Lecture Notes in Artificial Intelligence 6934

Subseries of Lecture Notes in Computer Science

Roberto Pirrone   Filippo Sorbello (Eds.)

AI*IA 2011:
Artificial Intelligence
Around Man and Beyond

XIIth International Conference
of the Italian Association for Artificial Intelligence
Palermo, Italy, September 15-17, 2011
Proceedings

 Springer

Series Editors

Randy Goebel, University of Alberta, Edmonton, Canada
Jörg Siekmann, University of Saarland, Saarbrücken, Germany
Wolfgang Wahlster, DFKI and University of Saarland, Saarbrücken, Germany

Volume Editors

Roberto Pirrone
University of Palermo
Department of Chemical, Management, Computer,
and Mechanical Engineering (DICGIM)
Viale delle Scienze, Edificio 6
90128 Palermo, Italy
E-mail: pirrone@unipa.it

Filippo Sorbello
University of Palermo
Department of Chemical, Management, Computer,
and Mechanical Engineering (DICGIM)
Viale delle Scienze, Edificio 6
90128 Palermo, Italy
E-mail: sorbello@unipa.it

ISSN 0302-9743   e-ISSN 1611-3349
ISBN 978-3-642-23953-3   e-ISBN 978-3-642-23954-0
DOI 10.1007/978-3-642-23954-0
Springer Heidelberg Dordrecht London New York

Library of Congress Control Number: 2011935741

CR Subject Classification (1998): I.2, F.1, F.4.1, I.2.6, H.2.8, I.5

LNCS Sublibrary: SL 7 – Artificial Intelligence

Typesetting: Camera-ready by author, data conversion by Scientific Publishing Services, Chennai, India

Printed on acid-free paper

Springer is part of Springer Science+Business Media (www.springer.com)

Preface

AI*IA 2011 was the 12th International Conference on Advances in Artificial Intelligence that is held bi-annually by the Italian Association for Artificial Intelligence (AI*IA).

This edition of the conference was entitled: "Artificial Intelligence Around Man and Beyond" owing to the strong focus of scientific contributions and invited speeches on the new frontiers of AI research.

Nowadays, AI-enabled technologies support a sort of "distributed intelligent environment." In this environment human beings can better manipulate information and more efficiently perform tasks involving a huge cognitive effort. All the foundation fields in AI are strongly oriented toward this goal: semantic information systems, natural language processing aimed to enhanced HCI, distributed gaming, and social networking are only a few application examples.

Moreover, such technologies shift theoretical speculation toward philosophical themes that are "beyond humans," such as the design of metacognitive artificial agents, machine consciousness, embodied robotics for seamless interaction with humans. The debate is no longer about merely intelligent or cognitive machines, but rather it focuses on their ability of being conscious and/or thinking about their cognition also in relation with humans.

AI*IA 2011 received 58 submissions from 18 countries. The conference accepted 31 oral presentations, and 13 posters, which covered broadly the many aspects of theoretical and applied artificial intelligence. They were grouped into seven sessions: "Machine Learning," "Distributed AI: Robotics and MAS," "Theoretical Issues: Knowledge Representation and Reasoning, Planning, Cognitive Modeling," "Natural Language Processing," "AI Applications I & II."

Moreover, *AI*IA 2011* hosted three invited speeches given by Stephen Grossberg, Hisroshi Ishiguro, and Roger Azevedo. A series of workshops dedicated to specific topics complemented the main conference program:

- Semantic Technologies in Enterprises
- AI and Cultural Heritage
- Intelligent Social and Service Robots
- Technological Challenges and Scenarios of the Ageing Society
- Learning by Reading in the Real World
- Analyzing, Modeling, Evaluating, and Fostering Metacognition with Intelligent Learning Environments
- Mining Complex Patterns
- DART 2011 - 5th International Workshop on New Challenges in Distributed Information Filtering and Retrieval

AI*IA 2011 stimulated debate among researchers and the birth of new groups devoted to particular investigations within the AI community. Moreover, many of the workshops promoted dialogue with the enterprise world: this is crucial for the growth of any scientific community.

The conference was made possible thanks to the support of several institutions and people that we want to acknowledge here.

We are very grateful to the AI*IA President, Paola Mello, for her precise suggestions and encouragements, and to the whole AI*IA Governing Board that selected Palermo as the venue for *AI*IA 2011*, providing us with continuous logistic support during the conference organization phases.

We also want to thank our supporting scientific institutions: the University of Palermo, CITC "Centro Interdipartimentale Tecnologie della Conoscenza," ICAR-CNR "Istituto di Calcolo e Reti ad Alte Prestazioni," and the *Artificial Intelligence Journal* published by Elsevier.

Moreover, let us thank Palermo Municipality, Provincia Regionale di Palermo, ARS Sicily Parliament Assembly, and the Sicily Governor Bureau for their general advice, Engineering Ingegneria Informatica S.p.A., Consorzio ARCA, and Informamuse s.r.l. for supporting the conference.

Last but not least, our thanks go to the AI Group at the DICGIM Dipartimento di Ingegneria Chimica Gestionale Informatica e Meccanica: Vincenzo Cannella, Arianna Pipitone, Alessandra De Paola, Valeria Seidita, Marco Ortloani, Orazio Gambino, Giuseppe Russo, Daniele Peri, and Agnese Augello. They supported the organization locally and concretely contributed to the success of *AI*IA 2011*.

July 2011   Roberto Pirrone
                                                             Filippo Sorbello

Organization

AI*IA 2011 was organized by DICGIM (Department of Chemical, Management, Computer, and Mechanical Engineering), University of Palermo, in cooperation with the Italian Association for Artificial Intelligence (AI*IA).

Executive Committee

Conference Chairs	Roberto Pirrone
	Filippo Sorbello
Program Chair	Salvatore Gaglio
Refereeing Chair	Edoardo Ardizzone
Workshop Chair	Antonio Chella
Sponsorship Chair	Antonio Gentile
Local Events Chair	Rosario Sorbello

Program Committee

Edoardo Ardizzone	University of Palermo, Italy
Roger Azevedo	McGill University, Canada
Stefania Bandini	University of Milan Bicocca, Italy
Roberto Basili	University of Rome Tor Vergata, Italy
Gautam Biswas	Vanderbilt University, USA
Gerhard Brewka	University of Leipzig, Germany
Ernesto Burattini	University of Naples Federico II, Italy
Stefano Cagnoni	University of Parma, Italy
Antonio Camurri	University of Genoa, Italy
Vincenzo Cannella	University of Palermo, Italy
Amedeo Cappelli	ISTI-CNR, Italy
Luigia Carlucci Aiello	University of Rome La Sapienza, Italy
Cristiano Castelfranchi	ISTC-CNR, Italy
Antonio Chella	University of Palermo, Italy
Cristina Conati	University of British Columbia, Canada
Vincenzo Conti	University of Enna Kore, Italy
Haris Dindo	University of Palermo, Italy
Floriana Esposito	University of Bari, Italy
Salvatore Gaglio	University of Palermo, Italy
Orazio Gambino	University of Palermo, Italy
Antonio Gentile	University of Palermo, Italy
Alfonso Gerevini	University of Brescia, Italy
Marco Gori	University of Siena, Italy
Marco La Cascia	University of Palermo, Italy

Evelina Lamma	University of Bologna, Italy
Maurizio Lenzerini	University of Rome La Sapienza, Italy
Leonardo Lesmo	University of Turin, Italy
Giuseppe Lo Re	University of Palermo, Italy
Bernardo Magnini	FBK, Italy
Sara Manzoni	University of Milan Bicocca, Italy
Francesco Mele	CIB-CNR, Italy
Paola Mello	University of Bologna, Italy
Daniele Nardi	University of Rome La Sapienza, Italy
Sergei Nirenburg	UMBC, USA
Andrea Omicini	University of Bologna, Italy
Marco Ortolani	University of Palermo, Italy
Maria Teresa Pazienza	University of Rome Tor Vergata, Italy
Daniele Peri	University of Palermo, Italy
Roberto Pirrone	University of Palermo, Italy
Fabrizio Riguzzi	University of Ferrara, Italy
Andrea Roli	University of Bologna, Italy
Francesca Rossi	University of Padova, Italy
Lorenza Saitta	University of Turin, Italy
Alexei Samsonovich	GMU, USA
Marco Schaerf	University of Rome La Sapienza, Italy
Giovanni Semeraro	University of Bari, Italy
Roberto Serra	University of Modena and Reggio-Emilia, Italy
Giovanni Soda	University of Florence, Italy
Rosario Sorbello	University of Palermo, Italy
Oliviero Stock	FBK, Italy
Pietro Torasso	University of Turin, Italy
Manuela Veloso	CMU, USA
Marco Villani	University of Modena and Reggio-Emilia, Italy
Salvatore Vitabile	University of Palermo, Italy

Local Organization

Vincenzo Cannella
Arianna Pipitone
Alessandra De Paola
Valeria Seidita
Marco Ortolani
Orazio Gambino
Giuseppe Russo
Daniele Peri
Agnese Augello
(University of Palermo, Italy)

Referees

V. Alcazar	C. D'Amato	F. Loebe	F. Riguzzi
E. Ardizzone	B. De Carolis	B. Magnini	A. Roli
R. Azevedo	A. De Paola	S. Manzoni	F. Rossi
L. Baltrunas	N. Di Mauro	A. Martelli	S. Rossi
S. Bandini	H. Dindo	A. Mazzei	A. Saetti
P. Basile	F. Esposito	F. Mele	L. Saitta
R. Basili	D. Ferraretti	P. Mello	A. Samsonovich
G. Biswas	S. Gaglio	M. Montali	M. Schaerf
G. Brewka	O. Gambino	M. Morana	G. Semeraro
E. Burattini	M. Gavanelli	D. Nardi	R. Serra
S. Cagnoni	A. Gentile	D. Nau	G. Soda
A. Camurri	A. Gerevini	S. Nirenburg	R. Sorbello
V. Cannella	M. Gori	A. Omicini	D. Sottara
A. Cappelli	M. Guerini	M. Ortolani	O. Stock
L. Carlucci Aiello	M. La Cascia	M. Palmonari	H. Strass
C. Castelfranchi	E. Lamma	M. Pazienza	P. Torasso
A. Chella	M. Lenzerini	D. Peri	M. Veloso
F. Chesani	L. Lesmo	R. Pirrone	R. Ventura
C. Conati	S. Liemhetcharat	D. Radicioni	M. Villani
V. Conti	G. Lo Re	G. Randelli	S. Vitabile

Sponsoring Institutions

University of Palermo (Italy)
Artificial Intelligence Journal by Elsevier (The Netherlands)
ICAR-CNR "Istituto di Calcolo e Reti ad Alte Prestazioni" (Italy)
CITC "Centro Interdipartimentale Tecnologie della Conoscenza" (Italy)
Engineering Ingegneria Informatica S.p.A. (Italy)
Informamuse s.r.l. (Italy)

Supporting Institutions

Italian Association for Artificial Intelligence (AI*IA)
Comune di Palermo
Provincia Regionale di Palermo
Assemblea Regionale Siciliana
Presidenza della Regione Siciliana
ARCA - "Consorzio per l'Applicazione della Ricerca e la Creazione di Aziende innovative" (Italy)

Table of Contents

Invited Talks

Machine Learning

Distributed AI: Robotics and MAS

Theoretical Issues: Knowledge Representation and Reasonoinng, Planning, Cognitive Modeling

Natural Language Processing

AI Applications I

AI Applications II

Poster Session

Foundations and New Paradigms of Brain Computing: Past, Present, and Future

Stephen Grossberg

Center for Adaptive Systems, Boston University,
677 Beacon Street, Boston, MA 02215
steve@cns.bu.edu

1 Introduction

A paradigm shift: autonomous adaptation to a changing world. There
has been rapid progress over the past fifty years in modeling how brains control
behavior; that is, in developing increasingly sophisticated and comprehensive
computational solutions of the classical mind/body problem. Not surprisingly,
such progress embodies a major paradigm shift, but one that is taking a long
time to fully take hold because it requires a synthesis of knowledge from multiple disciplines, including psychology, neuroscience, mathematics, and computer
science.

Linking brain to mind clarifies both brain *mechanisms* and behavioral *functions*. Such a linkage is needed to develop applications to computer science,
engineering, and technology, since mechanisms tell us how things work, whereas
functions tell us what they are for. Knowing how things work and what they
are for are both essential in any application. Such models represent a paradigm
shift because the brain is unrivaled in its ability to autonomously adapt in real
time to complex and changing environments. Models that embody adaptive autonomous intelligent responses to unexpected contingencies are just the sorts of
models that can fully realize the dream of artificial intelligence.

A method to link brain to mind. By what method can such models be
discovered? A successful method has been elaborated over the past fifty years.
The key is to begin with behavioral data, typically scores or even hundreds of
parametrically structured behavioral experiments in a particular problem domain. One begins with behavioral data because brain evolution needs to achieve
behavioral success. Any theory that hopes to link *brain* to behavior thus must
discover the computational level on which brain dynamics control behavioral
success. This level has proved to be the network and system level. That is why
the name *neural networks* is appropriate for these models.

Behavioral data provide a theorist with invaluable clues about the functional
problems that the brain is trying to solve. One starts with large amounts of data
because otherwise too many seemingly plausible hypotheses cannot be ruled out.
A crucial meta-theoretical constraint is to insist upon understanding the behavioral datawhich comes to us as static numbers or curves on a pageas the emergent

R. Pirrone and F. Sorbello (Eds.): AI*IA 2011, LNAI 6934, pp. 1–7, 2011.

properties of a dynamical process which is taking place moment-by-moment in an individual mind. One also needs to respect the fact that our minds can adapt on their own to changing environmental conditions without being told that these conditions have changed. One thus needs to frontally attack the problem of how an intelligent being can *autonomously adapt to a changing world*. Knowing how to do this, as with many other theoretical endeavors in science, is presently an art form. There are no known algorithms with which to point the way.

Whenever I have applied this method in the past, I have never used homunculi, or else the crucial constraint on *autonomous* adaptation would be violated. The result has regularly been the discovery of new organizational principles and mechanisms, which are then realized as a minimal model operating according to only locally defined laws that are capable of operating on their own in real time. The remarkable fact is that, when such a behaviorally-derived model has been written down, it has always been interpretable as a neural network. These neural networks have always included known brain mechanisms. The functional interpretation of these mechanisms has, however, often been novel because of their derivation from a behavioral analysis. The networks have also often predicted the existence of unknown neural mechanisms, and many of these predictions have been supported by subsequent neurophysiological, anatomical, and even biochemical experiments over the years.

Once this neural connection has been established by a top-down analysis from behavior, one can work both top-down from behavior and bottom-up from brain to exert a tremendous amount of conceptual pressure with which to better understand the current model and to discover design principles that have not yet been satisfied in it. Then the new design principles help to derive the next model stage. This Method of Minimal Anatomies acknowledges that one cannot derive "the brain" in one theoretical step. But one can do it incrementally in stages by carrying out a form of conceptual evolution. Applying this method, a sequence of self-consistent but evolving models can be derived, with each subsequent model capable of explaining and predicting more data than its ancestors.

A fundamental empirical conclusion can be drawn from many experiences of this type; namely, the brain as we know it can be successfully understood as an organ that is designed to achieve autonomous adaptation to a changing world. Although I am known as one of the founders of the field of neural networks, I have never tried to derive a neural network. Neural networks arise from a real-time behavioral analysis because they provide natural computational realizations with which to control autonomous behavioral adaptation to a changing world.

New paradigms: Complementary computing, laminar computing, and nano chips. How does the brain carry out autonomous adaptation to a changing world? What new computational paradigms are needed to accomplish this goal?

Complementary Computing clarifies the nature of brain specialization. It provides an alternative to the previous computer-inspired paradigm of independent modules. If there were independent modules in the brain, properties such as visual lightness, depth, and motion would be computed independently, which is not the case. Complementary Computing explains why the brain is specialized into parallel processing streams, and how these streams interact in specific ways to overcome their complementary deficiencies [8].

Laminar Computing clarifies how the ubiquitous organization of cerebral cortex into layered circuits can support, through variations of the same laminar architecture, such different aspects of biological intelligence as vision, speech, and cognition [1] [9] [11] [12].

These new computational paradigms promise to have a major impact on the design of future computers that increasingly embody aspects of human intelligence. For example, it is widely acknowledged that Moores Law will break down within ten years. Current Von Neumann chip designs cannot continue to become increasingly dense without becoming highly noisy and generating too much heat. The DARPA SyNAPSE program, among others, has responded to this challenge by supporting research to design new nano-scale VLSI chips that better embody properties of biological intelligence. The idea is for future computers to contain the fastest traditional chips, which can carry out many functions that human brains cannot, as well as brain-inspired chips whose successive generations can carry out increasingly complex types of characteristically human intelligence, notably autonomous adaptation to a changing world.

Nano-scale chips tend to be noisy chips, unlike the perfect chips in Von Neumann computers on which current AI builds. In order to generate less heat, the new nano-scale chips need to use discrete spikes in time to communicate between processing elements. In order to pack in the necessary processing, they may also need to be organized in processing layers. DARPA turned to the brain for design inspiration because the cerebral cortex, which supports all higher aspects of biological intelligence, provides a paradigmatic example of a noisy, layered, spiking intelligent device. That is why Laminar Computing is starting to change the way in which future chips are being designed. [2] have described, using the example of the 3D LAMINART model of 3D vision, a general method for converting fifty years of neural networks based on continuous rate-based signals into spiking neural networks that are amenable to being embodied in SyNAPSE-style chips.

*Research progress: towards autonomous adaptive agents.*Using this method, my colleagues and I have developed increasingly detailed and comprehensive neural models of vision and visual object recognition; audition, speech, and language; development; attentive learning and memory; cognitive information processing; reinforcement learning and motivation; cognitive-emotional interactions; navigation; sensory-motor control and robotics; and mental disorders. These models involve many parts of the brain, ranging from perception to action, and multiple levels of brain organization, ranging from individual spikes and their synchronization to cognition. My web page http://cns.bu.edu/~steve contains many downloadable articles that illustrate this progress. In my talk at AI*IA, I will summarize some recent theoretical progress towards designing autonomous mobile adaptive agents. One of these developments is summarized below.

2   What Is an Object? Learning Object Categories under Free Viewing Conditions

ARTSCAN model: What-Where stream coordination supports invariant object learning. What is an object? How can we learn what an object is without any external supervision? In particular, how does the brain learn to recognize a complex object from multiple viewpoints, and when it is seen at multiple positions and distances? Such a competence is essential in any mobile autonomous adaptive agent. Consider what happens when we first look at an object that is not instantly recognizable. We make scanning eye movements, directing our foveas to a variety of points of interest, or views, on the object. The objects retinal representations of these views are greatly distorted by cortical magnification in cortical area V1. The brain somehow combines several such distorted views into an object recognition category that is invariant to where we happen to be gazing at the moment. Future encounters with the same object can therefore lead to fast recognition no matter what view we happen to see.

How does the brain know that the views that are foveated on successive saccades belong to the same object? How does the brain avoid the problem of erroneously learning to classify parts of different objects together? Only views of the same object should be linked through learning to the same view-invariant object category. How does the brain know which views belong to the same object, even before it has learned a view-invariant category that can represent the object as a whole? How does the brain do this without an external teacher; that is, under the unsupervised learning conditions that are the norm during many object learning experiences *in vivo*?

My colleagues and I [3] [6] [10] have been developing a neural model to explain how spatial and object attention to coordinate the brains ability to learn representations of object categories that are seen at multiple positions, sizes, and viewpoints. Such invariant object category learning and recognition can be achieved using interactions between a hierarchy of processing stages in the visual brain. These stages include retina, lateral geniculate nucleus, and cortical areas V1, V2, V4, and IT in the brain's What cortical stream, as they interact with spatial attention processes within the parietal cortex of the Where cortical stream.

The model first was developed to explain view-invariant object category learning and recognition [6] [7] [10]. This version of the model is called ARTSCAN. ARTSCAN has been generalized to the *positional* ARTSCAN, or pARTSCAN, model which explains how view-, position-, and size-invariant object categories may be learned [3].

I predict that view-invariant object learning and recognition is achieved by the brain under free viewing conditions through the coordinated use of spatial and object attention. Many studies of spatial attention have focused on its spatial distribution and how it influences visual perception. I predict that spatial attention plays a crucial role in controlling view-invariant object category learning. In particular, several authors have reported that the distribution of spatial attention can configure itself to fit an objects form. Form-fitting spatial attention is sometimes called an *attentional shroud* [13]. ARTSCAN predicts how an objects pre-attentively formed surface representation can induce such a form-fitting attentional shroud. Moreover, while this attentional shroud remains active, I predict that it accomplishes two things.

First, it ensures that eye movements tend to end at locations on the objects surface, thereby enabling views of the same object to be sequentially explored. Second, it keeps the emerging view-invariant object category active while different views of the object are learned and associated with it. Thus, the brain avoids what would otherwise seem to be an intractable infinite regress: If the brain does not already know what the object is, then how can it, without external guidance, prevent views from several objects from being associated? My proposal is that the *pre-attentively formed surface representation of the object* provides the object-sensitive substrate that prevents this from happening, even before the brain has learned knowledge about the object. This hypothesis is consistent with a burgeoning psychophysical literature showing that 3D boundaries and surfaces are the units of pre-attentive visual perception, and that attention selects these units for recognition.

This proposed solution can be stated more formally as a temporally-coordinated cooperation between the brains What and Where cortical processing streams: The Where stream maintains an attentional shroud whose spatial coordinates mark the surface locations of a current "object of interest", whose identity has yet to be determined in the What stream. As each view-specific category is learned by the What stream, it focuses object attention via a learned top-down expectation on the critical features that will be used to recognize that view and its variations in the future. When the first such view-specific category is learned, it also activates a cell population at a higher cortical level that will become the view-invariant object category.

Suppose that the eyes or the object move sufficiently to expose a new view whose critical features are significantly different from the critical features that are used to recognize the first view. Then the first view category is reset, or inhibited. This happens due to the mismatch of its learned top-down expectation, or prototype of attended critical features, with the newly incoming view information [4] [5]. This top-down prototype focuses object attention on the incoming visual information. Object attention hereby helps to control which view-specific categories are learned by determining when the currently active view-specific category should be reset, and a new view-specific category should be activated. However, the view-invariant object category should *not* be reset every time a view-specific category is reset, or else it can never become view-invariant. This is what the attentional shroud accomplishes: It inhibits a tonically-active reset signal that would otherwise shut off the view-invariant category when each view-based category is reset. As the eyes foveate a sequence of object views through time, they trigger learning of a sequence of view-specific categories, and each of them is associatively linked through learning with the still-active view-invariant category.

When the eyes move off an object, its attentional shroud collapses in the Where stream, thereby disinhibiting the reset mechanism that shuts off the view-invariant category in the What stream. When the eyes look at a different object, its shroud can form in the Where stream and a new view category can be learned that can, in turn, activate the cells that will become the view-invariant category in the What stream.

The original archival articles show how these concepts can explain many psychological and neurobiological data about object category learning and recognition. In particular, the model mechanistically clarifies basic properties of attention shifts (engage, move, disengage) and inhibition of return. It simulates human reaction time data about object-based spatial attention shifts, and learns with 98.1% accuracy and a compression of 430 on a letter database whose letters vary in size, position, and orientation.

References

1. Cao, Y., Grossberg, S.: A laminar cortical model of stereopsis and 3D surface perception: closure and da Vinci stereopsis. Spatial Vision 18, 515–578 (2005)
2. Cao, Y., Grossberg, S.: Stereopsis and 3D surface perception by spiking neurons in laminar cortical circuits: A method for converting neural rate models into spiking models. Neural Networks (in press, 2011)
3. Cao, Y., Grossberg, S., Markowitz, J.: How does the brain rapidly learn and reorganize view- and positionally-invariant object representations in inferior temporal cortex? Neural Networks (in press, 2011)
4. Carpenter, G.A., Grossberg, S.: A massively parallel architecture for a self-organizing neural pattern-recognition machine. Computer Vision Graphics and Image Processing 37, 54–115 (1987)
5. Carpenter, G.A., Grossberg, S.: Pattern recognition by self-organizing neural networks. MIT Press, Cambridge (1991)
6. Fazl, A., Grossberg, S., Mingolla, E.: View-invariant object category learning, recognition, and search: how spatial and object attention are coordinated using surface-based attentional shrouds. Cognitive Psychology 58(1), 1–48 (2009)
7. Foley, N., Grossberg, S., Mingolla, E.: Neural dynamics of object-based multifocal visual spatial attention and priming: Object cueing, useful-field-of-vew, and crowding (submitted for publication, 2011)
8. Grossberg, S.: The complementary brain: Unifying brain dynamics and modularity. Trends in Cognitive Sciences 4, 233–246 (2000)
9. Grossberg, S.: How does the cerebral cortex work? Development, learning, attention, and 3D vision by laminar circuits of visual cortex. Behavioral and Cognitive Neuroscience Reviews 2, 47–76 (2003)
10. Grossberg, S.: Cortical and subcortical predictive dynamics and learning during perception, cognition, emotion, and action. Philosophical Transactions of the Royal Society of London, Special Issue Predictions in the Brain: Using our Past to Generate a Future 364, 1223–1234 (2009)
11. Grossberg, S., Kazerounian, S.: Laminar cortical dynamics of conscious speech perception: Neural model of phonemic restoration using subsequent context in noise. Journal of the Acoustical Society of America (in press, 2011)

12. Grossberg, S., Pearson, L.: Laminar cortical dynamics of cognitive and motor working memory, sequence learning and performance: Toward a unified theory of how the cerebral cortex works. Psychological Review 115, 677–732 (2008)
13. Tyler, C.W., Kontsevich, L.L.: Mechanisms of stereoscopic processing: stereoattention and surface perception in depth reconstruction. Perception 24, 127–153 (1995)

Robotics as the Hub of Various Research Areas

Hiroshi Ishiguro

Department of Systems Innovation, Osaka University and
ATR Hiroshi Ishiguro Laboratory
ishiguro@sys.es.osaka-u.ac.jp

Abstract. Application fields of robotics are shifting from industrial robots to everyday robots that provide various services in our daily life. For realizing the everyday robots, we need to study more on human. We, robotisist, have to develop the interactive robots in the open environment, although we do not have sufficient knowledge on human. That is, we develop the robots based on the knowledge on human and use the robots for verifying hypothesis for understanding human. This is a new interdisciplinary framework that connect various research areas, such as cognitive science, brain science, sensor networks, mechanical engineering, and artificial intelligence. Further, it also provide philosophical questions. In this talk, I will introduce a series of robots that I have developed in Osaka University and ATR Hiroshi Ishiguro Laboratory and discuss the related fundamental issues for understanding human.

R. Pirrone and F. Sorbello (Eds.): AI*IA 2011, LNAI 6934, p. 8, 2011.
© Springer-Verlag Berlin Heidelberg 2011

Can We Design Artificial Pedagogical Agents to Be Intelligent Enough to Detect, Model, and Foster Regulatory Learning Processes?

Roger Azevedo

Department of Educational & Counselling Psychology,
McGill University
roger.azevedo@mcgill.ca

Abstract. Self-regulated learning (SRL) involves a complex set of interactions between cognitive, metacognitive, motivational and affective processes. The key to understanding the influence of these self-regulatory processes on learning with open-ended, non-linear learning computer-based environments involves detecting, capturing, identifying, and classifying these processes as they temporally unfold during learning. Understanding the complex nature of these processes is key to building intelligent learning environments that are capable of adapting to the inherent fluctuations in learners' SRL processes and emerging understanding of the embedded educational content and related disciplinary knowledge. Recent developments in the use of and advances in the design of artificial pedagogical agents have begun to address these issues. However, we are still experiencing major theoretical, conceptual, methodological, and analytical challenges that may impede our ability to design more intelligent agents that are effectively and reliably able to detect, model, and foster learners SRL processes during learning. As such, the foci of this presentation are to: (1) introduce the complexity of SRL with current intelligent agent systems, (2) briefly present a hybrid theoretical model of SRL and describe how it can be used to analyze the temporally, unfolding sequences of processes during learning, (3) present and describe sample data to illustrate the nature and complexity of these processes and the various challenges they pose for designers and learners, and (4) present challenges for future research that combines several techniques and methods to design pedagogical agents and intelligent learning environments that effectively and reliably trace, model, and foster SRL.

R. Pirrone and F. Sorbello (Eds.): AI*IA 2011, LNAI 6934, p. 9, 2011.

A Comparative Study of Thresholding Strategies in Progressive Filtering

Andrea Addis, Giuliano Armano, and Eloisa Vargiu

University of Cagliari,
Department of Electrical and Electronic Engineering
{addis,armano,vargiu}@diee.unica.it
http://iasc.diee.unica.it

Abstract. Thresholding strategies in automated text categorization are an under-explored area of research. Indeed, thresholding strategies are often considered a post-processing step of minor importance, the underlying assumptions being that they do not make a difference in the performance of a classifier and that finding the optimal thresholding strategy for any given classifier is trivial. Neither these assumptions are true. In this paper, we concentrate on progressive filtering, a hierarchical text categorization technique that relies on a local-classifier-per-node approach, thus mimicking the underlying taxonomy of categories. The focus of the paper is on assessing TSA, a greedy threshold selection algorithm, against a relaxed brute-force algorithm and the most relevant state-of-the-art algorithms. Experiments, performed on Reuters, confirm the validity of TSA.

1 Introduction

Thresholding strategies can make the difference in the performance of a classifier. Moreover, how to find the best acceptance threshold for a classifier is a hard problem, especially in case of classifiers linked to taxonomic relationships. Nevertheless, in the literature, only few studies have been made on this issue [18], [9], [13], [7], so that thresholding strategies in automated Text Categorization (TC) are still an under-explored area of research [19].

In this paper, we perform a comparative experimental assessment of a greedy threshold selection algorithm, called TSA, aimed at finding a suboptimal combination of thresholds in the context of Progressive Filtering (PF), the Hierarchical Text Categorization (HTC) technique discussed in [2]. Experimental results, performed on the Reuters data collections, show that the proposed approach is able to find suboptimal solutions while maintaining a quadratic complexity, which allows to adopt the algorithm also for large taxonomies [4]. This paper extends our previous work [5] by reporting further experimental results and comparing them with those obtained by running a relaxed brute force approach and the algorithm proposed in [13].

The rest of the paper is organized as follows: Section 2 summarizes the main thresholding strategies proposed in the literature. Section 3 briefly recalls PF. Section 4 describes TSA, putting into evidence the theoretical background that allowed to devise the algorithm, together with its computational benefits. Experiments and results are illustrated in Section 5. Conclusions and future work are discussed in Section 6, which ends the paper.

R. Pirrone and F. Sorbello (Eds.): AI*IA 2011, LNAI 6934, pp. 10–20, 2011.

2 Thresholding Strategies

In TC, the three most commonly used thresholding strategies are RCut, PCut, and SCut [18]. For each document, RCut sorts categories by score and selects the t top-ranking categories, with $t \geq 1$ (however, as noted in [14], RCut is not a strict thresholding policy). For each category C_j, PCut sorts the documents by score and sets the threshold of C_j so that the number of documents accepted by C_j corresponds to the number of documents assigned to C_j. For each category C_j, SCut scores a validation set of documents and tunes the threshold over the local pool of scores, until a suboptimal, though satisfactory, performance of the classifier is obtained for C_j.

Few threshold selection algorithms have been proposed for HTC [9] [13] [7].

The algorithm proposed by D'Alessio et al. [9] tunes the thresholds by considering categories in a top-down fashion. For each category C_j, the search space is visited by incrementing the corresponding threshold with steps of 0.1. For each threshold value, the number of True Positives (TP) and False Positives (FP), i.e., the number of documents that would be correctly and incorrectly placed in C_j, is calculated. The utility function, i.e., the goodness measure that must be maximized for each threshold, is $TP - FP$[1].

Ruiz [13] selects thresholds that optimize the F_1 values for the categories. The whole training set is used to identify (sub)optimal thresholds. His expert-based system takes a binary decision at each expert gate and then optimizes the thresholds using only examples that reach leaf nodes. This task is performed by grouping experts into levels and finding the thresholds that maximize F_1. The best results are selected upon trials performed with each combination of thresholds from the vector $[0.005, 0.01, 0.05, 0.10]$ for level 1 and $[0.005, 0.01, 0.05, 0.10, 0.15, \ldots, 0.95]$ for levels 2, 3, 4.

The algorithm proposed by Ceci and Malerba [7] is based on a recursive bottom-up threshold determination. The algorithm takes as input a category C and the set of thresholds already computed for some siblings of C and their descendants. It returns the union of the input set with the set of thresholds computed for all descendants of C. In particular, if C' is a direct subcategory of C, the threshold associated to C' is determined by examining the sorted list of classification scores and by selecting the middle point between two values in the list, to minimize the expected error. The error function is estimated on the basis of the distance between two nodes in a tree structure (TD), the distance being computed as the sum of the weights belonging to the edges of the unique path connecting the two categories in the hierarchy (a unit weight is associated to each edge).

3 Progressive Filtering

A simple way to categorize the various proposals that have been made in HTC is to focus on the mapping between classifiers and the underlying taxonomy. According to [15], the approaches proposed in the literature can be framed as follows: (i) local classifier per node, (ii) local classifier per parent node, (iii) local classifier per level, and (iv) global classifier.

---

[1] In the event that the same value of the utility function occurs multiple times, the lowest threshold with that value is selected.

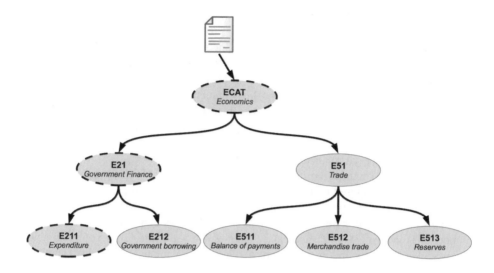

Fig. 1. An example of pipeline (highlighted with bold-dashed lines)

PF is a simple categorization technique framed within the local classifier per node approach, which admits only binary decisions, as each classifier is entrusted with deciding whether the input in hand can be forwarded or not to its children. The first proposals in which sequential boolean decisions are applied in combination with local classifiers per node can be found in [9], [10], and [16]. In [17], the idea of mirroring the taxonomy structure through binary classifiers is clearly highlighted; the authors call this technique "binarized structured label learning".

In PF, given a taxonomy, where each node represents a classifier entrusted with recognizing all corresponding positive inputs (i.e., interesting documents), each input traverses the taxonomy as a "token", starting from the root. If the current classifier recognizes the token as positive, it is passed on to all its children (if any), and so on. A typical result consists of activating one or more branches within the taxonomy, in which the corresponding classifiers have been activated by the token (see Figure 1).

A way to implement PF consists of unfolding the given taxonomy into pipelines of classifiers, as depicted in Figure 2. Each node of the pipeline represents a category that embeds a binary classifier able to recognize whether or not an input belongs to the category itself. Unfolding the taxonomy in pipelines gives rise to a set of new classifiers, each represented by a pipeline.

4   The Threshold Selection Algorithm

4.1   Motivations

According to classical text categorization, given a set of documents D and a set of labels C, a function $CSV_i : D \rightarrow [0,1]$ exists for each $c_i \in C$. The behavior of c_i is controlled by a threshold θ_i, responsible for relaxing or restricting the acceptance

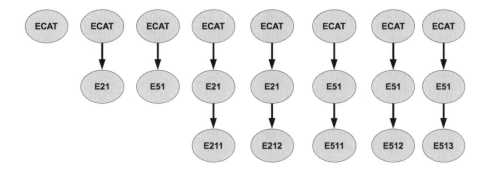

Fig. 2. The pipelines corresponding to the taxonomy in Figure 1

rate of the corresponding classifier. Let us recall that, with $d \in D$, $CSV_i(d) \geq \theta_i$ is interpreted as a decision to categorize d under c_i, whereas $CSV_i(d) < \theta_i$ is interpreted as a decision not to categorize d under c_i.

In PF, we assume that CSV_i exists, with the same semantics adopted by the classical setting. Considering a pipeline π, composed by n classifiers, the acceptance policy strictly depends on the vector of thresholds $\theta = \langle \theta_1, \theta_2, \ldots, \theta n \rangle$ that embodies the thresholds of all classifiers in π. In order to categorize d under π, the following constraint must be satisfied: for $k = 1 \ldots n$, $CSV_i(d) \geq \theta_k$. On the contrary, d is not categorized under c_i in the event that a classifier in π rejects it. Let us point out that we allow different behaviors for a classifier, depending on which pipeline it is embedded in. As a consequence, each pipeline can be considered in isolation from the others. For instance, given $\pi_1 = \langle C_1, C_2, C_3 \rangle$ and $\pi_2 = \langle C_1, C_2, C_4 \rangle$, the classifier C_1 is not compelled to have the same threshold in π_1 and in π_2 (the same holds for C_2). In so doing, the proposed approach performs a sort of "flattening", though preserving the information about the hierarchical relationships embedded in a pipeline. For instance, the pipeline $\langle C_1, C_2, C_3 \rangle$ actually represents the classifier C_3, although the information about the existing subsumption relationships are preserved (i.e., $C_3 \prec C_2 \prec C_1$, where "$\prec$" denotes the usual covering relation).

In PF, given a utility function[2], we are interested in finding an effective and computationally "light" way to reach a sub-optimum in the task of determining the best vector of thresholds. Unfortunately, finding the best acceptance thresholds is a difficult task. In fact, exhaustively trying each possible combination of thresholds (brute-force approach) is unfeasible, the number of thresholds being virtually infinite. However, the brute-force approach can be approximated by defining a granularity step that requires to check only a finite number of points in the range $[0, 1]$, in which the thresholds are permitted to vary with step δ. Although potentially useful, this "relaxed" brute force algorithm for calibrating thresholds (RBF for short) is still too heavy from a computational point of view. On the contrary, the threshold selection algorithm described in this paper is characterized by low time complexity while maintaining the capability of finding near-optimum solutions.

---

[2] Different utility functions (e.g., precision, recall, F_β, user-defined) can be adopted, depending on the constraint imposed by the underlying scenario.

4.2 Analysis and Implementation of TSA

Utility functions typically adopted in TC and in HTC are nearly-convex with respect to the acceptance threshold. Figure 3 depicts three typical trends of utility functions, i.e., precision, recall, and F_1.

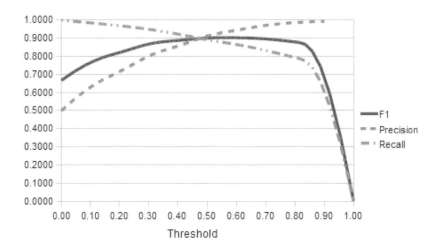

Fig. 3. Example of utility functions

Setting the threshold of a classifier to 0, no matter which utility function is adopted, forces the classifier to reach its maximum error in terms of false positives (FP). Conversely, setting the threshold to 1 forces the classifier to reach its maximum error in terms of false negatives (FN).

Due to the shape of the utility function and to its dependence on FP and FN, it becomes feasible to search its maximum around a subrange of $[0, 1]$. Bearing in mind that the lower the threshold the less restrictive is the classifier, we propose a greedy bottom-up algorithm for selecting decision threshold that relies on two functions [3]:

- *Repair* ($\mathcal{R}$), which operates on a classifier C by increasing or decreasing its threshold –i.e., $\mathcal{R}(up, C)$ and $\mathcal{R}(down, C)$, respectively– until the selected utility function reaches and maintains a local maximum.
- *Calibrate* ($\mathcal{C}$), which operates going downwards from the given classifier to its offspring by repeatedly calling $\mathcal{R}$. It is intrinsically recursive and, at each step, calls $\mathcal{R}$ to calibrate the current classifier.

Given a pipeline $\pi = \langle C_1, C_2, \ldots, C_L \rangle$, *TSA* is defined as follows (all thresholds are initially set to 0):

$$TSA(\pi) := for\ k = L\ downto\ 1\ do\ \mathcal{C}(up, C_k) \tag{1}$$

which asserts that $\mathcal{C}$ is applied to each node of the pipeline, starting from the leaf ($k = L$).

Under the assumption that p is a structure that contains all information about a pipeline, including the corresponding vector of thresholds and the utility function to be optimized, the pseudo-code of TSA is:

```
function TSA(p:pipeline):
    for k := 1 to p.length do p.thresholds[i] = 0
    for k := p.length downto 1do Calibrate(up,p,k)
    return p.thresholds
end TSA
```

The *Calibrate* function is defined as follows:

$$
\begin{aligned}
\mathcal{C}(up, C_k) &:= \mathcal{R}(up, C_k), \quad k = L \\
\mathcal{C}(up, C_k) &:= \mathcal{R}(up, C_k); \mathcal{C}(down, C_{k+1}), \quad k < L
\end{aligned}
\tag{2}
$$

and

$$
\begin{aligned}
\mathcal{C}(down, C_k) &:= \mathcal{R}(down, C_k), \quad k = L \\
\mathcal{C}(down, C_k) &:= \mathcal{R}(down, C_k); \mathcal{C}(up, C_{k+1}), k < L
\end{aligned}
\tag{3}
$$

where the ";" denotes a sequence operator, meaning that in "*a;b*" action a is performed *before* action b.

The pseudo-code of *Calibrate* is:

```
function Calibrate(dir:{up,down}, p:pipeline, level:integer):
    Repair(dir,p,level)
    if level < p.length then Calibrate(toggle(dir),p,level+1)
end Calibrate
```

where *toggle* is a function that reverses the current direction (from *up* to *down* and vice versa). The reason why the direction of threshold optimization changes at each call of *Calibrate* (and hence of *Repair*) lies in the fact that increasing the threshold θ_{k-1} is expected to forward less FP to C_k, which allows to decrease θ_k. Conversely, decreasing the threshold θ_{k-1} is expected to forward more FP to C_k, which must react by increasing θ_k. The pseudo-code of *Repair* is:

```
function Repair(dir:{up,down}, p:pipeline, level:integer):
    delta := (dir = up) ? p.delta : -p.delta
    best_threshold := p.thresholds[level]
    max_uf := p.utility_function()
    uf := max_uf
    while uf >= max_uf * p.sf and p.thresholds[level] in [0,1]
        do p.thresholds[level] += delta
            uf := p.utility_function()
            if uf < max_uf then continue
            max_uf := uf
            best_threshold := p.thresholds[level]
    p.thresholds[level] := best_threshold
end Repair
```

The scale factor ($p.sf$) is used to limit the impact of local minima during the search, depending on the adopted utility function (e.g., a typical value of $p.sf$ for F_1 is 0.8).

It is worth pointing out that, as also noted in [11], the sub-optimal combination of thresholds depends on the adopted dataset, hence the sub-optimal combination of thresholds need to be recalculated for each dataset.

4.3   Expected Runnning Time and Computational Complexity of TSA

Searching for a sub-optimal combination of thresholds in a pipeline π can be viewed as the problem of finding a maximum of a utility function F, the maximum being dependent on the corresponding thresholds θ. In symbols:

$$\theta^* = \underset{\theta}{argmax}\ F(\theta; \pi) \tag{4}$$

Unfortunately, the above task is characterized by high time complexity. In particular, two sources of intractability hold: (i) the optimization problem that involves the thresholds and (ii) the need of retraining classifiers after modifying thresholds. In this work, we concentrate on the former issue while deciding not to retrain the classifiers. In any case, it is clear that the task of optimizing thresholds requires a solution that is computationally light. To calculate the computational complexity of TSA, let us define a granularity step that requires to visit only a finite number of points in a range $[\rho_{min}, \rho_{max}]$, $0 \le \rho_{min} < \rho_{max} \le 1$, in which the thresholds vary with step δ. As a consequence, $p = \lfloor \delta^{-1} \cdot (\rho_{max} - \rho_{min}) \rfloor$ is the maximum number of points to be checked for each classifier in a pipeline. For a pipeline π of length L, the expected running time for TSA, say $T_{TSA}(\pi)$, is proportional to $(L + L^2) \cdot p \cdot (\rho_{max} - \rho_{min})$. This implies that TSA has complexity $O(L^2)$, quadratic with the number of classifiers embedded by a pipeline. A comparison between TSA and the brute-force approach is unfeasible, as the generic element of the threshold vector is a real number. However, a comparison between TSA and RBF is feasible, although RBF is still computationally heavy. Assuming that p points are checked for each classifier in a pipeline, the expected running time for RBF, $T_{RBF}(\pi)$, is proportional to p^L, which implies that its computational complexity is $O(p^L)$.

To show the drastic reduction of complexity brought by TSA, let us consider a pipeline π composed of 4 classifiers (i.e., $L = 4$), and $p = 100$. In this case, the orders of magnitude of $T_{RBF}(\pi)$ and $T_{TSA}(\pi)$ are 10^8 and 10^3, respectively. These values are confirmed by the experiments reported in Section 5. It is worth noting that, due its intrinsic complexity, the RBF approach can be applied in practice only setting p to a value much lower than the one applied to TSA. For instance, with $p_{TSA} = 2000$, $\rho_{max} = 1$, $\rho_{min} = 0$, and $L = 4$, $T_{TSA}(\pi) \propto 32,000$. To approximately get the same running time for RBF, $p_{RBF} \simeq 6.7$, which is two orders of magnitude lower than p_{TSA}.

5   Experimental Results

The Reuters Corpus Volume I (RCV1-v2) [12] has been chosen as benchmark dataset. In this corpus, stories are coded in four hierarchical groups: Corporate/Industrial (CCAT),

Table 1. Comparisons between TSA and RBF (in milliseconds), averaged on pipelines with $L = 4$.

Algorithm	t_{lev4}	t_{lev3}	t_{lev2}	t_{lev1}	F_1
experiments with $\delta = 0.1, (p = 10)$					
TSA	33	81	131	194	0.8943
RBF	23	282	3,394	43,845	0.8952
experiments with $\delta = 0.05, (p = 20)$					
TSA	50	120	179	266	0.8953
RBF	35	737	17,860	405,913	0.8958
experiments with $\delta = 0.01, (p = 100)$					
TSA	261	656	1,018	1,625	0.8926
RBF	198	17,633	3.1E+6	1.96E+8	0.9077

Economics (ECAT), Government/Social (GCAT), and Markets (MCAT). Although the complete list consists of 126 categories, only some of them have been used to test our hierarchical approach. The total number of codes actually assigned to the data is 93, whereas the overall number of documents is about 803,000. Each document belongs to at least one category and, on average, to 3.8 categories.

Reuters dataset has been chosen because it allows us to perform experiments with pipelines up to level 4 while maintaining a substantial number of documents along the pipeline. To compare the performance of TSA with respect to RBF and to the selected state-of-the-art algorithms, we used, about 2,000 documents per each leaf category. We considered the 24 pipelines of depth 4 yielding a total of about 48,000 documents.

Experiments have been performed on a SUN Workstation with two Opteron 280, 2Ghz+ and 8Gb Ram. The system used to perform benchmarks has been implemented using X.MAS [1], a generic multiagent architecture built upon JADE [6] and devised to make it easier the implementation of information retrieval/filtering applications.

Experiments have been carried out by using classifiers based on the *wk-NN* technology [8], which do not require specific training and are very robust with respect to noisy data. As for document representation, we adopted the bag of words approach, a typical method for representing texts in which each word from a vocabulary corresponds to a feature and a document to a feature vector. In this representation, all non-informative words such as prepositions, conjunctions, pronouns and very common verbs are disregarded by using a stop-word list. Moreover, the most common morphological and inflexional suffixes are removed by adopting a standard stemming algorithm. After determining the overall sets of features, their values are computed for each document resorting to the well-known TFIDF method. To reduce the high dimensionality of the feature space, we select the features that represent a node by adopting the information gain method.

During the training activity, each classifier has been trained with a balanced data set of 1000 documents, characterized by 200 features.

Experiments, performed on a balanced dataset of 2000 documents for each class, were focused on (i) comparing the running time and F_1 of TSA vs. RBF, and on (ii) comparing TSA with the selected state-of-the-art algorithms, i.e., those proposed by D'Alessio et al. [9], by Ruiz [13], and by Ceci and Malerba [7]. Let us note that the focus of this work is only on threshold selection algorithms, so that we selected the most representative algorithms proposed in the literature and we decided to not compare TSA with other types of heuristic text filtering algorithms.

TSA vs. RBF. Experiments performed to compare TSA with RBF have been carried out calculating the time (in milliseconds) required to set the optimal vector of thresholds for both algorithms, i.e., the one that reaches the optimal value in term of F_1, the selected utility function. Different values of δ (i.e., $0.1, 0.05, 0.01$) have been adopted to increment thresholds during the search. Each pair of rows in Table 1 reports the comparison in terms of the time spent to complete each calibrate step ($t_{lev4} \ldots t_{lev1}$), together with the corresponding F_1. Results clearly show that the cumulative running time for RBF tends to rapidly become intractable[3], while the values of F_1 are comparable.

TSA vs. state-of-the-art algorithms. As already pointed out, to compare TSA we considered the algorithms proposed by D'Alessio et al. [9], by Ruiz [13], and by Ceci and Malerba [7]. We used $\delta = 10^{-3}$ for TSA. Let us note that Ruiz uses the same threshold value for level 3 and level 4, whereas we let its algorithm to search on the entire space of thresholds. In so doing, the results in terms of utility functions cannot be worse than those calculated by means of the original algorithm. However, the running time is one order of magnitude greater than the original algorithm.

Table 2. Comparisons between TSA and three state-of-the-art algorithms (UF stands for Utility Function)

$UF\!:\!F1$	F1	TP-FP	TD	Time (s)
TSA	**0.9080**	814.80	532.24	1.74
Ceci & Malerba	**0.0927**	801.36	545.44	0.65
Ruiz	**0.8809**	766.72	695.32	29.39
D'Alessio et al.	**0.9075**	812.88	546.16	14.42
$UF\!:\!TP\text{-}FP$	F1	TP-FP	TD	Time (s)
TSA	0.9050	**813.36**	521.48	1.2
Ceci & Malerba	0.9015	**802.48**	500.88	1.14
Ruiz	0.8770	**764.08**	675.20	24.4
D'Alessio at al.	0.9065	**812.88**	537.48	11.77
$UF\!:\!TD$	F1	TP-FP	TD	Time (s)
TSA	0.8270	704.40	**403.76**	1.48
Ceci & Malerba	0.8202	694.96	**404.96**	0.62
Ruiz	0.7807	654.72	**597.32**	26.31
D'Alessio et al.	0.8107	684.78	**415.60**	13.06

---

[3] Note that 1.9E+8 millisecond are about 54.6 hours.

For each algorithm, we performed three sets of experiments in which a different utility function has been adopted. The baseline of our experiments is a comparison among the four algorithms. In particular, we used:

- $F1$, according to the metric adopted in our previous work on PF [2] and in [13];
- $TP - FP$, according to the metric adopted in [9];
- TD, according to the metric adopted in [7].

As reported in Table 2, for each experimental setting, we calculated the performance and the time spent for each selected metric.

Table 2 summarizes the results. For each experimental setting, the most relevant results (highlighted in bold in the table) are those which correspond to the metric used as utility function. As shown, TSA always performs better in terms of $F1$, $TP - FP$, and TD. As for the running time, the algorithm proposed by Ceci and Malerba shows the best performance.

6 Conclusions and Future Work

In this paper we described TSA, a threshold selection algorithm for hierarchical text categorization. The underlying insight is that relaxing the behavior of a classifier at level k-1 requires to constraint the behavior of classifier at level k, and so on. To assess the performance of TSA, we made experiments aimed at comparing TSA with a relaxed brute-force approach and with the most relevant state-of-the-art algorithms. Experimental results confirm the validity of TSA.

As for future work, considering that most real-world domains are characterized by high imbalance between positive and negative examples, we are currently performing comparative experiments considering also the input imbalance.

Acknowldegments. This research was partly sponsored by the RAS (Autonomous Region of Sardinia), through a grant financed with the "Sardinia POR FSE 2007-2013" funds and provided according to the L.R. 7/2007 "Promotion of the Scientific Research and of the Technological Innovation in Sardinia".

References

1. Addis, A., Armano, G., Vargiu, E.: From a generic multiagent architecture to multiagent information retrieval systems. In: AT2AI-6, Sixth International Workshop, From Agent Theory to Agent Implementation, pp. 3–9 (2008)
2. Addis, A., Armano, G., Vargiu, E.: Assessing progressive filtering to perform hierarchical text categorization in presence of input imbalance. In: Proceedings of International Conference on Knowledge Discovery and Information Retrieval, KDIR 2010 (2010)
3. Addis, A., Armano, G., Vargiu, E.: Experimental assessment of a threshold selection algorithm for tuning classifiers in the field of hierarchical text categorization. In: Proceedings of 17th RCRA International Workshop on Experimental Evaluation of Algorithms for Solving Problems with Combinatorial Explosion (2010)

4. Addis, A., Armano, G., Vargiu, E.: Using the progressive filtering approach to deal with input imbalance in large-scale taxonomies. In: Large-Scale Hierarchical Classification Workshop (2010)
5. Addis, A., Armano, G., Vargiu, E.: A comparative experimental assessment of a threshold selection algorithm in hierarchical text categorization. In: Clough, P., Foley, C., Gurrin, C., Jones, G.J.F., Kraaij, W., Lee, H., Mudoch, V. (eds.) ECIR 2011. LNCS, vol. 6611. Springer, Heidelberg (2011)
6. Bellifemine, F., Caire, G., Greenwood, D.: Developing Multi-Agent Systems with JADE. Wiley Series in Agent Technology. John Wiley and Sons, Chichester (2007)
7. Ceci, M., Malerba, D.: Classifying web documents in a hierarchy of categories: a comprehensive study. Journal of Intelligent Information Systems 28(1), 37–78 (2007)
8. Cost, R.S., Salzberg, S.: A weighted nearest neighbor algorithm for learning with symbolic features. Machine Learning 10, 57–78 (1993)
9. D'Alessio, S., Murray, K., Schiaffino, R.: The effect of using hierarchical classifiers in text categorization. In: Proceedings of of the 6th International Conference on Recherche Information Assiste par Ordinateur (RIAO), pp. 302–313 (2000)
10. Dumais, S.T., Chen, H.: Hierarchical classification of Web content. In: Belkin, N.J., Ingwersen, P., Leong, M.K. (eds.) Proceedings of SIGIR 2000, 23rd ACM International Conference on Research and Development in Information Retrieval, pp. 256–263. ACM Press, New York (2000)
11. Lewis, D.D.: Evaluating and optimizing autonomous text classification systems. In: SIGIR 1995: Proceedings of the 18th Annual International ACM SIGIR Conference on Research and Development in Information Retrieval, pp. 246–254. ACM, New York (1995)
12. Lewis, D.D., Yang, Y., Rose, T., Li, F.: RCV1: A new benchmark collection for text categorization research. Journal of Machine Learning Research 5, 361–397 (2004)
13. Ruiz, M.E.: Combining machine learning and hierarchical structures for text categorization. Ph.D. thesis (2001), supervisor-Srinivasan, Padmini
14. Sebastiani, F.: Machine learning in automated text categorization. ACM Computing Surveys (CSUR) 34(1), 1–55 (2002)
15. Silla, C., Freitas, A.: A survey of hierarchical classification across different application domains. Data Mining and Knowledge Discovery 22, 31–72 (2011), http://dx.doi.org/10.1007/s10618-010-0175-9, 10.1007/s10618-010-0175-9
16. Sun, A., Lim, E.: Hierarchical text classification and evaluation. In: ICDM 2001: Proceedings of the 2001 IEEE International Conference on Data Mining, pp. 521–528. IEEE Computer Society, Washington, DC (2001)
17. Wu, F., Zhang, J., Honavar, V.: Learning classifiers using hierarchically structured class taxonomies. In: Zucker, J.-D., Saitta, L. (eds.) SARA 2005. LNCS (LNAI), vol. 3607, pp. 313–320. Springer, Heidelberg (2005)
18. Yang, Y.: An evaluation of statistical approaches to text categorization. Information Retrieval 1(1/2), 69–90 (1999)
19. Yang, Y.: A study of thresholding strategies for text categorization. In: SIGIR 2001: Proceedings of the 24th Annual International ACM SIGIR Conference on Research and Development in Information Retrieval, pp. 137–145. ACM, New York (2001)

Semi-Supervised Multiclass Kernel Machines with Probabilistic Constraints

Stefano Melacci and Marco Gori

Department of Information Engineering,
University of Siena, 53100 - Siena, Italy
{mela,marco}@dii.unisi.it

Abstract. The extension of kernel-based binary classifiers to multi-class problems has been approached with different strategies in the last decades. Nevertheless, the most frequently used schemes simply rely on different criteria to combine the decisions of a set of independently trained binary classifiers. In this paper we propose an approach that aims at establishing a connection in the training stage of the classifiers using an innovative criterion. Motivated by the increasing interest in the semi-supervised learning framework, we describe a soft-constraining scheme that allows us to include probabilistic constraints on the outputs of the classifiers, using the unlabeled training data. Embedding this knowledge in the learning process can improve the generalization capabilities of the multiclass classifier, and it leads to a more accurate approximation of a probabilistic output without an explicit post-processing. We investigate our intuition on a face identification problem with 295 classes.

Keywords: Multiclass Support Vector Machines, Probabilistic Constraints, Semi-Supervised Learning.

1 Introduction

In multiclass classification problems we have a set of $k > 2$ classes, and the goal is to construct a classifier which correctly predicts the class to which an input point belongs. Although many real-world classification problems are multiclass, many of the most efficient classifiers are specifically designed for binary problems ($k = 2$), such as Support Vector Machines (SVMs) [16].

The simplest strategy to allow them to handle a larger number of classes is commonly referred to as "one-versus-all" (OVA), and it consists in independently training k classifiers to discriminate each class from the $k - 1$ remaining ones. Given an input instance, the class label corresponding to the classifier which outputs the maximum value is selected [14]. Even if some more sophisticated schemes have been proposed (based on directed acyclic graphs, on error correcting coding theory, or on combination of different strategies [13,5,4], for example) the OVA strategy is still one of the most popular approaches, since it has been shown to be as accurate as the most of the other techniques [14].

R. Pirrone and F. Sorbello (Eds.): AI*IA 2011, LNAI 6934, pp. 21–32, 2011.
© Springer-Verlag Berlin Heidelberg 2011

In this paper we propose to tackle the OVA multiclass problem for regularized kernel machines in an innovative fashion, enforcing the k outputs of the classification functions to fulfill a probabilistic relationship. In detail, the *probabilistic constraints* represent a domain information on the multiclass problem that we enforce on the available unlabeled training data, in a Semi-Supervised setting. As a matter of fact, the constraints introduce a dependency among the training stages of the k classifiers, encouraging an inductive transfer that may improve the generalization capabilities of the multiclass classifier.

For this reason our work is related to Multi-Task learning [3] and it shares some principles with approaches that post-process the output of an SVM to approximate posterior estimates [12,17]. It is substantially different from logistic regression models [15,7] that yield probabilistic outcomes based on a maximum likelihood argument. In particular, we focus on the improvements in terms of generalization performance that the proposed constraining can introduce, and not on the strict definition of a model that is guaranteed to produce a probabilistic output. Nevertheless, we show that it is satisfactorily approximated by our soft-constraining procedure.

We investigate our approach on a face identification problem with 295 classes, using the publicly available XM2VTS multimodal dataset. A detailed experimental analysis shows improvements in the quality of the classifier, successfully exploiting the interaction that is established by the probabilistic constraints.

This paper is organized as follows. In Section 2 the Semi-Supervised binary classifiers on which we focus are introduced. In Section 3 the probabilistic constraints are presented. Section 4 collects our experimental results, and, finally, in Section 5 some conclusions are drawn.

2   Multiclass Learning with Constraints

Given a set of objects in $\mathcal{X}$, let us suppose that each object $\boldsymbol{x}_i \in \mathcal{X} \subset \mathbb{R}^d$ is described by a d-dimensional vector of features. In a generic k-class classification problem, we want to infer the function $c : \mathcal{X} \to \mathcal{Y}$, where $\mathcal{Y}$ is a set of labels. We indicate with $y_i \in \mathcal{Y}$ the label associated to $\boldsymbol{x}_i$. Suppose that there is a probability distribution P on $\mathcal{X} \times \mathcal{Y}$, according to which data are generated.

In a Supervised classification problem, we have a labeled training set $\mathcal{L}$ of l pairs,

$$\mathcal{L} = \{(\boldsymbol{x}_i, y_i) | i = 1, \ldots, l, \ \boldsymbol{x}_i \in \mathcal{X}, \ y_i \in \mathcal{Y}\},$$

and the classifier is trained to estimate $c(\cdot)$ using the information in $\mathcal{L}$. A labeled validation set $\mathcal{V}$, if available, is used to tune the classifier parameters, whereas the generalization capabilities are evaluated on an out-of-sample test set $\mathcal{T}$, in a typical inductive setting.

In the Semi-Supervised learning framework, we have also a set $\mathcal{U}$ of u unlabeled training instances,

$$\mathcal{U} = \{\boldsymbol{x}_i | i = 1, \ldots, u, \ \boldsymbol{x}_i \in \mathcal{X}\},$$

that is exploited to improve the quality of the classifier. In a practical context, unlabeled data can be acquired relatively easily, whereas labeling requires the expensive work of one or more supervisors, so that frequently we have $u \gg l$. Unlabeled samples are drawn accordingly to the marginal distribution $P_{\mathcal{X}}$ of P, and the Semi-Supervised framework attempts to incorporate them into the learning process in different ways. Popular Semi-Supervised classifiers make specific assumptions on the geometry of $P_{\mathcal{X}}$, such as, for example, having the structure of a Riemannian manifold [9]. We indicate with n the total number of labeled and unlabeled training points collected in the set $\mathcal{S} = \mathcal{L} \cup \mathcal{U}$, $\mathcal{L} \cap \mathcal{U} = \emptyset$, where the union and intersection are intended to consider only the first element of each pair in $\mathcal{L}$ ($n = l$ in the supervised setting).

SVM-like kernel machines are specifically designed as binary classifiers. Their extension to multiclass classification in a "one-versus-all" (OVA) scheme, consists in the *independent* training of k binary classifiers that discriminate each class from the other $k - 1$ ones. We indicate with f_j the function learnt by the j-th classifier, $j = 1, \ldots, k$, and with y_{ij} the target of the sample $\boldsymbol{x}_i$ in the j-th binary problem. Following the Multi-Task learning framework [3], each function represents a specific "task", collected in the vector $\boldsymbol{f} = [f_1, \ldots, f_k]^T$. In the classical multiclass scenario, all the tasks are defined on the same set of points, and the decision function $c(\cdot)$ that determines the overall output of the classifier is

$$c(\boldsymbol{x}) = \arg\max_j f_j(\boldsymbol{x}). \tag{1}$$

When some prior knowledge on the correlation among the tasks is available, we propose to model it with a set of q constraints on $\{f_1(\boldsymbol{x}), \ldots, f_k(\boldsymbol{x})\}$, represented by the functions $\phi_h : \mathbb{R}^k \to \mathbb{R}$:

$$\phi_h(f_1(\boldsymbol{x}), \ldots, f_k(\boldsymbol{x})) \qquad h = 1, \ldots, q \tag{2}$$

that hold $\forall \boldsymbol{x} \in \mathcal{X}$.

Given a positive definite Kernel function $K : \mathbb{R}^d \times \mathbb{R}^d \to \mathbb{R}$, we indicate with $\mathcal{H}$ the Reproducing Kernel Hilbert Space (RKHS) corresponding to it, and with $\|\cdot\|_{\mathcal{H}}$ the norm of $\mathcal{H}$. Each f_j belongs to $\mathcal{H}^1$, and we formulate the learning problem in the risk minimization scheme, leading to

$$\min_{\boldsymbol{f}} \left(\sum_{j=1}^{k} \sum_{i=1}^{l} V(f_j(\boldsymbol{x}_i), y_{i,j}) + \sum_{j=1}^{k} \lambda_j \cdot \|f_j\|_{\mathcal{H}}^2 + C(\boldsymbol{f}) \right) \tag{3}$$

where

$$C(\boldsymbol{f}) = \sum_{h=1}^{q} \gamma_h \cdot \sum_{i=1}^{n=l+u} L_h(\phi_h(f_1(\boldsymbol{x}_i), \ldots, f_k(\boldsymbol{x}_i))).$$

In detail, the loss function $V(f_j(\boldsymbol{x}_i), y_{i,j})$ measures the fitting quality of each f_j with respect to the targets y_{ij}, and $\|f_j\|_{\mathcal{H}}^2$ is a smoothing factor weighted

---

[1] More generally, we can define each f_j in its own RKHS. We consider the case of a shared RKHS among the functions just to simplify the description of our idea.

by λ_j, that makes the learning problem well-posed[2]. Unlike the previous terms, $C(\cdot)$ is a penalty function that models a correlation among the tasks during the learning process, expressed by the constraints ϕ_h, $h = 1, \ldots, q$. The parameters $\{\gamma_h\}_{h=1}^{q}$ allow us to weight the contribution of each constraint, and the penalty loss function $L_h(\phi_h)$ is positive when the constraint is violated, otherwise it is zero. To simplify the notation we avoided additional scaling factors on the terms of the summation in Eq. 3.

In this soft-constraining scheme, there are no guarantees of ending up in a classifier that perfectly fulfills the relationships of ϕ_h, $h = 1, \ldots, q$, whereas some violations are tolerated. As a matter of fact, the solution of Eq. 3 is a trade-off among label fitting, smoothness on the entire input space, and problem specific constraints. Note that if $V(f_j(\boldsymbol{x}_i), y_{i,j})$ is a linear hinge loss, and we remove the $C(\cdot)$ term (i.e. $\gamma_h = 0$, $h = 1, \ldots, q$) we get SVM classifiers.

If the loss function V and the term $C(\cdot)$ are convex, the problem of Eq. 3 admits a unique minimizer. The optimal solution of Eq. 3 can be expressed as a kernel expansion, as stated in the following Representer Theorem.

Theorem 1. *Let us consider the minimization problem of Eq. 3, where the function $f_1, \ldots, f_k$ belong to a RKHS $\mathcal{H}$. Then the optimal solution f_j^*, $j = 1, \ldots, k$ is expressed as*

$$f_j^*(\boldsymbol{x}) = \sum_{i=1}^{n=l+u} \alpha_{ij} K(\boldsymbol{x}, \boldsymbol{x}_i), \quad j = 1, \ldots, k$$

where $K(\cdot, \cdot)$ is the reproducing kernel associated to $\mathcal{H}$, $\boldsymbol{x}_i \in \mathcal{S}$, and α_{ij} are n scalar values.

Proof: Using a simple orthogonality argument, the proof is a straightforward extension of the representer theorem for plain kernel machines [16]. Is is only sufficient to notice that V is measured on the l labeled training points only, whereas the penalty term $C(\cdot)$ involves a set of constraints evaluated on all the $n = l + u$ samples belonging to $\mathcal{S}$, so that the optimal solution lyes in the span of the n training points (both labeled and unlabeled), as in [1].   □

In the next section we will describe the probabilistic constraints using an instance of the described learning framework. Nevertheless this Semi-Supervised scheme is generic and it can be applied to model any kind of interaction among tasks that comes from a problem-dependent prior knowledge.

3   Probabilistic Constraints

For the j-th task, we select $y_{ij} \in \{0, 1\}$, where $y_{ij} = 1$ means that $\boldsymbol{x}_i$ belongs to class j while 0 indicates that it belongs to the other classes, and we penalize the

---

[2] We are assuming that the *kernel* function is not yielding to interaction among the different tasks, but the essence of what is proposed could be directly extended to the general case of multitask *kernel* functions [2].

label fitting with a squared loss $V(f_j(\boldsymbol{x}_i), y_{ij}) = (f_j(\boldsymbol{x}_i) - y_{ij})^2$. Note that using a hinge loss leads to the same classification accuracies, as investigated in [14], and it would not make any substantial differences with respect to the selected V due to the nature of the constraints that we will introduce in the following (that will enforce f_j in $[0, 1]$).

In its unconstrained and fully Supervised formulation, the OVA scheme does not guarantee that the output values $f_1(\boldsymbol{x}), \ldots, f_k(\boldsymbol{x})$, $\forall \boldsymbol{x} \in I\!\!R^d$ have the properties of a probability (i.e. that they are in $[0, 1]$ and they sum to one). In other words, the classifier do not fulfill what we refer to as the *probabilistic constraints*, that can be modeled with the following linear system,

$$\begin{cases} \sum_{j=1}^{k} f_j(\boldsymbol{x}) = 1 \\ f_j(\boldsymbol{x}) \geq 0 \quad j = 1, \ldots, k. \end{cases} \tag{4}$$

Clearly, for $\boldsymbol{x}_i \in \mathcal{L}$, this information is implicitly embedded on the targets y_{ij}, $j = 1, \ldots, k$. As a consequence, a hypothetic perfect fitting of labeled points would fulfill Eq. 4 $\forall \boldsymbol{x} \in \mathcal{L}$. On the other hand, each f_j is requested to be smooth in the RKHS, and a perfect fitting is generally not achieved.

Interestingly, Eq. 4 gives us a basic domain information on the problem that is supposed to hold in the entire input space. We exploit this information on the relationship among the functions f_j, $j = 1, \ldots, k$, to introduce an interaction among the tasks within their training stage. As a matter of fact Eq. 4 must hold also for points $\boldsymbol{x} \notin \mathcal{L}$, so that we can cast the problem in the Semi-Supervised setting described in Section 2, enforcing the probabilistic constraints also on the (largely available) unlabeled training data. Differently from approaches that estimate probabilities in a post-processing stage [12,17] or from kernel logistic regression [15,7], we do not aim at obtaining a perfectly fulfilled probabilistic output, but at improving the quality of the classifier by task interaction.

We can formulate the probabilistic constraints as a set of $q = k + 1$ linear functions that become zero when they are fulfilled,

$$\begin{cases} \phi_1^{sum}(f_1(\boldsymbol{x}), \ldots, f_k(\boldsymbol{x}))) = \sum_{j=1}^{k} f_j(\boldsymbol{x}) - 1 \\ \phi_h^{pos}(f_{h-1}(\boldsymbol{x})) = \max(-f_{h-1}(\boldsymbol{x}), 0) \quad h = 2, \ldots, k+1. \end{cases} \tag{5}$$

In particular, in this specific problem only ϕ_1^{sum} involves all the k tasks, whereas ϕ_h^{pos}, $h = 2, \ldots, k+1$ model a prior knowledge on the single binary functions. The paired interaction of ϕ_1^{sum} and ϕ_h^{pos} is introduced in the optimization problem of Eq. 3 by the following $C(\cdot)$ term,

$$C(\boldsymbol{f}) = \sum_{i=1}^{n} \Big(\gamma_1 L_1(\phi_1^{sum}(f_1(\boldsymbol{x}_i), \ldots, f_k(\boldsymbol{x}_i))) \\ + \sum_{h=2}^{k+1} \gamma_h L_h(\phi_h^{pos}(f_{h-1}(\boldsymbol{x}_i))) \Big). \tag{6}$$

In order to keep intact the squared nature of the problem, we select L_h, $h = 1, \ldots, q$ to be squared loss functions. A constraint violation is then quadratically penalized. Moreover, the γ_h, $h = 2, \ldots, k+1$ are set to the same value, that we will indicate with γ (without any subscripts), to equivalently weight the ϕ_h^{pos} constraint in each task, whereas γ_1 is set to $k \cdot \gamma$. As a matter of fact we want to emphasize the effect of ϕ_1^{sum} in the minimization procedure, since it encourages the interaction among the binary classifiers. The λ_j, $j = 1, \ldots, k$, coefficients of Eq. 3 are set to λ.

Enforcing the probabilistic constraints with non-linear kernel functions $K(\cdot, \cdot)$ appears the most natural choice. Following the Representer Theorem (Theorem 1), their combination can model highly non-linear f_j, $j = 1, \ldots, k$, allowing the classifier to efficiently alter the shape of each of them accordingly to the interaction with the other ones, to the labeled data fitting and to the smoothness constraint. Popularly used kernels, such as the Gaussian kernel or the polynomial kernel of degree ≥ 2, are well suited for this approach. Modeling the constraints with linear kernels yields to f_j solutions that can easily degenerate towards a constant value, as we experienced, in particular if the dimension of the input space is small. As a matter of fact, enforcing the probabilistic relationship tends to "over constrain" the linear f_j, $j = 1, \ldots, k$. More generally, our approach is well suited for problems with a large number of classes, that emphasize the importance of the task interaction.

3.1   Training the Multiclass Classifier

In order to devise a compact formulation of the minimization problem of Eq. 3, we assume that the n training points of S are ordered so that the first l are the labeled ones and the remaining u are the unlabeled samples. We overload the notation of K, so that it also indicates the Gram matrix associated to the selected kernel function $K(\cdot, \cdot)$ evaluated on the training data, $K = [K(\boldsymbol{x}_i, \boldsymbol{x}_j)]_{i,j=1,\ldots,n}$. Let $A \in \mathbb{R}^{n,k}$ be the matrix where the j-th column collects the n coefficients α_{ij} of the kernel expansion of the j-th task (from Theorem 1). As a result, each column of $KA \in \mathbb{R}^{n,k}$ collects the outputs of a f_j function evaluated on the training points. The subscript is used to refer to a column of a given matrix, so that, for example, A_j indicates the j-th column of A. In $Y \in \{0,1\}^{l,k}$ we collect the task-specific targets for labeled points, i.e. the entry in position (i,j) is y_{ij}. $\mathbf{1} \in \mathbb{R}^n$ is the vector of n elements equal to 1 while $J = [I, O] \in \mathbb{R}^{l,n}$ is a rectangular matrix composed by the identity matrix $I \in \mathbb{R}^{l,l}$ and by a matrix $O \in \mathbb{R}^{l,u}$ of zeros. Finally, the notation $(\boldsymbol{v})_+$ indicates that all the negative elements of the vector $\boldsymbol{v}$ are set to zero.

The instance of the problem of Eq. 3 that we want to minimize is then

$$
\begin{aligned}
A^* = \arg\min_A \Big(& \sum_{j=1}^k (JKA_j - Y_j)^T (JKA_j - Y_j) + \lambda \sum_{j=1}^k A_j^T K A_j \\
& + k\gamma (\sum_{j=1}^k KA_j - \mathbf{1})^T (\sum_{j=1}^k KA_j - \mathbf{1}) + \gamma \sum_{j=1}^k -A_j^T K(-KA_j)_+ \Big).
\end{aligned}
\tag{7}
$$

The objective function is continuous, piecewise quadratic (due to the piecewise linear ϕ_h^{pos}, $h = 2, \ldots, k+1$ and the quadratic loss functions V and L_h,

$h = 1, \ldots, k + 1$), and strictly convex. As recently investigated for the case of Laplacian SVMs [9], we can efficiently optimize it in its primal formulation using Preconditioned Conjugate Gradient (PCG). In our specific problem, the gradient $\nabla_j \in I\!R^n$ of Eq. 7 with respect to the A_j coefficients of the j-th function f_j is

$$\nabla_j = 2K \left(\ J^T(JKA_j - Y_j) + \lambda A_j + k\gamma(\textstyle\sum_{j=1}^{k} KA_j - \mathbf{1}) - \gamma(-KA_j)_+ \right) \quad (8)$$

and preconditioning by the matrix K comes at no additional cost, as discussed in [9].

In order to optimize the multiclass problem, all the ∇_j, $j = 1, \ldots, k$ must be computed. The computational cost of each PCG iteration is $O(kn^2)$, due to the KA product, and the complexity is reduced if K is sparse. Moreover, selecting $\alpha_{ij} = 0$, $i = 1, \ldots, n$, $j = 1, \ldots, k$, as initial point for the optimization procedure, each gradient iteration can be easily parallelized by computing in a separate process each ∇_j, and sharing the KA_j vectors ($j = 1, \ldots, k$) among the parallel processes at the end of the iteration.

4   Experimental Results

Face recognition involves a large number of classes, corresponding to the number of subject to be recognized. We evaluate the performances of the proposed approach in the traditional face identification scenario, where the identity of a given input face must be retrieved among a set of known subjects. If each input face is known to belong to such set, the problem is casted in a *winner-take-all* scenario, where the identity predicted with the highest confidence is selected as overall decision of the classifier, as in the described OVA scheme. SVM-like regularized classifiers have been widely applied to face recognition, focusing on different aspects of the problem [11,6].

We selected the XM2VTS multimodal database to test our system. It is a publicly available collection of face pictures, video sequences, speech recordings taken of 295 subjects, and distributed by the University of Surrey [10]. In particular, it collects 8 frontal view face pictures for each subject, acquired in four separate sessions uniformly distributed over a period of four months. Face images were acquired in controlled conditions (constant face-camera distance and lighting, uniform background) at the resolution of 720x576.

Data was preprocessed as in many popular eigenface-based face recognition approaches [18]. We cropped each image so that only the face region from eyebrows to the chin was kept; images were converted to gray scale and (uniformly) rescaled to 56x64, using the image height as a reference to compute the scaling factor; pixel values were rescaled to $[0, 1]$; Principal Component Analysis (PCA) was performed, and we kept the first 184 eigenfaces, describing 85% of the total variance. In Fig. 1 some examples of the cropped/scaled images are reported.

Following the data partitioning suggested in the second configuration of the so called "Lausanne" protocol (defined for face verification competitions on the

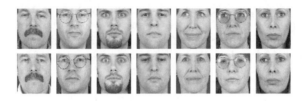

Fig. 1. Some examples of the cropped/scaled XM2VTS faces from two of the four sessions (top and bottom row)

XM2VTS data [8]) we split the available data as described in Table 1, where the details on the XM2VTS data are resumed. Moreover, the training set $\mathcal{S}$ was divided in the sets $\mathcal{L}$ and $\mathcal{U}$ of labeled and unlabeled points, respectively, simulating a Semi-Supervised scenario.

Table 1. The XM2VTS face dataset. For each subject there are 8 images (identified by the numbers $1, \ldots, 8$). The selected data splits follow the Lausanne protocol.

Dataset	Subjects	Total Images
XM2VTS	295	2360

Data Split		Image IDs (per subject)	Total Images
Training	$\mathcal{S} = \mathcal{L} \cup \mathcal{U}$	1, 2, 3, 4	1180
Validation	$\mathcal{V}$	5,6	590
Test	$\mathcal{T}$	7,8	590

We compared the proposed Multiclass Classifier with Probabilistic Constraints (MC-PC) with an unconstrained OVA Multiclass Support Vector Machines (MSVM) that it is one of the most popular approaches and it has been shown to be as accurate as the most of the other existing techniques [14]. Experiments have also been performed using a K-Nearest Neighbors (KNN) classifier with Euclidean distance, since it is frequently used in face recognition experiments.

For each experiment, and for all the compared algorithms, parameters were tuned by computing the error rate on the validation set $\mathcal{V}$ and selecting the best configuration. In the case of MC-PC, the optimal λ and γ were selected from the set $\{10^{-6}, 10^{-4}, 10^{-2}, 10^{-1}, 1, 10, 100\}$. The same range of values was used for the weight λ of the regularization term in MSVM. The number of neighbors in KNN was changed from 1 to 10.

Our analysis is aimed at showing the behavior of the proposed constraining scheme in variable conditions, using different kernel functions and different configurations of the available supervision. Hence, we selected Gaussian and polynomial kernels, due to their popularity in many classification problems. A Gaussian kernel $k(\boldsymbol{a}, \boldsymbol{b}) = \exp \frac{-\|\boldsymbol{a}-\boldsymbol{b}\|^2}{2\sigma^2}$ was tested with $\sigma \in \{5, 10, 20\}$ to assess the behavior of larger and tighter Gaussian functions (rbf). The polynomial kernel $k(\boldsymbol{a}, \boldsymbol{b}) = (\boldsymbol{a}^T \boldsymbol{b} + 1)^p$ was tested with a degree $p = 2$ and with $p = 3$ (poly).

We iteratively increased the fraction of labeled training points to evaluate the behavior of our Semi-Supervised approach as the size of $\mathcal{L}$ increases, and, consequently the amount of unlabeled training points in the set $\mathcal{U}$ decreases. We have $k = 295$ subjects, and for each of them the number of labeled images in $\mathcal{L}$ has been incrementally changed from 1 to 3 whereas $\mathcal{U}$ is reduced from 3 to 1 unlabeled points. The corresponding results are collected in Table 2, where the transductive and inductive configurations are evaluated.

Table 2. The *error rates* on the set $\mathcal{U}$ (transductive) and on the test set $\mathcal{T}$ (inductive). A one-vs-all Multiclass SVM (MSVM) and a Multiclass Classifier with the Probabilistic Constraints (MC-PC) are compared. Two kernel functions and a different numbers of labeled ($|\mathcal{L}|$) and unlabeled ($|\mathcal{U}|$) training points are used ($k = 295$). The variation of correctly identified faces between MSVM and MC-PC is reported in brackets.

Transductive Setting

Kernel	Classifier	$\|\mathcal{L}\| = 1 \cdot k$ $\|\mathcal{U}\| = 3 \cdot k$	$\|\mathcal{L}\| = 2 \cdot k$ $\|\mathcal{U}\| = 2 \cdot k$	$\|\mathcal{L}\| = 3 \cdot k$ $\|\mathcal{U}\| = 1 \cdot k$
rbf ($\sigma = 5$)	MSVM	34.35	28.47	8.81
	MC-PC	**31.53 (+25)**	**27.97 (+3)**	**8.14 (+2)**
rbf ($\sigma = 10$)	MSVM	31.98	26.10	7.46
	MC-PC	**31.30 (+6)**	**25.08 (+6)**	7.46
rbf ($\sigma = 20$)	MSVM	32.09	25.59	9.15
	MC-PC	**31.98 (+1)**	**25.41 (+1)**	**8.47 (+2)**
poly ($p = 2$)	MSVM	36.95	**30**	13.9
	MC-PC	**34.92 (+18)**	30.17 (-1)	**9.83 (+12)**
poly ($p = 3$)	MSVM	39.77	36.27	14.92
	MC-PC	**39.66 (+1)**	**35.93 (+2)**	**14.24 (+2)**
	KNN	*40.68*	*39.83*	*18.64*

Inductive Setting

Kernel	Classifier	$\|\mathcal{L}\| = 1 \cdot k$ $\|\mathcal{U}\| = 3 \cdot k$	$\|\mathcal{L}\| = 2 \cdot k$ $\|\mathcal{U}\| = 2 \cdot k$	$\|\mathcal{L}\| = 3 \cdot k$ $\|\mathcal{U}\| = 1 \cdot k$
rbf ($\sigma = 5$)	MSVM	41.36	30.17	19.32
	MC-PC	**39.32 (+12)**	**29.49 (+4)**	**18.98 (+2)**
rbf ($\sigma = 10$)	MSVM	36.93	24.92	15.25
	MC-PC	**36.44 (+3)**	**24.41 (+3)**	15.25
rbf ($\sigma = 20$)	MSVM	37.29	25.93	15.59
	MC-PC	**36.95 (+2)**	25.93	**15.25 (+2)**
poly ($p = 2$)	MSVM	43.56	31.19	20.34
	MC-PC	**42.37 (+7)**	**30.51 (+4)**	**19.15 (+7)**
poly ($p = 3$)	MSVM	48.81	39.83	28.64
	MC-PC	**47.12 (+10)**	**37.46 (+14)**	**26.78 (+11)**
	KNN	*50.64*	*42.03*	*32.71*

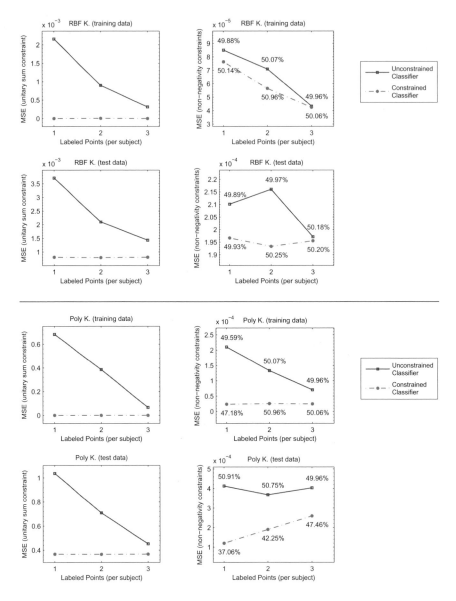

Fig. 2. The Mean Squared Error (MSE) of the unitary sum constraint (ϕ_1^{sum}) and of the non-negativity constraints (ϕ_h^{pos}, $h = 2, \ldots, k + 1$, where the reported MSE is averaged over the k measurements) on training and test data, in function of the number of labeled examples per subject. In the latter, the percentage of points for which $f(\boldsymbol{x}) < 0$ is also displayed. The two plots in each graph describe the behavior of a classifier in which such constraints were or were not enforced during the training stage. In the group of graphs on top, a radial basis function kernel (RBF) with $\sigma = 10$ is used, whereas the group on bottom refers to a polynomial kernel (Poly) of degree 3.

The experimental setup of Table 2 with $|\mathcal{L}| = 1 \cdot k$ and $|\mathcal{U}| = 3 \cdot k$ is the one that is closer to a truly Semi-Supervised setting, where a large amount of unlabeled points are available and just a few labels can be fed to the classifier. In roughly all the experiments (and with all the described kernel functions) the introduction of the probabilistic constraints improves the quality of the classifier, both in a transductive framework (on the set $\mathcal{U}$) and in the inductive one (on the set $\mathcal{T}$), showing an increment of the generalization capabilities.

As the number of labeled data increases, we move towards a fully Supervised setting and we reasonably expect a weaken impact of the probabilistic constraints, since the information that they carry is already included on training labels.

In particular, in the case of Gaussian kernel, the error rate on test data is improved mainly when the kernel width is small. As a matter of fact, due to the very local support of the kernel, the information on labeled points is not enough to fulfill the probabilistic constraints on the set $\mathcal{U}$, and the classifier can benefit from its explicit enforcement, even in this close-to-fully Supervised setup. When a polynomial kernel is used, the interaction among the 295 binary classifiers introduced by the probabilistic constraints keeps increasing the quality of the classifier, since they are far from being fulfilled in the whole space, and the action of our soft-constraining can be appreciated.

Those intuitions are confirmed by the graphs in Fig. 2, where we investigate "how strongly" the output values f_j, $j = 1, \ldots, k$ fulfill the probabilistic constraints in our Semi-Supervised scheme, with respect to the unconstrained case. The Mean Squared Error (MSE) of the unitary sum (ϕ_1^{sum}) and non-negativity (ϕ_h^{pos}, $h = 2, \ldots, k+1$) constraints on training and test data is reported. Thanks to our soft constraining procedure, the output values f_j, $j = 1, \ldots, k$ are very close to a probability. When only 1 labeled example per subject is used to train the classifier, the effect can be significantly appreciated, whereas as such number increases, the output of the unconstrained classifier becomes more similar to the constrained one, since labeled training data is the majority portion of $\mathcal{S}$.

The percentage of points for which $f(\boldsymbol{x}) < 0$ (reported over the plots) does not have significant changes when the constraints are introduced, due to the selected squared penalty approach that do not favor sparsity, whereas the fulfillment of such constraints is improved. Finally, we can see that in the case of Gaussian kernel the outputs f_j, $j = 1, \ldots, k$, of the unconstrained classifier are more similar to a probability than in the case of a polynomial kernel, where the importance of the explicit constraining is evident.

5   Conclusions

In this paper we presented an innovative approach to multiclass classification for popular kernel-based binary classifiers. We casted the "one-versus-all" k-class problem in the Semi-Supervised learning framework, where a set of probabilistic constraints is introduced among the outputs of the k classifiers, establishing an interaction in their training stages that biases an inductive transfer of information. The experiments on a face identification problem with 295 classes showed improvements in the generalization capabilities of the multiclass classifier, together with a more accurate approximation of a probabilistic output. Interestingly, the proposed constraining scheme is general, and it also applies to different categories of classifiers.

References

1. Belkin, M., Niyogi, P., Sindhwani, V.: Manifold regularization: A geometric framework for learning from labeled and unlabeled examples. Journal of Machine Learning Research 7, 2399–2434 (2006)
2. Caponnetto, A., Micchelli, C., Pontil, M., Ying, Y.: Universal multi-task kernels. Journal of Machine Learning Research 9, 1615–1646 (2008)
3. Caruana, R.: Multitask learning. Machine Learning 28(1), 41–75 (1997)
4. Crammer, K., Singer, Y.: On the algorithmic implementation of multiclass kernel-based vector machines. Journal of Machine Learning Research 2, 265–292 (2002)
5. Crammer, K., Singer, Y.: On the learnability and design of output codes for multiclass problems. Machine Learning 47(2), 201–233 (2002)
6. Heisele, B., Ho, P., Wu, J., Poggio, T.: Face recognition: component-based versus global approaches. Computer Vision and Image Understanding 91(1-2), 6–21 (2003)
7. Karsmakers, P., Pelckmans, K., Suykens, J.A.K.: Multi-class kernel logistic regression: a fixed-size implementation. In: Int. Joint Conf. on Neural Networks, pp. 1756–1761. IEEE, Los Alamitos (2007)
8. Matas, J., Hamouz, M., Jonsson, K., et al.: Comparison of face verification results on the XM2VTS database. In: Int. Conf. on Pattern Recognition, vol. 4, pp. 858–863. IEEE Computer Society, Los Alamitos (2000)
9. Melacci, S., Belkin, M.: Laplacian Support Vector Machines Trained in the Primal. Journal of Machine Learning Research 12, 1149–1184 (2011)
10. Messer, K., Matas, J., Kittler, J., Luettin, J., Maitre, G.: XM2VTSDB: The Extended M2VTS Database. In: Proc. of the Int. Conf. on Audio and Video-based Biometric Person Authentication, pp. 72–79 (1999)
11. Phillips, P.: Support vector machines applied to face recognition. Advances in Neural Information Processing Systems, 803–809 (1999)
12. Platt, J.C.: Probabilistic outputs for support vector machines and comparison to regularized likelihood methods. Advances in Kernel Methods Support Vector Learning, 61–74 (2000)
13. Platt, J., Cristianini, N., Shawe-Taylor, J.: Large margin DAGs for multiclass classification. Advances in NIPS 12(3), 547–553 (2000)
14. Rifkin, R., Klautau, A.: In defense of one-vs-all classification. Journal of Machine Learning Research 5, 101–141 (2004)
15. Roth, V.: Probabilistic discriminative kernel classifiers for multi-class problems. In: Radig, B., Florczyk, S. (eds.) DAGM 2001. LNCS, vol. 2191, pp. 246–253. Springer, Heidelberg (2001)
16. Scholkopf, B., Smola, A.: Learning with Kernels. MIT Press, Cambridge (2001)
17. Wu, T., Lin, C., Weng, R.: Probability estimates for multi-class classification by pairwise coupling. Journal of Machine Learning Research 5, 975–1005 (2004)
18. Zhao, W., Chellappa, R., Phillips, P.J., Rosenfeld, A.: Face recognition: A literature survey. ACM Computing Surveys 35(4), 399–458 (2003)

Plugging Numeric Similarity in First-Order Logic Horn Clauses Comparison

S. Ferilli[1,2], T.M.A. Basile[1], N. Di Mauro[1,2], and F. Esposito[1,2]

[1] Dipartimento di Informatica, Università di Bari
{ferilli,basile,ndm,esposito}@di.uniba.it
[2] Centro Interdipartimentale per la Logica e sue Applicazioni, Università di Bari

Abstract. Horn clause Logic is a powerful representation language exploited in Logic Programming as a computer programming framework and in Inductive Logic Programming as a formalism for expressing examples and learned theories in domains where relations among objects must be expressed to fully capture the relevant information. While the predicates that make up the description language are defined by the knowledge engineer and handled only syntactically by the interpreters, they sometimes express information that can be properly exploited only with reference to a suitable background knowledge in order to capture unexpressed and underlying relationships among the concepts described. This is typical when the representation includes numerical information, such as single values or intervals, for which simple syntactic matching is not sufficient.

This work proposes an extension of an existing framework for similarity assessment between First-Order Logic Horn clauses, that is able to handle numeric information in the descriptions.

The viability of the solution is demonstrated on sample problems.

1   Introduction

First-Order Logic (*FOL* for short) is a powerful representation language that allows to express relationships among objects, which is often an unnegligible requirement in real-world and complex domains. Logic Programming [7] is a computer programming framework based on a FOL sub-language, which allows to perform reasoning on knowledge expressed in the form of Horn clauses. Inductive Logic Programming (ILP) [8] aims at learning automatically logic programs from known examples of behaviour, and has proven to be a successful Machine Learning approach in domains where relations among objects must be expressed to fully capture the relevant information. Many AI tasks can take advantage from techniques for descriptions comparison: subsumption procedures (to converge more quickly), flexible matching, instance-based classification techniques or clustering, generalization procedures (to focus on the components that are more likely to correspond to each other). In FOL, this is a particularly complex task due to the problem of *indeterminacy* in mapping portions of one formula onto portions of another.

R. Pirrone and F. Sorbello (Eds.): AI*IA 2011, LNAI 6934, pp. 33–44, 2011.

Usually, predicates that make up the description language used to tackle a specific problem are defined by the knowledge engineer that is in charge of setting up the reasoning or learning task, and are handled as purely syntactic entities by the systems. However, the use of uninterpreted predicates and terms (meaning by 'interpretation' their mapping onto meaningful objects, concepts and relationships) is often too limiting for an effective application of this kind of techniques to real-world problems. Indeed, in the real world there are a huge number of implicit connections and inter-relationships between items that would be ignored by the system. For limited and simple domains, only a few of these relationships are actually significant, and must be expressed not to prevent finding a solution. In these cases, they can be provided in the form of a background knowledge. However, if the amount of relevant information to be expressed as background knowledge grows, this becomes infeasible manually and requires the support of readily available resources in the form of explicit knowledge items or computational procedures.

This work builds on previous results concerning a framework for similarity assessment between FOL Horn clauses, where the overall similarity depends on the similarity of the pairs of literals associated by the least general generalization, the similarity of two literals in turn depends on the similarity of their corresponding arguments (i.e., terms), and the similarity between two terms is computed according to the predicates and positions in which they appear. Following a previous paper in which the framework was extended to handle taxonomic knowledge, here, a novel and general approach to the assessment of similarity between numeric values and intervals is proposed, and its integration as a corresponding further extension of the similarity framework for clauses including numeric information is described.

The rest of this paper is organized as follows. Section 3 introduces the basic formula and framework for the overall assessment of similarity between Horn clauses. Section 4 proposes an application of the same formula to compute the numeric similarity between values and/or intervals, and introduces it in the previous framework. Section 5 shows experiments that suggest the effectiveness of the proposed approach. Lastly, Section 6 concludes the paper and outlines future work directions.

2   Background

Let us preliminary recall some basic notions involved in Logic Programming. The *arity* of a predicate is the number of arguments it takes. A *literal* is an n-ary predicate, applied to n terms, possibly negated. *Horn clauses* are logical formulæ usually represented in Prolog style as $l_0 \text{ :- } l_1, \ldots, l_n$ where the l_i's are *literals*. It corresponds to an implication $l_1 \wedge \cdots \wedge l_n \Rightarrow l_0$ to be interpreted as "l_0 (called *head* of the clause) is true, provided that l_1 and ... and l_n (called *body* of the clause) are all true". Datalog [2] is, at least syntactically, a restriction of Prolog in which, without loss of generality [9], only variables and constants (i.e., no functions) are allowed as terms. A set of literals is *linked* if and only if each literal in the set has at least one term in common with another literal in the set. We will deal with the case of linked Datalog clauses. In the following, we will call *compatible* two sets or sequences of literals that can be mapped onto each other without yielding inconsistent term associations (i.e., a term in one formula cannot correspond to different terms in the other formula).

Real-world problems, and the corresponding descriptions, often involve numeric features, that are to be expressed in the problem formalization and handled by the inferential procedures. For instance, when describing a bicycle we would like to say that the front wheel diameter is 28 inches, or when defining the title block in a scientific paper we would like to say that it must be placed in a range going from 5% to 20% of the page height from the top. Let us call this kind of features (such as size, height in the above examples) *numeric attributes*. Clearly, to be properly handled such a kind of information needs to be suitably interpreted according to a background knowledge consisting of the mathematical models of numbers and their ordering relationships. Unfortunately, the purely logical setting ignores such a background knowledge, and considers each single value as completely unrelated to all other values. This problem has been tackled in two different ways.

Keeping the purely logical setting ensures general applicability of the logical representation and inference techniques: a classical solution in this case has been discretization of the range of numeric values allowed for a given attribute into pre-defined intervals, each of which can be associated to a symbolic descriptor (e.g., size_small, size_large, ...; position_top, position_middle, ...). Thus, the original descriptions are to be pre-processed to turn all instances of numeric attributes into the corresponding discretized descriptors. What is a useful number of intervals in which splitting the range of values allowed for a numeric attribute? How to choose the cut points between intervals? Both choices are crucial, since once it is determined even points that are very close to each other (e.g., 4.999 and 5.001 for a cut point placed at 5) will be considered as two completely different entities. Techniques for (semi-)automatic definition of the intervals given samples of values for the attribute have been proposed (e.g., [1]), based on statistics on the values occurrence and distribution, although a (partial) manual intervention is often required to provide and/or fix their outcome. In any case, if the intervals are not to be considered as completely distinct entities, the problem is simplified but not completely solved, since additional background knowledge must be provided to express the ordering relationships between intervals (requiring a number of items that is quadratic in the number of intervals, to express which one precedes which other for all possible pairs) or progressive levels of aggregations of groups of adjacent intervals into wider ones (requiring, for all possible combinations, a number of items that is exponential in the number of intervals).

As another option, plugging the ability to handle numeric information directly in the inference engine somehow 'spoils' its behavior and adds complexity (reducing efficiency). A problem (in both cases) is the fact that the specific way in which numeric information is to be handled is strictly domain-dependent: Are values 15 and 300 close or distant (and how much are they)? This question cannot be answered in general (a difference of, say, 215 meters might be meaningful when comparing two fields, but completely insignificant when comparing planets according to their size.

3 Similarity Framework

In this section, a short description of the similarity framework in its current status, borrowed from [5], will be provided. The original framework for computing the similarity between two Datalog Horn clauses has been provided in [4]. Intuitively, the evaluation of similarity between two items i' and i'' might be based both on parameters expressing the amounts of common features, which should concur in a positive way to the similarity evaluation, and of the features of each item that are not owned by the other (defined as the *residual* of the former with respect to the latter), which should concur negatively to the whole similarity value assigned to them [6]:

n , the number of features owned by i' but not by i'' (*residual* of i' wrt i'');
l , the number of features owned both by i' and by i'';
m , the number of features owned by i'' but not by i' (*residual* of i'' wrt i').

A similarity function that expresses the degree of similarity between i' and i'' based on the above parameters, and that has a better behaviour than other formulæ in the literature in cases in which any of the parameters is 0, is [4]:

$$sf(i', i'') = \mathrm{sf}(n, l, m) = 0.5\frac{l+1}{l+n+2} + 0.5\frac{l+1}{l+m+2} \tag{1}$$

It takes values in $]0, 1[$, which resembles the theory of probability and hence can help human interpretation of the resulting value. When $n = m = 0$ it tends to the limit of 1 as long as the number of common features grows. The full-similarity value 1 is never reached, being reserved to two items that are exactly the same ($i' = i''$), which can be checked in advance. Consistently with the intuition that there is no limit to the number of different features owned by the two descriptions, which contribute to make them ever different, it is also always strictly greater than 0, and will tend to such a value as long as the number of non-shared features grows. Moreover, for $n = l = m = 0$ the function evaluates to 0.5, which can be considered intuitively correct for a case of maximum uncertainty. Note that each of the two terms refers specifically to one of the two items under comparison, and hence they could be weighted to reflect their importance.

In FOL representations, usually terms denote objects, unary predicates represent object properties and n-ary predicates express relationships between objects; hence, the overall similarity must consider and properly mix all such components. The similarity between two clauses C' and C'' is guided by the similarity between their structural parts, expressed by the n-ary literals in their bodies, and is a function of the number of common and different objects and relationships between them, as provided by their least general generalization

$C = l_0 \text{ :- } l_1, \ldots, l_k$. Specifically, we refer to the θ_{OI} generalization model [3]. The resulting formula is the following:

$$\text{fs}(C', C'') = \text{sf}(k' - k, k, k'' - k) \cdot \text{sf}(o' - o, o, o'' - o) + \text{avg}(\{\text{sf}_s(l'_i, l''_i)\}_{i=1,\ldots,k})$$

where k' is the number of literals and o' the number of terms in C', k'' is the number of literals and o'' the number of terms in C'', o is the number of terms in C and $l'_i \in C'$ and $l''_i \in C''$ are generalized by l_i for $i = 1, \ldots, k$. The similarity of the literals is smoothed by adding the overall similarity in the number of overlapping and different literals and terms.

The similarity between two compatible n-ary literals l' and l'', in turn, depends on the multisets of n-ary predicates corresponding to the literals directly linked to them (a predicate can appear in multiple instantiations among these literals), called *star*, and on the similarity of their arguments:

$$\text{sf}_s(l', l'') = \text{sf}(n_s, l_s, m_s) + \text{avg}\{\text{sf}_o(t', t'')\}_{t'/t'' \in \theta}$$

where θ is the set of term associations that map l' onto l'' and S' and S'' are the stars of l' and l'', respectively:

$$n_s = |S' \setminus S''| \qquad l_s = |S' \cap S''| \qquad m_s = |S'' \setminus S'|$$

Lastly, the similarity between two terms t' and t'' is computed as follows:

$$\text{sf}_o(t', t'') = \text{sf}(n_c, l_c, m_c) + \text{sf}(n_r, l_r, m_r)$$

where the former component takes into account the sets of properties (unary predicates) P' and P'' referred to t' and t'', respectively:

$$n_c = |P' \setminus P''| \qquad l_c = |P' \cap P''| \qquad m_c = |P'' \setminus P'|$$

and the latter component takes into account how many times the two objects play the same or different roles in the n-ary predicates; in this case, since an object might play the same role in many instances of the same relation, the *multisets* R' and R'' of roles played by t' and t'', respectively, are to be considered:

$$n_r = |R' \setminus R''| \qquad l_r = |R' \cap R''| \qquad m_r = |R'' \setminus R'|$$

This general (uninterpreted) framework was extended [5] to consider taxonomic information (assuming that there is some way to distinguish taxonomic predicates from ordinary ones). Using the same similarity function, the parameters for taxonomic similarity between two concepts are determined according to their associated sets of ancestors (say I_1 and I_2) in a given heterarchy of concepts: their intersection (i.e., the number of common ancestors) is considered as common information (yielding $l_t = |I_1 \cap I_2|$), and the two symmetric differences as residuals (yielding $n_t = |I''_1 - I'_1|$ and $m_t = |I''_2 - I'_2|$). Since the taxonomic predicates represent further information about the objects involved in a description, in addition to their properties and roles, term similarity is the component where the corresponding similarity can be introduced in the overall framework:

$$\text{sf}_o(t', t'') = \text{sf}(n_c, l_c, m_c) + \text{sf}(n_r, l_r, m_r) + \text{sf}(n_t, l_t, m_t)$$

where the additional component refers to the number of common and different ancestors of the two concepts associated to the two terms, as specified above.

4   Numeric Similarity

Given the above considerations, it is clear that, in an extended framework considering interpreted predicates and constants in addition to simple syntactic entities, the ability to handle numeric information is at least as fundamental as considering conceptual ones. Hence, the motivation for this work on the extension of the similarity-related technique. While observations usually described specific cases using single constants, models typically refer to allowed ranges of values: thus, we are interested in handling all the following cases:

– comparison between two intervals
– comparison between an interval and a value
– comparison between two values

To ensure a smooth integration of the new components, a first requirement is preserving the same basic similarity function, whence the need to specify how to extract parameters l, n and m from numeric comparisons. In this respect, let us start our discussion from the case of comparison between two intervals, say $[i'_1, i'_2]$ and $[i''_1, i''_2]$. Two intuitive approaches are available (assume, without loss of generality, that $i'_1 \leq i'_2$):

– basing the comparison on the distance between the interval **extremes**: then, we take $n = |i''_1 - i'_1|$, $m = |i''_2 - i'_2|$ and $l = \min(i'_2, i''_2) - \max(i'_1, i''_1)$ if non-negative (or 0 otherwise). Note that this solution does not take into account the actual distance between the two intervals when they are disjoint, but modifying the function to take into account this distance as a negative value would spoil uniformity of the approach and make the function definition more complex.
– considering the intervals as **sets**, and exploiting set operators and interpretation: l would be the width (expressed as $|| \cdot ||$) of the overlapping part $(l = ||[i'_1, i'_2] \cap [i''_1, i''_2]||)$, and n, m their symmetric differences, respectively $(n = ||[i'_1, i'_2] \setminus [i''_1, i''_2]||, m = ||[i''_1, i''_2] \setminus [i'_1, i'_2]||)$.

Both strategies can be straightforwardly applied also to the comparison of an interval to a single value, considered as an interval in which both extremes are equal. However, the l parameter would always be zero in this case.

Let us now check and evaluate the behavior of the two candidate approaches by applying them on a set of sample intervals, as shown in Table 1. Overall, both approaches seem reasonable. As expected, their outcome is the same for partially overlapping intervals, so that case is not a discriminant to prefer either over the other. Different behavior emerges in the cases of disjoint intervals or of inclusion of intervals (which is always the case of an interval compared to a single value). In the former, the extreme-based approach ensures more distinction power, because the distance between the intervals is taken into account. While this behavior seems reasonable (according to the intuition that the farther two intervals, the more different they are), on the other hand, in the case of an interval being a sub-interval of the other it is not. Indeed, the set-based approach charges the whole

Table 1. Similarity values between sample intervals

Intervals		Extreme-based				Set-based			
I_1	I_2	n	l	m	similarity	similarity	n	l	m
$[10,15]$	$[11,16]$	1	4	1	$0.\overline{714285}$	$0.\overline{714285}$	1	4	1
$[10,15]$	$[10,15]$	0	5	0	$0.\overline{857142}$	$0.\overline{857142}$	0	5	0
$[21,25]$	$[1,5]$	20	0	20	$0.0\overline{45}$	$0.1\overline{6}$	4	0	4
$[1,5]$	$[21,25]$	20	0	20	$0.0\overline{45}$	$0.1\overline{6}$	4	0	4
$[1,5]$	$[6,10]$	5	0	5	$0.\overline{142857}$	$0.1\overline{6}$	4	0	4
$[1,5]$	$[0,6]$	1	4	1	$0.\overline{714285}$	$0.7291\overline{6}$	0	4	2

Table 2. Similarity values between sample interval-value pairs

Intervals		Extreme-based				Set-based			
I	v	n	l	m	similarity	similarity	n	l	m
$[1,5]$	$1 \equiv [1,1]$	0	0	4	$0.\overline{3}$	$0.\overline{3}$	4	0	0
$[1,5]$	2	1	0	3	$0.2\overline{6}$	$0.\overline{3}$	4	0	0
$[1,5]$	3	2	0	2	0.25	$0.\overline{3}$	4	0	0
$[1,5]$	4	3	0	1	$0.2\overline{6}$	$0.\overline{3}$	4	0	0
$[1,5]$	5	4	0	0	$0.\overline{3}$	$0.\overline{3}$	4	0	0
$[1,5]$	$6 \equiv [6,6]$	5	0	1	0.238095	$0.\overline{3}$	0	0	4
$[1,5]$	21	20	0	16	$0.0\overline{5}$	$0.\overline{3}$	4	0	0
$[7,10]$	11	4	0	1	0.25	0.35	3	0	0
$[7,10]$	14	7	0	4	$0.13\overline{8}$	0.35	3	0	0
$[7,10]$	3	4	0	7	$0.13\overline{8}$	0.35	3	0	0
$6 \equiv [6,6]$	$10 \equiv [10,10]$	4	0	4	$0.1\overline{6}$	0.5	0	0	0
$10 \equiv [10,10]$	$10 \equiv [10,10]$	0	0	0	0.5	0.5	0	0	0

difference to the residual of the larger interval, which complies with the intuition that it has more 'different stuff' that the other does not have; conversely, the extreme-based approach splits such a difference on both parameters n and m, resulting in a smaller similarity value.

This is even more evident looking at the behavior of the two approaches in the case of an interval and a single value, as shown in Table 2. Only the case where the first item is an interval and the second one is a value is reported, due to the similarity function being symmetric providing the same result also in the opposite case. Here, given a value included in an interval, their similarity according to the set-based approach is constant, while in the extreme-based approach it is affected by the position of the former within the latter: the closer the value to the middle of the interval, the smaller their similarity; conversely, as long as the value approaches the interval extremes, their similarity grows up to the same similarity as the set-based approach. If we assume that the interval just specifies an allowed range, with no reason to prefer particular regions within that range, the set-based approach is clearly more intuitive. Again, in the case of a value outside the interval (corresponding to disjoint intervals) an opposite evaluation holds: the actual distance of the value from the interval is considered by the extreme-based approach, affecting its evaluation, but not by the set-based

Table 3. Similarity values between sample pairs of values

Values		Extreme-based				Specific
v_1	v_2	n	l	m	similarity	similarity
1	1	0	0	0	0.5	1
1	2	1	0	1	$0.\overline{3}$	0.5
1	3	2	0	2	0.25	$0.\overline{3}$
1	4	3	0	3	0.2	0.25
1	5	4	0	4	$0.1\overline{6}$	0.2
1	10000	9999	0	9999	$0.\overline{00009999}$	0.0001
6	4	2	0	2	0.25	$0.\overline{3}$

approach, where the absurd that a value outside a range has a larger similarity than a value falling in the range happens. In both approaches, the larger the interval, the smaller the similarity (which is consistent with the intuition that a value is more likely to fall in a wider range than in a narrower one).

When comparing two values, in particular, the set-based approach returns maximum uncertainty about their similarity (0.5) due to all parameters being zero, and hence it is not applicable. The extreme-based approach evaluates their similarity according to how close to each other they are on the real-valued axis, but loses expressive power (because any pair of values yields $n = m$), and has the additional drawback that when comparing a value to itself it yields $n = l = m = 0$ and hence similarity 0.5 (whereas we would expect to get 1 as a perfect matching). Thus, a different approach is needed. The similarity assessment should be independent of the different ranges of values used in the specific domain (e.g., the range for describing the length of a pen is incomparable to that for describing the width of a building). We propose the following formula:

$$\text{sf}_n(v_1, v_2) = \frac{1}{|v_1 - v_2| + 1}$$

Let us briefly examine the behavior of such a function. It is clearly symmetric. When the difference between the two values approaches zero it approaches 1, and becomes actually 1 for $v_1 = v_2$, as expected (differently from the previous cases in the logical setting, one is sure that two equal values denote exactly the same entity). As long as the difference increases, the function monotonically approaches 0, but never reaches that value (according to the intuition that a larger difference can be always thought of, requiring a smaller similarity value). The rate at which 0 is approached decreases as long as the difference takes larger and larger values, consistently with the intuition that for very large distances one does not care small variations. Of course, if the descriptions are consistent, only values referred to the same kind of entities/attributes will be compared to each other, and hence the corresponding ranges of similarities should be consistent and comparable to each other.

As to the similarity between values, some sample comparisons are reported in Table 3 (both the specific strategy and the extreme-based one are symmetric).

As desired, identity of values yields similarity 1, and wider distances among the two values result in smaller similarity values (independently of the actual values).

Summing up, a specific strategy is needed when comparing two values. When at least an interval is involved, both the set-based and the extreme-based strategies are equivalent in the case of partially overlapping intervals. Otherwise, the former is better in the case of an interval or value being included in another interval, because it better fits the concept on which the similarity function parameters are based. Conversely, the extreme-based strategy is able to consider the actual distance from the interval and/or value extremes in the case of disjoint intervals, which affects the residual parameters. Overall, a cooperation of the three strategies is desirable to fully match the spirit of the similarity function parameters. In this case, a deeper empirical study is adviceable, and is planned as future work, to establish if and how a smooth combination thereof can be obtained, ensuring meaningful overall results and comparable similarity assessments even when determined by different approaches (e.g., the similarity for two distinct values should not be larger than the similarity between a value and an interval it belongs to).

A final issue is where to embed the new similarity component in the overall First-Order Logic similarity framework. Without loss of generality, we can assume that numeric attributes (those associating a numeric value to a given entity) apply to single terms in the overall description, and thus are represented by binary predicates $a(t, v)$ meaning that "object t has (numeric) value v for attribute a". Indeed, the case of relationships associated with numeric information (e.g., the weight of arcs in a graph), can be easily handled by reifying the relationship in a term and then associating the value to the new term: e.g., $arc\_weight(n_1, n_2, v)$ would become $arc(n_1, n_2, a), weight(a, v)$. In this setting, the numeric predicates represent further information about the objects involved in a description, in addition to their properties and roles (and taxonomic position, in case), and hence term similarity is the component where the corresponding similarity can be introduced in the overall framework.

Of course, we assume that there is some way to distinguish numeric predicates from ordinary ones, so that they can be handled separately by the procedures (the numeric values themselves should be sufficient for this). The overall similarity between two terms becomes:

$$\mathrm{sf}_o(t', t'') = \mathrm{sf}(n_c, l_c, m_c) + \mathrm{sf}(n_r, l_r, m_r)[+\mathrm{sf}(n_t, l_t, m_t)] + \mathrm{sf}_n(N_1, N_2)$$

where the components can be weighted differently if needed, and the additional component refers to the numeric similarity between intervals and/or single values, as applicable, specified above.

5   Evaluation

To fully evaluate the effectiveness of the proposed approach to numeric similarity embedded in the wider First-Order Logic Horn-clause similarity framework, a dataset involving numeric attributes and including both intervals and specific values should be exploited. Unfortunately, the available datasets typically fill each numeric attribute with a single number expressing its value for the object being described, while intervals are usually exploited in general models rather than in observations. Thus, in this work we will evaluate only the comparison of specific values embedded in the overall structural similarity. We are currently working at building datasets including both intervals and specific values, and an extended evaluation of the full-fledged numeric similarity strategy is planned as future work. Specifically, we focus on the classical problem of Mutagenesis [10], as a most famous success in ILP and as a dataset involving both relational and numeric descriptors. It should be noted that this is a very stressing dataset for our technique, because it is known for being a problem where the key to success is in the relational part, rather than in the attribute/value (numeric) one. Indeed, it was exploited to demonstrate that ILP can learn predictive theories from a dataset on which classical attribute-value techniques based on regression failed. Thus, the numeric values are more likely to act as noise in the proper similarity assessment, rather than as useful information to be leveraged (in other word, one might expect that ignoring numeric information would improve effectiveness). It aims at learning when a chemical compound is active, and when it is not, with respect to mutagenicity. Molecules in this dataset are represented according to the atoms that make them up, and to the bonds linking those atoms, so that the 3D structure of the molecule is completely described. Atom substances are represented by symbolic predicates, while numeric features are used for atom weights and bond charges. The dataset includes 188 examples, of which 63 are positive (class *active*) and 125 are negative (class *not_active*).

The clustering procedure was stopped as soon as a loop was detected: after a few steps, the k-means algorithm started oscillating between two sets of clusters such that applying a further distribution on the former yields the latter, and *vice versa*. Thus, both can be considered as candidate solutions, there being no reason to prefer either over the other. The purity for the two candidate clusters was, however, so close that either choice might be adopted: 76.59% and 76.06%, respectively.

Another experiment, more suitable for evaluation of the correctness of the proposed approach, concerns the problem of classification of documents according to their layout description. In this case, the same dataset as in [3] was exploited, made up of 353 examples of scientific papers from 4 different series: Elsevier journals, Springer-Verlag Lecture Notes (SVLN), Machine Learning Journal (MLJ), Journal of Machine Learning Research (JMLR). The descriptions involve both relational descriptors (for the mutual position and alignment among layout blocks) and attribute-value descriptors for each layout block. In particular, 4 numeric attributes are present: horizontal/vertical position in the page of the block, and width/height thereof. Thus, differently from the Mutagenesis dataset, the expectation is that these descriptors are very significant to class discrimination. Previous applications on this dataset [3] were carried out by discretizing the allowed values for these descriptors into (manually-defined) intervals, and assigning a symbolic descriptor to each such interval. Here, the aim is checking whether

Table 4. Dispersion matrix for the Document Clustering problem

	Elsevier	MLJ	JMLR	SVLN	Total	Errors	Accuracy
Cluster 1	50	1	0	1	52	2	96.15
Cluster 2	7	84	1	0	92	8	91.30
Cluster 3	0	30	99	0	129	30	76.74
Cluster 4	4	7	0	69	80	11	86.25
Total	61	122	100	70	353	51	85.55
Missed	11	38	1	1	–	–	–

the introduction of the numeric similarity component, able to handle directly the original observations (and hence avoiding the need for human intervention to provide a discretization knowledge), can provide effective results. Again, a k-means algorithm is run, asked to find 4 groups in the observations obtained by hiding the class information from the above examples. After 100656.767 seconds needed to compute the similarity matrix among all pairs of observations, the resulting clusters are shown in Table 4.

It clearly emerges which clusters represent which classes, due to a predominance of the corresponding elements: Cluster 1 corresponds to Elsevier, including 50 out of its 61 correct elements (plus 2 wrong elements from other classes), Cluster 2 corresponds to MLJ, Cluster 3 corresponds to JMLR and Cluster 4 to SVLN. Given this correspondence, the *purity* of each cluster with respect to the associated class can be computed, as the ratio of elements from that class over the total elements in the cluster. There are 51 errors overall, yielding an overall 85.55% accuracy, that increases to 87.61% taking the average accuracy of the various classes/clusters. Compared to the 92.35% purity reported in [4] it can be considered a satisfactory result, considering that here no help is provided to the system, while there a manual discretization carried out by experts was provided to turn numeric values into symbolic ones (the reference value of *supervised* learning on the same dataset, using the experts' discretization, is 98% accuracy, that can be considered as the counterpart of purity). The worst performing class is MLJ, that is also the largest one however. It has the largest number of missed items (most of which fall in the JMLR class/cluster, and all clusters include at least one element from this class. Indeed, by observing its layout, it turns out that it is in some way at the crossing of the other classes, and in particular the 30 documents in JMLR are actually very similar to real JMLR ones (the Authors blocks are in the same place, under the title, both have a heading at the top of the page — although it is narrower in JMLR). This suggests that these kinds of blocks are the most significant to discriminate different classes in this dataset.

6   Conclusions

Horn clause Logic is a powerful representation language for automated learning and reasoning in domains where relations among objects must be expressed to fully capture the relevant information. While the predicates in the description language are defined by the knowledge engineer and handled only syntactically by the interpreters, they sometimes express information that can be properly exploited only with reference to a background knowledge in order to capture unexpressed and underlying relationships among the concepts described. After a previous work aimed at extending a general similarity framework for First-Order Logic Horn clauses with the ability to deal with taxonomic information, in this paper we presented another extension concerning numeric descriptors involving intervals and/or numeric values. A composite approach to the numeric similarity assessment, compliant with the general framework, has been proposed, discussed and evaluated. Clustering experiments on a typical dataset mixing relational and numeric information show that the proposal is effective.

Future work will concern deeper empirical evaluation of the behavior of the proposed approaches to numeric computation in the case of intervals, and of its integration in the general similarity framework (e.g., to determine what weight it should have with respect to the other similarity parameters). Then, experiments aimed at application of the framework to other real-world problems are planned. Finally, another research direction concerns the exploitation of the proposed technique as a support to refinement operators for incremental ILP systems.

References

[1] Biba, M., Esposito, F., Ferilli, S., Di Mauro, N., Basile, T.M.A.: Unsupervised discretization using kernel density estimation. In: IJCAI 2007, pp. 696–701 (2007)
[2] Ceri, S., Gottlöb, G., Tanca, L.: Logic Programming and Databases. Springer, Heidelberg (1990)
[3] Esposito, F., Fanizzi, N., Ferilli, S., Semeraro, G.: A generalization model based on oi-implication for ideal theory refinement. Fundamenta Informaticæ 47(1-2), 15–33 (2001)
[4] Ferilli, S., Basile, T.M.A., Biba, M., Di Mauro, N., Esposito, F.: A general similarity framework for horn clause logic. Fundamenta Informaticæ 90(1-2), 43–46 (2009)
[5] Ferilli, S., Biba, M., Mauro, N., Basile, T.M., Esposito, F.: Plugging taxonomic similarity in first-order logic horn clauses comparison. In: Serra, R., Cucchiara, R. (eds.) AI*IA 2009. LNCS, vol. 5883, pp. 131–140. Springer, Heidelberg (2009)
[6] Lin, D.: An information-theoretic definition of similarity. In: Proc. 15th International Conf. on Machine Learning, pp. 296–304. Morgan Kaufmann, San Francisco (1998)
[7] Lloyd, J.W.: Foundations of Logic Programming, 2nd edn. Springer, Heidelberg (1987)
[8] Muggleton, S.: Inductive logic programming. New Generation Computing 8(4), 295–318 (1991)
[9] Rouveirol, C.: Extensions of inversion of resolution applied to theory completion. In: Inductive Logic Programming, pp. 64–90. Academic Press, London (1992)
[10] Srinivasan, A., Muggleton, S., King, R., Sternberg, M.: Mutagenesis: ILP experiments in a non-determinate biological domain. In: Wrobel, S. (ed.) Proceedings of the 4th International Workshop on Inductive Logic Programming. GMD-Studien, vol. 237, pp. 217–232 (1994)

Learning to Tag Text from Rules and Examples

Michelangelo Diligenti, Marco Gori, and Marco Maggini

Dipartimento di Ingegneria dell'Informazione, Università di Siena, Italy
{diligmic,marco,maggini}@dii.unisi.it

Abstract. Tagging has become a popular way to improve the access to resources, especially in social networks and folksonomies. Most of the resource sharing tools allow a manual labeling of the available items by the community members. However, the manual approach can fail to provide a consistent tagging especially when the dimension of the vocabulary of the tags increases and, consequently, the users do not comply to a shared semantic knowledge. Hence, automatic tagging can provide an effective way to complete the manual added tags, especially for dynamic or very large collections of documents like the Web. However, when an automatic text tagger is trained over the tags inserted by the users, it may inherit the inconsistencies of the training data. In this paper, we propose a novel approach where a set of text categorizers, each associated to a tag in the vocabulary, are trained both from examples and a higher level abstract representation consisting of FOL clauses that describe semantic rules constraining the use of the corresponding tags. The FOL clauses are compiled into a set of equivalent continuous constraints, and the integration between logic and learning is implemented in a multi-task learning scheme. In particular, we exploit the kernel machine mathematical apparatus casting the problem as primal optimization of a function composed of the loss on the supervised examples, the regularization term, and a penalty term deriving from forcing the constraints resulting from the conversion of the logic knowledge. The experimental results show that the proposed approach provides a significant accuracy improvement on the tagging of bibtex entries.

1 Introduction

Tagging consists in associating a set of terms to resources (e.g. documents, pictures, products, blog posts, etc.) with the aim of making it easier their search and organization. Tags reflecting semantic properties of the resources (e.g. categories, keywords summarizing the content, etc.) are effective tools for searching the collection. Therefore, high consistency and precision in the tag assignment would allow the development of more sophisticated information retrieval mechanisms, as the ones typically provided in search-by-keyword applications. In the context of folksonomies or Web directories, tagging is manually performed and the vocabulary of tags is usually unrestricted and freely chosen by the users. Beside semantic tags, the users often use tags to represent meta–information to be attached to each item (e.g. dates, the camera brand for pictures, etc.). Unfortunately, a manual collective tagging process has many limitations. First, it

R. Pirrone and F. Sorbello (Eds.): AI*IA 2011, LNAI 6934, pp. 45–56, 2011.

is not suited for very large and dynamic collections of resources, where response time is crucial. Furthermore, the collective tagging process does not provide any guarantee of consistency of the tags across the different items, creating many issues for the subsequent use of the tags [8]. This problem is especially crucial in the context of social networks and folksonomies where there is not a standardized semantic knowledge shared by the users, since tags are independently chosen by each user without any restriction.

Automatic text tagging is regarded as a way to address, at least partially, these limitations [7]. Automatic text tagging can be typically seen as a classical text categorization task [10], where each tag corresponds with a different category. Differently to many categorization tasks explored in the literature, the number of tags is typically in the order of hundreds to several thousands, and the tags are not mutually exclusive, thus yielding a multi-class classification task. Automatic text categorization and tagging has been approached with either pure ontology-based approaches [12,6] or with learning-from-examples techniques based on machine learning [1,11].

Manually inserted tags can be used to effectively train an automatic tagger, which generalizes the tags to new documents [9]. However, when an automatic text tagger is trained over the tags inserted by the users of a social network, it may inherit the same inconsistencies of the training data.

The approach presented in this paper bridges the knowledge-based and machine learning approaches, where a text categorizer and the reasoning process defined via a formal logic language are jointly implemented via kernel machines. The formal logic language enforces the tag consistency without depending on specially trained human taggers. In particular, higher level abstract representations consist of FOL clauses that constrain the configurations assumed by the task predicates, each one associated to a tag. The FOL clauses are then compiled into a set of equivalent continuous constraints, and the integration between logic and learning is implemented in a multi-task learning scheme where the kernel machine mathematical apparatus makes it possible to cast the problem as primal optimization of an objective function combining the loss on the supervised examples, the regularization term, and a penalty term deriving from forcing the constraints resulting from the conversion of the FOL knowledge base.

One main contribution of this paper is the definition of a novel approach to convert the FOL clauses into a set of constraints. Unlikely previously solutions, the proposed conversion process guarantees that each configuration satisfying the FOL rules corresponds to a minimum of the cost function. Furthermore, the paper provides an experimental evaluation of the proposed technique on a real world text tagging dataset. The paper is organized as follows. The next section describes how constraints can be embedded as penalties in the kernel machine framework. Section 3 shows how constraints provided as a FOL knowledge base can be mapped to a set of continuous penalty functions. Finally, section 4 reports the experimental results on a bibtex tagging benchmark.

2 Learning to Tag with Constraints

We consider an alphabet T of tags, whose size we indicate as $|T|$. Therefore, a set of multivariate functions $\{f_k(\boldsymbol{x}), k = 1, \ldots, |T|\}$ must be estimated, where the k-th function approximates a predicate determining whether the k-th tag should be assigned to the input text represented by the feature vector $\boldsymbol{x} \in \mathcal{D}$. We consider the case when the task functions f_k have to meet a set of functional constraints that can be expressed as

$$\phi_h(f_1, \ldots, f_{|T|}) = 0 \quad h = 1, \ldots, H \tag{1}$$

for any valid choice of the functions $f_k : k = 1, \ldots, |T|$ defined on the input domain $\mathcal{D}$. In particular, in the next section we will show how appropriate functionals can be defined to force the function values to meet a set of first-order logic constraints.

Once we assume that all the functions f_k can be approximated from an appropriate Reproducing Kernel Hilbert Space $\mathcal{H}$, the learning procedure can be cast as a set of $|T|$ optimization problems that aim at computing the optimal functions $f_1, \ldots, f_{|T|}$ in $\mathcal{H}$. In the following, we will indicate by $\boldsymbol{f} = [f_1, \ldots, f_{|T|}]'$ the vector collecting these functions. In general, it is possible to consider a different RKHS for each function given some a priori knowledge on its properties (i.e. we may decide to exploit different kernels for the expansion).

We consider the classical learning formulation as a risk minimization problem. Assuming that the correlation among the input features $\boldsymbol{x}$ and the desired k-th task function output y is provided by a collection of supervised input-target examples

$$\mathcal{L}_k = \left\{ (\boldsymbol{x}_k^i, y_k^i) | i = 1, \ldots, \ell_k \right\}, \tag{2}$$

we can measure the risk associated to a hypothesis $\boldsymbol{f}$ by the empirical risk

$$R[\boldsymbol{f}] = \sum_{k=1}^{|T|} \frac{\lambda_k^e}{|\mathcal{L}_k|} \sum_{\left(\boldsymbol{x}_k^j, y_k^j\right) \in \mathcal{L}_k} L^e\left(f_k(\boldsymbol{x}_k^j), y_k^j\right) \tag{3}$$

where $\lambda_k^e > 0$ is the weight of the risk for the k-th task and $L^e(f_k(\boldsymbol{x}), y)$ is a loss function that measures the fitting quality of $f_k(\boldsymbol{x})$ with respect to the target y. Common choices for the loss function are the quadratic function especially for regression tasks, and the hinge function for binary classification tasks. Different loss functions and λ_k^e parameters may be employed for the different tasks.

As for the regularization term, unlike the general setting of multi-task kernels [2], we simply consider scalar kernels that do not yield interactions amongst the different tasks, that is

$$N[\boldsymbol{f}] = \sum_{k=1}^{|T|} \lambda_k^r \cdot ||f_k||_{\mathcal{H}}^2, \tag{4}$$

where $\lambda_k^r > 0$ can be used to weigh the regularization contribution for the k-th task function.

Clearly, if the tasks are uncorrelated, the optimization of the objective function $R[\boldsymbol{f}] + N[\boldsymbol{f}]$ with respect to the $|T|$ functions $f_k \in \mathcal{H}$ is equivalent to $|T|$ stand-alone optimization problems. However, if we consider the case in which some correlations among the tasks are known a priori and coded as rules, we can enforce also these constraints in the learning procedure. Following the classical penalty approach for constrained optimization, we can embed the constraints by adding a term that penalizes their violation. In general we can assume that the functionals $\phi_h(\boldsymbol{f})$ are strictly positive when the related constraint is violated and zero otherwise, such that the overall degree of constraint violation of the current hypothesis $\boldsymbol{f}$ can be measured as

$$V[\boldsymbol{f}] = \sum_{h=1}^{H} \lambda_h^v \cdot \phi_h(\boldsymbol{f}) \ , \tag{5}$$

where the parameters $\lambda_h^v > 0$ allow us to weigh the contribution of each constraint. It should be noticed that, differently from the previous terms considered in the optimization objective, the penalty term involves all the functions and, thus, explicitly introduces a correlation among the tasks in the learning statement. In general, the computation of the functionals $\phi_h(\boldsymbol{f})$ implies the evaluation of some property[1] of the functions f_k on their whole input space $\mathcal{D}$. However, we can approximate the exact computation by considering an empirical realization of the input distribution. We assume that beside the labeled examples in the supervised sets $\mathcal{L}_k$, a collection of unsupervised examples $\mathcal{U} = \{\boldsymbol{x}^i | i = 1, \ldots, u\}$ is also available. If we define the set $\mathcal{S}_k^L = \{\boldsymbol{x}_k | (\boldsymbol{x}_k, y_k) \in \mathcal{L}_k\}$ containing the inputs for the labeled examples for the k-th task, in general we can assume that $\mathcal{S}_k^L \subset \mathcal{U}$, i.e. all the available points are added to the unsupervised set. Even if this is not always required, it is clearly reasonable when the supervisions are partial, i.e. a labeled example for a task k-th is not necessarily contained in the supervised learning sets for the other tasks. This assumption is important for tagging tasks, where the number of classes is very high and providing an exhaustive supervision over all the classes is generally impossible. Finally, the exact constraint functionals will be replaced by their approximations $\hat{\phi}_h(\mathcal{U}, \boldsymbol{f})$ that exploit only the values of the unknown functions $\boldsymbol{f}$ computed for the points in $\mathcal{U}$. Therefore, in eq. (5) we can consider $\phi_h(\boldsymbol{f}) \approx \hat{\phi}_h(\mathcal{U}, \boldsymbol{f})$, such that the objective function for the learning algorithm becomes

$$E[\mathbf{f}] = R[\mathbf{f}] + N[\mathbf{f}] + \sum_{h=1}^{H} \lambda_h^v \hat{\phi}_h(\mathcal{U}, \boldsymbol{f}) \ . \tag{6}$$

The solution to the optimization task defined by the objective function of equation (6) can be computed by considering the following extension of the Representer Theorem.

---

[1] As shown in the following, the functional may imply the computation of the function maxima, minima, or integral on the input domain $\mathcal{D}$.

Theorem 1. *Given a multitask learning problem for which the task functions* $f_1, \ldots, f_{|T|}$, $f_k : I\!\!R^d \to I\!\!R$, $k = 1, \ldots, |T|$, *are assumed to belong to the Reproducing Kernel Hilbert Space* $\mathcal{H}$ *and the problem is formulated as*

$$[f_1^*, \ldots, f_{|T|}^*] = argmin_{f_k \in \mathcal{H}, k=1, \ldots, |T|} E[f_1, \ldots, f_{|T|}]$$

where $E[f_1, \ldots, f_{|T|}]$ *is defined as in equation (6), then each function* f_k^* *in the solution can be expressed as*

$$f_k^*(\boldsymbol{x}) = \sum_{\boldsymbol{x}^i \in \mathcal{S}_k} w_{k,i}^* K(\boldsymbol{x}^i, \boldsymbol{x})$$

where $K(\boldsymbol{x}', \boldsymbol{x})$ *is the reproducing kernel associated to the space* $\mathcal{H}$, *and* $\mathcal{S}_k = \mathcal{S}_k^L \cup \mathcal{U}$ *is the set of the available samples for the k-th task function.*

The proof is a straightforward extension of that for the original Representer Theorem and is based on the fact that the objective function only involves values of the functions f_k computed on the finite set of supervised and unsupervised points.

3   Translation of First-Order Logic Clauses into Real-Valued Constraints

We consider knowledge-based descriptions given by first-order logic (FOL–KB). While the framework can be easily extended to arbitrary FOL predicates, in this paper we will focus on formulas containing only unary predicates to keep the notation simple. In the following, we indicate by $\mathcal{V} = \{v_1, \ldots, v_N\}$ the set of the variables used in the KB, with $v_s \in \mathcal{D}$. Given the set of predicates used in the KB $\mathcal{P} = \{p_k | p_k : \mathcal{D} \to \{true, false\}, k = 1, \ldots, |T|\}$, the clauses will be built from the set of atoms $p(v) : p \in \mathcal{P}, v \in \mathcal{V}$.

Any FOL clause has an equivalent version in *Prenex Normal form* (PNF), that has all the quantifiers $(\forall, \exists)$ and their associated quantified variables at the beginning of the clause. Standard methods exist to convert a generic FOL clause into its corresponding PNF and the conversion can be easily automated. Therefore, with no loss of generality, we restrict our attention to FOL clauses in the PNF form. We assume that the task functions f_k are exploited to implement the predicates in $\mathcal{P}$ and that the variables in $\mathcal{V}$ correspond to the input $\boldsymbol{x}$ on which the functions f_k are defined. Different variables are assumed to refer to independent values of the input features $\boldsymbol{x}$. In this framework, the predicates yield a continuous real value that can be interpreted as a *truth degree*. The FOL–KB will contain a set of clauses corresponding to expressions with no free variables (i.e. all the variables appearing in the expression are quantified) that are assumed to be *true* in the considered domain. These clauses can be converted into a set of constraints as in eq. (1) that can be enforced during the kernel based learning process. The conversion process of a clause into a constraint functional consists of the following steps:

I. CONVERSION OF THE PROPOSITIONAL EXPRESSION: conversion of the quantifier-free expression using a mixture of Gaussians.
II. QUANTIFIER CONVERSION: conversion of the universal and existential quantifiers.

Logic expressions in a continuous form

This subsection describes a technique for the conversion of an arbitrary propositional logic clause into a function, yielding a continuous truth value in $[0, 1]$. Previous work in the literature [3] concentrated on a conversion schema based on t-norms [5]. In this paper a different approach based on mixtures of Gaussians was pursued. This approach has the advantage of not making any independence assumption among the variables like it happens when using t-norms. In particular, let us assume to have available a propositional logic clause composed by n logic variables. The logic clause is equivalent to its truth table containing 2^n rows, each one corresponding to a configuration of the input variables.

The continuous function approximating the clause is based on a set of Gaussian functions, each one centered on a configuration corresponding to a true value in the truth table. These Gaussians, basically corresponding to minterms in a disjunctive expansion of the clause, can be combined by summing all the terms as

$$t(x_1, \ldots, x_n) = \sum_{[c_1, \ldots, c_n] \in \mathcal{T}} \exp\left(-\frac{|[x_1, \ldots, x_n]' - [c_1, \ldots, c_n]'|^2}{2\sigma^2}\right), \quad (7)$$

where $x_1, \ldots, x_n$ are the input variables generalizing the Boolean values on a continuous domain, and $c_1, \ldots, c_n$ span the set $\mathcal{T}$ of all the possible configurations of the input variables which correspond to the true values in the table. In the following, we indicate as *mixture-by-sum* this conversion procedure.

Another possibility for combining the minterm functions is to select the one with the highest value for a given input configuration (*mixture-by-max*). This latter solution selects the configuration closest to the true value in the truth table as

$$t(x_1, \ldots, x_n) = \max_{[c_1, \ldots, c_n] \in \mathcal{T}} \exp\left(-\frac{|[x_1, \ldots, x_n]' - [c_1, \ldots, c_n]'|^2}{2\sigma^2}\right). \quad (8)$$

For example, let us consider the simple OR of two atoms $A \vee B$. The mixture-by-max corresponding to the logic clause is the continuous function $t : \mathbb{R}^2 \Rightarrow \mathbb{R}$

$$t(x_1, x_2) = \max(e^{-\frac{|[x_1, x_2] - [1,1]'|^2}{2\sigma^2}}, e^{-\frac{|[x_1, x_2] - [1,0]'|^2}{2\sigma^2}}, e^{-\frac{|[x_1, x_2] - [0,1]'|^2}{2\sigma^2}}),$$

where $\boldsymbol{x} = [x_1, x_2]$ collects the continuous values computed for the atoms A and B, respectively. Instead, for the mixture-by-sum, the clause is converted as

$$t(x_1, x_2) = e^{-\frac{|[x_1, x_2] - [1,1]'|^2}{2\sigma^2}} + e^{-\frac{|[x_1, x_2] - [1,0]'|^2}{2\sigma^2}} + e^{-\frac{|[x_1, x_2] - [0,1]'|^2}{2\sigma^2}}.$$

If $x_1, \ldots, x_n$ is equal to a configuration verifying the clause, in the case of the mixture-by-sum it holds $t(x_1, \ldots, x_n) \geq 1$ whereas for the mixture-by-max $t(x_1, \ldots, x_n) = 1$. Otherwise, the value of $t()$ will decrease depending on the distance from the closest configuration verifying the clause. The variance σ^2 is a parameter which can be used to determine how quickly $t(x_1, \ldots, x_n)$ decreases when moving away from a configuration verifying the constraint. Please note that each configuration verifying the constraint is always a global maximum of t when using the mixture-by-max.

Quantifier conversion

The quantified portion of the expression is processed recursively by moving backward from the inner quantifier in the PNF expansion.

Let us consider the universal quantifier first. The universal quantifier requires that the expression must hold for any realization of the quantified variable v_q. When considering the real–valued mapping of the original boolean expression, the universal quantifier can be naturally converted measuring the degree of non-satisfaction of the expression over the domain $\mathcal{D}$ where the feature vector $\boldsymbol{x}$, corresponding to the variable v_q, ranges. In particular, it can be proven that if $E(v, \mathcal{P})$ is an expression with no quantifiers, depending on the variable v, and $t_E(v, \boldsymbol{f})$ is its translation into the mixture-by-max representation, given that $f_k \in \mathbb{C}^0$, $k = 1, \ldots, T$, then $\|1 - t_E(v, \boldsymbol{f})\|_p = 0$ if and only if $\forall v \; E(v, \mathcal{P})$ is *true*. Hence, in general, the satisfaction measure can be implemented by computing the overall distance of the penalty associated to the expression E, depending on the set of variables $\boldsymbol{v}_E$, i.e. $\varphi_E(\boldsymbol{v}_E, \boldsymbol{f}) = 1 - t_E(\boldsymbol{v}_E, \boldsymbol{f})$ for mixture-by-max, from the constant function equal to 0 (the value for which the constraint is verified), over the values in the domain $\mathcal{D}$ for the input $\boldsymbol{x}$ corresponding to the quantified variable $v_q \in \boldsymbol{v}_E$. In the case of the infinity norm we have

$$\forall v_q E(\boldsymbol{v}_E, \mathcal{P}) \to \|\varphi_E(\boldsymbol{v}_E, \boldsymbol{f})\|_\infty = \sup_{v_q \in \mathcal{D}} |\varphi_E(\boldsymbol{v}_E, \boldsymbol{f})| \,, \tag{9}$$

where the resulting expression depends on all the variables in $\boldsymbol{v}_E$ except v_q. Hence, the result of the conversion applied to the expression $E_q(\boldsymbol{v}_{E_q}, \mathcal{P}) = \forall v_q \; E(\boldsymbol{v}_E, \mathcal{P})$ is a functional $\varphi_{E_q}(\boldsymbol{v}_{E_q}, \boldsymbol{f})$, assuming values in $[0, 1]$ and depending on the set of variables $\boldsymbol{v}_{E_q} = \boldsymbol{v}_E \setminus \{v_q\}$. The variables in $\boldsymbol{v}_{E_q}$ need to be quantified or assigned a specific value in order to obtain a constraint functional depending only on the functions $\boldsymbol{f}$.

In the conversion of the PNF representing a FOL constraint without free variables, the variables are recursively quantified until the set of the free variables is empty. In the case of the universal quantifier we apply again the mapping described previously. The existential quantifier can be realized by enforcing the De Morgan law to hold also in the continuous mapped domain. The De Morgan law states that $\exists v_q \; E(\boldsymbol{v}_E, \mathcal{P}) \iff \neg \forall v_q \; \neg E(\boldsymbol{v}_E, \mathcal{P})$. Using the conversion of the universal quantifier defined in eq. (9), we obtain the following conversion for the existential quantifier

$$\exists v_q \; E(\boldsymbol{v}_E, \mathcal{P}) \to \inf_{v_q \in \mathcal{D}} \varphi_E(\boldsymbol{v}_E, \boldsymbol{f}) \,.$$

The conversion of the quantifiers requires to extend the computation on the whole domain of the quantified variables. Here, we assume that the computation can be approximated by exploiting the available empirical realizations of the feature vectors. If we consider the examples available for training, both supervised and unsupervised, we can extract the empirical distribution $\mathcal{S}_k$ for the input to the k-th function. Hence, the universal quantifier exploiting the infinity norm is approximated as

$$\forall v_q \ E(\boldsymbol{v}_E, \mathcal{P}) \rightarrow \max_{v_q \in \mathcal{S}_{k(q)}} |\varphi_E(\boldsymbol{v}_E, \boldsymbol{f})| \ .$$

Similarly, for the existential quantifier it holds

$$\exists v_q \ E(\boldsymbol{v}_E, \mathcal{P}) \rightarrow \min_{v_q \in \mathcal{S}_{k(q)}} |\varphi_E(\boldsymbol{v}_E, \boldsymbol{f})| \ .$$

It is also possible to select a different functional norm to convert the universal quantifier. For example, when using the $\|\cdot\|_1$ norm, and the empirical distribution for the $\boldsymbol{x}$, we have

$$\forall v_q \ E(\boldsymbol{v}_E, \mathcal{P}) \rightarrow \frac{1}{|\mathcal{S}_{k(q)}|} \sum_{v_q \in \mathcal{S}_{k(q)}} |\varphi_E(\boldsymbol{v}_E, \boldsymbol{f})| \ .$$

Please note that $\varphi_E([], \boldsymbol{f})$ will always reduce to the following form, when computed for an empirical distribution of data for any selected functional norm

$$\varphi_E([], \boldsymbol{f}) = O_{v_{s(1)} \in \mathcal{S}_{k(s(1))}} \cdots O_{v_{s(Q)} \in \mathcal{S}_{k(s(Q))}} t_{E_0}(\boldsymbol{v}_{E_0}, \boldsymbol{f}) \ ,$$

where E_0 is the quantifier free expression, $O_{v_q \in \mathcal{S}_{x_{j(q)}}}$ specifies the aggregation operator to be computed on the sample set $\mathcal{S}_{k(q)}$ for each quantified variable v_q. In the case of the infinity norm, $O_{v_q \in \mathcal{S}_{k(q)}}$ is either the minimum or maximum operator over the set $\mathcal{S}_{k(q)}$. Therefore, the presented conversion procedure implements the logical constraint depending on the realizations of the functions over the data point samples. For this class of constraints, Theorem 1 holds and the optimal solution can be expressed as a kernel expansion over the data points. In fact, since the constraint is represented by $\varphi_E([], \boldsymbol{f}) = 0$ in the definition of the learning objective function of eq. (6) we can substitute $\hat{\phi}(\mathcal{U}, \boldsymbol{f}) = \varphi_E([], \boldsymbol{f})$.

When using the minimum and/or maximum operators for defining $\hat{\phi}(\mathcal{U}, \boldsymbol{f})$, the resulting objective function is continuous with respect to the parameters $w_{k,i}$ defining the RKHS expansion, since it is obtained by combining continuous functions. However, in general, its derivatives are no more continuous, but, in practice, this is not a problem for gradient descent based optimization algorithms once appropriate stopping criteria are applied. In particular, the optimal minima can be located also in configurations corresponding to discontinuities in the gradient values, i.e. when a maximum or minimum operator switches its choice among two different points in the dataset.

As an example of the conversion procedure, let $a(\cdot), b(\cdot)$ be two predicates, implemented by the functions $f_a(\cdot), f_b(\cdot)$. The clause $\forall v_1 \ \forall v_2 \ (a(v_1) \wedge \neg b(v_2)) \vee$

Table 1. A sample of the semantic rules used in training the kernel machines

$\forall x$ phase(x) $\wedge$ transition(x) $\Rightarrow$ physics(x)
$\forall x$ chemistry(x) $\Rightarrow$ science(x)
$\forall x$ immunoelectrode(x) $\Rightarrow$ physics(x) $\vee$ biology(x)
$\forall x$ semantic(x) $\wedge$ web20(x) $\Rightarrow$ knowledgemanagement(x)
$\forall x$ rdf(x) $\Rightarrow$ semanticweb(x)
$\forall x$ software(x) $\wedge$ visualization(x) $\Rightarrow$ engineering(x)
$\forall x$ folksonomy(x) $\Rightarrow$ social(x)
$\forall x$ mining(x) $\wedge$ web(x) $\Rightarrow$ informationretrieval(x)
$\forall x$ mining(x) $\wedge$ information(x) $\Rightarrow$ datamining(x)
$\forall x$ computer(x) $\wedge$ science(x) $\Rightarrow$ engineering(x)

$(\neg a(v_1) \wedge b(v_2))$ is converted starting with the conversion of the quantifier free expression $E_0([v_1, v_2], \{a(\cdot), b(\cdot)\}) = (a(v_1) \wedge \neg b(v_2)) \vee (\neg a(v_1) \wedge b(v_2))$ as

$$t_{E_0}([v_1, v_2], [f_a, f_b]) =$$
$$= \max \left(e^{-\frac{(f_a(v_1)-1)^2 + f_b(v_2)^2}{2\sigma^2}}, e^{-\frac{f_a(v_1)^2 + (f_b(v_2)-1)^2}{2\sigma^2}} \right) .$$

Then, the quantifier free expression $E_0([v_1, v_2], \{a(\cdot), b(\cdot)\})$ is converted into the distance measure and the two universal quantifiers are converted using the infinity norm over their empirical realizations, yielding the constraint

$$\varphi_{E_0}([], [f_a, f_b]) = \max_{v_1 \in S_a} \max_{v_2 \in S_b} (1 - t_{E_0}([v_1, v_2], [f_a, f_b])) = 0 .$$

4 Experimental Results

The dataset used in the experiments consists of 7395 bibtex entries that have been tagged by users of a social network[2] using 159 tags. Each bibtex entry contains a small set of textual elements representing the author, the title, and the conference or journal name. The text is represented as a bag-of-words, with a feature space with dimensionality equal to 1836. The training set was obtained by sampling a subset of the entries (10%-50%), leaving the remaining ones for the test set. Like previous studies in the literature on this dataset (see e.g. [4]), the F1 score was employed to evaluate the tagger performances.

A knowledge base containing a set of 106 rules, expressed by FOL, has been created to express semantic relationships between the categories (see Table 1). The experiments tested the prediction capabilities of the classifiers, when considering all tags or only the 25 most popular tags in the dataset as output categories. In this second case, only the 8 logic rules fully defined over the subset of tags have been exploited.

---

[2] The dataset can be freely downloaded at:
http://mulan.sourceforge.net/datasets.html

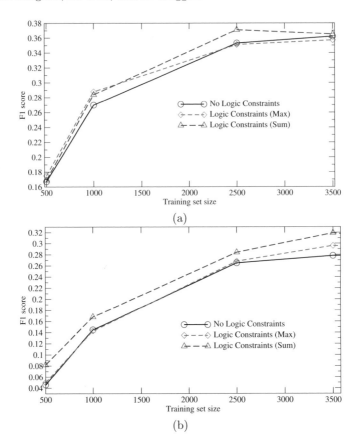

Fig. 1. F1 scores considering (a) the top 25 and (b) all the tags on the test set, when using or not using the knowledge base

The rules correlate the tags and, after their conversion into the continuous form, they have been used to train the kernel machines according to the procedure described in the previous sections. The same experiments have been performed using a kernel machine trained using only the supervised examples. All the kernel machines used in this experiment have been based on a linear kernel, as they have been assessed as the best performers on this dataset in a first round of experiments (not reported in this paper).

Figure 1 reports the F1 scores computed over the test set provided by the 25 and 159 tag predictors, respectively. The logic constraints implemented via the mixture-by-sum, overperformed the logic constraints implemented as a mixture-by-max. Enforcing the logic constraints during learning was greatly beneficial with respect to a standard text classifier based on a kernel machine, as the F1 scores are improved by 2-4% when all tags are considered. This gain is consistent for the training set sizes that have been tested. When restricting the attention to the top 25 tags, the gains obtained by enforcing the mixture-by-sum constraints

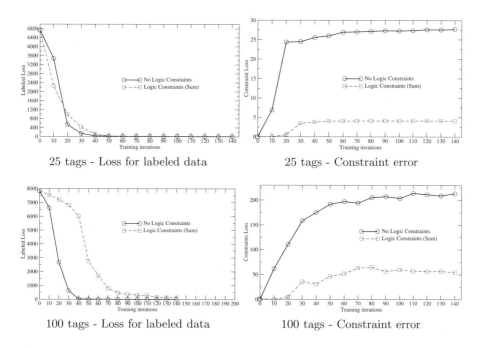

25 tags - Loss for labeled data   25 tags - Constraint error

100 tags - Loss for labeled data   100 tags - Constraint error

Fig. 2. Loss term on labeled data and on constraints deriving from the rules over the test set (generalization) for the 25 and all tag datasets

are smaller, ranging around 0-2%. These smaller gains are due to the significantly smaller number of logic rules that are defined over the 25 tags.

Figures 2 plots the loss on the labeled data and on the constraints for the test set (generalization) at the different iterations of the training for the 25 and 159 tag classifiers. In all the graphs, the loss on the constraint is low at the beginning of the training. This happens because all the provided rules are in the form of implications, which are trivially verified if the precondition is not true. Since no tags are yet assigned at the beginning of the training, the precondition of the rules is false, making the constraint verified. The figures show how the introduction of the constraints does not change significantly the loss on the labeled data at convergence, whereas the constraint loss is strongly reduced at the end of the training process. This means that the final weights of the kernel machine implement tagging functions that are able to fit much better the prior knowledge on the task.

5   Conclusions and Future Work

This paper presented a novel approach to text tagging, bridging pure machine learning approaches to knowledge based annotators based on logic formalisms. The approach is based on directly injecting logic knowledge compiled as continuous constraints into the kernel machine learning algorithm. While previous work concentrated on t-norms, this paper presents a new approach to convert FOL clauses into constraints using mixtures of Gaussians. The experimental results show how the approach can over-perform a text annotator based on classical kernel machines by a significant margin. In this paper, we assumed that a logic predicate corresponds to each tag and, as a consequence, to a function to estimate. Therefore, the logic rules are defined directly over the tags. As future work we plan to test this approach to the case where new logic predicates can be added to the knowledge base. For example, ontology-based taggers use regular expressions to define partial tagging rules [6]. In our approach this would be implemented by assigning a new FOL predicate to each regular expression. After converting the logic rules as explained in this paper, it would be possible to train the kernel machines to learn jointly from examples and the KB.

References

1. Bengio, S., Weston, J., Grangier, D.: Label embedding trees for large multi-class tasks. Advances in Neural Information Processing Systems 23, 163–171 (2010)
2. Caponnetto, A., Micchelli, C., Pontil, M., Ying, Y.: Universal Kernels for Multi-Task Learning. Journal of Machine Learning Research (2008)
3. Diligenti, M., Gori, M., Maggini, M., Rigutini, L.: Multitask Kernel-based Learning with Logic Constraints. In: Proceedings of the 19th European Conference on Artificial Intelligence, pp. 433–438. IOS Press, Amsterdam (2010)
4. Katakis, I., Tsoumakas, G., Vlahavas, I.: Multilabel text classification for automated tag suggestion. ECML PKDD Discovery Challenge 75 (2008)
5. Klement, E.P., Mesiar, R., Pap, E.: Triangular Norms. Kluwer Academic Publishers, Dordrecht (2000)
6. Laclavik, M., Seleng, M., Gatial, E., Balogh, Z., Hluchy, L.: Ontology based text annotation. In: Proceedings of the 18th International Conference on Information Modelling and Knowledge Bases, pp. 311–315. IOS Press, Amsterdam (2007)
7. Liu, D., Hua, X., Yang, L., Wang, M., Zhang, H.: Tag ranking. In: Proceedings of the 18th International Conference on World Wide Web, pp. 351–360. ACM, New York (2009)
8. Matusiak, K.: Towards user-centered indexing in digital image collections. OCLC Systems & Services 22(4), 283–298 (2006)
9. Peters, S., Denoyer, L., Gallinari, P.: Iterative Annotation of Multi-relational Social Networks. In: Proceedings of the International Conference on Advances in Social Networks Analysis and Mining, pp. 96–103. IEEE, Los Alamitos (2010)
10. Sebastiani, F.: Machine learning in automated text categorization. ACM Computing Surveys (CSUR) 34(1), 1–47 (2002)
11. Weinberger, K., Saul, L.: Distance metric learning for large margin nearest neighbor classification. The Journal of Machine Learning Research 10, 207–244 (2009)
12. Zavitsanos, E., Tsatsaronis, G., Varlamis, I., Paliouras, G.: Scalable Semantic Annotation of Text Using Lexical and Web Resources. Artificial Intelligence: Theories, Models and Applications, 287–296 (2010)

Comparison of Dispersion Models by Using Fuzzy Similarity Relations

Angelo Ciaramella[1], Angelo Riccio[1], Stefano Galmarini[2],
Giulio Giunta[1], and Slawomir Potempski[3]

[1] Department of Applied Science,
University of Naples "Parthenope", Isola C4, Centro Direzionale, I-80143,
Napoli (NA), Italy
{angelo.ciaramella,angelo.riccio,giulio.giunta}@uniparthenope.it
[2] European Commission, DG Joint Research Centre,
Institute for Environment and Sustainability,
21020 Ispra (VA), Italy
stefano.galmarini@jrc.ec.europa.eu
[3] Institute of Atomic Energy, Otwock-Swierk, Poland
slawek@cyf.gov.pl

Abstract. Aim of this work is to introduce a methodology, based on the combination of multiple temporal hierarchical agglomerations, for model comparisons in a multi-model ensemble context. We take advantage of a mechanism in which hierarchical agglomerations can easily combined by using a transitive consensus matrix. The hierarchical agglomerations make use of fuzzy similarity relations based on a generalized Łukasiewicz structure. The methodology is adopted to analyze data from a multi-model air quality ensemble system. The models are operational long-range transport and dispersion models used for the real-time simulation of pollutant dispersion or the accidental release of radioactive nuclides in the atmosphere. We apply the described methodology to agglomerate and to individuate the models that characterize the predicted atmospheric pollutants from the ETEX-1 experiment.

Keywords: Fuzzy Similarity, Hierarchical Agglomeration, Ensemble Models, Air Pollutant Dispersion.

1 Introduction

Clustering is an exploratory tool in data analysis that arises in many different fields such as data mining, image processing, machine learning, and bioinformatics. One of the most popular and interesting clustering approaches is the hierarchical agglomerative clustering. In this work we introduce a novel methodology based on fuzzy similarity relations that permits to combine multiple temporal hierarchical agglomerations.

This methodology has been applied to data concerning the real-time forecasting of atmospheric compounds from the *ENSEMBLE* system [5,6,7]. ENSEMBLE is a web-based system aiming at assisting the analysis of multi-model data

R. Pirrone and F. Sorbello (Eds.): AI*IA 2011, LNAI 6934, pp. 57–67, 2011.

provided by many national meteorological services and environmental protection agencies worldwide for the real-time forecasting of deliberate/accidental releases of harmful radionuclides (e.g. Fukushima, Chernobyl).

In previous works [15,12] an approach for the statistical analysis of multi-model ensemble results has been presented. The authors used a well-known statistical approach to multimodel data analysis, i.e., *Bayesian Model Averaging*, which is a standard method for combining predictive distributions from different sources. Moreover, similarities and differences between models were explored by means of correlation analysis. In [13] the authors investigate some basic properties of multi-model ensemble systems, which can be deduced from general characteristics of statistical distributions of the ensemble membership. Cluster-based approaches [1,2,3] have also been developed and applied. These approaches discriminate between data that are less dependent (in the statistical sense), so that "redundant" information can be more easily discarded and equivalent performance can be achieved with a considerable lower number of models.

In this paper we generalize these clustering approaches, by introducing a new methodology based on fuzzy similarity relations that allows to combine multiple hierarchical agglomerations, each for a different forecasting leading time.

We conjecture that this framework is amenable to easily incorporate observations that may become available during the course of the event, so as to improve the forecast by "projecting" observations onto the hierarchical combination of clusters.

The paper is organized as follows. In Sections 2 and 3 some fundamental concepts on t-norms and fuzzy similarity relations are given. The proposed methodology is detailed in Section 3.3. Finally, in Section 4 some experimental results obtained by applying this methodology on an ensemble of prediction models are described. Conclusions and future remarks are given in Section 5.

2   Norms and Residuum

In this Section we introduce some basic terminologies and successively we outline the minimum requirements a fuzzy relation should satisfies in order to correspond to a dendrogram and in what cases dendrograms can be aggregated into a consensus matrix.

The popularity of *fuzzy logic* comes mainly from many applications, where linguistic variables are suitably transformed in fuzzy sets, combined via the conjunction and disjunction operations by using continuous triangle norms or co-norms, respectively. Moreover, it offers the possibility of soft clustering, in contrast with algorithms that output hard (crisp or non-fuzzy) clustering of data.

A fundamental concept in fuzzy logic is that of *norm* [9]. A *triangular norm* (*t-norm* for short) is a binary operation t on the unit interval $[0, 1]$, i.e., a function $t : [0, 1]^2 \rightarrow [0, 1]$, such that for all $x, y, z \in [0, 1]$ the following four axioms are satisfied:

$$\begin{array}{llll}
t(x,y) & = & t(y,x) & (commutativity) \\
t(x,t(y,z)) & = & t(t(x,y),z) & (associativity) \\
t(x,y) & \leq & t(x,z) \quad \text{whenever } y \leq z & (monotonicity) \\
t(x,1) & = & x & (boundary\ condition)
\end{array} \qquad (1)$$

Several parametric and non-parametric t-norms have been introduced [9] and recently a their generalized version has been studied [4]. The four basic t-norms are $t_{\mathbf{M}}$, $t_{\mathbf{P}}$, $t_{\mathbf{L}}$ and $t_{\mathbf{D}}$ given by, respectively:

$$\begin{array}{lll}
t_{\mathbf{M}}(x,y) = & \min(x,y) & (minimum) \\
t_{\mathbf{P}}(x,y) = & x \cdot y & (product) \\
t_{\mathbf{L}}(x,y) = & \max(x+y-1,0) & (\text{Ł}ukasiewicz\ t\text{-}norm) \\
t_{\mathbf{D}}(x,y) = & \begin{cases} 0 & \text{if } (x,y) \in [0,1]^2 \\ \min(x,y) & \text{otherwise} \end{cases} & (drastic\ product)
\end{array} \qquad (2)$$

In the following, we concentrate on the $t_{\mathbf{L}}$ norm.

The union and intersection of two unit interval valued fuzzy sets are essentially lattice operations. In many applications, however, lattice structure alone is not rich enough to model fuzzy phenomena. An important concept is the residuated lattice and its structure appears, in one form or in another, in practically all fuzzy inference systems, in the theory of fuzzy relations and fuzzy logic. One important operator here is the *residuum* $\rightarrow_t$ defined as

$$x \rightarrow_t y = \bigvee \{z | t(z,x) \leq y\} \qquad (3)$$

where $\bigvee$ is the *union* operator and for the left-continuous basic t-norm $t_{\mathbf{L}}$ is given by

$$x \rightarrow_{\mathbf{L}} y = \min(1-x+y,1) \quad (\text{Ł}ukasiewicz\ implication) \qquad (4)$$

Moreover, let p a fixed natural number in a *generalized Łukasiewicz structure*, we have

$$t_{\mathbf{L}}(x,y) = \sqrt[p]{\max(x^p + y^p - 1, 0)} \qquad (5)$$

$$x \rightarrow_{\mathbf{L}} y = \min(\sqrt[p]{1 - x^p + y^p}, 1)$$

Finally, we define as *bi-residuum* on a residuated lattice the operation

$$x \leftrightarrow_t y = (x \rightarrow_t y) \wedge (y \rightarrow_t x) \qquad (6)$$

where $\wedge$ is the *meet*.

For the left-continuous basic t-norm $t_{\mathbf{L}}$ we have

$$x \leftrightarrow_{\mathbf{L}} y = 1 - \max(x,y) + \min(x,y) \qquad (7)$$

3   Fuzzy Similarity

A binary *fuzzy relation* R on $U \times V$ is a fuzzy set on $U \times V$ ($R \subseteq U \times V$). *Similarity* is a fuzzy relation $S \subseteq U \times U$ such that, for each $u, v, w \in U$

$$S\langle u, u \rangle \qquad\qquad = \quad 1 \qquad (\text{everthing is similar to itself})$$

$$S\langle u, v \rangle \qquad\qquad = S\langle v, u \rangle \qquad\qquad (\text{symmetric}) \qquad (8)$$

$$t(S\langle u, v \rangle, S\langle v, w \rangle) \leq S\langle u, w \rangle \qquad (\text{weakly transitive})$$

It is also well known that a fuzzy set with membership function $\mu : X \to [0, 1]$ generate a fuzzy similarity S defined as $S\langle a, b \rangle = \mu(a) \leftrightarrow_t \mu(b)$ for all $a, b \in X$. We also note that, let $t_{\mathbf{L}}$ be the Lukasiewicz product, we have that S is a fuzzy equivalence relation on X with respect to $t_{\mathbf{L}}$ iif $1 - S$ is a *pseudo-metric* on X. Further, a main result is the following [17,16]:

Proposition 1. *Consider n Lukasiewicz valued fuzzy similarities S_i, $i = 1, \ldots, n$ on a set X. Then*

$$S\langle x, y \rangle = \frac{1}{n} \sum_{i=1}^{n} S_i \langle x, y \rangle \qquad (9)$$

is a Lukasiewicz valued fuzzy similarity on X.

3.1   Min-transitive Closure

If a similarity relation is *min-transitive* ($t = \min$ in (8)), it is called a *fuzzy-equivalence relation*. Each fuzzy-equivalence relation can be graphically described by a *dendrogram* [10]. Therefore, the requirement for the existence of the dendrogram, for a similarity matrix, is the transitivity.

The methodology introduced in this paper uses a min-transitive closure [11]. The transitive closure is obtained by computing a sufficiently high power of the given similarity matrix. Let n the dimension of a relation matrix, the transitive closure R^T of R is calculated by

$$R^T = \bigcup_{i=1}^{n-1} R^i \qquad (10)$$

where R^{i+1} is defined as

$$R^{i+1} = R^i \circ R \qquad (11)$$

The composition $R \circ S$ of fuzzy relations R and S is a fuzzy relation defined by

$$R \circ S\langle x, y \rangle = \text{Sup}_{z \in X}\{R\langle x, z \rangle \wedge S\langle z, y \rangle\} \qquad (12)$$

$\forall x, y \in X$ and where $\wedge$ stands for a t-norm (e.g., min operator) [11]. Using this methodology the min-transitive closure R^T can be computed by the algorithm described in Algorithm 1.

Algorithm 1. Min-transitive closure

1: **Input** R the input relation
2: **Output** R^T the output transitive relation
3: 1. Calculate $R^* = R \cup (R \circ R)$
    2. if $R^* \neq R$ replace R with R^* and go to step 1
    else $R^T = R^*$ and the algorithm terminates.

The transitive property of binary relations is closely related to the theory of the graphs. In other words, if a relation is represented as a directed graph, then computation of transitive closure of this relation is equivalent to finding the tightest path between each pair of vertices. The strength of a path is determined by the minimum of the weights on that path.

3.2 Dendrogram Description Matrices

As previously described, any dendrogram could be associated with a fuzzy equivalence relation and, equivalently, with its matrix representation if the min-transitive closure property is satisfied. The elements of a fuzzy equivalence matrix describe the similarity between objects. Moreover we have that [11]

Lemma 1. *Letting R be a similarity relation with the elements $R\langle x, y \rangle \in [0, 1]$ and letting D be a dissimilarity relation, which is obtained from R by*

$$D(x, y) = 1 - R\langle x, y \rangle \tag{13}$$

then D is ultrametric iif R is min-transitive.

There is a one-to-one correspondence between min-transitive similarity matrices and dendrogram. The correspondence between ultrametric dissimilarity matrices and dendrograms is also on-to-one. In other words, a dendrogram could be generated corresponding to a dissimilarity matrix if it is *ultrametric*.

3.3 Agglomerative Methodology

We remark that the aim is to agglomerate, by an unsupervised methodology, the distributions obtained by the ensemble models at different times. Substantially a hierarchical tree (dendrogram), that permits to cluster models that have similar behavior, must be obtained. We calculate the similarity (or dissimilarity) matrix between the distributions of the models by using the fuzzy similarity described in equation 9. Successively the algorithm described in Algorithm 1 is applied to obtain the min-transitive closure.

We also may express the information by *fuzzy set*. A simple way is to describe the *membership functions* by the following equation [17]

$$\mu(\mathbf{x}_i) = \frac{\mathbf{x}_i - \min(\mathbf{x}_i)}{\max(\mathbf{x}_i) - \min(\mathbf{x}_i)} \tag{14}$$

Algorithm 2. Combination of dendrograms

1: **Input** $S^{(i)}$, $1 \leq i \leq L$ L input similarity matrices (dendrograms)
2: **Output** S the resulted similarity matrix (dendrogram)
  1. Aggregate the similarity matrices to a final similarity matrix $S = Aggregate(S^{(1)}, S^{(2)}, \ldots, S^{(L)})$
  a. Let S^* be the identity matrix
  b. For each $S^{(i)}$ calculate e $S^* = S^* \cup (S^* \circ S^{(i)})$
  c. c. If S^* is not changed $S = S^*$ and goto step 3 else goto step 1.b
3: Create the final dendrogram from the S

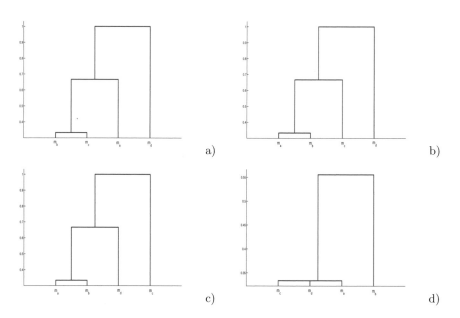

Fig. 1. Combination algorithm: a-b-c) input dendrograms; d) combined hierarchy

where $\mathbf{x}_i = [x_1^i, x_2^i, \ldots, x_L^i]$ is the i-th observation vector of the L models.

Successively, we apply the agglomerative hierarchical clustering approach to obtain the dendrogram. A consensus matrix that it is representative of all dendrograms is obtained by combining the transitive closure and equation 12 (i.e., max-min) [11]. The algorithm to obtain the final dendrogram is described in Algorithm 2.

In Figure 1 we show a realistic agglomeration result. In Figures 1a-b-c three input hierarchies to be combined are plotted. Four models are considered, namely m_a, m_b, m_c and m_d, respectively. In Figure 1d we show the final result obtained calculating the dendrogram on the similarity matrix. The result seems to be rational, because the output hierarchy contains the clusters (m_a, m_b, m_c) and (m_a, m_b, m_c, m_d) at different levels, and each of these clusters are repeated at least in two out of the three input dendrograms [11].

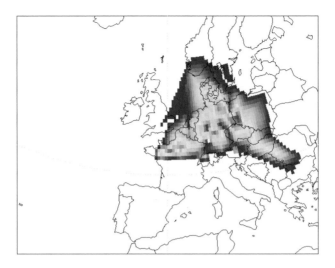

Fig. 2. ETEX-1 integrated (in time) observations

4   Experimental Results

In this Section we propose the results obtained applying the described methodology to compare mathematical operational long-range transport and dispersion models used for the real-time simulation of pollutant dispersion.

In [7,15] the authors analyzed the output of multi-model ensemble results for the ETEX-1 experiment. They already showed that the *"Median Model"* provided a more accurate reproduction of the concentration trend and estimate of the cloud persistence at sampling locations.

The ETEX-1 [8] experiment concerned the release of pseudo-radioactive material on 23 October 1994 at 16:00 UTC from Monterfil, southeast of Rennes (France). Briefly, a steady westerly flow of unstable air masses was present over central Europe. Such conditions persisted for the 90 h that followed the release with frequent precipitation events over the advection area and a slow movement toward the North Sea region. In Figure 2 we show the integrated concentration after 78 hours from release. Several independent groups worldwide tried to forecast these observations. The ensemble is composed by 25 members. Each simulation, and therefore each ensemble member, is produced with different atmospheric dispersion models and is based on weather fields generated by (most of the time) different *Global Circulation Models* (GCM). All the simulations relate to the same release conditions. For details on the groups involved in the exercise and the model characteristics, refer to [7] and [8].

Now we describe the phases needed to analyze this dataset. The first step is the *fuzzification*. Namely, equation 14 is used on the estimated model concentrations at each time level. Successively, a similarity matrix (dendrogram) is obtained for the concentrations at different times (by using equation 9, Łukasiwicz

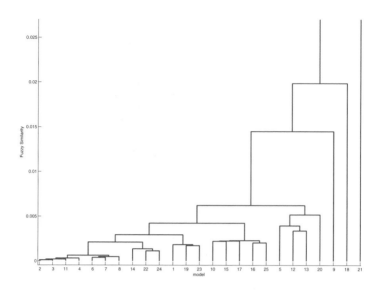

Fig. 3. Combined fuzzy similarity dendrogram

with $p = 1$). Finally the representative similarity matrix is estimated making use of Algorithm 2. A particular of the dendrogram obtained on the integrated concentrations after 78 hours is plotted in Figure 3. In this figure the information on the abscissa are related to the models and those on the ordinate are related to the model data similarities obtained by using the fuzzy similarity. As an example, in Figure 4 we show some distributions of the models. The distributions in Figures 4a-b are very close and this is confirmed by the dendrogram. Instead the model in Figure 4c has a diffusive distribution far from the other distributions and also confirmed by the dendrogram.

The hierarchical mechanism permits to clusterize the observations in a fixed number of clusters. A *Mean Square Error* (MSE) between each model and the median value of the cluster where it belongs is determined. For each cluster the model with the minimum MSE is considered. Finally the *median model* of these selected models is calculated and it is compared with the real observation by using the RMSE. Moreover, varying the number of clusters the models that have the best approximation of the real observation can be defined (see [15] and [3] for more details). In Figure 5 we show the *Root Mean Square Error (RMSE)* obtained varying the number of clusters. In this case the best approximation is obtained by using 6 clusters.

As can be inferred from the analysis of this figure, a lower RMSE does not necessarily corresponds to the use of a large number of models; similar (or even better) performance can be achieved with a few models; even more interestingly, since the selection framework is not based on the prior knowledge of experimental values, the satisfactory comparison of selected subset of models with

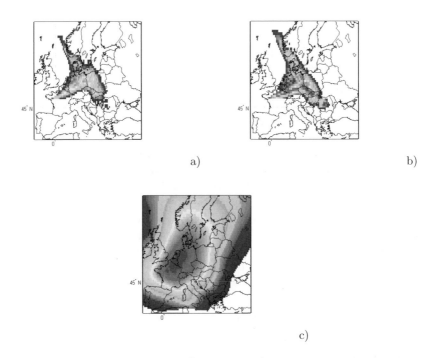

a)
b)
c)

Fig. 4. Model distributions: a) model 22; b) model 24; model 21

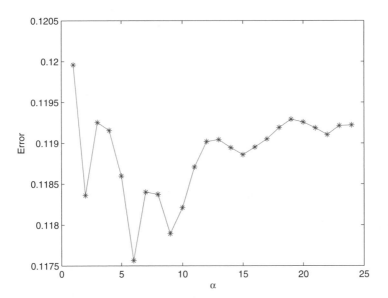

Fig. 5. RMSE varying the clustering number

experimental values suggest promising perspectives for the systematic reduction of ensemble data complexity. Furthermore, comparing this new methodology with the previous one, by using the consensus matrix we add temporal information that permits to obtain a more robust and realistic analysis.

5 Conclusions

In this work we introduced a methodology, based on the combination of multiple temporal hierarchical agglomerations, for model comparison in a multi-model ensemble context. Here we suggest to use fuzzy similarity relations in a Lukasiewicz structure. We remark that further studies can be made by using also different fuzzy similarities (e.g., [14]). Moreover, we take advantage of a mechanism in which hierarchical agglomerations can be easily combined by using a transitive consensus matrix. The proposed methodology is able to combine multiple temporal hierarchical agglomerations of dispersion models used for the real-time simulation of pollutant dispersions. The results show that this methodology is able to discard redundant temporal information and equivalent performance can be achieved considering a lower number of models reducing, the data complexity. In the next future, further studies could be conducted on real pollutant dispersions (e.g., Fukushima) and on the structure utilized in the fuzzy similarity relations.

References

1. Ciaramella, A., Cocozza, S., Iorio, F., Miele, G., Napolitano, F., Pinelli, M., Raiconi, G., Tagliaferri, R.: Interactive data analysis and clustering of genomic data. Neural Networks 21(2-3), 368–378 (2008)
2. Napolitano, F., Raiconi, G., Tagliaferri, R., Ciaramella, A., Staiano, A., Miele, G.: Clustering and visualization approaches for human cell cycle gene expression data analysis. International Journal of Approximate Reasoning 47(1), 70–84 (2008)
3. Ciaramella, A., Giunta, G., Riccio, A., Galmarini, S.: Independent Data Model Selection for Ensemble Dispersion Forecasting. In: Okun, O., Valentini, G. (eds.) Applications of Supervised and Unsupervised Ensemble Methods. SCI, vol. 245, pp. 213–231. Springer, Heidelberg (2009)
4. Ciaramella, A., Pedrycz, W., Tagliaferri, R.: The Genetic Development of Ordinal Sums. Fuzzy Sets and Systems 151, 303–325 (2005)
5. Galmarini, S., Bianconi, R., Bellasio, R., Graziani, G.: Forecasting consequences of accidental releases from ensemble dispersion modelling. J. Environ. Radioactiv. 57, 203–219 (2001)
6. Galmarini, S., Bianconi, R., Klug, W., Mikkelsen, T., Addis, R., Andronopoulos, S., Astrup, P., Baklanov, A., Bartniki, J., Bartzis, J.C., Bellasio, R., Bompay, F., Buckley, R., Bouzom, M., Champion, H., D'Amours, R., Davakis, E., Eleveld, H., Geertsema, G.T., Glaab, H., Kollax, M., Ilvonen, M., Manning, A., Pechinger, U., Persson, C., Polreich, E., Potemski, S., Prodanova, M., Saltbones, J., Slaper, H., Sofiev, M.A., Syrakov, D., Sorensen, J.H., Van der Auwera, L., Valkama, I., Zelazny, R.: Ensemble dispersion forecasting–Part I: concept, approach and indicators. Atmos. Environ. 38, 4607–4617 (2004a)

7. Galmarini, S., Bianconi, R., Addis, R., Andronopoulos, S., Astrup, P., Bartzis, J.C., Bellasio, R., Buckley, R., Champion, H., Chino, M., D'Amours, R., Davakis, E., Eleveld, H., Glaab, H., Manning, A., Mikkelsen, T., Pechinger, U., Polreich, E., Prodanova, M., Slaper, H., Syrakov, D., Terada, H., Van der Auwera, L.: Ensemble dispersion forecasting–Part II: application and evaluation. Atmos. Environ. 38, 4619–4632 (2004b)
8. Girardi, F., Graziani, G., van Veltzen, D., Galmarini, S., Mosca, S., Bianconi, R., Bellasio, R., Klug, W.: The ETEX project. EUR Report 181-43 EN. Office for official publications of the European Communities, Luxembourg, p. 108 (1998)
9. Klement, E.P., Mesiar, R., Pap, E.: Triangular norms. Kluwer Academic Publishers, Dordrecht (2001)
10. Meyer, H.D., Naessens, H., Baets, B.D.: Algorithms for computing the min-transitive closure and associated partition tree of a symmetric fuzzy relation. Eur. Journal Oper. Res. 155(1), 226–238 (2004)
11. Mirzaei, A., Rahmati, M.: A novel Hierarchical-Clustering-Combination Scheme based on Fuzzy-Similarity Relations. IEEE Transaction on Fuzzy Systems 18(1), 27–39 (2010)
12. Potempski, S., Galmarini, S., Riccio, A., Giunta, G.: Bayesian model averaging for emergency response atmospheric dispersion multimodel ensembles: Is it really better? How many data are needed? Are the weights portable? Journal of Geophysical Research 115 (2010), doi:10.1029/2010JD014210
13. Potempski, S., Galmarini, S.: Est modus in rebus: analytical properties of multi-model ensembles. Atmospheric Chemistry and Physics 9(24), 9471–9489 (2009)
14. Rezaei, H., Emoto, M., Mukaidono, M.: New Similarity Measure Between Two Fuzzy Sets. Journal of Advanced Computational Intelligence and Intelligent Informatics 10(6) (2006)
15. Riccio, A., Giunta, G., Galmarini, S.: Seeking for the rational basis of the median model: the optimal combination of multi-model ensemble results. Atmos. Chem. Phys. 7, 6085–6098 (2007)
16. Sessa, S., Tagliaferri, R., Longo, G., Ciaramella, A., Staiano, A.: Fuzzy Similarities in Stars/Galaxies Classification. In: Proceedings of IEEE International Conference on Systems, Man and Cybernetics, pp. 494–4962 (2003)
17. Turunen, E.: Mathematics Behind Fuzzy Logic. In: Advances in Soft Computing. Springer, Heidelberg (1999)

An Interaction-Oriented Agent Framework for Open Environments

Matteo Baldoni[1], Cristina Baroglio[1], Federico Bergenti[4], Elisa Marengo[1],
Viviana Mascardi[3], Viviana Patti[1], Alessandro Ricci[2], and Andrea Santi[2]

[1] Università degli Studi di Torino
{baldoni,baroglio,emarengo,patti}@di.unito.it
[2] Università degli Studi di Bologna
{a.ricci,a.santi}@unibo.it
[3] Università degli Studi di Genova
mascardi@disi.unige.it
[4] Università degli Studi di Parma
federico.bergenti@unipr.it

Abstract. The aim of the work is to develop formal models of interaction and of the related support infrastructures, that overcome the limits of the current approaches. We propose to represent explicitly not only the agents but also the computational environment in terms of rules, conventions, resources, tools, and services, that are functional to the coordination and cooperation of the agents. These models will enable the verification of the interaction in the MAS, thanks to the introduction of a novel social semantics of interaction based on commitments and on an explicit account of the regulative rules.

Keywords: Commitment-based protocols, High-level environment models, Direct and mediated communication, Agent-oriented infrastructure.

1 Introduction

The growing pervasiveness of computer networks and of Internet is an important catalyst pushing towards the realization of *business-to-business, cross-business solutions* or, more generally, of *open environment systems*. Interaction, coordination, and communication acquire in this context a special relevance since they allow the involved groups, often made of heterogeneous and antecedently existing entities, to integrate their capabilities, to interact according to some agreed contracts, to share best practices and agreements, to cooperatively exploit resources and to facilitate the identification and the development of new products. Multi-Agent Systems (MASs) seem to be the most proper abstraction for developing cross-business systems, since they share with them the same characteristics (autonomy, heterogeneity) and the same issues (interaction, coordination, communication). These issues received a lot of attention by the research community working on agent models and platforms but, in our opinion, no proposal tackles them in a seamless and integrated way.

R. Pirrone and F. Sorbello (Eds.): AI*IA 2011, LNAI 6934, pp. 68–79, 2011.

The first limit of existing agent frameworks and platforms concerns the *forms of allowed interactions*. Most of them, such as [8,9], only supply the means for realizing agent interaction through *direct communication* (message exchange). This feature is common to all those approaches which foresee *agents* as the only available abstraction, thereby leading to message exchange as the only natural way agents have to interact. However, think of a business interaction between two partners. Once the client has paid for some item into the paypal account of the merchant, it should not be necessary that the client also sends a message to the merchant about the payment. Indeed, it is sufficient for the merchant to check his/her account to be informed about the transaction. The interaction already occurred.

In general, indirect communication fosters the collaboration and the coordination inside open systems, in that it allows anonymous, non-specialized interfaces that invite participation of new knowledge sources [21]. Indirect communication can be realized by means of persistent observable state changes, as it is for instance done by stigmergic approaches. In the literature there are models that allow the MAS designer to cope with a wider range of communication kinds by explicitly providing abstractions that support also the realization of *indirect forms* of interaction. In particular, the Agents & Artifacts model (A&A) [44,28,38] provides the explicit abstractions of *environment* and *artifact*, that are entities which can be acted upon, observed, perceived, notified, and so forth. Unfortunately, the A&A *misses a semantics* of communication, which would instead be required to achieve the seamless and coherent integration of the interaction, coordination, communication issues that we look forward. Other frameworks lack a satisfactory *semantics of interaction*. For instance, many of them, e.g. [8,9], adopt a *mentalistic* semantics which does not allow the agents, that are involved in the interaction, to verify that their partners respect the agreements. This, however, is a crucial aspect when one models business relationships.

In this respect, one interesting possibility would be to use an approach based on a *social* and *observational* semantics, like the commitment-based approach [30]. In this context, the agents' own behaviors remain private but agents agree on the meaning of a set of social actions: since interactions are observable and their semantics is shared, each agent is able to draw conclusions concerning the behavior of their partners as well as of the system as a whole. The advantage is that it becomes possible to detect violations to the agreed interaction and identify responsibilities. Proposals of this kind, that overcome the limitations of a pure mentalistic approach, seem particularly suitable to give to the A&A meta-model the semantics of communication that it still misses. Moreover, an interaction-oriented framework must supply the means for representing laws, rules, habits, which have strong implications on how agents can interact. These can be modeled by means of temporal regulations, along the line of [6,22].

This paper proposes a new agent-programming framework, Mercurio, that supports interaction-oriented computing and is currently under development. The paper is organized as follows. Section 2 describes the proposal. Section 3 discusses how the proposal advances the state of art. Final remarks in Section 4 conclude the paper.

2   The Mercurio Framework

Open systems are made of heterogeneously designed and pre-existing parties
that are assembled with some aim, which none of them can pursue alone. In
order to allow for a fruitful cooperation, the interaction that each agent carries
on with the others, in the context of the assembled system, must respect some
rules. In other words, the regulation of interaction is a decisive factor. The proposals that can be found in the literature, concerning the formation of and the
interaction within decentralized structures, are still incomplete. For instance,
electronic institutions [15,2] regulate interaction, tackle open environments, and
their semantics allows the verification of properties but they only cope with direct communication protocols, based on speech acts. On the other hand, most of
the models and architectures for "environments", which also allow indirect communications, prefigure simple/reactive agent models without defining semantics
that are comparable to the ones for ACL. This lack hinders agents to reason
about their partners' actions and to detect possible violations to the agreed behaviors [25]. In general, no proposal can yet capture all the requirements posed
by interaction-oriented computation in open environments. Here, in fact, it is
necessary to coordinate autonomous and heterogeneous agents and it is not
possible to assume mutual trust among them. It is necessary to have an unambiguous semantics allowing the verification of interaction properties both before
the interaction takes place [35] as well as during the interaction [1], preserving
at the same time the privacy of the implemented policies.

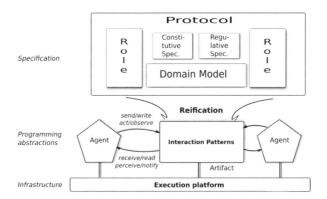

Fig. 1. The Mercurio architecture

Figure 1 draws the overall picture of the proposal. We distinguish three levels: the *specification* level, the level of the *programming abstractions*, and the
infrastructure. As we will see, this setting enables forms of verification that encompass both global interaction properties and specific agent properties, such
as interoperability and conformance [4].

2.1 Specification Level

The specification level allows the designer to shape the interactions that will characterize the system. Open systems involve autonomous partners with heterogeneous software designs and implementations; hence, the need of identifying high-level abstractions that allow modeling them in a natural way. In order to minimize the effort needed to define proper interfaces and to minimize the altering of internal implementations, Telang and Singh [42] propose that such abstractions should capture the contractual relationships among the partners, which are well-known to the business analysts and motivate the specification of the processes, and identify in commitment-based approaches [41,46] the appropriate features for performing this task. These, in fact, allow capturing the business intent of the system, leaving aside implementative issues. Other existing approaches (e.g. BPEL, WS-CDL) rely on the specification of control and business flows, imposing unnecessarily restrictive orderings of the interactions but supply the means for capturing neither the business relationships nor the business intent. This limitation is highlighted also by authors like Pesic and van der Aalst [31,26], who, in particular, show the advantages of adopting a declarative rather than a procedural representation.

By relying on an *observational semantics*, commitment-based approaches can cope both with direct forms of communication (by supporting the implementation of communicative acts), and with forms of interaction, that are mediated by the environment (by supporting the implementation of non-communicative acts, having a social meaning). Agents can not only send and receive messages but they can also act upon or perceive the social state.

Traditional commitment-based approaches, nevertheless, do not suit well those situations where the evolution of the social state is constrained by conventions, laws, and the like, because they do not supply the means for specifying legal *patterns of interaction*. This kind of constraints characterizes, however, many practical situations. Recent proposals, like [6,22], solve the problem by enriching commitment protocols with temporal regulations. In particular, [6] proposes a decoupled approach that separates a constitutive and a regulative specification [40]. A clear separation brings about many advantages, mostly as a consequence of the obtained modularity: easier re-use of actions, easier customization, easier composition. Roughly speaking, constitutive rules, by identifying certain behaviors as foundational of a certain type of activity, create that activity. They do so by specifying the semantics of actions. Regulative rules contingently constrain a previously constituted activity. In other words, they rule the "flow of activity", by capturing some important characteristics of how things should be carried on in *specific contexts* of interaction [12].

In [6] the constitutive specification defines the meaning of actions based on their effects on the social state, the regulative specification reinforces the regulative nature of commitment by adding a set of behavioral rules, given in terms of *temporal constraints* among commitments (and literals). These constraints define schemes on how commitments must be satisfied, regulating the evolution of the social state independently from the executed actions: the constitutive specification defines *which* engagement an agent has to satisfy, whereas the regulative specification defines *how* it must achieve what promised. Since interactions are observable and their semantics is shared, each agent is able to draw conclusions concerning the behavior of the partners or concerning the system as a whole. The specification level of Mercurio relies on the representation described in [6] for defining the set of legal interactions.

2.2   Programming Abstractions Level

This level realizes at a programming language level the abstractions defined above. This is done by incorporating interaction protocols based on commitments, patterns of interaction, forms of direct and indirect communication and coordination between agents (such as stigmergic coordination) inside the programmable environments envisaged by the A&A meta-model [44,28,38]. The resulting programmable environments will provide flexible *communication channels* that are specifically suitable for open systems. *Agents*, on the other hand, are the abstraction used to capture the interacting partners.

The notion of environment has always played a key role in the context of MAS; recently, it started to be considered as a first-class abstraction useful for the design and the engineering of MAS [44]. A&A follows this perspective, being a meta-model rooted upon Activity Theory and Computer Support Cooperative Work that defines the main abstractions for modeling a MAS, and in particular for modeling the environment in which a MAS is situated. A&A promotes a vision of an endogenous environment, that is a sort of software/computational environment, part of the MAS, that encapsulates the set of tools and resources useful/required by agents during the execution of their activities. A&A introduces the notion of artifact as the fundamental abstraction used for modeling the resources and the tools that populate the MAS environment. In particular, the fact of relying on computational environments provides features that are important from a Software Engineering point of view:

abstraction - the main concepts used to define application environments, i.e. artifacts and workspaces, are first-class entities in the agents world, and the interaction with agents is built around the agent-based concepts of action and perception (use and observation);

modularity and encapsulation - it provides an explicit way to modularize the environment, since artifacts are components representing units of functionality, encapsulating a partially-observable state and operations;

extensibility and adaptation - it provides a direct support for environment extensibility and adaptation, since artifacts can be dynamically constructed (instantiated), disposed, replaced, and adapted by agents;

reusability - it promotes the definition of types of artifact that can be reused, such as in the case of coordination artifacts empowering agent interaction and coordination, blackboards and synchronizers.

2.3   Infrastructure Level

In the state of the art numerous applications of the endogenous environments, i.e. environments used as a computational support for the agents' activities, have been explored, including coordination artifacts [29], artifacts used for realizing argumentation by means of proper coordination mechanisms [27], artifacts used for realizing stigmergic coordination mechanisms [36], organizational artifacts [19,33]. Our starting point is given by works that focus on the integration of agent-oriented programming languages with environments [37] and by the CArtAgO framework [39]. The CArtAgO framework provides the basis for the engineering of MAS environments on the base of: (i) a proper computational model and (ii) a programming model for the design and the development of the environments on the base of the A&A meta-model.

Even if CArtAgO can be considered a framework sufficiently mature for the concrete developing of software/computational MAS environments it can not be considered "complete" yet. Indeed at this moment the state of the art and in particular the CArtAgO framework are still lacking: (i) a reference standard on the environment side comparable to the existing standards in the context of the agents direct communications (FIPA ACL), (ii) the definition of a rigorous and formal semantics, in particular related to the artifact abstraction, (iii) an integration with the current communication approaches (FIPA ACL, KQML, etc.), and finally (iv) the support of semantic models and ontologies.

The infrastructure level will be developed by taking as reference JaCaMo http://jacamo.sourceforge.net, which is a platform for *multi-agent programming*, that successfully integrates BDI agents (developed in Jason [10]), artifact-based environments (CArtAgO) and organisations (specified by MOISE [20]).

2.4   Mediation as a Foundation for Interaction

One of the tasks, usually assigned to the environment, is that of *medium* of communication [44,28]. In our opinion, a programmable environment could play the important role of *arbitrator*. In other words, besides supplying the functionalities, needed for communicating, it can also be entrusted to check that the interaction is evolving in obedience to the rules as well as to reify the social state. The key is to develop artifacts that: (i) implement interaction specifications, and in particular the social expectations (given as commitments) entailed by them, (ii) monitor on-going interactions in order to detect possible incorrect executions and manage them. This perspective allows interaction, that is mediated by programmable environments, to be declined so as to capture different features and structures that are currently studied by as different researches as those concerning organizational theory [24,7,19,20], normative MAS [47], and e-institutions [15,2].

Figure 2 sketches an e-institution in the setting that we are proposing [23]. It is composed of three artifacts: one for the social state, one acting as a catalog of social actions, and the last one encoding the constraints upon the interaction. We adopt the definition of artifact in [44,28,38]: so, each artifact provides a set

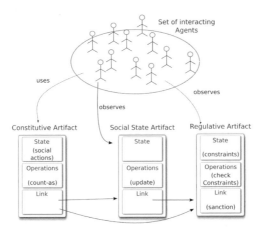

Fig. 2. E-institution representation by means of agents and artifacts

of *operations*, a set of *observable properties* and a set of *links*. Using an artifact involves two aspects: (1) being able to execute operations, that are listed in its usage interface, and (2) being able to perceive the artifact observable information. From an agent's standpoint, artifact operations represent the actions, that are provided by the environment. Moreover, artifacts can be linked together so as to enable one artifact to trigger the execution of operations over another artifact. The *social state artifact* contains the commitments and the facts as envisioned by commitment-based specification. It is updated after each execution of a social action. Agents can observe its contents. The *constitutive artifact* contains the set of the institutional actions together with their semantics, given in terms of effects on the social state. It must be observable by the agents. The link to the social state represents the fact that the execution of an institutional action triggers the update of the social state. The *regulative artifact* contains the constraints on the evolution of the social state. Violations can be detected automatically when the state is updated by checking whether some constraint is unattended. The check is triggered by the social state itself which, in turn, triggers the regulative artifact. The Mercurio framework accommodates equally well the solution where checks are performed by an arbiter agent. We propose the one based on artifacts because, as observed in [17], agents are autonomous and, therefore,they are free to choose whether to sanction a violation or not. Artifacts guarantee that violations are always detected.

Notice that the modularity entailed by the separation of the constitutive and of the regulative specifications, which characterizes the model, allows also for dinamically changing the agreed rules of the interaction, i.e. the regulative specification, at run-time, along the line of [3]. Indeed, it would be sufficient to include appropriate meta-actions to the constitutive artifact, whose effect is to change some of the constraints defined in the regulative artifact.

Organizations can be realized in a similar way. The focus, here, is on the structure of a certain reality, in terms of roles and groups. Each role will have an artifact associated with it, implementing the actions that the role player can execute on the social state. Organizations can be used to regiment the execution of the protocol [18]. More details in [23].

3 Discussion

We think that Mercurio will give significant contributions to industrial applicative contexts; in particular, to those companies working on software development in large, distributed systems and in service-oriented architectures. Among the most interesting examples are the integration and the cooperation of e-Government applications (services) spread over the nation. In this context, the aim is to verify the adherence of bureaucratic procedures, of the public administration, to the current laws (e.g. http://www.ict4law.org). Another interesting application regards (Web) services. Some of fundamental aspects promoted by the SOA model, such as autonomy and decoupling, are addressed in a natural way by the agent-oriented paradigm. The development and analysis of service-oriented systems can benefit from the increased level of abstraction offered by agents, by reducing the gap between the modeling, design, development, and implementation phases. In this context, it is necessary to deploy complex interactions having those characteristics of flexibility that agents are able to guarantee.

Nowadays, the development of MASs is based on two kinds of tools: agent platforms and BDI (or variations) development environments. The former, e.g. JADE and FIPA-OS provide only a transport layer and some basic services. Moreover, they lack support for semantic interoperability because they do not take into account the semantics of the ACL they adopt. The available BDI development environments, such as Jadex [11] and 2APL [14], support only syntactic interoperability because they do not integrate the semantics of the adopted ACL. Our proposal is compatible with the programming of BDI agents, in that the way agents are realized is totally transparent to the interaction media supplied by Mercurio. As such, Mercurio allows the interaction of *any kind* of agent. At the same time, by directly implementing the interaction specifications into an "intelligent" medium, the social semantics of the provided interactive actions is *guaranteed*; moreover, the medium supplies additional features, in particular concerning the verification of interactions.

Although our proposal is more general than e-institutions, it is interesting to comment some work carried on in this context. The current proposals for electronic institutions do not supply yet all of the solutions that we need: either they do not account for indirect forms of communication or they lack mechanisms for allowing the a priori verification of global properties of the interaction. As [16,43] witness, there is an emerging need of defining a more abstract notion of action, which is not limited to direct speech acts, whose use is not always natural. For instance, for voting in the human world, people often raise their hands rather than saying the name corresponding to their choice. If the environment were represented explicitly it would be possible to use a wider range of instrumental actions, that can be perceived by the other agents through the environment that acts as a medium.

Moreover, for what concerns the abstract architecture, e-institutions, e.g. Ameli [15], place a middleware composed of governors and staff agents between participating agents and an agent communication infrastructure. The notion of environment is dialogical: it is not something agents can sense and act upon but a conceptual one that agents, playing within the institution, can interact with by means of norms and laws, based on specific ontologies, social structures, and language conventions. Agents communicate with each other by means of speech acts and, behind the scene, the middleware mediates such communication. Finally, there are two significant differences between artifacts and e-institutions [2]: (i) e-institutions are tailored to a particular, though large, family of applications while artifacts are more generic; (ii) e-institutions are a well established and proven technology that includes a formal foundation, and advanced engineering and tool support, while for artifacts, these features are still in a preliminary phase. Mercurio gives to artifacts the formal foundation in terms of commitments and interaction patterns; engineering tools are currently being developed.

For what concerns organizations, instead, there are some attempts to integrate them with artifacts, e.g. ORA4MAS [19]. Following the A&A perspective, artifacts are concrete bricks used to structure the agents' world: part of this world is represented by the organizational infrastructure, part by artifacts introduced by specific MAS applications, including entities/services belonging to the external environment. In [19] the organizational infrastructure is based on $Moise^+$, which allows both for the enforcement and the regimentation of the rules of the organization. This is done by defining a set of conditions to be achieved and the roles that are permitted or obliged to perform them. The limit of this approach is that it cannot capture contexts in which regulations are, more generally, norms because norms cannot be restricted to achievement goals.

Finally, for what concerns commitment-based approaches, the main characteristic of the Mercurio proposal stands in the realization of the commitment stores as artifacts/environments which are capable to monitor the interaction, taking into account also the regulative specification. Such kind of artifact is a first-class object of the model, and is available to the designer. These features are advantageous w.r.t. approaches like [13], where these elements reside in the middleware, and are therefore shielded from the agents and from the designer. A negative consequence of the lack of appropriate programming abstraction is the difficulty to verify whether the system corresponds to the specification.

4   Conclusion

The Mercurio framework, presented in this paper, represents the core of a research project proposal, that is currently waiting for approval. The formal model of Mercurio relies on the commitment-based approach, presented in [6,5]. The proposals coming from Mercurio conjugate the flexibility and openness features that are typical of MAS with the needs of modularity and compositionality that are typical of design and development methodologies. The adoption of commitment protocols makes it easier and more natural to represent (inter)actions that are not limited to communicative acts but that include interactions mediated by the environment, namely actions upon the environment and the detection of variations of the environment.

For what concerns the infrastructure, a first result is the definition of environments based on A&A and on CArtAgO, that implement the formal models and the interaction protocols mentioned above. A large number of environments, described in the literature and supporting communication and coordination, have been stated considering purely reactive architectures. In Mercurio we formulate environment models that allow goal/task-oriented agents (those that integrate pro-activities and re-activities) the participation to MAS. Among the specific results related to this, we foresee an advancement of the state of the art with respect to the definition and the exploitation of forms of stigmergic coordination [36] in the context of intelligent agent systems. A further contribution regards the flexible use of artifact-based environments by intelligent agents, and consequently the reasoning techniques that such agents may adopt to take advantage of these environments. First steps in this direction, with respect to agents with BDI architectures, have been described in [34,32].

The Mercurio project aims at putting forward a proposal for a language that uses the FIPA ACL standard as a reference but integrates forms of interactions, that are enabled and mediated by the environment, with direct forms of communication. This will lead to an explicit representation of environments as first-class entities (in particular endogenous environments based on artifacts) and of the related model of actions/perceptions. As future work, we plan to implement a prototype of the reference infrastructural model. The prototype will be developed upon and integrate existing open-source technologies, among which: JADE, the reference FIPA platform, CArtAgO, the reference platform and technology for the programming and execution of environments, as well as agent-oriented programming languages such as Jason and 2APL.

References

1. Alberti, M., Chesani, F., Gavanelli, M., Lamma, E., Mello, P., Torroni, P.: Verifiable Agent Interaction in Abductive Logic Programming: The SCIFF Framework. ACM Trans. Comput. Log. 9(4) (2008)
2. Arcos, J.L., Noriega, P., Rodríguez-Aguilar, J.A., Sierra, C.: E4MAS Through Electronic Institutions. In: Weyns, et al. (eds.) [45], pp. 184–202
3. Artikis, A.: A Formal Specification of Dynamic Protocols for Open Agent Systems. CoRR, abs/1005.4815 (2010)
4. Baldoni, M., Baroglio, C., Chopra, A.K., Desai, N., Patti, V., Singh, M.P.: Choice, Interoperability, and Conformance in Interaction Protocols and Service Choreographies. In: Proc. of AAMAS, pp. 843–850 (2009)
5. Baldoni, M., Baroglio, C., Marengo, E.: Behavior-oriented Commitment-based Protocols. In: Proc. of ECAI, pp. 137–142 (2010)
6. Baldoni, M., Baroglio, C., Marengo, E., Patti, V.: Constitutive and Regulative Specifications of Commitment Protocols: a Decoupled Approach. In: ACM TIST, Special Issue on Agent Communication (2011)

7. Baldoni, M., Boella, G., van der Torre, L.: Bridging Agent Theory and Object Orientation: Agent-like Communication among Objects. In: Bordini, R.H., Dastani, M.M., Dix, J., El Fallah Seghrouchni, A. (eds.) PROMAS 2006. LNCS (LNAI), vol. 4411, pp. 149–164. Springer, Heidelberg (2007)
8. Bellifemine, F., Bergenti, F., Caire, G., Poggi, A.: JADE - A Java Agent Development Framework. In: Multi-Agent Progr.: Lang., Plat. and Appl. MAS, Art. Soc., and Sim. Org., vol. 15, pp. 125–147. Springer, Heidelberg (2005)
9. Bellifemine, F., Caire, G., Poggi, A., Rimassa, G.: JADE: A Software Framework for Developing Multi-agent Applications. Lessons learned. Information & Software Technology 50(1-2), 10–21 (2008)
10. Bordini, R.H., Hübner, J.F., Vieira, R.: Jason and the Golden Fleece of Agent-Oriented Programming. In: Multi-Agent Progr.: Lang., Plat. and Appl. MAS, Art. Soc., and Sim. Org., vol. 15, pp. 3–37. Springer, Heidelberg (2005)
11. Braubach, L., Pokahr, A., Lamersdorf, W.: Jadex: A BDI Agent System Combining Middleware and Reasoning. In: Software Agent-Based Applications, Platforms and Development Kits. Birkhauser Book, Basel (2005)
12. Cherry, C.: Regulative Rules and Constitutive Rules. The Philosophical Quarterly 23(93), 301–315 (1973)
13. Chopra, A.K., Singh, M.P.: An Architecture for Multiagent Systems: An Approach Based on Commitments. In: Braubach, L., Briot, J.-P., Thangarajah, J. (eds.) ProMAS 2009. LNCS, vol. 5919. Springer, Heidelberg (2010)
14. Dastani, M.: 2APL: a Practical Agent Programming Language. Autonomous Agents and Multi-Agent Systems 16(3), 214–248 (2008)
15. Esteva, M., Rosell, B., Rodríguez-Aguilar, J.A., Arcos, J.L.: AMELI: An Agent-Based Middleware for Electronic Institutions. In: AAMAS, pp. 236–243 (2004)
16. Fornara, N., Viganò, F., Colombetti, M.: Agent Communication and Artificial Institutions. JAAMAS 14(2), 121–142 (2007)
17. Fornara, N., Viganò, F., Verdicchio, M., Colombetti, M.: Artificial Institutions: a Model of Institutional Reality for Open Multiagent Systems. Artif. Intell. Law 16(1), 89–105 (2008)
18. Hübner, J.F., Boissier, O., Bordini, R.H.: From Organisation Specification to Normative Programming in Multi-agent Organisations. In: Dix, J., Leite, J., Governatori, G., Jamroga, W. (eds.) CLIMA XI. LNCS (LNAI), vol. 6245, pp. 117–134. Springer, Heidelberg (2010)
19. Hubner, J.F., Boissier, O., Kitio, R., Ricci, A.: Instrumenting Multi-agent Organisations with Organisational Artifacts and Agents: Giving the Organisational Power Back to the Agents. In: Proc. of AAMAS, vol. 20 (2009)
20. Hübner, J.F., Sichman, J.S., Boissier, O.: Developing Organised Multiagent Systems Using the MOISE. IJAOSE 1(3/4), 370–395 (2007)
21. Keil, D., Goldin, D.: Modeling Indirect Interaction in Open Computational Systems. In: Proc. of TAPOCS, pp. 355–360. IEEE Press, Los Alamitos (2003)
22. Marengo, E., Baldoni, M., Baroglio, C., Chopra, A.K., Patti, V., Singh, M.P.: Commitments with Regulations: Reasoning about Safety and Control in REGULA. In: Proc. of AAMAS (2011)
23. Marengo, E.: Designing and Programming Commitment-based Service-oriented Architectures on top of Agent and Environment Technologies. Technical Report RT 129/2010, Dip. di Informatica, Univ. di Torino (2010)
24. Masolo, C., Vieu, L., Bottazzi, E., Catenacci, C., Ferrario, R., Gangemi, A., Guarino, N.: Social Roles and their Descriptions. In: Proc. of KR 2004, pp. 267–277. AAAI Press, Menlo Park (2004)
25. McBurney, P., Parsons, S.: Games That Agents Play: A Formal Framework for Dialogues between Autonomous Agents. JLLI 11(3), 315–334 (2002)

26. Montali, M., Pesic, M., van der Aalst, W.M.P., Chesani, F., Mello, P., Storari, S.: Declarative Specification and Verification of Service Choreographies. ACM TWEB 4(1) (2010)

27. Oliva, E., McBurney, P., Omicini, A.: Co-argumentation Artifact for Agent Societies. In: Rahwan, I., Parsons, S., Reed, C. (eds.) Argumentation in Multi-Agent Systems. LNCS (LNAI), vol. 4946, pp. 31–46. Springer, Heidelberg (2008)

28. Omicini, A., Ricci, A., Viroli, M.: Artifacts in the A&A Meta-model for Multi-agent Systems. JAAMAS 17(3), 432–456 (2008)

29. Omicini, A., Ricci, A., Viroli, M., Castelfranchi, C., Tummolini, L.: Coordination Artifacts: Environment-Based Coordination for Intelligent Agents. In: Proc. of AAMAS, pp. 286–293 (2004)

30. Singh, M.P.: An Ontology for Commitments in Multiagent Systems. Artif. Intell. Law 7(1), 97–113 (1999)

31. Pesic, M., van der Aalst, W.M.P.: A Declarative Approach for Flexible Business Processes Management. In: Dustdar, S., Fiadeiro, J.L., Sheth, A.P. (eds.) BPM 2006. LNCS, vol. 4102, pp. 169–180. Springer, Heidelberg (2006)

32. Piunti, M., Ricci, A.: Cognitive Use of Artifacts: Exploiting Relevant Information Residing in MAS Environments. In: Meyer, J.-J.C., Broersen, J. (eds.) KRAMAS 2008. LNCS, vol. 5605, pp. 114–129. Springer, Heidelberg (2009)

33. Piunti, M., Ricci, A., Boissier, O., Hübner, J.F.: Embodying Organisations in Multi-agent Work Environments. In: Proc. of IAT, pp. 511–518 (2009)

34. Piunti, M., Ricci, A., Braubach, L., Pokahr, A.: Goal-Directed Interactions in Artifact-Based MAS: Jadex Agents Playing in CARTAGO Environments. In: Proc. of IAT, pp. 207–213. IEEE, Los Alamitos (2008)

35. Rajamani, S.K., Rehof, J.: Conformance Checking for Models of Asynchronous Message Passing Software. In: Brinksma, E., Larsen, K.G. (eds.) CAV 2002. LNCS, vol. 2404, pp. 166–179. Springer, Heidelberg (2002)

36. Ricci, A., Omicini, A., Viroli, M., Gardelli, L., Oliva, E.: Cognitive Stigmergy: Towards a Framework based on Agents and Artifacts. In: [45], pp. 124–140

37. Ricci, A., Piunti, M., Acay, D.L., Bordini, R.H., Hübner, J.F., Dastani, M.: Integrating Heterogeneous Agent Programming Platforms within Artifact-based Environments. In: Proc. of AAMAS, vol. (1), pp. 225–232 (2008)

38. Ricci, A., Piunti, M., Viroli, M.: Environment Programming in MAS – An Artifact-Based Perspective. JAAMAS

39. Ricci, A., Piunti, M., Viroli, M., Omicini, A.: Environment Programming in CArtAgO. In: Multi-Agent Prog. II: Lang., Plat. and Appl. (2009)

40. Searle, J.R.: The Construction of Social Reality. Free Press, New York (1995)

41. Singh, M.P.: A Social Semantics for Agent Communication Languages. In: Dignum, F.P.M., Greaves, M. (eds.) Issues in Agent Communication. LNCS, vol. 1916, pp. 31–45. Springer, Heidelberg (2000)

42. Telang, P.R., Singh, M.P.: Abstracting Business Modeling Patterns from RosettaNet. In: SOC: Agents, Semantics, and Engineering (2010)

43. Verdicchio, M., Colombetti, M.: Communication Languages for Multiagent Systems. Comp. Intel. 25(2), 136–159 (2009)

44. Weyns, D., Omicini, A., Odell, J.: Environment as a First Class Abstraction in Multiagent Systems. JAAMAS 14(1), 5–30 (2007)

45. Weyns, D., Van Dyke Parunak, H., Michel, F. (eds.): E4MAS 2006. LNCS (LNAI), vol. 4389. Springer, Heidelberg (2007)

46. Yolum, p., Singh, M.P.: Commitment machines. In: Meyer, J.-J.C., Tambe, M. (eds.) ATAL 2001. LNCS (LNAI), vol. 2333, pp. 235–247. Springer, Heidelberg (2002)

47. Zambonelli, F., Jennings, N.R., Wooldridge, M.: Developing Multiagent Systems: The Gaia Methodology. ACM Trans. Softw. Eng. Methodol. 12(3), 317–370 (2003)

Fuzzy Conformance Checking of Observed Behaviour with Expectations

Stefano Bragaglia[1], Federico Chesani[1], Paola Mello[1],
Marco Montali[2], and Davide Sottara[1]

[1] DEIS, University of Bologna,
Viale Risorgimento, 2 - 40136 Bologna, Italy
`name.surname@unibo.it`
[2] KRDB Research Centre, Free University of Bozen-Bolzano,
Piazza Domenicani, 3 - 39100 Bolzano, Italy
`montali@inf.unibz.it`

Abstract. In some different research fields a research issue has been to establish if the external, observed behaviour of an entity is conformant to some rules/specifications/expectations. Research areas like Multi Agent Systems, Business Process, and Legal/Normative systems, have proposed different characterizations of the same problem, named as the conformance problem. Most of the available systems, however, provide only simple yes/no answers to the conformance issue.

In this paper we introduce the idea of a gradual conformance, expressed in fuzzy terms. To this end, we present a system based on a fuzzy extension of Drools, and exploit it to perform conformance tests. In particular, we consider two aspects: the first related to fuzzy ontological aspects, and the second about fuzzy time-related aspects. Moreover, we discuss how to conjugate the fuzzy contributions from these aspects to get a single, fuzzy score representing a conformance degree.

Keywords: fuzzy conformance, production rule systems, expectations, time reasoning

1 Introduction

In the last ten years there has been a flourishing of models and technologies for developing, deploying, and maintaining ICT systems based on (heterogeneous, distributed) components. Paradigms such as Service Oriented Architectures, Cloud Computing, Business Process Workflows, have been exploited from the industry: nowadays, mature standards and solutions are available to the average customer, covering many of the ICT needs within the industry.

However, the complexity of such systems has grown pair-wise with the availability of such standards and tools. At the same time, the adoption of standards has fostered the use of heterogeneous (software/hardware) components. As a consequence, assuring the correct behaviour of such systems has become an important issue. To this end, approaches based on the notion of *conformance* have been proposed. Roughly speaking, the *expected* behaviour of the system is specified a-priori, by means of some formal language. Then, the complex system is observed at run-time, and the externally observed

R. Pirrone and F. Sorbello (Eds.): AI*IA 2011, LNAI 6934, pp. 80–91, 2011.

behaviour is confronted with the *expectations*. In case the expectations are not met by the observations, some alarms and/or managing procedures are triggered. With the term *conformance test* we refer to the process of evaluating if the observed behaviour matches the expectations, i.e., if the observed behaviour is *conformant* with the expectations.

Notably, the *conformance*-based approach has been object of intense research activity in many different application fields, especially when considering it in its most abstract way. In Multi-Agent Systems (MAS), for example, social approaches specify the agents' allowed interactions as expected behaviours (externally observed), and define violations in terms of deviance from what is expected. The framework SCIFF [1], as defined by Alberti and colleagues, is mainly focused on the notion of *expectations* and their violations. Commitments, as deeply investigated by authors such as Singh [21,9,25] or Colombetti and Fornara [13], concentrate on promises that arise as consequences of agent interactions: a debtor agent becomes committed towards a creditor (i.e., it is *expected*) to bring about (make true) a certain property.

In the Business Process field, for example, van der Aalst and colleagues have proposed declarative languages to focus on the properties that the system should exhibit: in the DecSerFlow language [18] the users can specify which are the business activities that are (not) *expected* to be executed, as consequence of previously (not) executed activities. Within the field of legal reasoning and normative systems, authors like Governatori and Rotolo [14] have proposed logic frameworks and languages to represent legal contracts between parties: the focus is on the *compliance* problem, and they evaluate it by establishing if the possible executions of a system are conformant with the legal aspects.

Most of the approaches investigated so far provide a boolean answer to the conformance problem. If the question is *"Is the observed behaviour conformant with what is expected"*, most of the systems generate only a simple answer of the type *yes/no*. However, taking inspiration from our everyday experiences, we argue that in many cases it is required a richer, more informative answer. Indeed, yes/no answers tend to oversimplify and to collapse the conformance check to only two possible values, while real situations would require an answer with some degree. In this sense, a score (a value comprised in the interval $[0, 1]$) would be a reasonable desiderata.

Let us consider, for example, an internet book seller who delivers items by mail. To reduce its costs, the seller often delegates the packaging and the shipping of the items to smaller book retailers, following a commercial agreement. The delivers must be conformant to some criteria established within the agreement. Then, the book seller performs a continuous monitoring of the delivery process. In particular, in this example we consider two aspects: 1) the quality of the packaging; and 2) the timing of the delivery. Both aspects contribute to establish if the delivering has been conformant with the expectations established in the business contract.

Evaluating the quality of the packaging means to take into consideration several different aspects, such as the use of a box of the right dimension, the water-sealing, the material used within the package, the care with which the items have been packaged, and many others[1]. Although a yes/no answer is still possible, it is reasonable to assign a score to the overall quality, thus capturing "how well" the packaging was done. Moreover, the evaluation criteria should be known a-priori (for example, they could be part of

---

[1] Ultimately, also the customer feedback would contribute to evaluate the packaging quality.

a business agreement). Such criteria would explicitly define the concept of "good package". Evaluating the quality of a particular packaging would consist of establishing if that package belongs, and with which degree, to the category of "good packages".

Similar observations could be done also for timing aspects. Again, answering with a yes/no answer to the question "was the package delivered in time?" is correct. However, in case of delays, there would be no way to evaluate how "big" was the delay. Depending on the business agreement (that sets the expectations), few or many days of delay would have a different impact when evaluating the conformance of that particular delivery (w.r.t. to the expectations).

All these examples suggest that a conformance test could be significantly enriched with evaluation scores. Given the *vague* and *gradual* definition of the desired constraints, it would be natural to exploit *fuzzy logic* for defining the notion of conformance. In this paper, we present a prototypical system, based on a fuzzy extension of the Drools rule-based framework [16], to evaluate in a fuzzy manner how much an observed event, fact or object, is conformant to a certain expectation. In particular, we focus on two different aspects: on one side, we tackle the problem of establishing if a certain event matches the expectation, by considering fuzzy ontological aspects on *what* happens and *what* is expected. On the other side, we concentrate on *temporal*-related aspects, and show how it is possible within Drools to define custom, fuzzy, time-related operators. Finally, we show also how it is possible to conjugate both the aspects (ontological and temporal ones) to get a single fuzzy evaluation. Such evaluation not only allows to provide a fuzzy answer to the conformance issue, but intrinsically supports also the ranking of the observed events/facts (w.r.t. expectations) on the base of the conformance criteria.

The actual idea behind this work is to show the feasibility of a hybrid rule-based and semantic approach to the conformance evaluation problem, with fuzziness added on top. We do not aim to define some new fuzzy theory, and we do not provide any new contribution to the fuzzy research field. Rather, we investigate how fuzzy logics can be used to characterize the conformance problem, and we do this by exploiting an existing (fuzzy) rule-based framework. While indeed trivial from a "fuzzy point of view", our approach is quite new, to the best of our knowledge, in other research fields like, e.g., Multi-Agent Systems.

2   Background on the Drools Framework and Its Support to Fuzzy Reasoning

Drools [16] is an open source system with the aim of becoming a "Knowledge modelling and integration platform". At its core, among several other tools, it includes a reactive production rule engine, which is is based on an object-oriented implementation of the RETE algorithm [12].

From a user perspective, the system offers a blackboard-like container, called *Working Memory* (WM), where the facts describing the "state of the world" can be insert-ed, updated or retracted. The rules, then, are *activated* accordingly whenever the WM is modified. A rule is an IF-THEN like construct, composed of a premise (Left Hand Side, LHS for short) and a consequence (Right Hand Side, RHS). The LHS part is composed by one or more *patterns*, which must be matched by one or more facts in the WM for the rule to become active. An active rule is then eligible to be fired, executing the actions defined in the RHS, which may either be logical actions on the WM or side effects. A pattern is a sequence of constraints a fact must satisfy in order to *match* with that pattern. Since facts are objects, in Drools, the first constraint is a *class* constraint, while the following ones are boolean expressions involving one or more object's *fields*.

2.1 Fuzzy Reasoning Capabilities

Drools' language is rich and expressive, but the core expressions that can be used in a LHS are equivalent to a boolean formula where the atomic constraints are linked using simple logic connectives such as and, or, not and the quantifiers forall and exists. The use of boolean logic accounts for high efficiency, but limits the expressiveness of the system.

In [23] an extension of Drools, namely Drools Chance, has been proposed to support fuzzy reasoning. The RETE engine has been extended to support (among other frameworks) fuzzy logics "in a narrow sense", i.e., to support special many-valued logics addressing vagueness, via truth degrees taken from an ordered scale. Under this assumption, the constraints are no longer evaluated to true or false, but generalized using an abstract and pluggable representation, called *degree*. Degrees can be concretely implemented using, among others, real numbers in the interval $[0, 1]$, intervals, fuzzy numbers, ... Exploiting the fact that operators (*evaluators* in the Drools terminology) can be externally defined, the engine has been extended to allow the evaluators to return degrees in place of booleans. Moreover, logic connectives and quantifiers have also been made "pluggable" (externally defined) and configurable. Given many valued predicates, connectives and quantifiers, Drools Chance can evaluate the LHS of a rule in the context of a "narrow" fuzzy logic. Further extensions allow to support fuzzy logics in a "broader sense", and in particular to support linguistic variables and fuzzy sets in constraints, as introduced in [26].

2.2 Fuzzy Ontological Reasoning

In [5], a further extension has been built on top of Drools Chance, to support fuzzy semantic reasoning within the Drools rule system: in particular, the implementation of a fuzzy tableaux-based reasoner [3] is presented. This extension allows to define and reason on knowledge bases whose expressiveness is equivalent to the $\mathcal{ALC}$ fragment of the family of Description Logics.

The resulting framework (Drools, Drools Chance and the extensions for ontological reasoning) allows to import ontologies defined in a fuzzy manner,within the Drools WM. Moreover, it allows to write rules where the LHS pattern matching mechanism is extended with some (fuzzy) ontological evaluators. The framework allows to integrate, in a unified formalism and model, both ontological and rule-based reasoning, and both the aspects are treated in a fuzzy manner.

During the evaluation of the LHS of a rule, each (fuzzy ontological) statement in the LHS is associated with an *interval*, whose bounds are fully included into the range $[0, 1]$. The interval defines lower and upper bounds of the truth degree of the statement. Such

Listing 1.1. A rule using fuzzy and semantic statements

```
rule "Dispatching an order"
when
  $o : Order
  $c : Contact( this ~isA "retailer" )
       Score( retailer == $c,
              this ~seems "high" )
then
  /* Suggests $c for $o */
end
```

bounds can be respectively considered as the *necessity* (lower bound) and the *possibility* values (upper bound) of the truth degree for its related statement. The use of an interval (instead of a single value) has been chosen by the authors to overcome the dichotomy between Open World Assumption (OWA) and Closed World Assumption (CWA). Indeed, ontological reasoning is often performed assuming OWA semantics, while rule based systems such as Drools adopt the CWA semantics: such important difference makes impossible to define a unified semantics for both the reasoning paradigms.

The solution to the OWA/CWA issue proposed in [5] consists of assigning to each ontological statement an interval delimited by two fuzzy values, that are interpreted as the necessity and the possibility of the truth degree of the statement. In our interpretation, the necessity value (lower bound of the interval) corresponds to the OWA hypothesis, while the possibility value (upper bound of the interval) corresponds to the CWA hypothesis. To easy the use of the ontological statements and their truth value within a rule, two operators, POS and NEC, are provided to the user to extract OWA and / or CWA degree, respectively.

An example of rules integrating ontological statements and rule-based reasoning is shown in Listing 1.1. The meaning of such rule is the following: anytime she receives an order, the seller should choose a retailer to dispatch it. To this aim, assuming that her address book is an ontology where contacts are organized into (not necessarily disjoint) subclasses including "retailer" and that the system has a distinct component that returns a score on the service of each retailer according to customers' feedback, the rule browses through the seller's address book identifying all the contacts that qualify as retailers and whose score seems high. All the retailers so identified are suggested to the seller who can choose then the most appropriate retailer for the given order.

2.3 Exploiting Drools and Drools Chance to Manage Fuzzy Time Aspects

To implement the fuzzy reasoning capabilities (see Section 2.1), the authors of Drools Chance [23] have exploited a powerful feature of Drools: the possibility given to the users to provide their own definition of new evaluators. We have chosen to follow the same approach: on this line, we have defined new evaluators for treating temporal related aspects. Currently, we have not implemented any particular time-related formalism, although we are aware there are many in the literature. Probably the most famous logic about temporal aspects is the one proposed by Allen in [2], where a (crisp) logic for reasoning over temporal intervals, together with some operators, is provided.

Our current choice has been to implement ad-hoc, on-purpose time-related evaluators, to address the specific needs of our test-cases. Probably, for very simple cases, the same expressivity could be achieved directly using Drools fuzzy rules, or by means of a fuzzy ontology defining some basic time-related concepts. It is out of the scope of this paper to establish when some knowledge is better represented using a formalism rather than another one. Here we will stress only the fact that the notion of conformance has been always referred to *what* is expected towards *what* is observed. Recent works within the MAS research community, such as [24], have also stressed the importance of temporal aspects (such as, for example, deadlines). Our choice then has been to focus on ontological and time related aspects explicitly, thus making these two aspects "first class entities" in the notion of conformance.

3   Checking Fuzzy Conformance

Generally speaking, the conformance problem amounts to establish if the externally observed behaviour of a system/entity does respect (satisfy/fulfill) some given *expectations*. Thus, the notion of conformance is strictly related (and depends on) to the notion of expectation. Depending on the research field, conformance and expectations have been given different characteristics and flavours.

In our system we do not restrict to a particular notion of conformance and/or of expectation. We rather assume that the answer will be always a real number in the interval $[0, 1]$, and that such a number can be interpreted as *the extent to which* the observed behaviour is conformant to the expectation. Such generalisation allows to consider also previous "yes/no" approaches as particular cases.

Moreover, we consider the conformance test as the result of combining many, different aspects, each one contributing with its own fuzzy result to the final conformance degree. The number of aspects to be taken into consideration, and the methods/algorithms to evaluate each aspect, depend on the application domain and its modelling. In this paper we present some examples of conformance based on two different aspects: more precisely, the ontological aspect and the temporal one. Many other different aspects could be considered, such as, for example, geographical information.

Our notion of fuzzy conformance then is a two-level process, where at the lower level many single components (the evaluators) provide a fuzzy conformance degree related to each single aspect, while at the higher level each fuzzy contribution is combined to achieve a single fuzzy degree representing the overall conformance. To support such notion of conformance, then, many user-defined evaluators are needed. Single aspects need ad-hoc, specific evaluators properly designed for a particular aspect of the domain. For the evaluation of fuzzy ontological aspects, we resorted to use an existing extension of Drools Chance (see Section 2.2). For the time-related aspects instead, we provided our own implementation of the needed evaluators.

In the remain of this section we will discuss the example introduced in Section 1, and will show a possible implementation of our notion of fuzzy conformance. Briefly recapping the example, in the context of a business agreement, some local book stores perform the packaging and the delivering of items on behalf of an internet book seller. The seller continuously checks if the delivering of the packages is conformant with the commercial agreement. Indeed, such contract establishes which are the *expectations* about the items delivery. The local stores provides in a log file the description of the packaging, and the time it was delivered.

3.1 Conformance as Ontological Fuzzy Evaluation

Exploiting the extensions introduced in Section 2.2, our system supports the definition of fuzzy ontologies[2], i.e. of ontologies where, for example, individuals are instances of certain classes with a fuzzy degree. We can easily imagine then an ontology where the concept of GoodPackaging is defined as those individuals of the domain that are Packages (intended as atomic concept), and that have been water sealed, filled with bubble-wrap paper and carefully prepared. Of course, there would be cases of packages sealed without water-proof scotch tape, or packages where the items inside have not been rolled up with bubble-wrap. Such packages would belong with a low degree to the category of GoodPackaging. On the contrary, packages responding to all the requisites would be classified with the highest score to belonging to the GoodPackaging class.

From a practical view point, we aim to write a rule where in the LHS there is a pattern that evaluates how much a particular packaging was well-done. I.e., we want to add in the LHS a (fuzzy) ontological statement about a particular package being instance of the class GoodPackaging. The evaluation of the LHS would then compute the truth value of such statement, and such truth value could be then used within the rule itself (in the LHS as well as in the RHS). However, to exploit the support of fuzzy ontologies presented in [5], we have to deal with the fact that an interval bounded by two fuzzy values is given as answer, when evaluating fuzzy ontological statements. Such values represent the truth degree of the statement under the OWA and the CWA hypotheses: consequently, we must resort, on a domain-basis, to the OWA semantics (lower bound), or to the CWA semantics (upper bound).

Listing 1.2. A rule with a fuzzy ontological statement

```
rule "Fuzzy Ontological Matching"
when
    $p: Package( $p nec ~isA "GoodPackaging" )
then
    println($p.id + " isA GoodPackaging: " +
      Drools.degree);
end
```

Within the context of our example, we have depicted a quite simple fuzzy ontology: in this particular case using OWA rather than CWA would make no difference, since under both the semantics the statement would be evaluated with the same, exact score. In more complex situations however, where more complex representations of the domain would be taken into consideration, the choice of which semantics should be adopted is not a

---

[2] See [4] for an introduction to Fuzzy Description Logics, their representation, and many examples of fuzzy ontologies.

trivial task, and would largely depend on the particular application domain. Generally speaking, since we are applying such ontological reasoning to evaluate conformance, Open World Assumption semantics seems to be a safer choice, since it would support a "lazier" evaluation of the conformance. CWA semantics would support a more stricter notion of conformance, but with the risk of some "false positive" results.

In Listing 1.2 an example of a rule with an ontological statement in its LHS is presented. The LHS is evaluated every time an object representing a packaging in inserted in the working memory. The Drools engine then evaluates the ontological statement, where ∼isA is a shortcut for the classical instanceOf ontological operator; the "∼" symbols indicates that the statement is indeed evaluated within the fuzzy domain. The rules print out on the console the the truth value of the statement "p instanceOf GoodPackaging". Note also the use of the "nec" operator to select the lower bound (necessity) of the fuzzy ontological evaluation, corresponding to the OWA semantics.

3.2 Fuzzy Temporal Evaluation

We have already discussed in Section 2.3 how it is possible to easily extend DROOLS with new operators. Exploiting such possibility, it is possible to create fuzzy time-related evaluators on the basis of the needs for representing the domain. Let us consider again the example of the internet book store. In a crisp evaluation setting, the conformance would depend on the promised and the effective delivery date. If the latter follows the former, we could conclude that the deadline has not been respected, and the expedition is not conformant.

However, it makes sense to consider the expedition process with a larger perspective. One day of delay could be insignificant in certain situations, while could be a terrific problem in other situations. For example, if one-business day delivery is expected, one day of delay has a huge impact on the notion of conformance. Differently, if the delivery is expected within thirty days, one or two more days would have a more little impact on the evaluation of concept.

To support our example, we have defined a new simple evaluator that provides a fuzzy evaluation score of how much a deadline has been met or not. We named such evaluator ∼InTime, and it takes two parameters: a) the difference t_d between the expected delivery date and the effective delivery date (zero if the delivery met the deadline); and b) the duration of the interval of time t_e expected for the delivery (such expectation established, for example, when a customer finalized the order to the internet book seller). A possible definition of our evaluator could be the following:

$$InTime(t_d, t_e) = \left\lfloor 1 - \frac{t_d}{t_e} \right\rfloor_0$$

An example of a rule considering time-related aspects is presented in Listing 1.3.

3.3 Combining Different Fuzzy Contributions

Once the different aspects have been evaluated, the problem of deciding how to combine such different fuzzy contributions arises. Typically, such choice would be highly

Listing 1.3. A Drools rule evaluating in a fuzzy manner how much a temporal dealine has been respected

```
rule "Fuzzy evaluation of the delivery delay "
when
  Order ($e: expectedInDays)
  DeliveryLog( $d: delay ~InTime $e)
then
  println("Delivery is conformant with the" +
          " temporal deadline with score: " +
          Drools.degree);
end
```

dependent by the modelled domain. Again, Drools Chance offers the possibility of user--defining logical operators. Moreover, the most common fuzzy logic operators are natively supported by the framework, hence providing the user a vast range of possibilities. When defining new user fuzzy operators, a particular attention must be paid to define them in terms of simpler, available operators that ensure the truth functionality property: the aim is to guarantee that the resulting system has still an underlying semantics based on a (infinitely many-valued) fuzzy logic. Of course, if the chosen operators do not guarantee the truth-functionality property, then also the resulting system will not exhibit such feature.

Listing 1.4. A rule evaluating conformance in a fuzzy manner

```
rule "Fuzzy evaluation of conformance"
when
  Order ($e: expectedInDays )
  DeliveryLog(
    $d: delay ~InTime $e
    , @imperfect(kind=="userOp")
    $p: packaging nec ~isA "GoodPackaging")
then
  println("Degree of Delivery Conformance: " +
          Drools.degree);
end
```

In Listing 1.4 an example of a rule evaluating the conformance in a fuzzy manner is presented. The rule exploits the operators defined previously, and show how it is possible to evaluate the conformance in a fuzzy manner by means of Drools Chance and its extensions. In particular, we have considered here only aspects related to *what* is expected/observed, and *when* it is expected/observed. In a similar manner, the framework can easily extended to support many different aspects, adapting to the needs of the modelled domain.

4   Discussion and Conclusions

In this paper we have presented our idea of *fuzzy conformance*, motivated by the fact that the usual crisp notion of conformance might result too poor for capturing the vagueness and uncertain characteristics of real application domains. The use of (infinitely) many--valued fuzzy logic has been a natural choice for supporting such idea. Moreover, we propose to split the notion of conformance into the process of evaluating the conformance on many, different aspects, each providing its own fuzzy conformance degree, and then to combine such contributions into a single (fuzzy) conformance value. In particular, we focussed our attention on ontological and on time-related aspects. Nevertheless, we acknowledge that such a choice could result as too restrictive: for example, spatial/geographical aspects could be of a great interest when evaluating conformance in particular domains. Our approach is easily extendible towards such directions. From the practical viewpoint, we have shown how it is possible to easily exploit existing tools to implement such notion of conformance. In particular, we have used the Drools rule--based system, together with the Drools Chance extension, and a recent extension that supports fuzzy ontological reasoning. Then, we have discussed how, within such framework, it is possible to write rules that evaluates the conformance in a fuzzy manner.

The implementation of the presented framework is still in a prototypical stage, and further refinements will be addressed in the future. For example, we support conjunction of atomic expectations, combined with the temporal dimension: disjunctions of expectations are only partially supported, but we deem them as a fundamental feature. Also the expressivity of the supported ontology language will be addressed: our current implementation is based on [5], i.e. it supports $\mathcal{ALC}$ fragment, while we aim to exploit [4] for a greater expressivity. Moreover, a better assessment of the approach, using a real and complex scenario, together with an evaluation of the performances and a comparison with other solutions, is planned for the near future.

An important (open) issue is about using which one, and when, the right modelling formalism. For simple examples like the one used in the paper, it can happen that more than one formalism would be expressive enough for modelling that part of knowledge. E.g., the ontology described in Section 3.1 could have been represented also by means of custom-tailored rules, instead of a fuzzy ontology reasoner. Discussing which formalism is better for which situation is far behind the scope of this paper. However, we have adopted the following thumb rule: fuzzy ontologies have been used to describe the static structure and properties of the domain, while more dynamic aspects such as temporal (or spatial proximity) issues have been modelled by using/extending the rules.

Our work has many contacts with studies on the evaluation of constraints, temporal or not, under the generic notion of uncertainty. The concept can be applied at various levels of abstraction, from value-constrained measurements (e.g. [7]) to time--constrained interactions (e.g. [6], where the interactions are modelled by state transitions in automata). Many works on the evaluation of semantic constraints comes also from the Semantic Web area, especially to address the problems of service selection/ranking in general and of QoS evaluation in particular. In [20], the conformance is given a probabilistic connotation, while in [22] a fuzzy evaluation is performed. In [15], [17], finally, both semantic and fuzzy aspects about the QoS parameters are defined using an ontology, and evaluated using fuzzy predicates. Such solution, however, is based on a loosely coupling: examples of tighter levels of integration can be found for example in [19], where matchings are performed for recommendation instead of conformance check purposes.

With respect to the fuzzy research field, our work has many conjunctions with the well known idea of fuzzy pattern matching. In particular, our prototype share some similarities with the idea of weighted/tolerant fuzzy pattern matching [11,10], where a data item (composed of many features) is matched against an and/or formula (the pattern) of conditions. While in the fuzzy pattern matching both the pattern and the data can be vague/imprecise, in our scenario the data is certain and precise, since it results from the observation of the systems at run-time. Moreover, necessity and probability in our approach are mainly referred to the results of fuzzy ontological evaluation, and are a direct consequence of the ontology definition, while in the fuzzy pattern matching problem these terms refer to the vagueness introduced by the data (the observation in our case). Finally, researchers in the fuzzy area have focused on defining the operators so that the final resulting degree of matching will exhibit some properties, like for example preserving the semantics of necessity and possibility. In our prototype we leave the implementation of the operators to the domain modeller; we do not support directly the weights associated to each pattern's component, leaving the user the possibility of defining custom operators and, possibly, supporting multi-criteria evaluation policies.

We are aware that the presented approach juxtaposes with many other contiguous research fields, such as User Preferences specification or Recommender Systems. The possibility of having a degree of conformance allows to create a ranking of the observed behaviour, ordered on the basis of such conformance degree. If the evaluation of conformance is applied upon possible events/course of actions, instead of observed ones, we could get a sort of recommendation system. Investigating the relations with such existing research fields will be matter of future work.

Acknowledgements. This work has been partially support by the DEPICT Project, DEIS, and by the CIRI Life Sciences and Health Technologies, Univ. of Bologna.

References

1. Alberti, M., Chesani, F., Gavanelli, M., Lamma, E., Mello, P., Torroni, P.: Verifiable agent interaction in abductive logic programming: The SCIFF framework. ACM Trans. Comput. Logic 9(4), 1–43 (2008)
2. Allen, J.F.: Maintaining knowledge about temporal intervals. Commun. ACM 26(11), 832–843 (1983)
3. Bobillo, F., Straccia, U.: fuzzydl: An expressive fuzzy description logic reasoner. In: IEEE International Conference on Fuzzy Systems, FUZZ-IEEE 2008 (IEEE World Congress on Computational Intelligence), pp. 923–930 (2008)
4. Bobillo, F., Straccia, U.: Fuzzy ontology representation using owl 2. International Journal of Approximate Reasoning (to appear, 2011)
5. Bragaglia, S., Chesani, F., Mello, P., Sottara, D.: A rule-based implementation of fuzzy tableau reasoning. In: Dean, et al. (eds.) [8], pp. 35–49
6. Crespo, F., de la Encina, A., Llana, L.: Fuzzy-timed automata. In: Hatcliff, J., Zucca, E. (eds.) FMOODS 2010. LNCS, vol. 6117, pp. 140–154. Springer, Heidelberg (2010), doi:10.1007/978-3-642-13464-712

7. De Capua, C., De Falco, S., Liccardo, A., Morello, R.: A technique based on uncertainty analysis to qualify the design of measurement systems. In: Proceedings of the 2005 IEEE International Workshop on Advanced Methods for Uncertainty Estimation in Measurement, pp. 97–102 (May 2005)

8. Dean, M., Hall, J., Rotolo, A., Tabet, S. (eds.): RuleML 2010. LNCS, vol. 6403. Springer, Heidelberg (2010)

9. Desai, N., Chopra, A.K., Singh, M.P.: Representing and reasoning about commitments in business processes. In: AAAI, pp. 1328–1333. AAAI Press, Menlo Park (2007)

10. Dubois, D., Prade, H.: Tolerant fuzzy pattern matching: An introduction. In: Bosc, P., Kacprzyk, J. (eds.) Fuzziness in Database Management Systems, pp. 42–58. Physica-Verlag, Heidelberg (1995)

11. Dubois, D., Prade, H., Testemale, C.: Weighted fuzzy pattern matching. Fuzzy Sets Syst. 28, 313–331 (1988)

12. Forgy, C.: Rete: A fast algorithm for the many patterns/many objects match problem. Artif. Intell. 19(1), 17–37 (1982)

13. Fornara, N., Colombetti, M.: A commitment-based approach to agent communication. Applied Artificial Intelligence 18(9-10), 853–866 (2004)

14. Governatori, G., Rotolo, A.: Norm compliance in business process modeling. In: Dean, et al. (eds.) [8], pp. 194–209

15. HongKang, Z., XueLi, Y., GuangPing, Z., Kun, H.: Research on services matching and ranking based on fuzzy qos ontology. In: 2010 International Conference on Computational Aspects of Social Networks (CASoN), pp. 579–582 (September 2010)

16. JBossL. JBoss Drools 5.0 - Business Logic Integration Platform (2010)

17. Lee, C.-H.L., Liu, A., Hung, J.-S.: Service quality evaluation by personal ontology. J. Inf. Sci. Eng. 25(5), 1305–1319 (2009)

18. Montali, M., Pesic, M., Aalst, W.M.P.v.d., Chesani, F., Mello, P., Storari, S.: Declarative specification and verification of service choreographiess. TWEB 4(1) (2010)

19. Ragone, A., Straccia, U., Noia, T.D., Sciascio, E.D., Donini, F.M.: Fuzzy matchmaking in e-marketplaces of peer entities using datalog. Fuzzy Sets Syst. 160, 251–268 (2009)

20. Rosario, S., Benveniste, A., Haar, S., Jard, C.: Probabilistic qos and soft contracts for transaction-based web services orchestrations. IEEE Trans. Serv. Comput. 1, 187–200 (2008)

21. Singh, M.P., Chopra, A.K., Desai, N.: Commitment-based service-oriented architecture. IEEE Computer 42(11), 72–79 (2009)

22. Sora, I., Lazar, G., Lung, S.: Mapping a fuzzy logic approach for qos-aware service selection on current web service standards. In: 2010 International Joint Conference on Computational Cybernetics and Technical Informatics (ICCC-CONTI), pp. 553–558 (May 2010)

23. Sottara, D., Mello, P., Proctor, M.: A configurable rete-oo engine for reasoning with different types of imperfect information. IEEE Trans. Knowl. Data Eng. 22(11), 1535–1548 (2010)

24. Torroni, P., Chesani, F., Mello, P., Montali, M.: Social commitments in time: Satisfied or compensated. In: Baldoni, M., Bentahar, J., van Riemsdijk, M.B., Lloyd, J. (eds.) DALT 2009. LNCS, vol. 5948, pp. 228–243. Springer, Heidelberg (2010)

25. Torroni, P., Chesani, F., Yolum, P., Gavanelli, M., Singh, M.P., Lamma, E., Alberti, M., Mello, P.: Modelling Interactions via Commitments and Expectations. In: Handbook of Research on Multi-Agent Systems: Semantics and Dynamics of Organizational Models, pp. 263–284. IGI Global (2009)

26. Zadeh, L.A.: Fuzzy sets. Information and Control 8(3), 338–353 (1965)

Dealing with Crowd Crystals in MAS-Based Crowd Simulation: A Proposal

Stefania Bandini[1,2], Lorenza Manenti[1,2], Luca Manzoni[1], and Sara Manzoni[1,2]

[1] CSAI - Complex Systems & Artificial Intelligence Research Center,
Universita' di Milano-Bicocca,
Viale Sarca 336, 20126, Milano, Italy
[2] Centre of Research Excellence in Hajj and Omrah,
Umm Al-Qura University,
Makkah, Saudi Arabia
{bandini,manenti,luca.manzoni,sara.manzoni}@disco.unimib.it

Abstract. The paper presents an agent-based model for the explicit representation of groups of pedestrians in a crowd. The model is the result of a multidisciplinary research (CRYSTALS project) where multicultural dynamics and spatial and socio-cultural relationships among individuals are considered as first class elements for the simulation of crowd of pilgrims taking to the annual pilgrimage towards Makkah. After an introduction of advantages of Multi-Agent System approach for pedestrian dynamics modelling, a formal description of the model is proposed. The scenario in which the model was developed and some examples about modelling heterogeneous groups of pedestrians are described.

Keywords: crowd, groups, agent-based model, proxemics.

1 Introduction

Models for the simulation of pedestrian dynamics and crowds of pedestrians have been proposed and successfully applied to several scenarios and case studies: these models are based on physical approach, Cellular Automata approach and Multi-Agent System approach (see [1] for a state of the art). In this work, we refer to the Multi-Agent System (MAS) approach according to which crowds are studied as complex systems whose dynamics results from local behaviour of individuals and the interactions with their surrounding environment. A MAS is a system composed of a set of autonomous and heterogeneous entities distributed in an environment, able to cooperate and coordinate with each other [2,3]. Many research areas contribute to the development of tools and techniques based on MAS for the modelling and simulation of complex systems, as crowds of pedestrians are. In particular, Artificial Intelligence (AI) has contributed in different ways [4]. At the very beginning, AI researchers mainly worked towards encapsulating intelligence in agent behaviours. Other main aspects which AI researchers recently investigated concern modeling and computational tools to deal with interactions [5,6]. The result of this line of research is that we currently can exploit sounding tools that are flexible, adaptable, verifiable, situated

R. Pirrone and F. Sorbello (Eds.): AI*IA 2011, LNAI 6934, pp. 92–103, 2011.

and distributed. Due to the suitability of agents and of MAS-approach to deal with heterogeneity of complex systems, several examples of its application in the pedestrian dynamics area are presented in the literature [7,8,9].

Despite simulators can be found on the market and they are commonly employed by end-user and consultancy companies to provide suggestions to crowd managers and public events organizers about questions regarding space management (e.g. positioning signals, emergency exits, mobile structures), some main open issues in Pedestrian Dynamics community are highlighted as specific modelling requirements. For instance, theoretical studies and empirical evidences demonstrated that the presence of groups strongly modifies the overall dynamics of a crowd of pedestrians [10,11].

In this paper, we propose an agent-based model for the explicit representation and modelling of groups of pedestrians, starting from some fundamental elements we derived from theories and empirical studies from sociology [12], anthropology [13] and direct observations gathered during experiments in collective environments [14]. This work is the result of CRYSTALS project, a multidisciplinary research project where multicultural dynamics and spatial and social relationships among individuals are considered as first class elements for the simulation of crowd of pilgrims taking to the annual Hajj (the annual pilgrimage towards Makkah). In modelling groups, considering the differences in the agent-based tools before mentioned, we provide a general platform-independent model, without an explicit description of space, time, perception functions and behavioural functions which are usually strictly related to the development of the tool. On the contrary, we focus on the organization of pedestrians and on the study of relationships among individuals and the relative group structure. The main contribution of the approach we presented concerns the expressiveness of modelling. Considering the explicit representation of relationships among pedestrians, it is moreover possible to apply methods of network analysis, in particular regarding the identification of relevant structures (i.e. borders and spatially located groups [15,16]).

Differently, other proposals about group modelling presented in the pedestrian dynamics literature do not explicitly investigate the whole concept of group (both from static and dynamic way) and do not consider elements derived from anthropological and sociological studies: in [17] e.g. a proposal in which the concept of group is related to the idea of attraction force applied among pedestrians is presented as an extension of social force field model; [18] proposes a model of pedestrian group dynamics using an agent-based approach, based on utility theory, social comparison theory and leader-follower model; in [19] a MAS-based analysis in which social group structures is presented, exploiting inter and intra relationships in groups by means of the creation of static influence weighted matrices not depending on the evolution of the system.

The paper is organized as follows: we focus on the description of basic elements of the model and on the description of agent behavioural rules. First, in section 2 the scenario of the CRYSTALS project in which the presence of heterogeneous groups is particularly evident is explained. Conclusion and future directions are then presented.

2   The Scenario of Arafat I Station on Mashaer Line

In this section we describe a case of study in which model requirements have
been developed with the study of affluence and entrance on Arafat I station of
new Mashaer train line (Fig. 1) during Hajj 2010, the annual Pilgrimage towards
Makkah. Hajj is a phenomenon in which millions of pilgrims organized in groups
come from all the continents and stay and live together for a limited period of
time. In this situation, a lot of groups with different cultural characteristics live
together and create the whole crowd of the Pilgrimage.

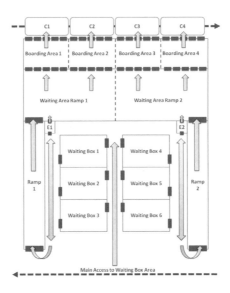

Fig. 1. A representation of the scenario of Mashaer train station in Arafat I

An analysis focused on the presence of groups according to cultural relationships highlighted that four main types of groups can be identified within Hajj
pilgrims crowds:

1. *primary groups*, the basic units social communities are built on consisting in
   small units whose members have daily direct relationships (e.g. families);
2. *residential groups*, characterized by homogeneous spatial localization and
   geographical origin;
3. *kinship groups*, based on descent;
4. *functional groups*, "artificial" groups which exist only to perform a specific
   functions (i.e. executive, control, expressive function). Relationships among
   members are only based on the fulfilment of a goal.

To model groups during Hajj, four kinds of static relationships have to be considered: *primary, residential, kinship, functional*. Moreover, every group can be
characterized by a set of features like the country of origin, the language, the

Fig. 2. Figure on the left shows a group of people following a domestic flag: this group is a residential group, in which people are characterized by the same geographical origin. Figure on the right shows some primary and kinship groups, composed of few people interconnected by means of descent relationships.

social rank. Differently, every pedestrian can be characterized by personal features like the gender, the age, the marital status. In Fig. 2 and 3 some examples on the previously presented groups are shown.

3   Crowd Crystals: A Formal Model

In this model, we refer to some considerations about organizing structures related to particular patterns of pedestrians such as *crystals of crowd*. This concept is directly derived by the theory of Elias Canetti [12]:

> *Crowd Crystals are the small, rigid groups of men, strictly delimited and of great constancy, which serve to precipitate crowds. Their structure is such that they can be comprehended and taken in at a glance. Their unity is more important than their size. The crowd crystal is a constant: it never changes its size.*

Starting from this definition, a crowd can be seen as a set of crystals (i.e. groups of agents); a crowd of crystals is a system formally described as:

$$S = \langle \mathcal{A}, \mathcal{G}, \mathcal{R}, \mathcal{O}, \mathcal{C} \rangle$$

where:

- $\mathcal{A} = \{a_1, \ldots, a_n\}$ is the population of agents;
- $\mathcal{G} = \{G_1, \ldots, G_m\}$ is a finite set of groups;
- $\mathcal{R} = \{r_1, \ldots, r_l\}$ is a finite set of static binary relationships defined on the system;

Fig. 3. These figures show the situation in a waiting box in which a lot of people are waiting to enter the station. Considering the whole group of people who are waiting, we can identify it as a functional group: they are interconnected by a functional relationship, based on the goal of the group (i.e. enter the station).

- $\mathcal{O} = \{o_1, \ldots, o_k\}$ is a finite set of goals present in the system;
- $\mathcal{C} = \{C_1, C_2, \ldots, C_s\}$ is a family of features defined on the system regarding the groups where each C_i is a set of possible values that the ith feature can assume.

In the next sections we formally define groups and agents.

3.1 Crystals

We define the concept of group in a crowd starting from the previously presented definition of crystals of crowd. Every group is defined by a set of agents and by a relationship that defines the membership of agents to the group. We derive the importance and the connection between the notion of group and the notion of relationship by multidisciplinary studies: informally, a group is *a whole of individuals in a relationship with a common goal and/or a common perceived identity*.

Every group is defined a priori by a set of agents: this set has a *size* (i.e. the cardinality of the group) and the composition of members can not change. Moreover, among group members, a static relationship already exists: the kind of relationship determines the type of group, e.g. a family, a group of friends, a working group and so on.

In order to characterize pedestrian groups, it is possible to identify a set of features, shared among all groups in a system: these features allow to analyse and describe in more detail different aspects which is necessary to take into account in the modelling of the system due to a their potential influence in the simulation. On the basis of this assumption, a vector with the values of features is associated to every group. These values are shared and homogeneous on agents belonging to the same group. In the same way, every group has a goal that is shared among all the group members. In fact, every agent belonging to a group inherits from it the global attributes of the group and the goal. The latter idea is not a restriction: following multidisciplinary studies, people involve in a group share the same objective or project. The problem to mediate the goal associated to the group and the "local" goal associated to agent as single entity is not dealt with in this first proposal.

We define a group G_i as a 4-tuple:

$$G_i = \langle A_i, z_i, r_i, o_i \rangle$$

where:

- $A_i \subseteq \mathcal{A}$ is a finite set of agents belonging to G_i;
- $z_i \in C_1 \times C_2 \times \ldots \times C_s$ is a vector with the values of features related to G_i group;
- $r_i \in \mathcal{R}$ is a static irreflexive, symmetric relationship among agents which belong to the group G_i and such that for all $a, b \in A_i$ with $a \neq b$, the pair (a, b) is in the transitive closure of r_i. This means that the graph given by r_i is undirected and connected without self-loops. Note that r_i can be defined on a superset of $A_i \times A_i$. To overcome this difficulty, we can simply consider $r_i | A_i \times A_i$ (i.e. the restriction of r_i on $A_i \times A_i$);
- $o_i \in \mathcal{O}$ is the goal associated to the group G_i.

In this first proposal, we assume that agents can not belong to two different groups at the same time:

$$A_i \bigcap A_j = \emptyset \; \forall i, j = 1, \ldots, m \text{ and } i \neq j$$

This constraint is certainly a restriction for the generalization of the model. Future works are related to the extension of the model to lead with this aspect. We can also describe the population of agents $\mathcal{A}$ as the union the populations of every group:

$$\mathcal{A} = \bigcup_{i=1}^{m} A_i$$

Visually, we can represent each group as a graph $GA_i = (A_i, E_i)$ where A_i is the set of agents belonging to G_i and E_i is the set of edges given by the relationship r_i. We require that GA_i is a non-oriented and connected graph (i.e. for every pair of distinct nodes in the graph there is a path between them).

3.2 Agents

Another fundamental element besides groups is the agent population $\mathcal{A}$ in which every agent represents a pedestrian in a crowd. In order to introduce characteristics related the pedestrians, we introduce $\mathcal{L} = \{L_1, \ldots, L_q\}$ as a family of agent features where every L_i is a set of possible values that the ith feature can assume. Every agent can have different values related to a set of characteristics $\mathcal{L}$:

$$a = \langle w_a \rangle$$

where $w_a \in L_1 \times L_2 \times \ldots \times L_q$ is a vector with the values of features related to agent a.

3.3   Agent Behavioural Rules

After the characterisation of the main elements of the system, we now focus on behavioural rules of pedestrians belonging to a group in a crowd.

We deeply focus on two behavioural rules: the fact that pedestrians tend to maintain a minimum distance from pedestrians belonging the other groups (i) and the fact that pedestrians in a group tend to keep a maximum distance from other agents belonging to the same group (ii).

These rules are directly derived by *Proxemics* a theory first introduced by E.T. Hall [13] and related to the study of the set of measurable distances between people as they interact. The core of this theory is the fact that different persons perceive the same distance in different way, due to personal attitude. In order to develop these rules, it is necessary to introduce a set of functions to measure distances among agents in the case of a pedestrian inside and outside a group, depending on the semantic of space.

On $\mathcal{A}$ we define a pseudo-semi-metric:

$$p : \mathcal{A} \times \mathcal{A} \mapsto \mathcal{D},$$

that is a function that measures distances between agents, such that, given two agents $a, b \in \mathcal{A}$, $p(a, b) = p(b, a)$ (i.e., p is symmetric) and $p(a, a) = 0_{\mathcal{D}}$, where $\mathcal{D}$ is a domain of distances, described as a totally ordered set with $0_{\mathcal{D}}$ as a minimal element. We introduce $\mathcal{D}$ with the scope to not restrict the definition of the environment in a spatial domain: different simulation tools describe space both in a continuous and discrete way. In order to be platform-independent, in this work, we do not explicitly define the environment and, i.e., distances, in a spatial domain.

From p we derive, for any specific agent $a \in \mathcal{A}$, a function $p_a : \mathcal{A} \mapsto \mathcal{D}$ that associates to a its distance from any other agents in $\mathcal{A}$. Given two agents $a, b \in \mathcal{A}$, $p_a(b) = p_b(a)$.

Moreover, for every group G_i we introduce another pseudo-semi-metric:

$$v_i : A_i \times A_i \mapsto \mathcal{D}$$

that denotes the distance between two different agents belonging to the same group G_i. Given two agents $a, b \in G_i$, $v_i(a, b) = v_i(b, a)$ (i.e., v_i is symmetric) and $v_i(a, a) = 0_{\mathcal{D}}$. From v_i we derive, for any specific agent $a \in G_i$ a function $v_{i_a} : A_i \mapsto \mathcal{D}$ that associates to the agent its distance from any other agents in G_i. Given two agents $a, b \in G_i$, $v_{i_a}(b) = v_{i_b}(a)$.

In fact, we introduce two different functions p and v_i due to a potential difference in their semantic from a theoretical point of view. Actually, considering scenarios of crowd simulations, this distinction is not necessary: in this sense, we assume that p and v_i functions have the same semantic $\forall G_i$. A simplification is possible:

$$\forall G_i, \; v_i(a, b) = p(a, b) \; \forall a, b \in \mathcal{A}$$

In the next section we will use p in order to calculate the distance among agents and to guide the behaviours of agents inside and outside groups. As previously written, we have introduced the distance domain $\mathcal{D}$ in order to allow us to not restrict the definition of distance to a spatial domain. Obviously, all crowd simulations are situated in a particular environment in which distances can be measured in $\mathbb{R}^+$: thinking about a spatially located or binary (true/false) systems simulating pedestrians, only positive real values are admissible. For this reason, we can reduce the complexity of $\mathcal{D}$ and admit that $\mathcal{D} \subset \mathbb{R}^+$, in which also binary values are included (i.e. false=0 and true=1).

3.3.1   Safe Proxemic Rule. The first rule we want to introduce is related to the behaviour during interaction between a pedestrian and other pedestrians belonging to a different group. From this point of view, in order to introduce the importance of personal differences derived, for instance, by cultural attitude and social context, in the pedestrian simulating context, we associate to every agent $a \in \mathcal{A}$ belonging to a group G_i a personal distance $d_a \in \mathcal{D}$.

We introduce a function da that, considering the feature values associate to the agent and to its group, derives d_a as follows:

$$da : \left(\prod_{C \in \mathcal{C}} C \right) \times \left(\prod_{L \in \mathcal{L}} L \right) \mapsto \mathcal{D}$$

Given an agent $a \in G_i$, with $a = \langle w_a \rangle$ and its group G_i with features z_i, its personal distance is $da(z_i, w_a) = d_a$. This distance derives both from the global characteristics of group (i.e. z_i) and from the local characteristics of agent (i.e. w_a) we are considering.

Considering the distance among a and the other agents not belonging to its group, we require that $a \in G_i$ is in a safe proxemic condition if the distance $p_a(b)$ is above d_a for all $b \in \mathcal{A} \setminus A_i$.

Formally, we define that an agent $a \in G_i$ is in a *safe proxemic condition* iff:

$$\nexists b \in \mathcal{A} \setminus A_i \; : p_a(b) \leq d_a$$

This first rule represents the fact that pedestrians tend to maintain a minimum distance from pedestrians belonging the other groups; if the safe proxemic condition is violated, agents tend to restore the condition of proxemic safeness.

3.3.2   Safe Group Rule. Every group G_i is characterized by a private defined distance $\delta_{G_i} \in \mathcal{D}$ that depends on the values of group features z_i. We introduce a function dg that calculates δ_{G_i} as follows:

$$dg : \prod_{C \in \mathcal{C}} C \mapsto \mathcal{D}$$

Given a group G_i, $dg(z_i) = \delta_{G_i}$.

Previously, the introduced relationships $\mathcal{R}$ were called static relationships. The introduction of time into the model gives the possibility to define relationships that are time dependent: due to the fact that time can be modelled in a continuous or discrete way, the proposed model is defined in a way applicable to both continuous and discrete modelling. Considering a particular time $t \in \mathcal{T} \subseteq \mathbb{R}$ and t_0 as the starting time, the evolution of the system is given by a map $\varphi : \mathcal{S} \times \mathcal{T} \mapsto \mathcal{S}$, where $\mathcal{S}$ is the space of possible systems. The state of the system at time t is $\varphi(S_0, t)$, where S_0 is the state of the system at time t_0. We use the definition of time in order to introduce a new kind of relationship time-dependent (differently from the previous one). We call *dynamic relationship* a function $\mathfrak{r}$ such that $\mathfrak{r}_t$ is a dynamic irreflexive, symmetric relationship among agents which belong to the group G_i. $\mathfrak{r}_t$ represents the relation at time t that is dependent on the whole evolution of the system from time t_0 to time t. For each group G_i at time t it is possible to consider the graph given by the relation $\mathfrak{r}_t$. In particular, to model the proximity relationship between agents, a possible definition of $\mathfrak{r}_t$ is the following:

$$\forall a, b \in G_i, \ (a, b) \in \mathfrak{r}_t \text{ iff } p(a, b) \leq \delta_{G_i}$$

recalling that v_i is potentially different for each $\varphi(S_0, t)$ since it is defined into the system.

It is possible to define a group as having the *safe group condition* at time t on the basis of the history of the evolution of the graph structure given by $\mathfrak{r}_t$. Let $\mathfrak{S}$ be the function that defines the presence or absence of the safe group condition. In other words $\mathfrak{S}\left(\langle \mathfrak{r}_j \mid j \leq t \rangle\right) \in \{0, 1\}$. The fact that $\mathfrak{S}$ is dependent on the whole history of the graph structure is motivated by the necessity to take care of particular conditions that can temporary change the graph structure but that can be quickly recovered. By using the whole history we can avoid to consider unsafe (respect to safe) a group that is, in fact, in a safe (respect to unsafe) condition. For instance, considering a simulation placed into two rooms separated by a turnstile. The passage of a group through the turnstile can divide the group: in fact the group is not in an unsafe condition if we can detect that the passage through the turnstile is a temporary condition.

The safe group rule represents the fact that pedestrians in a group tend to keep a maximum distance from other agents belonging to the same group: if the safe group condition is violated, agents tend to restore the condition of group safeness.

In the next section we show how groups in the scenario of Arafat I Station can be modelled following the above presented formalization.

4   Modelling Groups in the Scenario of Arafat I Station

Considering the scenario of Arafat I station, in this section we exploit the model above presented to describe groups of pedestrians in the process of entering the station depicted in Fig. 1. In particular, considering Fig. 2 on the right, we can define the system $S = \langle \mathcal{A}, \mathcal{G}, \mathcal{R}, \mathcal{O}, \mathcal{C} \rangle$ as follows:

- $\mathcal{A} = \{a_1, \ldots, a_{52}\}$ is the set of the pilgrims in the waiting boxes;
- $\mathcal{G} = \{G_1, \ldots, G_4\}$ is the set of groups of pilgrims we consider;
- $\mathcal{R} = \{$primary, kindship, residential, functional$\}$ represents the types of groups;
- $\mathcal{O} = \{C\}$ is the set of possible goals, i.e. to reach the train carriages;
- $\mathcal{C} = \{$country, language, social rank$\}$ is the family of features regarding groups.

In this scenario, the four groups identified (numbered from left to right in Fig. 2) can be defined as the 4-tuple $G_i = \langle A_i, z_i, r_i, o_i \rangle$ and, in particular:

- $G_1 = \langle A_1 \subseteq \mathcal{A}, ($Saudi Arabia, Arabic, low$)$, kindship, C$\rangle$;
- $G_2 = \langle A_2 \subseteq \mathcal{A}, ($Saudi Arabia, Arabic, medium$)$, primary, C$\rangle$;
- $G_3 = \langle A_3 \subseteq \mathcal{A}, ($Saudi Arabia, Arabic, medium$)$, primary, C$\rangle$;
- $G_4 = \langle A_4 \subseteq \mathcal{A}, ($Saudi Arabia, Arabic, high$)$, primary, C$\rangle$.

Regarding the definition of characteristics of agents, a plausible family of features can be $\mathcal{L} = \{$gender, age, marital status, impaired status$\}$. From this point of view, an agent $a \in \mathcal{A}$ can be defined, for example, as follows:

$$a_i = \langle \text{male, adult, married, not impaired} \rangle$$

All the information about the features of groups are inherited from the group $G_j \ni a_i$. These information are derived by the multidisciplinary study performed during the CRYSTALS project and by the collaboration with Hajj experts.

5   Conclusions and Future Directions

In this paper, we proposed an agent-based model for the explicit representation and modelling of groups in a crowd, focusing on the organization of pedestrians and on the study of relationships among individuals and the relative group structure including the analysis of internal states and transitions as shown in Fig. 4.

Future directions are related to the development of simulation in the presented scenario in order to test and validate the model, and in the application of methods for network analysis on the group structures, in order to identify and study, for example, the presence of recursive patterns.

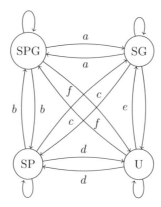

Fig. 4. A Finite state automata representing states and transitions of an agent in a system S: *Safe Proxemic and Group state* (SPG): an agent is in this state iff both the safe proxemic and safe group condition are verified; *Safe Proxemic state* (SP): an agent is in this state iff only the safe proxemic condition is verified; *Safe Group state* (SG): an agent is in this state iff only the safe group condition is verified; *Unsafe state* (U): an agent is in this safe iff neither the safe proxemic nor safe group conditions are verified.

Acknowledgment. This research was fully supported by the Center of Research Excellence in Hajj and Omrah, Umm Al-Qura University, Makkah, Saudi Arabia, grant title "Crystal Proxemic Dynamics of Crowd & Groups".

References

1. Bandini, S., Manzoni, S., Vizzari, G.: Crowd Behaviour Modeling: From Cellular Automata to Multi-Agent Systems. In: Multi-Agent Systems: Simulation and Applications. CRC Press, Boca Raton (2009)
2. Ferber, J.: Multi-Agent Systems: An Introduction to Distributed Artificial Intelligence, 1st edn. Addison-Wesley Longman Publishing, Boston (1999)
3. Wooldridge, M.: An introduction to multiagent systems. Wiley, Chichester (2002)
4. Baldoni, M., Baroglio, C., Mascardi, V., Omicini, A., Torroni, P.: Agents, multi-agent systems and declarative programming: What, when, where, why, who, how? In: 25 Years GULP, pp. 204–230 (2010)
5. Omicini, A., Piunti, M., Ricci, A., Viroli, M.: Agents, intelligence and tools. In: Artificial Intelligence: An International Perspective, pp. 157–173 (2009)
6. Torroni, P., Chesani, F., Mello, P., Montali, M.: Social commitments in time: Satisfied or compensated. In: Baldoni, M., Bentahar, J., van Riemsdijk, M.B., Lloyd, J. (eds.) DALT 2009. LNCS, vol. 5948, pp. 228–243. Springer, Heidelberg (2010)
7. Klugl, F., Rindsfuser, G.: Large-scale agent-based pedestrian simulation. In: Proceedings of the 5th German Conference on Multiagent System Technologies, pp. 145–156 (2007)

8. Toyama, M., Bazzan, A., Da Silva, R.: An agent-based simulation of pedestrian dynamics: from lane formation to auditorium evacuation. In: Proceedings of the Fifth International Joint Conference on Autonomous Agents and Multiagent Systems, pp. 108–110. ACM Press, New York (2006)
9. Vizzari, G., Manzoni, S., Bandini, S.: Simulating and Visualizing Crowd Dynamics with Computational Tools based on Situated Cellular Agents. Pedestrian Behavior: Models, Data Collection, and Applications (2009)
10. Rogsch, C., Klingsch, W., Schadschneider, A.: Pedestrian and evacuation dynamics 2008. Springer, Heidelberg (2010)
11. Challenger, R., Clegg, C., Robinson, M.: Understanding Crowd Behaviours, vol. 1. Cabinet Office (2009)
12. Canetti, E.: Crowds and Power. In: Gollancz, V, ed. (1962)
13. Hall, E.: The hidden dimension. Doubleday, New York (1966)
14. Manenti, L., Manzoni, S., Vizzari, G., Kazumichi, O., Shimura, K.: Towards an Agent- Based Proxemic Model for Pedestrian and Group Dynamic. In: Proceedings of the 11th Italian Workshop From Objects to Agents, WOA 2010. CEUR, vol. 631 (2010)
15. Lembo, P., Manenti, L., Manzoni, S.: Towards patterns of comfort: A multilayered model based on situated multi-agent systems. In: Bandini, S., Manzoni, S., Umeo, H., Vizzari, G. (eds.) ACRI 2010. LNCS, vol. 6350, pp. 439–445. Springer, Heidelberg (2010)
16. Manenti, L., Manzoni, L., Manzoni, S.: Towards an application of graph structure analysis to a mas-based model of proxemic distances in pedestrian systems. CEUR, vol. 631 (2010)
17. Moussaïd, M., Helbing, D., Garnier, S., Johansson, A., Combe, M., Theraulaz, G.: Experimental study of the behavioural mechanisms underlying self-organization in human crowds. Proceedings, Biological Sciences / The Royal Society 276, 2755–2762 (2009)
18. Qiu, F., Hu, X.: Modeling dynamic groups for agent-based pedestrian crowd simulations. In: 2010 IEEE/WIC/ACM International Conference on Web Intelligence and Intelligent Agent Technology, vol. 2, pp. 461–464 (2010)
19. Qiu, F., Hu, X., Wang, X., Karmakar, S.: Modeling Social Group Structures in Pedestrian Crowds. Simulation Modelling Practice and Theory (SIMPAT) 18, 190–205 (2010)

An Agent Model of Pedestrian and Group Dynamics: Experiments on Group Cohesion

Stefania Bandini[1,2], Federico Rubagotti[1],
Giuseppe Vizzari[1,2], and Kenichiro Shimura[3]

[1] Complex Systems and Artificial Intelligence (CSAI) research center,
Università degli Studi di Milano - Bicocca, Viale Sarca 336/14, 20126 Milano, Italy
{bandini,vizzari}@csai.disco.unimib.it, f.rubagotti@campus.unimib.it
[2] Crystal Project, Centre of Research Excellence in Hajj and Omrah (Hajjcore),
Umm Al-Qura University, Makkah, Saudi Arabia
[3] Research Center for Advanced Science & Technology,
The University of Tokyo, Japan
shimura@tokai.t.u-tokyo.ac.jp

Abstract. The simulation of pedestrian dynamics is a consolidated area of application for agent–based based models; however generally the presence of groups and particular relationships among pedestrians is treated in a simplistic way. This work describes an innovative agent–based based approach encapsulating in the pedestrian's behavioural model effects representing both proxemics and a simplified account of influences related to the presence of groups in the crowd. The model is tested in a simple scenario to evaluate the effectiveness of mechanisms to preserve groups cohesion maintaining a plausible overall crowd dynamic.

1 Introduction

Crowds of pedestrians can be safely considered as complex entities; various phenomena related to crowds support this statement: pedestrian behaviour shows a mix of competition for the space shared and collaboration due to the (not necessarily explicit) social norms; the dependency of individual choices on the past actions of other individuals and on the current perceived state of the system (that, in turn, depends on the individual choices of the comprised agents); the possibility to detect self-organization and emergent phenomena. The definition of models for explaining or predicting the dynamics of a complex system is a challenging scientific effort; nonetheless the significance and impact of human behaviour, and especially of the movements of pedestrians, in built environment in normal and extraordinary situations motivated a prolific research area focused on the study of pedestrian and crowd dynamics. The impact the results of these researches on activities of architects, designers and urban planners is apparent (see, e.g., [1] and [2]), especially considering dramatic episodes such as terrorist attacks, riots and fires, but also due to the growing issues in facing the organization and management of public events (ceremonies, races, carnivals, concerts, parties/social gatherings, and so on) and in designing naturally crowded places

R. Pirrone and F. Sorbello (Eds.): AI*IA 2011, LNAI 6934, pp. 104–116, 2011.

(e.g. stations, arenas, airports). These research efforts led to the realization of commercial, off-the-shelf simulators often adopted by firms and decision makers to elaborate what-if scenarios and evaluate their decisions with reference to specific metrics and criteria.

Cellular Automata have been widely adopted as a conceptual and computational instrument for the simulation of complex systems (see, e.g., [3]); in this specific context several CA based models (see, e.g., [4,5]) they have been adopted as an alternative to particle-based approaches [6], and they also influenced new approaches based on autonomous situated agents (see, e.g., [7,8,9]). The main aim of this work is to present an approach based on reactive autonomous situated agents derived by research on CA based model for pedestrian and crowd dynamics for a multidisciplinary investigation of the complex dynamics that characterize aggregations of pedestrians and crowds. This work is set in the context of the Crystals project[1], a joint research effort between the Complex Systems and Artificial Intelligence research center of the University of Milano–Bicocca, the Centre of Research Excellence in Hajj and Omrah and the Research Center for Advanced Science and Technology of the University of Tokyo. The main focus of the project is to investigate how the presence of heterogeneous groups influences emergent dynamics in the context of the Hajj and Omrah. This point is an open topic in the context of pedestrian modeling and simulation approaches: the implications of particular relationships among pedestrians in a crowd are generally not considered or treated in a very simplistic way by current approaches. In the context of the Hajj, the yearly pilgrimage to Mecca, the presence of groups (possibly characterized by an internal structure) and the cultural differences among pedestrians represent two fundamental features of the reference scenario.

The paper breaks down as follows: the following Sect. introduces the agent–based pedestrian and crowd model considering the possibility of pedestrians to be organized in groups, while Sect. 3 summarizes the results of the application of this model in a simple simulation scenario. Conclusions and future developments will end the paper.

2   The GA-Ped Model

The Group–Aware Pedestrian (GA- Ped) model is a reactive agents based model characterized by an environment that is discrete both in space and in time. The model employs floor fields (see, e.g., [10]) to support pedestrian navigation in the environment. In particular, each relevant final or intermediate target for a pedestrian is associated to a floor field, a sort of gradient indicating the most direct way towards the associated point of interest. Our system is represented by the triple: $Sys = \langle Env, Ped, Rules \rangle$ whose elements will be now introduced.

---

[1] http://www.csai.disco.unimib.it/CSAI/CRYSTALS/

2.1   Space and Environment

The representation of the space in our model is derived from the Cellular Automata theory: space is discretized into small cells which may be empty or occupied by exactly one pedestrian. At each discrete time step it is possible to analise the state of the system by observing the state of each cell (and, consequently, the position of each pedestrian into the environment). The environment is defined as $Env = \langle Space, Fields, Generators \rangle$ where the $Space$ is a physical, bounded bi-dimensional area where pedestrians and objects are located; the size of the space is defined as a pair of values($xsize, ysize$) and it is specified by the user. In our model we consider only rectangular-shaped scenarios (but it is possible to shape the scenario defining non-walkable areas). The space in our model is modeled using a three-layer structure: $Space = \langle l_1, l_2, l_3 \rangle$ where each layer represents details related to a particular aspect of the environment. Each layer is a rectangular matrix sharing the same size of the other two. The first layer (l_1), contains all the details about the geometry of the environment and the properties of each cell. A cell may be a generating spot (i.e. a cell that can generate new pedestrians according to the simulation parameter), and can be walkable or not. A cell is thus characterized by a $cellID$, an unique key for each cell, it can be associated to a $generator$ if the cell can generate pedestrians, it can be walkable or not (e.g. the cell contains a wall). The $second\ layer$, denoted as l_2, contains information about the values of the floor fields of each cell. Values are saved as pairs ($floorID, value$). Data saved into the second layer concerns targets and the best path to follow to reach them. The $third\ layer$, l_3, is made up of cells that may be empty or occupied by one pedestrian. This layer stores the position of each pedestrian.

Generators and Targets – Information about generators and targets are saved into the first and second layer. A target is a location in the environment that the pedestrians may desire to reach. Examples of targets in a train station are ticket machines, platforms, exits and so on. A traveller may have a complex schedule composed of different targets like: (a) I have to buy a ticket, then (b) I want to drink a coffee and (c) reach platform number 10 to board the train to Berlin. This plan can be translated in the following schedule: (i) ticket machine, (ii) lounge, (iii) platform 10. From now on the words $schedule$ and $itinerary$ are used interchangeably. We will describe how pedestrians will be able to move towards the target later on.

Generators are cells that, at any iteration, may generate new pedestrians according to predetermined rules. $Generating\ spots$ are groups of generator cells located in the same area and driven by the same set of rules of generation. In our model a $generating\ spots$ is defined as $spot = \langle spotID, maxPed, positions, groups\ itineraries, frequency \rangle$ where $spotID$ is an identifier for the generator; $maxPed$ is the maximum amount of pedestrians that the spot can generate during the entire simulation; $positions$ indicate the cells belonging to that generating spot (a spot in fact may contain different cells); $groups$ being the set of group types that can be generated, each associated with a frequency of generation; $itineraries$ that can be assigned to each pedestrian, considering the fact that group members share the same schedule but that different groups may have different schedules, each associated with a frequency; $frequency$ is a value between 0 and 100, specifying the frequency of pedestrian generation (0 means never generate pedestrians, 100 means generate pedestrians at each iteration, if free space is available and if the desired maximum density has not been reached.).

Information about generators are stored in the first layer, on the contrary, targets are represented in the second layer, specified with their floor field values. In fact, every target has a position and it is associated to a floor field that guides pedestrians to it.

Floor Fields – As stated previously, the floor field can be thought of as a grid of cells underlying the primary grid of the environment. Each target has a floor field and the values are saved into the $l2$ of the environment. A floor field contains information suggesting the shortest path to reach the destination. In our model each cell contains information about every target defined in the model. Given the cell at position (x, y), the corresponding floor field values are saved into l_2, the content of $l_2(c_{x,y})$ is a list of pairs with the following structure: $(floorID, value)$. Values of a floor field are integers between 0 (no indication on how to reach the target) and 256 (target is present in the cell); the value of a floor field decreases when the distance from the target grows (e.g. according to the Manhattan distance). The GA-Ped model only comprises *static* floor fields, specifying the shortest path to destinations and targets. Interactions between pedestrians, that in other models are described by the use of *dynamic floor fields*, in our model are managed through a perception model based on the idea of *observation fan*, which will be introduced in Section 2.2.

Time and Environment Update Type. Our model is a discrete-time dynamical system, and update rules are applied to all pedestrians following an update method called *shuffled sequential update* [11]. At each iteration, pedestrians are updated following a random sequence. This choice was made in order to implement our method of collision avoidance based on cell reservation. In the shuffled sequential update, a pedestrian, when choosing the destination cell, has to check if this cell has been reserved by another pedestrian within the same time step. If not, the pedestrian will reserve that cell, but moving into at the end of the iteration. If the cell is already reserved, an alternative destination cell can be chosen. Each iteration corresponds to an amount of time directly proportional to the size of the cells of the environment and to the reaction time: given a squared cell of $40 \times 40cm^2$, the corresponding timescale is approximately of $0.3sec$, obtained by transposing the empirically observed value of average velocity of a pedestrian, that is $1.3m/s$ to the maximal walking speed of one cell per time step.

2.2   Pedestrians

Pedestrians are modeled as simple reactive agents situated in a bidimensional grid. Each pedestrian is provided with some attributes describing details like group membership, ID, schedules. Each pedestrian is also endowed with a set of *observation fans* that determine how he sees and evaluates the environment. Attributes, internal state and environment influence the behavior of our pedestrians: a pedestrian can move in one of the cells belonging to its Moore neighborhood, and to any possible movement is associated a revenue value, called *likability*, representing the desirability of moving into that position given the state of the pedestrian.

Pedestrian Characterization – a pedestrian is characterized by a simple set of attributes and in particular *pedestrian* = ⟨*pedID*, *groupID*, *schedule*⟩ with *pedID* being an identifier for each pedestrian, *groupID* (possibly null, in case of individuals) the group the pedestrian belongs to and *schedule* a list of goals to be accomplished by the pedestrian (one of the above introduced itineraries).

Perception model – In the GA-Ped model every pedestrian has the capability to observe the environment around him, looking for other pedestrians, walls and objects by means of an *observation fan*. An *observation fan* can be thought as the formalization of physical perceptive capabilities combined with the evaluation of the perception of relevant entities: it determines how far a pedestrian can see and how much importance has to be given to the presence of obstacles and other pedestrians. An *observation fan* is defined as follows:

$$\zeta = \langle type, xsize, ysize, weight, xoffset, yoffset \rangle$$

Where *type* identifies the direction of the fan: it can be 1 for diagonal directions and 2 for straight directions (the fan has different shapes and it may be asymmetric). Sizes and offsets are defined as shown in figure 1. Sizes (*xsize* and *ysize*) define the maximum distance to which the pedestrian can see. The shape of the fan is influenced by both the direction and the sizes. The offsets are used to define if the pedestrian can see backward and the size of the lateral view (only type 2, see Fig 1.c). The parameter *weight* is a matrix of values $w_{x,y} \in \mathbb{R}_+$ defined in the interval $[0, 1]$. These values determine the relationship between the *thing* that has been observed and the distance (e.g. given a wall, its distance influences differently the movement of a pedestrian).

For each class of groups is possible to define multiple *observation fans*; each fan can be applied when evaluating walls, pedestrians belonging to the same group, to other groups or, lastly, to particular groups. For instance, this feature is useful when modeling situations like football matches: it is possible to define two classes of groups, one made of supporters of the first team and the other of supporters of the second team. Groups belonging to the first class will interact differently if dealing with other groups belonging to the first class or belonging to the second one.

Behavior and Transition Rules – The behavior of a pedestrian is represented as a flow made up of four stages: *sleep, context evaluation, movement evaluation, movement*. When a new iteration is started, each pedestrian is in a sleeping state. This state is the only possible in this stage, and the pedestrian does

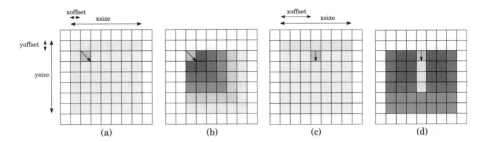

Fig. 1. Example of the shape of an observation fan for a diagonal direction (in this case south-east) and for a straight direction (in this case south): (a and c) in light cyan the cells that are observable by the pedestrian and are used for the evaluation, in green the observable backward area; (b and d) the weight matrix applied for the evaluation, in this case objects or pedestrians near the pedestrian have more weight that farther ones (e.g. this fan is useful for evaluating walls).

nothing but waits for a trigger signal from the system. The system wakes up each pedestrian once per iteration and, then, the pedestrian passes to a new state of context evaluation. In this stage, the pedestrian tries to collect all the information necessary to obtain spatial awareness. When the pedestrian has collected enough data about the environment around him, it reaches a new state. In this state behavioral rules are applied using the previously gathered data and a movement decision is taken. When the new position is notified to the system, the pedestrian returns to the initial state and waits for the new iteration.

In our model, pedestrian active behavior is limited to only two phases: in the second stage pedestrians collect all the information necessary to recognize the features of the environment around him and recall some data from their internal state about last actions and desired targets. A first set of rules determine the new state of the pedestrian. The new state, belonging to the stage of movement evaluation, depicts current circumstances the pedestrian is experiencing: e.g. the situation may be normal, the pedestrian may be stuck in a jam, it may be compressed in a dense crowd or lost in an unknown environment (i.e. no valid floor field values associated to the desired destination). This state of awareness is necessary to the choice of the movement as different circumstances may lead to different choices: a pedestrian stuck in a jam may try to go in the opposite direction in search for an alternative path, a lost pedestrian may start a random walk or look for other significant floor fields.

Pedestrian movement – Direction and speed of movement At each time step, pedestrians can change their position along nine directions (keeping the current position is considered a valid option), into the cells belonging to their Moore neighborhood of range $r = 1$. Each possible movement has a value called *likability* that determines how much the move is *good* in the terms of the criteria previously introduced. In order to keep our model simple and reduce complexity, we do not consider multiple speed. At each iteration a pedestrian can move only in the cells belonging to the Moore neighborhood, reaching a speed value of 1 or can maintain the position (in this case speed is 0)[2].

Functions and notation – In order to fully comprehend the pedestrian behavior introduced in the following paragraphs, it is necessary to premise the notational conventions and the functions we have introduced in our modelization:

- $c_{x,y}$ defines the cell with (valid) coordinates (x, y);
- *Floors* is the set of the targets instantiated during the simulation. Each target has a floor field and they share the same *floorID* (i.e. with $t \in Floors$ we define both the target and the associated floor field);
- *Groups* the set containing the *groupID* of the groups instantiated during the simulation dynamics;
- *Classes* is the set containing all the group classes declared when defining the scenario;
- *Directions* is the set of the possible directions. Are nine, defined using cardinal directions: $\{N, NE, E, SE, S, SW, W, NW, C\}$.

Given $x \in [0, xsize - 1]$ and $y \in [0, ysize - 1]$, we define some functions useful to determine the characteristics and the status of the cell $c_{x,y}$:

cell walkability, function $l_1(c_{x,y})$: this function determines if the cell $c_{x,y}$ is walkable or not (e.g. if there is a wall). If the cell is walkable the function returns the value 1, otherwise it returns 0. We assume that this function does not depend on time (i.e. the structure of the environment does not change during the simulation).

floor field value, function $l_2(c_{x,y}, t)$: this function determines the value of the floor field t in the cell $c_{x,y}$.

presence of pedestrians belonging to a given group, function $l_3(c_{x,y}, g)$: this function determines if in the cell $c_{x,y}$ contains a pedestrian belonging to a particular group g specified as input. If a pedestrian belonging to that group is contained in the cell, the function returns 1, otherwise it returns 0.

In addition to these functions, we also define the **observation fan** as $\zeta_{x,y,d}$, the set of cells that are observable according to the characteristics of the observation fan ζ, used by a pedestrian located in the cell at coordinates (x, y) and looking in the direction d.

The overall *likability* of a possible solution can be thought as the desirability of one of the neighboring cells. The more a cell is desirable, the higher is the probability that a pedestrian will choose to move into that position. In our model the *likability* is determined by the evaluation of the environment and it is defined as a composition of the following sequence of characteristics: (i) *goal driven component*, (ii) *group cohesion*, (iii) *proxemic repulsion*, (iv) *geometrical repulsion*, (v) *stochastic contribution*.

Formally, given a pedestrian belonging to the group class $g \in Groups$, in the state $q \in Q$ and reaching a goal $t \in Floors$, the *likability* of a neighbouring

---

[2] Our pedestrians can move only to the cells with distance 1 according to the Tchebychev distance.

cell $c_{x,y}$ is defined as $li(c_{x,y})$ and is obtained evaluating the maximum benefit the pedestrian can achieve moving into this cell (following the direction $d \in Directions$) using the observation fan ζ for the evaluation. The value of the characteristics that influence the likability are defined as follows:

goal driven component: it is the pedestrian wish to quickly reach its destination and is represented with the floor field. Our model follows the least effort theory: pedestrians will move on the shortest path to the target which needs the least effort. This component is defined as $l_2(c_{x,y}, t)$: it is the value of the floor field in the cell at coordinates (x, y) for the target t;

group cohesion: it is the whish to keep the group cohese, minimizing the distances between the members of the group. It is defined as the pedestrians belonging to the same group in the observation fan ζ, evaluated according to the associated weight matrix:

$$\zeta(group, d, (x, y), g) = \sum^{c_{i,j} \in \zeta_{x,y,d}} w^{\zeta}_{i,j} \cdot l_3(c_{i,j}, g) \tag{1}$$

geometrical repulsion: it represents the presence of walls and obstacles. Usually a pedestrian wishes to avoid the contact with these object and the movement is consequently influenced by their position. This influence is defined as the presence of walls (located in layer l_1) inside the observation fan ζ, according to the weight matrix for *walls* specified in the same observation fan:

$$\zeta(walls, d, (x, y)) = \sum^{c_{i,j} \in \zeta_{x,y,d}} w^{\zeta}_{i,j} \cdot l_1(c_{i,j}) \tag{2}$$

proxemic repulsion: it is the repulsion due to presence of pedestrians, alone or belonging to other groups (e.g. strangers). A pedestrian whishes to maintain a *safe* distance from these pedestrians and this desire is defined as the sum of these people in the observation fan ζ, according to the weight matrix for the group of these pedestrians:

$$\zeta(strangers, d, (x, y), g) = \sum^{c_{i,j} \in \zeta_{x,y,d}} w^{\zeta}_{i,j} \cdot (1 - l_3(c_{i,j}, g)); \tag{3}$$

stochasticity: similarly to some traffic simulation models (e.g. [12]), in order to introduce more realism and to obtain a non deterministic model, we define $\epsilon \in [0, 1]$ as a a random value that is different for each *likability* values and introduces stochasticity in the decision of the next movement.

The overall *likability* of a movement is thus defined as follows:

$$li(c_{x,y}, d, g, t) = j_w \zeta(walls, d, (x, y)) + j_f field(t, (x, y)) -$$
$$j_g \zeta(group, d, (x, y), g) - j_n \zeta(strangers, d, (x, y), g) + \epsilon. \tag{4}$$

Group cohesion and floor field are positive components because they positively influence a decision as a pedestrian wishes to reach the destination quickly, keeping the group cohese at the same time. On the contrary, the presence of obstacles and other pedestrians has a negative impact as a pedestrian usually tends to avoid this contingency. The formula 4 summarizes the evaluation of the aspects that characterize the *likability* of a solution. A pedestrian for each possible movement *opens* an observation fan and examines the environment in the corresponding directions, evaluating elements that may make that movement opportune (e.g. the presence of other pedestrians belonging to the same group or an high floor field value and data that may discourage as the presence of walls or pedestrians belonging to other groups).

3   Simulation Scenario

The simulated scenario is a rectangular corridor, $5m$ wide and $10m$ long. We assume that the boundaries are open and that walls are present in the north and south borders. The width of the cells is $40cm$ and the sizes of the corridor are represented with 14 cells and 25 cells respectively. Pedestrians are generated at the east and west borders and their goal is to reach the opposite exit.

Group dispersion – Since we are simulating groups of pedestrians, observing how different group sizes and overall densities affect the dispersion of groups through their movement in the environment is a central issue of our work. We considered three different approaches to the definition of such a metric: (i) dispersion as an area, (ii) dispersion as a distance from a centroid and (iii) dispersion as summation of the edges of a connected graph. The formulas of group dispersion for each approach are defined as follows:

$$(a)\ \Xi(C)^I = \frac{A_C}{|C|} \qquad (b)\ \Xi(C)^{II} = \frac{\sum_{i=1}^{|C|} d(c, p_i)}{|C|} \qquad (c)\ \Xi(C)^{III} = \frac{\sum_{i=1}^{|E|} w(e_i)}{|V|}$$

where C is the group of pedestrians (each member is enumerated using the notation p_i), A_C is its area, c as the centroid of the group, d is defined as the euclidean distance in $\mathbb{R}^2$ and $v_i \in V$ is a set of vertices (each pedestrian is a vertex) of a connected graph $G = \{V, E\}$. We tested the approaches using a set of over fifty different configurations of groups, representatives of significant situations for which we have an intuitive idea of the degree of dispersion of comprised groups. The results highlighted that the first and third approaches capture complementary aspects of our intuitive idea of dispersion, while the second one provides results similar to the third. We decided to combine the first and third approaches by means of a linear combination, allow obtaining a fairly unbiased measurement of dispersion. First of all, we normalized the metrics in the closed range $[0, 1]$ using the function $\Xi(C) = (\tanh(\Xi(C)^{\frac{1}{\eta}}))^{\varpi}$, with Ξ as the value generated by one of the three metrics previously introduced, and ϖ, η as normalization parameters. Then we combined two approaches in the following metric:

$$\overline{\Xi}(C) = \Xi(C)^I w^I + \Xi(C)^{III} w^{III}, \tag{5}$$

this function returns values the real range $[0, 1]$, with two weights w^I and w^{III} such that $w^I + w^{III} = 1$. In the preovious formula $\Xi(C)^I$ is the normalized value of the group dispersion for the group C, obtained with the first metric and $\Xi(C)^{III}$ the value obtained with the third metric.

Large group vs small group counterflow – We were interest in studying the dynamics of friction and avoidance that are verified when two groups with different size, traveling in opposite directions, are facing each others in a rectangular shaped corridor. We simulated the $5m \times 10m$ corridor with one large group traveling from the left (west) to the right (east), opposed to one small group traveling in the opposite direction. The aim of this particular set up was to investigate the differences in the dispersion of the smaller group with respect of the size of the large group and the overall time necessary to walk through the corridor. From now on we call the small group as the *challenging* group and the large group as the *opponent* group.

We considered opponent group of five different sizes: 10, 20, 30, 40 and 50. Challenging groups were defined with only two sizes: 3 and 5. The results are consistent with the observable phenomena as the model can simulate all the three possible cases that can be spotted in the real world: (i) the challenging group remains compact and moves around the opponent group; (ii) one or more members of the challenging group moves around the larger group in the other side with respect to the other members of the group; (iii) one or more members of the challenging group remain stuck in the middle of the opponent group and then the small group temporarily breaks up.

It is also interesting to point out that in our model, if a split is verified in the challenging group, when their members overcome the opponent group, they aim to form again a compact configuration. The actual size of the simulation scenario is however too small to detect this *reforming* of the group[3]. Figure 2 presents screenshots of the simulation at different time steps. It is possible to observe the range of different circumstance that our model is able to simulate: for example, in the simulation #3 the challenging groups can overcome the opponent one simply by moving around it, the same situation is represented in simulation #1 but the challenging group experiences more friction generated by the opponents. Simulations #2 and #4 show a challenging group that splits in two and their members moving around the opponent group on both the two sides. We investigated the relationships between the time necessary to the members of the challenging group to reach the opposite end of the corridor in relation with the size of the opponent group. As expected, and in tune with the previous observations, the larger the size of the opponent group, the higher time necessary to the members of the challenging group to reach their destination is. The difference of size in the challenging group only slightly influences the performances: it is

---

[3] We carried out additional simulations in larger environments and we qualitatively observed the group re-union.

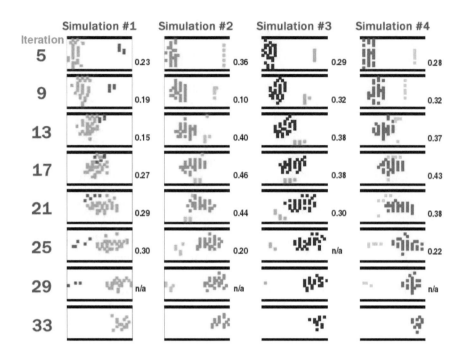

Fig. 2. Images representing the state of the simulation taken at different time steps. The opponent group is composed of 30 pedestrians, while the challenging group size is 5. The small number on the right of each state is the dispersion value of the challenging group.

easier to remain stuck in the opponent group but the difference between three and five pedestrians is insufficient to obtain significant differences.

Dispersion in counterflow – In addition to verifying the plausibility and validity of the overall system dynamics generated by our model [13], we tried to evaluate qualitatively and quantitatively the effectiveness of mechanism preserving group cohesion. We investigated the relationship between the the size of the opponent group and the average dispersion of the challenging group. As expected, the larger the size of the opponent group, the higher the dispersion of the challenging group is. This happens mainly for two reasons: (i) it is easier for the challenging group to remain stuck in the middle of a spatially wide opponent group and (ii) if the challenging group splits in two, the separation between the two sub-groups when moving around is higher. We also observed that small groups are more stable as they can maintain their compactness more frequently. It is also interesting to focus the attention on the high variability of plausible phenomena our model is able to reproduce: increasing the size of the opponent group usually increases the friction between the two groups and the consequently the possibility that the challenging group loses its compactness is higher. It is interesting to notice that there are situations in which dispersion does not present a monotonic behavior: in simulation #4 (Fig. 2) the challenging group firstly disperses, splitting in two parts, but eventually re-groups.

4   Conclusions and Future Developments

The paper presented an agent based pedestrian model considering groups as a fundamental element influencing the overall system dynamics. The model adopts a simple notion of group (i.e. a set of pedestrians sharing the destination of their movement and the tendency to stay close to each other) and it has been applied to a simple scenario in which it was able to generate plausible group dynamics, in terms of preserving when appropriate the cohesion of the group while also achieving a quantitatively realistic pedestrian simulation. Validation against real data is being conducted and preliminary results show a promising correspondence between simulated and observed data.

Acknowledgments. This work is a result of the Crystal Project, funded by the Centre of Research Excellence in Hajj and Omrah (Hajjcore), Umm Al-Qura University, Makkah, Saudi Arabia.

References

1. Batty, M.: Agent based pedestrian modeling (editorial). Environment and Planning B: Planning and Design 28, 321–326 (2001)
2. Willis, A., Gjersoe, N., Havard, C., Kerridge, J., Kukla, R.: Human movement behaviour in urban spaces: Implications for the design and modelling of effective pedestrian environments. Environment and Planning B 31(6), 805–828 (2004)
3. Weimar, J.R.: Simulation with Cellular Automata. Logos Verlag, Berlin (1998)
4. Schadschneider, A., Kirchner, A., Nishinari, K.: CA approach to collective phenomena in pedestrian dynamics. In: Bandini, S., Chopard, B., Tomassini, M. (eds.) ACRI 2002. LNCS, vol. 2493, pp. 239–248. Springer, Heidelberg (2002)
5. Blue, V.J., Adler, J.L.: Cellular automata microsimulation for modeling bi-directional pedestrian walkways. Transp. Research Part B 35(3), 293–312 (2001)
6. Helbing, D., Molnár, P.: Social force model for pedestrian dynamics. Phys. Rev. E 51(5), 4282–4286 (1995)
7. Dijkstra, J., Jessurun, J., Timmermans, H.J.P.: A Multi-Agent Cellular Automata Model of Pedestrian Movement. In: Pedestrian and Evacuation Dynamics, pp. 173–181. Springer, Heidelberg (2001)
8. Henein, C.M., White, T.: Agent-based modelling of forces in crowds. In: Davidsson, P., Logan, B., Takadama, K. (eds.) MABS 2004. LNCS (LNAI), vol. 3415, pp. 173–184. Springer, Heidelberg (2005)
9. Bandini, S., Federici, M.L., Vizzari, G.: Situated cellular agents approach to crowd modeling and simulation. Cybernetics and Systems 38(7), 729–753 (2007)
10. Burstedde, C., Klauck, K., Schadschneider, A., Zittartz, J.: Simulation of pedestrian dynamics using a two-dimensional cellular automaton. Physica A 295(3-4), 507–525 (2001)

11. Klüpfel, H.L.: A Cellular Automaton Model for Crowd Movement and Egress Simulation. PhD thesis, Universität Duisburg-Essen (July 2003)
12. Rickert, M., Nagel, K., Schreckenberg, M., Latour, A.: Two lane traffic simulations using cellular automata. Physica A 231(4), 534–550 (1996)
13. Bandini, S., Rubagotti, F., Vizzari, G., Shimura, K.: A Cellular Automata Based Model for Pedestrian and Group Dynamics: Motivations and First Experiments. In: Parallel Computing Technologies - PaCT 2011. LNCS. Springer, Heidelberg (in press, 2011)

An Architecture with a Mobile Phone Interface for the Interaction of a Human with a Humanoid Robot Expressing Emotions and Personality

Antonio Chella[1], Rosario Sorbello[1], Giovanni Pilato[2], Giorgio Vassallo[1], Giuseppe Balistreri[1], and Marcello Giardina[1]

[1] DICGIM Università degli Studi di Palermo, RoboticsLab, Viale delle Scienze, 90128 Palermo, Italy
[2] ICAR, CNR Viale delle Scienze, 90128, Palermo, Italy

Abstract. In this paper is illustrated the cognitive architecture of a humanoid robot based on the proposed paradigm of Latent Semantic Analysis (LSA). This paradigm is a step towards the simulation of an emotional behavior of a robot interacting with humans. The LSA approach allows the creation and the use of a data driven high-dimensional conceptual space. We developed an architecture based on three main areas: Sub-conceptual, Emotional and Behavioral. The first area analyzes perceptual data coming from the sensors. The second area builds the sub-symbolic representation of emotions in a conceptual space of emotional states. The last area triggers a latent semantic behavior which is related to the humanoid emotional state. The robot shows its overall behavior also taking into account its "personality". We implemented the system on a Aldebaran NAO humanoid robot and we tested the emotional interaction with humans through the use of a mobile phone as an interface.

Keywords: Humanoid Robot, Emotions, Personality, Latent Semantic Analysis.

1 Introduction

The concept to have a humanoid robot expressing emotions appear mind-boggling to the human beings; but it is possible to try to reproduce some emotional behaviors inside the robots in order to improve the interaction with people. In the past few years, many efforts have been made experimenting human robot social interaction [10], [3] through the emotions [1], [2], [4], [5]. Lun et al. [6] describe an emotional model and affective space of a humanoid robot and establish their studies on the psycho-dynamics psychological energy and affective energy conservation law. ARKIN et al. [7], [8], [9] show a robotic framework called TAME for human-robot interaction that tries to connect together in the same system affective phenomena as attitudes, emotions, moods and trait. Miwa et al. [1] have illustrated a mechanism for humanoid robot WE-4RII to express human-like emotions in a natural way. Research by Breazeal [2] has shown a humanoid

R. Pirrone and F. Sorbello (Eds.): AI*IA 2011, LNAI 6934, pp. 117–126, 2011.
© Springer-Verlag Berlin Heidelberg 2011

robots emotion model as an important basis in social interaction with humans. The interaction of an humanoid robots with humans and the environment should not be mechanical and deterministic in order to avoid a predictable and trivial behavior. To reach this goal we have modeled a Cognitive Architecture that drives the behavior of a humanoid robot taking into account its own personality", the context in which it is immersed, the present stimulus and the most recent tasks executed by the robot. This cognitive architecture of a humanoid robot is based on the paradigm of Latent Semantic Analysis [11], [12]. This paradigm is a step forward towards the simulation of an emotional behavior of a robot interacting with humans [13], [14], [15]. The approach presented integrates traditional knowledge representation and associative capabilities provided by geometric and sub symbolic information modeling. The paper is structured as follows: In section 2, we will describe the creation of the emotional space of the robot. Section 3 reports on the cognitive architecture of the humanoid robot expressing emotions and personality. Finally section 4, shows the emotional interaction system with humans through the use of a mobile phone as an interface.

2   The Emotional Space Creation

The architecture of the robot is based on the creation of a probabilistic emotional conceptual space automatically induced from data. The LSA technique has been widely explained and details can be found in [16], [14], [17]; for completeness we briefly recall it here.

The approach is based on the application of the Truncated Singular Value Decomposition (TSVD) technique, preceded and followed by specific pre- and post-processing phases that allow to give a probabilistic interpretation of the induced vector space.

The methodology is inspired to the Latent Semantic Analysis (LSA) approach, which is a technique based on the vector space paradigm, used to extract and represent the meaning of words through statistical computations applied to a large corpus of texts. The paradigm defines a mapping between words and documents belonging to the corpus into a continuous vector space S, where each word, and each document is associated to a vector in S [18], [19]. Even if Truncated SVD (TSVD) has been traditionally applied in the text classification and information retrieval fields [20], it can be successfully applied to any kind of dyadic domain. Let us consider a corpus made of N text chunks, each one expressing an emotion, and let M be the number of words taken into consideration in the whole data-set. The goal is to realize a mapping between the M words and the N text chunks verbally describing emotions into a continuous vector space S, where each word, as well as each emotion is associated to a emotional *knoxel* in S. Like knoxels in conceptual spaces, from the mathematical point of view, an emotional knoxel is a vector in the probabilistic emotional space; from the conceptual point of view, it is the epistemologically basic element at the level of analysis considered. Let B be the $M \times N$ matrix whose $(i, j) - th$ entry is the square root of the sample probability of the $i - th$ word belonging to the $j - th$ text chunk. The Singular Value Decomposition of the matrix B is performed, so that B is decomposed in the product of three matrices: a column-orthonormal $M \times N$ matrix U, a column-orthonormal $N \times N$ matrix V and a $N \times N$ diagonal matrix , whose elements are called singular values of B.

$$\mathbf{B} = \mathbf{U}\boldsymbol{\Sigma}\mathbf{V}^T$$

Let us suppose that B's singular values are ranked in decreasing order. Let R be a positive integer with $R < N$, and let U_R be the $M \times R$ matrix obtained from U by suppressing the last N-R columns, $_R$ the matrix obtained from by suppressing the last N-R rows and the last N-R columns and V_R be the $N \times R$ matrix obtained from V by suppressing the last N-R columns. Then

$$\mathbf{B}_R = \mathbf{U}_R \boldsymbol{\Sigma}_R \mathbf{V}_R^T$$

is a $M \times N$ matrix of rank R. The matrix B_R is then post-processed in order to obtain a new matrix Ψ in this manner:

$$[\psi]_{ij} = \begin{cases} 0 & \text{if } [b_R]_{ij} < 0 \\ \dfrac{[b_R]_{i,j}}{\sqrt{\sum [b_R]_{ij}^2 < 0}} & \text{otherwise} \end{cases} \tag{1}$$

It can be shown that the illustrated procedure can be seen as a statistical estimation process, being Ψ the probability amplitude estimated starting from the sample probability amplitude B.

3   The Cognitive Architecture

The architecture of the system presented is inspired to the approach illustrated in [21], [22], and it is organized, as shown in figure 1, in three main areas:

- The Sub-conceptual Area processes the perceptual data that come from the sensors.
- The Emotional Area is constituted by a "conceptual" space where emotions, behaviors, personality of the robot, together with the perceptual data are mapped as emotional knoxels. The coding of the personality distorts the perception of reality of the robot.
- The Behavioral Area reasons about the environment, and chooses the most adequate behaviors in order to react to the external stimuli emphatically. In this process, the linguistic area takes into account both the personality of the robot and the behaviors adopted in the recent past.

3.1   The Sub-conceptual Area

The sub-conceptual area is aimed at receiving stimuli from the different modalities of the robot and controlling its actuators at a low level also. Two main modules compose this area:

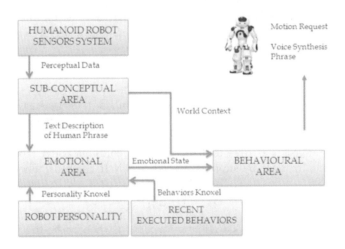

Fig. 1. The Emotional Humanoid Robot Architecture

- The *MotionModule* controls the actuators of the robot in order to perform the movement requested by the Behavioral Area.
- The *PerceptualModule* processes the raw data coming from robot sensors in order to obtain information about the environment, and the external stimuli.

The perceptual information sensed by the processing of the raw sensor data is associated to their English description: as an example, if the system recognizes a "red hammer", it will be associated to the sentence "I can see a red hammer". The use of a verbal description of the information retrieved from the environment allows its easy mapping into the emotional area. These natural language descriptions associated to each modality will be the input of the conceptual area.

3.2 The Emotional Area

The emotional area enables the robot able to find emotional analogies between the current status and the previous knowledge stored in the robot using the semantic space of emotional states. The associative area is built up in order to reflect and encode not only emotions and objects that provoke emotions, but also the personality of the robot. This is true in a twofold manner:

- documents used to induce the associative space characterize its dimensions and concepts organization beneath the space
- personality is also encoded as a set of knoxels that play the role of attractors of perceived objects beneath the space

A corpus of documents dealing with emotional states has been chosen in order to infer the space which plays therefore the role of a "probabilistic emotional space". Emotional states have been coded as "emotional knoxels" in this space using verbal description of situations capable to evoke feelings and reactions.

Environmental incoming stimuli are encoded using natural language and then are mapped in the space, to find analogies with the emotions encoded in it. A large amount of documents has been selected from several publicly available on-line sources. The excerpts have been organized in homogeneous paragraphs both for text length and emotion. The set of documents used has been obtained through an accurate selection of excerpts associated to feelings. We have selected the following emotional expressions: sadness, fear, anger, joy, surprise, love and a neutral state.

A matrix has been organized where the 6 emotional states and the neutral state have also been coded according to the procedure illustrated in the previous section that leads to the construction of a probabilistic emotional space.

A corpus of 1000 documents, equally distributed among the seven states, including other documents characterizing the personality of the robot, has been built. This set of documents represents the affective knowledge base of the robot. Each document has been processed in order to remove all words in literature named "stopwords" that do not carry semantic information like articles, prepositions and so on.

According to the technique outlined in section two, a 87×1000 terms documents matrix (B) has been created where M=80+7 is the number of words plus the emotional states and N=1000 is the number of excerpts. The generic entry $b_{i,j}$ of the matrix is the square root of the sample probability of the $i - th$ word belonging to the $j - th$ document. The TSVD technique, with K=150, has been applied to B in order to obtain its best probability amplitude approximation Ψ. This process leads to the construction of a K=150 dimensional conceptual space of emotions S. The axes of S represent the "fundamental" emotional concepts automatically induced by TSVD procedure arising from the data. In the obtained space S, a subset of n_i documents for each emotional state corresponding to one of the six "basic emotion" E_i has been projected in S using the folding-in technique. According to this technique each excerpt is coded as the sum of the vectors representing the terms composing it. As a result, the $j - th$ excerpt belonging to the subset corresponding to the emotional state E_i is represented in S by an associated vector $em_j^{(i)}$ and the emotional state E_i is represented by the set of vectors $\{\mathbf{em}_j^{(i)} : j = 1, 2, Kn_i\}$

The personality of the robot is encoded as a set of knoxels derived from documents describing the fundamental personality characteristics of the robot. As an example, if the robot has a "shy" personality, a set of documents dealing with shyness, bashfulness, diffidence, sheepishness, reserve, discretion, introversion, reticence, timidity, and so on are used to construct a cloud of "personality knoxels" that represent attraction points for the external stimuli that cause emotions in the robot. Therefore the $i - th$ excerpt belonging to the subset corresponding to the personality characteristic is mapped as a vector p_i in S. The inputs from the sense channels are coded in natural language words or sentences describing them and projected in the conceptual space using the folding-in technique. These vectors, representative of the inputs from the channels, are merged together as a weighted sum in a single vector $stim(t)$ that synthesizes the inputs stimuli from environment at instant t:

The input stimuli are therefore biased through the computation of the contribution of personality attractors in the space as the weighted sum of the knoxels p_i representing the personality of the robot, and therefore are called "personality knoxels". Each personality knoxel can be weighted with coefficients w_i (with $0 \leq w_i \leq 1$) in order to fine tune the personality influence upon the robot's behaviors.

$$\hat{stim}(t) = \frac{stim(t) + \sum w_i p_i}{||stim(t) + \sum w_i p_i||}$$

This procedure represents the common process that arises in human beings, when reality is "filtered" and interpreted by personality. The emotional semantic similarity between the vector $\hat{stim}(t)$ and the knoxels that code the six emotions in S, plus the "neutral" state, can be evaluated using the cosine similarity measure between each $em_j^{(i)}$ and $\hat{stim}(t)$:

$$sim(\hat{stim}(t), em_j^{(i)}) = \frac{\hat{stim}(t) \cdot em_j^{(i)}}{||\hat{stim}(t)|| \cdot ||em_j^{(i)}||}$$

A higher value of $sim(\hat{stim}(t), em_j^{(i)})$ corresponds to a higher value of similarity between the emotion evoked from the input and the emotion E_i associated with the vector $em_j^{(i)}$. The semantic similarity measure is calculated between $\hat{stim}(t)$ and each $em_j^{(i)}$. The vector $em_j^{(i)}$, which maximizes the quantity expressed in the formula, will be the inferred emotional state E_i. This process will activate the emotional stimulus "i" with a given intensity given by:

$$I_i(j) = sim(\hat{stim}(t), em_j^{(i)})$$

3.3   The Behavioral Area

The purpose of the Behavioral Area (figure 2) is to manage and execute behaviors coherently with the emotional state inferred by the Emotional Area, the personality of the robot, the environmental status, the stimulus given by the user and the recent past behavior adopted by the robot. A behavior is described as a sequence of primitive actions sent directly to the robot actuators. Each emotional state E_i is related with different behaviors $b_k^{(E_i)}$ in order to give the robot a non-monotonous, non-deterministic, and non-boring response. The choice of the behavior is a function $g()$ of the emotion aroused in the robot as a function of the environment, the stimulus perceived and the personality of the robot, and a function of the recent behaviors adopted in the past by the robot. Among the behaviors associated to the emotional state E_i, the behavior $b_k^{(E_i)}$ is selected through the evaluation of a score associated to each one of them.

$$Bbi(k, t) = \mu_k r - \frac{\lambda_k}{t}$$

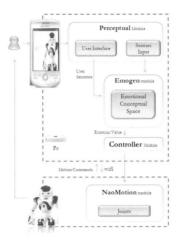

Fig. 2. An high-level description of the architecture

where r is a random value ranging from 0 to 1, $t(t > 0)$ is the time elapsed by the instant at which the $k - th$ behavior $b_k^{(E_i)}$ associated to the emotional state E_i has been executed and the instant at which this assessment is made; μ_k and λ_k are the weights assigned to the random value and to time elapsed respectively. The response with the highest weight is chosen and executed. Since the emotional stimulus is also weighted, thanks to the I "intensity" parameter. Thus, the reaction will be executed with the same intensity: movements of the parts of the body will be quicker, faster or slower. Summing up, the resulting behavior is therefore a composite function:

$$b_k^{(E_i)} = F(g(emotion, personality, stimulus), Bb(k,t), I_i(j))$$

If the emotional state is classified as "neutral", a standard behavior (lie down, sleep, and so on) is randomly selected.

4   Expected Results

The Architecture of Emotional Humanoid Robot NAO, as shown in the figure 3, was modeled using the cognitive architecture described in the previous section. Complex Humanoid Behaviors have been developed to express each of these seven emotions: sadness, fear, anger, joy, surprise, love, and a neutral. A human being can interact by talking to the robot through its voice recognition system. The Architecture of the robot elaborates the user sentence considered it as present stimulus and the image information of the camera system and generates the correct behavior taking into account also its own "personality", the context in which it is immersed, and the most recent behaviors executed by the robot. As a preliminary test bed, we considered a human being talking to the

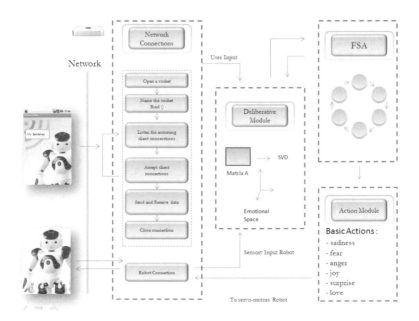

Fig. 3. The Architecture of Emotional Humanoid Robot NAO

Table 1. Samples of results of interaction: in bold it is highlighted the empathic behavior triggered by the user sentence. PE stands for the "Prevalent Emotion".

Sentence	Sadness	Anger	Fear	Joy	Surprise	Love	PE
I found myself confuse by sudden feelings of astonishment	-0,09	-0.09	-0.12	+0.08	**+0.50**	+0.08	Surprise
I don't know if my feeling for you is just pleasure or bond	-0.15	-0.33	-0.23	+0.50	+0.28	**+0.51**	Love
I want to share with you the delight of ecstasy for your help	-0.30	-0.25	-0.13	**+0.50**	+0.08	+0.47	Joy
I notice the sense of trepidation and anxiety in your eyes	+0.12	+0.32	**-0.50**	-0.13	-0.12	-0.11	Fear
I am irate and I come to you because the work you did	+0.15	**+1.00**	+0.65	-0.50	-0.17	-0.43	Anger
I am under depression for the trouble that I encountered	**+0.33**	+0.05	0.08	-0.20	-0.06	-0.15	Sadness

humanoid robot through a mobile phone. If the human being, for example, sends the following sentence to the robot as a stimulus,: "I found myself confused by sudden feelings of astonishment", the dominant emotional state activated in the LSA space of the robot with an "open" personality is "SURPRISE", and the robot executes a "SURPRISE" behavior.

The experiments, summarized in table 1, have been conducted in order to test the emotional capabilities of the robot by reproducing various emotive situations. A set of 100 sentences has been evaluated in order to understand if the behavior of the robot was accurate. While the 81% of the set was considered almost correct, in 19% of the trials the behavior of the robot was judged as being below the performance expected.

5   Conclusion and Future Works

The results shown in this paper demonstrate the possibility for a humanoid robot to generate emotional behaviors recognizable by humans. In the future we want to focus our attention on increasing the emotional interaction with humans.

References

1. Miwa, H., Itoh, K., Matsumoto, M., Zecca, M., Takanobu, H., Roccella, S., Carrozza, M.C., Dario, P., Takanishi, A.: Effective Emotional Expressions with Emotion Expression Humanoid Robot WE-4RII Integration of Humanoid Robot Hand RCH-1. In: IEEE/RSJ International Conference on Intelligent Robots and Systems, Sendai International Center, Sendai, Japan, September 28-October 2, vol. 3, pp. 2203–2208 (2004)
2. Breazeal, C.: Emotion and Sociable Humanoid Robots. International Journal Human-Computer Studies 59, 119–155 (2003)
3. Bruce, A., Nourbakhsh, I., Simmons, R.: "The Role of Expressiveness and Attention in Human-Robot Interaction" AAAI Technical Report FS-01-02 (2001)
4. Liu, Z., Pan, Z.G.: An Emotion Model of 3D Virtual Characters in Intelligent Virtual Environment. In: Tao, J., Tan, T., Picard, R.W. (eds.) ACII 2005. LNCS, vol. 3784, pp. 629–636. Springer, Heidelberg (2005)
5. Monceaux, J., Becker, J., Boudier, C., Mazel, A.: Demonstration: First Steps in Emotional Expression of the Humanoid Robot Nao. In: International Conference on Multimodal Interfaces, Cambridge, Massachusetts, USA, pp. 235–236 (2009)
6. Xie, L., Wang, Z.-L., Wang, W., Yu, G.-C.: Emotional gait generation for a humanoid robot. International Journal of Automation and Computing. Inst. of A.Chinese Ac. of Sc. (2010)
7. Moshkina, L., Arkin, R.C.: Beyond Humanoid Emotions: Incorporating Traits, Attitudes, and Moods. In: Kobe, J.P. (ed.) Proc. 2009 IEEE Workshop on Current Challenges and Future Perspectives of Emotional Humanoid Robotics (May 2009)
8. Arkin, R.C., Fujita, M., Takagi, T., Hasegawa, R.: An Ethological and Emotional Basis for Human-Robot Interaction. Robotics and Autonomous Systems 42, 191–201 (2003)
9. Arkin, R., Fujita, M., Takagi, T., Hasegawa, R.: Ethological Modeling and Architecture for an Entertaiment Robot. In: IEEE Int. Conf. on Robotics & Automation, Seoul, pp. 453–458 (2001)

10. Chella, A., Barone, R.E., Pilato, G., Sorbello, R.: Workshop An Emotional Storyteller Robot. In: AAAI 2008 Spring Symposium on Emotion, Personality and Social Behavior, March 26-28. Stanford University, Stanford (2008)

11. Prendinger, H., Ullrich, S., Nakasone, A., Ishizuka, M.: MPML3D: Scripting Agents for the 3D Internet. IEEE Transactions on Visualization and Computer Graphics 17(5), 655–668 (2011)

12. Neviarouskaya, A., Prendinger, H., Ishizuka, M.: SentiFul: A Lexicon for Sentiment Analysis. IEEE Transactions on Affective Computing 2(1), 22–36 (2011)

13. Anzalone, S.M., Cinquegrani, F., Sorbello, R., Chella, A.: An Emotional Humanoid Partner. In: Linguistic and Cognitive Approaches To Dialog Agents (LaCATODA 2010) At AISB 2010 Convention, Leicester, UK (April 2010)

14. Chella, A., Pilato, G., Sorbello, R., Vassallo, G., Cinquegrani, F., Anzalone, S.M.: An Emphatic Humanoid Robot with Emotional Latent Semantic Behavior. In: Carpin, S., Noda, I., Pagello, E., Reggiani, M., von Stryk, O. (eds.) SIMPAR 2008. LNCS (LNAI), vol. 5325, pp. 234–245. Springer, Heidelberg (2008)

15. Menegatti, E., Silvestri, G., Pagello, E., Greggio, N., Cisternino, A., Mazzanti, F., Sorbello, R., Chella, A.: 3D Models of Humanoid Soccer Robot in USARSim and Robotics Studio Simulators. International Journal of Humanoids Robotics (2008)

16. Agostaro, F., Augello, A., Pilato, G., Vassallo, G., Gaglio, S.: A Conversational Agent Based on a Conceptual Interpretation of a Data Driven Semantic Space. In: Bandini, S., Manzoni, S. (eds.) AI*IA 2005. LNCS (LNAI), vol. 3673, pp. 381–392. Springer, Heidelberg (2005)

17. Pilato, G., Vella, F., Vassallo, G., La Cascia, M.: A Conceptual Probabilistic Model for the Induction of Image Semantics. In: Proc. of the Fourth IEEE International Conference on Semantic Computing (ICSC 2010), September 22-24. Carnegie Mellon University, Pittsburgh (2010)

18. Landauer, T.K., Foltz, P.W., Laham, D.: Introduction to Latent Semantic Analysis. Discourse Processes 25, 259–284 (1998)

19. Colon, E., Sahli, H., Baudoin, Y.: CoRoBa, a Multi Mobile Robot Control and Simulation Framework. Int. Journal of Advanced Robotic Systems (2006)

20. Thagard, P., Shelley, C.P.: Emotional analogies and analogical inference. In: Gentner, D., Holyoak, K.H., Kokinov, B.K. (eds.) The Analogical Mind: Perspectives from Cognitive Science, pp. 335–362. MIT Press, Cambridge (2001)

21. Chella, A., Frixione, M., Gaglio, S.: An Architecture for Autonomous Agents Exploiting Conceptual Representations. Robotics and Autonomous Systems 25, 231–240 (1998)

22. Chella, A., Frixione, M., Gaglio, S.: A cognitive architecture for robot self-consciousness. Artificial Intelligence in Medicine 44(2), 147–154 (2008)

Checking Safety of Neural Networks with SMT Solvers: A Comparative Evaluation*

Luca Pulina[1] and Armando Tacchella[2]

[1] DEIS, University of Sassari, Piazza Università 11, Sassari, Italy
lpulina@uniss.it
[2] DIST, University of Genova, Viale F. Causa 13, Genova, Italy
armando.tacchella@unige.it

Abstract. In this paper we evaluate state-of-the-art SMT solvers on encodings of verification problems involving Multi-Layer Perceptrons (MLPs), a widely used type of neural network. Verification is a key technology to foster adoption of MLPs in safety-related applications, where stringent requirements about performance and robustness must be ensured and demonstrated. While safety problems for MLPs can be attacked solving Boolean combinations of linear arithmetic constraints, the generated encodings are hard for current state-of-the-art SMT solvers, limiting our ability to verify MLPs in practice.

Keywords: Empirical evaluation of SMT solvers, Applications of Automated Reasoning, Formal Methods for adaptive systems.

1 Introduction

SMT solvers [1] have enjoyed a recent widespread adoption to provide reasoning services in various applications, including interactive theorem provers like Isabelle [2], static checkers like Boogie [3], verification systems, e.g., ACL2 [4], software model checkers like SMT-CBMC [5], and unit test generators like CREST [6]. Research and development of SMT algorithms and tools is a very active research area, as witnessed by the annual competition, see e.g. [7]. It is fair to say that SMT solvers are the tool of choice in automated reasoning tasks involving Boolean combinations of constraints expressed in decidable background theories.

This paper is motivated by the fact that SMT solvers can also be used to solve formal verification problems involving neural networks [8]. Verification is indeed a standing challenge for neural networks which, in spite of some exceptions (see e.g., [9]), are confined to non-safety related equipment. The main reason is the lack of general, automated, yet effective safety assurance methods for neural-based systems, whereas existing mathematical methods require manual effort and ad-hoc arguments to justify safety claims [9].

In this paper we consider the problem of verifying a specific kind of neural network known as Multi-Layer Perceptron [10] (MLP). The main feature of MLPs is that, even

---

* This research has received funding from the European Community's Information and Communication Technologies Seventh Framework Programme [FP7/2007-2013] under grant agreement N. 215805, the CHRIS project.

R. Pirrone and F. Sorbello (Eds.): AI*IA 2011, LNAI 6934, pp. 127–138, 2011.
© Springer-Verlag Berlin Heidelberg 2011

with a fairly simple topology, they can in principle approximate any non-linear mapping $f : \mathbb{R}^n \rightarrow \mathbb{R}^m$ with $n, m \geq 1$ – see [11]. We consider two different kinds of safety conditions. The first is checking that the output of an MLP always ranges within stated safety thresholds – a "global" safety bound. The second is checking that the output of an MLP is close to some known value or range of values modulo the expected error variance – a "local" safety bound. Both conditions can be tested by computing an abstraction, i.e., an overapproximation, of the concrete MLP and solving a satisfiability problem in Quantifier Free Linear Arithmetic over Reals (QF_LRA in [12]) using some SMT solver. Conservative abstraction guarantees that safety of the abstract MLP implies safety of the concrete one. On the other hand, realizable counterexamples demonstrate that the concrete MLP is unsafe, whereas spurious counterexamples call for a refinement of the abstraction.

The goal of this paper is to compare state-of-the-art SMT solvers on challenging test cases derived from verification problems involving MLPs – see Section 3. In particular, we consider HySAT [13] a solver based on interval-arithmetic; MATHSAT [14], the winner of SMTCOMP 2010 in the QF_LRA category; and YICES [15] the winner of SMTCOMP 2009 in the same category. All the solvers above are tested extensively on different families of instances related to satisfiability checks in QF_LRA. These instances are obtained considering neural-based estimation of internal forces in the arm of a humanoid robot [16] – see Section 2. In particular, in Section 4 we describe three groups of experiments on the selected solvers. The first is a competition-style evaluation considering different safety flavours (local and global), different types and sizes of MLPs, and different degrees of abstraction "grain". The second is an analysis of scalability considering satisfiable and unsatisfiable encodings and varying the parameters which mostly influence performances in this regard. The last one is verification of the MLPs proposed in [16] with different solvers as back-ends. This experimental analysis shows that current state-of-the-art SMT solvers have the capability of attacking several non-trivial (sub)problems in the MLP verification arena. However, the overall verification process, particularly for networks of realistic size and fine grained abstractions, remains a standing open challenge.

2   Verification of MLPs: Case Study and Basic Concepts

All the encodings used in our analysis are obtained considering verification problems related to a control subsystem described in [16] to detect potentially unsafe situations in a humanoid robot. The idea is to detect contact with obstacles or humans, by measuring external forces using a single force/torque sensor placed along the kinematic chain of the arm. Measuring external forces requires *compensation* of the manipulator dynamic, i.e., the contribution of internal forces must be subtracted from sensor readings. Neural networks are a possible solution, but the actual network is to be considered safety-related equipment: An "incorrect" approximation of internal forces, may lead to either undercompensation – the robot is lured to believe that an obstacle exists – or overcompensation – the robot keeps moving even when an obstacle is hit. Internal forces estimation in [16] relies on an MLP. In the following, we introduce the main technical aspects of MLPs and their usage to estimate internal forces in the arm of the humanoid robot described in [16].

We consider fully-connected feed-forward MLPs composed of three layers, namely *input*, *hidden*, and *output*. Given an MLP $\nu : \mathbb{R}^n \to \mathbb{R}^m$, we have n neurons in the input layer, h neurons in the hidden layer, and m neurons in the output layer. The input of the j-th hidden neuron is defined as

$$r_j = \sum_{i=1}^{n} a_{ji} x_i + b_j \qquad j = \{1, \ldots, h\} \tag{1}$$

where a_{ji} is the *weight* of the connection from the i-th neuron in the input layer to the j-th neuron in the hidden layer, and the constant b_j is the *bias* of the j-th neuron. The activation function σ_h of hidden neurons is chosen to be a non-constant, bounded and continuous function of its input. As long as the activation function is differentiable everywhere, MLPs with *only one* hidden layer can, in principle, approximate any real-valued function with n real-valued inputs [11]. Commonly used activation functions (see, e.g., [8]) are sigmoidal non-linearities such as the *hyperbolic tangent* (tanh) and the *logistic* (logi) functions defined as follows:

$$\tanh(r) = \frac{e^r - e^{-r}}{e^r + e^{-r}} \qquad \mathrm{logi}(r) = \frac{1}{1 + e^{-r}} \tag{2}$$

where $\tanh : \mathbb{R} \to (-1, 1)$ and $\mathrm{logi} : \mathbb{R} \to (0, 1)$. The MLP suggested in [16] uses hyperbolic tangents, but our encodings are obtained using the logistic function instead.[1] The input received by an output neuron is

$$s_k = \sum_{j=1}^{h} c_{kj} \sigma_h(r_j) + d_k \qquad k = \{1, \ldots, m\} \tag{3}$$

where c_{kj} denotes the weight of the connection from the j-th neuron in the hidden layer to the k-th neuron in the output layer, while d_k represents the bias of the k-th output neuron. The output of the MLP is a vector $\nu(\underline{x}) = \{\sigma_o(s_1), \ldots, \sigma_o(s_m)\}$, where $\sigma_o = \sigma_h$, so that each output of ν is constrained to the range of σ_o.

We train MLPs considering a training set (X, Y) where $X = \{\underline{x}_1, \ldots, \underline{x}_t\}$ are the *patterns* and $Y = \{\underline{y}_1, \ldots, \underline{y}_t\}$ are the corresponding *labels*. In particular, we consider angular positions and velocities of two shoulder and two elbow joints as input patterns, i.e., for each $\underline{x} \in X$, we have $\underline{x} = \langle q_1, \ldots, q_4, \dot{q}_1, \ldots, \dot{q}_4 \rangle^2$ ($n = 8$). Labels $\underline{y} \in Y$ are corresponding values of internal forces and torques – denoted by f and τ, respectively – in a Cartesian space, i.e., $\underline{y} = \langle f_1, f_2, f_3, \tau_1, \tau_2, \tau_3 \rangle$ ($m = 6$). The unknown relation $\varphi : \mathbb{R}^8 \to \mathbb{R}^6$ is the one tying joint positions and velocities to internal forces, and takes into account components from gravity, Coriolis forces, and manipulator inertia.

3   SMT Encodings to Verify MLPs

We consider two verification problems involving MLPs, and an approach to solve them using SMT encodings – more precisely, encodings in Quantifier Free Linear Arithmetic

---

[1] From a practical standpoint, the impact of our choice is negligible, since the logistic function has the same "shape" of the hyperbolic tangent, and they are often used interchangeably.

[2] This is standard control-theory notation, where q represents the angular position of the joint, and $\dot{q}$ the angular velocity, i.e., the derivative of q with respect to time.

over Reals (QF_LRA) as defined in [12]. In the following we will always consider MLPs with $n \geq 1$ inputs and $m \geq 1$ outputs in which the *input domain* is a Cartesian product $\mathcal{I} = D_1 \times \ldots \times D_n$ where $D_i = [a_i, b_i]$ is a closed interval bounded by $a_i, b_i \in \mathbb{R}$ for all $1 \leq i \leq n$; analogously, the *output domain* is a Cartesian product $\mathcal{O} = E_1 \times \ldots \times E_m$ where $E_i = [c_i, d_i]$ is a closed interval bounded by $c_i, d_i \in \mathbb{R}$ for all $i \leq i \leq m$.[3]

3.1   Global Safety

Checking for global safety of an MLP $\nu : \mathcal{I} \to \mathcal{O}$ amounts to prove that

$$\forall \underline{x} \in \mathcal{I}, \ \forall k \in \{1, \ldots, m\} : \nu_k(\underline{x}) \in [l_k, h_k] \tag{4}$$

where $\nu_k(\underline{x})$ denotes the k-th output of ν, and $l_k, h_k \in E_k$ are *safety thresholds*, i.e., constants defining an interval wherein the k-th component of the MLP output is to range, given all acceptable input values. Condition (4) can be checked on a *consistent abstraction* of ν, i.e., a function $\tilde{\nu}$ such that if the property corresponding to (4) is satisfied by $\tilde{\nu}$ in a suitable abstract domain, then it must hold also for ν. The key point is that verifying condition (4) in the abstract domain can be encoded to a satisfiability check in QF_LRA. Clearly, abstraction is not sufficient per se, because of *spurious counterexamples*, i.e., abstract counterexamples that do not correspond to concrete ones. A spurious counterexample calls for a refinement of the abstraction which, in turn, generates a new satisfiability check in QF_LRA. Therefore, a global safety check for a single output of ν may generate several logical queries to the underlying SMT solver. In practice, we hope to be able to either verify ν or exhibit a counterexample within a reasonable number of refinements.

In the following, we briefly sketch how to construct consistent abstractions and related refinements to check for property (4). Given a concrete domain $D = [a, b]$, the corresponding abstract domain is $[D] = \{[x, y] \mid a \leq x \leq y \leq b\}$, i.e., the set of intervals inside D, where $[x]$ is a generic element. We can naturally extend the abstraction to Cartesian products of domains, i.e., given $\mathcal{I} = D_1 \times \ldots \times D_n$, we define $[\mathcal{I}] = [D_1] \times \ldots \times [D_n]$, and we denote with $[\underline{x}] = \langle [x_1], \ldots, [x_n] \rangle$ the elements of $[\mathcal{I}]$ that we call *interval vectors*. Given a generic MLP ν – the *concrete* MLP – we construct the corresponding *abstract* MLP by assuming that σ_h is the logistic function logi : $\mathbb{R} \to (0, 1)$ as defined in (2), and that σ_o is the identity function. Since no ambiguity can arise between σ_h and σ_o, we denote σ_h with σ for the sake of simplicity. Given an abstraction parameter $p \in \mathbb{R}^+$, the *abstract activation function* $\tilde{\sigma}^p$ can be obtained by considering the maximum increment of σ over intervals of length p. Since σ is a monotonically increasing function, and the tangent to σ reaches a maximum slope of $1/4$ we have that

$$\forall x \in \mathbb{R} : 0 \leq \sigma(x + p) - \sigma(x) \leq \frac{p}{4} \tag{5}$$

---

[3] In the definitions above, and throughout the rest of the paper, a closed interval $[a, b]$ bounded by $a, b \in \mathbb{R}$ is the set of real numbers comprised between a and b, i.e, $[a, b] = \{x \mid a \leq x \leq b\}$ with $a \leq b$.

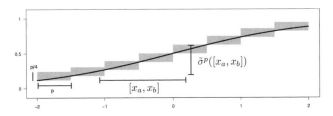

Fig. 1. Activation function $\sigma(x)$ and its abstraction $\tilde{\sigma}^p(x)$ in the range $x \in [-2, 2]$. The solid line denotes σ, while the boxes denote $\tilde{\sigma}^p$ with $p = 0.5$.

for any choice of the parameter $p \in \mathbb{R}^+$. Now let x_0 and x_1 be the values that satisfy $\sigma(x_0) = p/4$ and $\sigma(x_1) = 1 - p/4$, respectively, and let $p \in (0, 1)$. We define $\tilde{\sigma}^p :$ $[\mathbb{R}] \rightarrow [[0, 1]]$ as follows

$$
\tilde{\sigma}_p([x_a, x_b]) = \begin{cases}
[0, p/4] & \text{if } x_b \leq x_0 \\
[0, \sigma(p\lfloor \frac{x_b}{p} \rfloor) + \frac{p}{4}] & \text{if } x_a \leq x_0 \text{ and } x_b < x_1 \\
[\sigma(p\lfloor \frac{x_a}{p} \rfloor), \sigma(\lfloor \frac{x_b}{p} \rfloor) + \frac{p}{4}] & \text{if } x_0 \leq x_a \text{ and } x_b \leq x_1 \\
[\sigma(p\lfloor \frac{x_a}{p} \rfloor), 1] & \text{if } x_0 < x_a \text{ and } x_1 \leq x_b \\
[1 - p/4, 1] & \text{if } x_a \geq x_1 \\
[0, 1] & \text{if } x_a \leq x_0 \text{ and } x_1 \leq x_b
\end{cases}
\tag{6}
$$

Figure 1 gives a pictorial representation of the above definition.

According to (6) we can control how much $\tilde{\sigma}^p$ over-approximates σ, since large values of p correspond to coarse-grained abstractions, whereas small values of p correspond to fine-grained ones. We can now define $\tilde{\nu}^p : [\mathcal{I}] \rightarrow [\mathcal{O}]$ as

$$
\tilde{\nu}_k^p([\underline{x}]) = \sum_{j=1}^h c_{kj} \tilde{\sigma}^p(\tilde{r}_j([\underline{x}])) + d_k \qquad k = \{1, \ldots, m\}
\tag{7}
$$

where $\tilde{r}_j([\underline{x}]) = \sum_{i=1}^n a_{ji}[x_i] + b_j$ for all $j = \{1, \ldots, h\}$, and we overload the standard symbols to denote products and sums among interval vectors, e.g., we write $x + y$ to mean $x\tilde{+}y$ when $x, y \in [\mathbb{R}]$. Since $\tilde{\sigma}^p$ is a consistent abstraction of σ, and products and sums on intervals are consistent abstractions of the corresponding operations on real numbers, defining $\tilde{\nu}^p$ as in (7) provides a consistent abstraction of ν. This means that our original goal of proving the safety of ν according to (4) can be now recast, modulo refinements, to the goal of proving its abstract counterpart

$$
\forall [\underline{x}] \in [\mathcal{I}], \forall k \in \{1, \ldots, m\} : \tilde{\nu}_k^p([\underline{x}]) \sqsubseteq [l_k, h_k]
\tag{8}
$$

where "$\sqsubseteq$" stands for the usual containment relation between intervals, i.e., given two intervals $[a, b] \in [\mathbb{R}]$ and $[c, d] \in [\mathbb{R}]$ we have that $[a, b] \sqsubseteq [c, d]$ exactly when $a \geq c$ and $b \leq d$, i.e., $[a, b]$ is a subinterval of – or it coincides with – $[c, d]$.

3.2 Local Safety

The need to check for global safety can be mitigated using sigmoidal non-linearities in the output neurons to "squash" the response of the MLP within an acceptable range, modulo rescaling. A more stringent, yet necessary, requirement is represented by local safety. Informally speaking, we can say that an MLP ν trained on a dataset (X, Y) of t patterns is "locally safe" whenever given an input pattern $\underline{x}^*$ it turns out that $\nu(\underline{x}^*)$ is "close" to $\underline{y}_j \in Y$ as long as $\underline{x}^*$ is "close" to $\underline{x}_j \in X$ for some $j \in \{1, \ldots, t\}$. Local safety cannot be guaranteed by design, because the range of acceptable values varies from point to point. Moreover, it ensures that the error of an MLP never exceeds a given bound on yet-to-be-seen inputs, and it ensures that the response of an MLP is relatively stable with respect to small perturbations in its input.

To formalize local safety, given an MLP $\nu : \mathcal{I} \to \mathcal{O}$, and a training set (X, Y) consisting of t elements, we introduced the following concepts. Given two patterns $\underline{x}, \underline{x}' \in X$ their *distance along the i-th dimension* is defined as $\delta_i(\underline{x}, \underline{x}') = |x'_i - x_i|$. Given some $\underline{x} \in X$, the function $N_i^q : X \to 2^X$ maps $\underline{x}$ to the set of *q-nearest-neighbours along the i-th dimension*, i.e., the first q elements of the list $\{\underline{x}' \in X \mid \underline{x}' \neq \underline{x}\}$ sorted in ascending order according to $\delta_i(\underline{x}, \underline{x}')$. Given some $\underline{x} \in X$, the function $\delta_i^q : X \to \mathbb{R}$ maps $\underline{x}$ to the *q-nearest-distance along the i-th dimension*, i.e.,

$$\delta_i^q(\underline{x}) = \max_{\underline{x}' \in N_i^q(\underline{x})} \delta_i(\underline{x}, \underline{x}')$$

The *q-n-polytope* $\mathcal{X}_j^q$ corresponding to $\underline{x}_j \in X$ for some $j \in \{1, \ldots, t\}$ is the region of space comprised within all the $2n$ hyper-planes obtained by considering, for each dimension i, the two hyper-planes perpendicular to the i-th axis and intersecting it in $(x_i - \delta_i^q(\underline{x}))$ and $(x_i + \delta_i^q(\underline{x}))$, respectively. The above definitions can be repeated for labels, and thus $\mathcal{Y}_j^q$ denotes a q-m-polytope associated to $\underline{y}_j \in Y$ for some $j \in \{1, \ldots, t\}$. In the following, when the dimensionality is understood from the context, we use $\mathcal{X}$ to denote 1-n-polytopes, and $\mathcal{Y}$ to denote 1-m-polytopes. In the following, we refer to q as the *neighborhood size*.

Given an MLP $\nu : \mathcal{I} \to \mathcal{O}$ with training set (X, Y) consisting of t patterns we consider, for all $j \in \{1, \ldots, t\}$, the set of *input polytopes* $\{\mathcal{X}_1, \ldots, \mathcal{X}_t\}$ associated with each pattern $\underline{x}_j \in X$, and the set of *output polytopes* $\{\mathcal{Y}_1^q, \ldots, \mathcal{Y}_t^q\}$ associated with each label $\underline{y}_j \in Y$, for a fixed value of q. We say that ν is locally safe if the following condition is satisfied

$$\forall \underline{x}^* \in \mathcal{I}, \exists j \in \{1, \ldots, t\} : \underline{x}^* \in \mathcal{X}_j \to \nu(\underline{x}^*) \in \mathcal{Y}_j^q \tag{9}$$

Notice that this condition is trivially satisfied by all the input patterns $\underline{x}^*$ such that $\underline{x}^* \in \mathcal{I}$ but $\underline{x}^* \notin \mathcal{X}_j$ for all $j \in \{1, \ldots, t\}$. Indeed, these are patterns which are "too far" from known patterns in the training set, and for which we simply do not have enough information in terms of local (un)safety. Also, we always consider 1-n-polytopes on the input side of ν whereas we can vary the size of neighbourhoods on the output side by increasing q. Clearly, the larger is q, the larger is the neighbourhood considered in the output, and the less stringent condition (9) becomes. This additional degree of freedom is important in order to "tune" the safety condition according to the expected variance in the network error. Intuitively, assuming that we obtained a network whose expected error mean and variance is satisfactory, if we try to certify such network on safety bounds which imply a smaller error variance, we will invariably generate feasible counterexamples. The abstraction-refinement approach to check local safety is similar to the one

Table 1. Evaluation results at a glance. We report the number of encodings solved within the time limit ("#") and the total CPU time ("Time") spent on the solved encodings. Total number of formulas solved ("Total") is also split in satisfiable and unsatisfiable formulas ("Sat") and ("Unsat"), respectively. Solvers are sorted according to the number of encodings solved. A dash means that a solver did not solve any encoding in the related group.

Solver	Total		Sat		Unsat	
	#	Time	#	Time	#	Time
YICES	809	27531.90	639	22141.03	170	5390.87
HYSAT	698	17641.61	561	13089.82	137	4551.79
MATHSAT	683	41975.01	544	35043.15	139	6931.86
CVC	–	–	–	–	–	–

described for global safety. In particular, the abstract network $\tilde{\nu}^p$ is obtained as shown previously, i.e., by abstracting the activation function and thus the whole network to compute interval vectors. The abstract local safety condition corresponding to (9), for any fixed value of k, is

$$\forall [\underline{x}^*] \in [\mathcal{I}], \exists i \in \{1, \ldots, t\} : [\underline{x}^*] \in \mathcal{X}_i \to \tilde{\nu}([\underline{x}^*]) \sqsubseteq \mathcal{Y}_i^k \tag{10}$$

It can be shown that the above condition implies local safety of the concrete network ν, and, as in the case of (8), verifying it can be encoded into a QF_LRA satisfiability check.

4   Challenging SMT Solvers to Verify MLPs

The experiments detailed in this section are carried out on a family of identical Linux workstations comprised of 10 Intel Core 2 Duo 2.13 GHz PCs with 4GB of RAM. Unless otherwise specified, the resources granted to the solvers are 600s of CPU time and 2GB of memory. The solvers involved in the evaluation are CVC [17] (v. 3.2.3, default options), HYSAT [13] (v. 0.8.6, $\varepsilon = 10^{-5}$ and $\delta = 10^6$ options), MATHSAT [14] (v. 4, -no_random_decisions option), and YICES [15] (v. 2, default options).

To compare the solvers we use encodings considering both global and local safety, different types and sizes of MLPs, and different degrees of abstraction grain. We classify the encodings in "Suites" and "Families". The former distinction is about global vs. local safety, from which we obtain two suites, namely GLOBAL and LOCAL. For each suite, we group encodings in families differing for the number of hidden neurons and the numbers of output neurons. The family HN-XX_ON-YY denotes encodings with XX hidden and YY output neurons, respectively. We vary the number of hidden neurons in the range $\{5, 10, 20\}$ and we consider either one or six output neurons. For each suite, we produce all possible encodings obtained combining different values of these parameters. Finally, for each family, we encode formulas from $p = 0.5$, and decreasing it by a rate $r = 1.3$, i.e., $p = p/r$ at each refinement step. We stop when we obtain 20 encodings for each family. Moreover, in the case of LOCAL, we compute encodings with different neighborhood sizes, i.e. $q = \{1, 10, 20, 50, 100, 199\}$.

Table 2. Solver-centric view of the results. Columns "Suite" and "Family" report suite and family name of the encodings, respectively. The remainder of the table is organized similarly to Table 1, with the exception of column ("Unique"), that shows data about uniquely solved encodings.

Suite	Family	Solver	Total		Sat		Unsat		Unique	
			#	Time	#	Time	#	Time	#	Time
GLOBAL (120)	HN-5_ON-1 (20)	YICES	20	23.889	20	23.889	–	–	–	–
		MATHSAT	20	281.56	20	281.56	–	–	–	–
		HYSAT	12	288.91	12	288.91	–	–	–	–
	HN-5_ON-6 (20)	YICES	20	28.90	20	28.90	–	–	–	–
		HYSAT	20	712.30	20	712.30	–	–	–	–
		MATHSAT	19	548.96	19	548.96	–	–	–	–
	HN-10_ON-1 (20)	YICES	19	493.71	19	493.701	–	–	3	115.68
		HYSAT	17	816.59	17	816.59	–	–	1	478.53
		MATHSAT	9	731.21	9	731.21	–	–	–	–
	HN-10_ON-6 (20)	HYSAT	19	902.85	19	902.85	–	–	1	4.68
		YICES	19	1188.10	19	1188.10	–	–	–	–
		MATHSAT	15	1112.26	15	1112.26	–	–	–	–
	HN-20_ON-1 (20)	MATHSAT	20	1178.47	20	1178.47	–	–	2	326.58
		YICES	18	1637.82	18	1637.82	–	–	–	–
		HYSAT	17	1066.20	17	1066.20	–	–	–	–
	HN-20_ON-6 (20)	MATHSAT	20	1034.77	20	1034.77	–	–	2	406.29
		HYSAT	14	961.40	14	961.40	–	–	–	–
		YICES	13	2633.74	13	2633.74	–	–	–	–
LOCAL (720)	HN-5_ON-1 (120)	YICES	120	910.77	82	451.68	38	459.09	13	592.63
		MATHSAT	106	5103.72	74	2860.42	32	2243.30	–	–
		HYSAT	98	1639.34	70	393.22	28	1246.11	–	–
	HN-5_ON-6 (120)	HYSAT	120	20.57	100	19.53	20	1.04	–	–
		YICES	120	1225.49	100	1134.06	20	91.43	–	–
		MATHSAT	113	4714.75	93	4272.69	20	442.06	–	–
	HN-10_ON-1 (120)	YICES	120	2808.87	83	1603.70	37	1205.17	22	2194.79
		HYSAT	94	5382.65	67	3469.57	27	1913.08	–	–
		MATHSAT	94	6802.15	67	4412.35	27	2389.80	–	–
	HN-10_ON-6 (120)	YICES	120	4047.18	100	3760.97	20	286.20	9	1729.51
		HYSAT	106	1346.47	86	1344.97	20	1.49	–	–
		MATHSAT	104	6272.87	84	5728.48	20	544.39	–	–
	HN-20_ON-1 (120)	YICES	115	6014.31	80	3563.78	35	2450.53	23	4274.22
		HYSAT	90	2651.61	67	1263.40	23	1388.22	–	–
		MATHSAT	76	6420.13	56	5796.93	20	623.19	–	–
	HN-20_ON-6 (120)	YICES	105	6519.13	85	5620.70	20	898.43	11	3238.09
		HYSAT	91	1852.71	72	1850.86	19	1.85	1	3.03
		MATHSAT	87	7774.15	67	7085.04	20	689.11	–	–

In Table 1 we report a global picture of the evaluation results. In the following, when we say that "solver A *dominates* solver B" we mean that the set of problems solved by B is a subset of those solved by A. Looking at the result, we can see that all solvers but CVC, were able to solve at least 70% of the test set. CVC exhausts memory resources before reaching the time limit, and it is able to solve no encodings, thus we drop it from the analysis. Still looking at Table 1, we can see that YICES outperforms the other solvers conquering 96% of the test set, while HYSAT and MATHSAT were able to solve 83% and 81%, respectively. Despite the very similar performance of HYSAT and MATHSAT – only 15 encodings separate them – we can see that HYSAT spends about 42% of the CPU time spent by MATHSAT. We also report that HYSAT, has the best average time per encoding (about 25s) with respect to both YICES (about 34s) and MATHSAT (about 61s). Finally, we report no discrepancies in the satisfiability result of the evaluated solvers.

Table 3. Encoding-centric view of the results. The table consists of seven columns where for each family of encodings we report the name of the family in alphabetical order (column "Family"), the number of encodings included in the family, and the number of encodings solved (group "Overall", columns "N", "#", respectively), the CPU time taken to solve the encodings (column "Time"), the number of easy, medium and medium-hard encodings (group "Hardness", columns "EA", "ME", "MH").

Family	Overall		Time	Hardness			Family	Overall		Time	Hardness		
	N	#		EA	ME	MH		N	#		EA	ME	MH
GLOBAL_HN-5_ON-1	20	20	23.89	12	8	–	LOCAL_HN-5_ON-1	120	120	823.53	97	10	13
GLOBAL_HN-5_ON-6	20	20	28.91	19	1	–	LOCAL_HN-5_ON-6	120	120	20.57	113	7	–
GLOBAL_HN-10_ON-1	20	20	744.19	9	7	4	LOCAL_HN-10_ON-1	120	120	2563.94	90	8	22
GLOBAL_HN-10_ON-6	20	20	345.42	14	5	1	LOCAL_HN-10_ON-6	120	120	2137.99	99	12	9
GLOBAL_HN-20_ON-1	20	20	906.36	17	1	2	LOCAL_HN-20_ON-1	120	115	4811.98	74	18	23
GLOBAL_HN-20_ON-6	20	20	827.05	9	9	2	LOCAL_HN-20_ON-6	120	106	4484.83	83	11	12

Table 2 shows the results of the evaluation dividing the encodings by suites and families. As we can see, in terms of number of encodings solved, YICES is the strongest solver. Concerning the suite GLOBAL, it leads the count with 109 solved encoding (91% of the test set), while concerning the LOCAL suite, it solves 700 encodings (97% of the test set). Focusing on the suite GLOBAL, 10 encodings separate the strongest solver from the weakest one – HYSAT, that solves 99 encodings (82% of the test set). If we consider the problems that are uniquely solved, then we see that no solver is dominated by the others. Now focusing on the suite LOCAL, the first thing to observe is that the difference between the strongest and the weakest solver is increased: 101 encodings separate YICES and MATHSAT, that was able to solve 580 encodings (about 80% of the test set). We also report that MATHSAT is dominated by YICES.

In Table 3 we show the classification of encodings included in the test set. In the table, the number of encodings solved and the cumulative time taken for each family is computed considering the "SOTA solver", i.e., the ideal solver that always fares the best time among all considered solvers. An encoding is thus solved if at least one of the solvers solves it, and the time taken is the best among all times of the solvers that solved the encoding. The encodings are classified according to their hardness with the following criteria: easy encodings are those solved by all the solvers, medium encodings are those non easy encodings that could still be solved by at least two solvers, medium-hard encodings are those solved by one reasoner only, and hard encodings are those that remained unsolved.

According to the data summarized in Table 3, the test set consisted in 840 encodings, 821 of which have been solved, resulting in 636 easy, 97 medium, 88 medium-hard, and 19 hard encodings. Focusing on families comprised in the suite GLOBAL, we report that all 120 encodings were solved, resulting in 80 easy, 31 medium, and 9 medium-hard encodings. Considering the families in LOCAL, we report that 821 encodings (out of 840) were solved, resulting in 636 easy, 97 medium, and 88 medium-hard encodings.

Finally, we report the contribution of each solver to the composition of the SOTA solver. Focusing on the suite GLOBAL, YICES contributed to the SOTA solver 77 times (out of 120), while MATHSAT and HYSAT 29 and 14 times, respectively. If we consider now the suite LOCAL, the picture is quite different: HYSAT contributed 509 times,

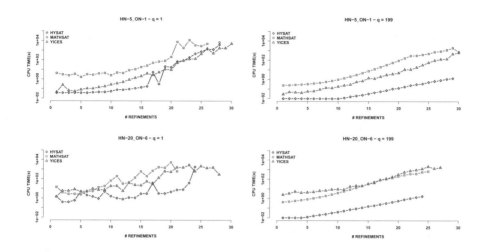

Fig. 2. Scalability test on the evaluated solvers. For each plot, in x axis is shown the refinement step, while in the y axis (in logarithmic scale) the related CPU time (in seconds). HYSAT performance is depicted by blue diamonds, while MATHSAT and YICES results are denoted by red boxes and green triangles, respectively. Plots in the same column are related to encodings having tha same satisfiability result, i.e. SAT (left-most column) and UNSAT (right-most column). Plots in the same row are related to encodings having tha same MLP architecture, i.e. HN-5_ON-1 (top) and HN-20_ON-6 (bottom).

while YICES and MATHSAT 188 and 4 times, respectively. Considering all 821 solved encodings, we report that HYSAT was the main contributor (64%), despite the fact that it is not the best solver in terms of total amount of solved encodings.

Our next experiment aims to draw some conclusions about the scalability of the evaluated solvers. In order to do that, we compute a pool of encodings satisfying the following desiderata: (1) Consider values of the abstraction parameter p which correspond to increasingly fine-grained abstractions; (2) Consider different MLP size in terms of hidden neurons; and (3) Consider the satisfiability result of the computed encodings. To cope with (1), we generate encodings related to 30 refinement steps. To take care the potentially increasing difficulty of such encodings, we set the time limit to 4 CPU hours. In order to satisfy desiderata (2), we compute encodings both considering HN-5_ON-1 and HN-20_ON-6 MLP architectures. Finally, in order to cover (3), we focus on the suite LOCAL, selecting the encodings related to $q = 1$, and $q = 199$. The former encodings are almost always satisfiable, i.e., an abstract counterexamples is easily found, and, conversely, the latter encodings are almost always unsatisfiable.

As a result of the selection above, we obtain 4 groups of encodings, and Figure 2 shows the results of experimenting with them. Looking at Figure 2 (top-left), we can see that HYSAT is the best solver along the first 17 refinement steps. After this point, its performance is comparable to YICES, but with the noticeable difference that the latter is able to solve all the encodings, while HYSAT exhausts CPU time resources

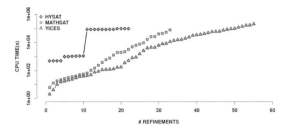

Fig. 3. CPU cumulative time (y axis) vs. number of refinement steps (x axis) with different back-engines. The plot is organized similarly to the plots in Figure 2.

trying to solve the two encodings having the smallest value of p. The CPU time spent by MATHSATon each of the first 24 refinement steps is at least one order of magnitude higher than HYSAT and YICES. Considering now the same safety problem, but related to a larger MLP architecture, we can see a different picture. From Figure 2 (bottom-left), we can see that no solver is able to solve all the encodings within 4 CPU hours. In particular, MATHSAT stops at the 21st step (out of 30), while YICES is able to solve all encodings but the last two. While the performances of MATHSAT and YICES seems to have a smooth increasing trend, HYSAT is less predictable: In the first 22 refinement steps it is up to one order of magnitude faster than YICES– with the noticeable exception of four "peaks" – and for the following two steps it is two order of magnitude slower than YICES.

Considering now the plots in Figure 2 (right), we can see that the trend in solver's performance is much smoother than the plots in Figure 2 (left). Looking at Figure 2 (top-right), we can see that, excluding the last encoding, HYSAT is one order of magnitude faster than YICES, that in turn is one order of magnitude faster that MATHSAT. Looking now at the last plot, we can see that we have to main differences with respect to the picture resulting from the previous plot. First, the encodings are more challenging, because no solver was able to solve all the pool within the CPU time limit. Second, there is no noticeable difference – excluding the last two solved encodings – between MATHSAT and YICES.

Our last experiment concerns the analysis of various solvers as back-ends. We experiment with a local safety problem with $q = 75$, $p = 0.5$, and $r = 1.1$ about an MLP having a HN-20_ON-6 architecture. In Figure 3 we report the results of such experiment. Also if we are not able to conclude about the safety of the considered MLP because all solvers exhaust their memory resources, looking at the figure we can see that YICES clearly outperforms both MATHSAT and HYSAT. YICES enabled allowed to refine 54 times, while MATHSAT and HYSAT stopped to 32 and 21 refinements, respectively. Concerning the cumulative CPU time, the performances of MATHSAT and YICES are in the same ballpark until the 12th step, and they increase smoothly until the end of the computation. Concluding our analysis, we report for HYSAT a very similar behaviour to the one shown in Figure 2 (bottom-left).

References

1. Barrett, C., Sebastiani, R., Seshia, S.A., Tinelli, C.: Satisfiability modulo theories. In: Handbook of Satisfiability, pp. 825–885. IOS Press, Amsterdam (2009)
2. Fontaine, P., Marion, J.Y., Merz, S., Nieto, L., Tiu, A.: Expressiveness+ automation+ soundness: Towards combining SMT solvers and interactive proof assistants. In: Hermanns, H. (ed.) TACAS 2006. LNCS, vol. 3920, pp. 167–181. Springer, Heidelberg (2006)
3. DeLine, R., Leino, K.R.M.: BoogiePL: A typed procedural language for checking object-oriented programs (2005)
4. Ray, S.: Connecting External Deduction Tools with ACL2. In: Scalable Techniques for Formal Verification, pp. 195–216 (2010)
5. Armando, A., Mantovani, J., Platania, L.: Bounded model checking of software using SMT solvers instead of SAT solvers. International Journal on Software Tools for Technology Transfer (STTT) 11(1), 69–83 (2009)
6. Hoang, T.A., Binh, N.N.: Extending CREST with Multiple SMT Solvers and Real Arithmetic. In: 2010 Second International Conference on Knowledge and Systems Engineering (KSE), pp. 183–187. IEEE, Los Alamitos (2010)
7. Barrett, C., de Moura, L., Stump, A.: SMT-COMP: Satisfiability Modulo Theories Competition. In: Etessami, K., Rajamani, S. (eds.) CAV 2005. LNCS, vol. 3576, pp. 20–23. Springer, Heidelberg (2005)
8. Bishop, C.M.: Neural networks and their applications. Review of Scientific Instruments 65(6), 1803–1832 (2009)
9. Schumann, J., Liu, Y. (eds.): Applications of Neural Networks in High Assurance Systems. SCI, vol. 268. Springer, Heidelberg (2010)
10. Haykin, S.: Neural networks: a comprehensive foundation. Prentice Hall, Englewood Cliffs (2008)
11. Hornik, K., Stinchcombe, M., White, H.: Multilayer feedforward networks are universal approximators. Neural Networks 2(5), 359–366 (1989)
12. Cok, D.R.: The SMT-LIBv2 Language and Tools: A Tutorial (2011), http://www.grammatech.com/resources/smt/
13. Franzle, M., Herde, C., Teige, T., Ratschan, S., Schubert, T.: Efficient solving of large nonlinear arithmetic constraint systems with complex boolean structure. Journal on Satisfiability, Boolean Modeling and Computation 1, 209–236 (2007)
14. Bruttomesso, R., Cimatti, A., Franzén, A., Griggio, A., Sebastiani, R.: The MathSAT 4 SMT Solver. In: Gupta, A., Malik, S. (eds.) CAV 2008. LNCS, vol. 5123, pp. 299–303. Springer, Heidelberg (2008)
15. Dutertre, B., De Moura, L.: A fast linear-arithmetic solver for DPLL (T). In: Ball, T., Jones, R.B. (eds.) CAV 2006. LNCS, vol. 4144, pp. 81–94. Springer, Heidelberg (2006)
16. Fumagalli, M., Gijsberts, A., Ivaldi, S., Jamone, L., Metta, G., Natale, L., Nori, F., Sandini, G.: Learning to Exploit Proximal Force Sensing: a Comparison Approach. In: Sigaud, O., Peters, J. (eds.) From Motor Learning to Interaction Learning in Robots. SCI, vol. 264, pp. 149–167. Springer, Heidelberg (2010)
17. Barrett, C., Tinelli, C.: CVC3. In: Damm, W., Hermanns, H. (eds.) CAV 2007. LNCS, vol. 4590, pp. 298–302. Springer, Heidelberg (2007)

A Dual Association Model for Acquisition and Extinction

Ashish Gupta[1] and Lovekesh Vig[2]

[1] Google Inc., MountainView, CA 94043, USA
[2] School of Information Technolgy,
Jawaharlal Nehru University, New Delhi 110067, India

Abstract. Phenomena like faster reacquisition of learned associations after extinction were initially explained via residual synaptic plasticity in the relevant neural circuits. However, this account cannot explain many recent behavioral findings. This includes phenomena like savings in extinction, reinstatement, spontaneous recovery and renewal. These phenomena point to the possibility that extinction isn't a mere reversal of the associations formed during acquisition. It instead involves the superimposition of some separate decremental process that works to inhibit the previously learned responses. We have explored this dual-pathway account using a neurocomputational model of conditioning. In our model, associations related to acquisition and extinction are maintained side by side as a result of the interaction between general neural learning processes and the presence of lateral inhibition between neurons. The model captures relevant behavioral phenomena that prompted the hypothesis of separate acquisition and extinction pathways. It also shows how seemingly complex behavior can emerge out of relatively simple underlying neural mechanisms.

Keywords: Extinction, Conditioning, Leabra, Savings, Reinstatement, Renewal.

1 Introduction

The relationship between the learning of an association and the unlearning of that same association is commonly thought to involve a singular representation of the strength of association, with that strength rising during learning and falling during unlearning. In animal conditioning, this view suggests that the extinction of a behavior involves reversing the synaptic modifications made during the initial acquisition of that behavior, causing the animal to stop producing the conditioned response (CR) [7]. While this theory is simple and elegant, it is not consistent with a growing body of behavioral findings.

Evidence from numerous studies points to the possibility that extinction isn't a mere reversal of the associations formed during acquisition [4,3]. Phenomena like savings, reinstatement, spontaneous recovery and renewal suggest that extinction training involves the superimposition of some separate decremental process that works to inhibit previously learned response, leaving most of the originally acquired CS-US association intact. The phenomenon of savings [11] involves the relatively small amount of reacquisition training needed to restore the response after extinction training. In reinstatement [3,12], the response is restored through the presentation of US, alone. Renewal [4]

R. Pirrone and F. Sorbello (Eds.): AI*IA 2011, LNAI 6934, pp. 139–150, 2011.

is said to occur when a shift in environmental context away from that in which extinction training took place results in renewed responding.

Recognition of retained association knowledge, even after responding has been extinguished, has led to theories involving residual synaptic plasticity and sub-threshold responding [7]. These theories hold that extinction training does not completely reverse synaptic changes made during initial acquisition, but only reverses these changes enough to effectively inhibit responding. When presented with the CS after extinction, the neural system involved in producing a response continues to become somewhat active, but not sufficiently active to produce an actual response. Thus, only small changes in association strength are needed to return this system to a state in which responding to the CS is robust.

However theories based on residual synaptic plasticity cannot account for some important additional observations. In particular, there is evidence that, just as extinction does not remove associations built up during previous acquisition training; subsequent reacquisition training does not remove the inhibitory force built up during previous extinction training. For example, animals continue to show spontaneous recovery - a phenomenon that only arises after extinction training - even if they experience a subsequent period of reacquisition that removes the behavioral impact of the previous extinction process [11]. Also, just as reacquisition after extinction is faster than initial acquisition, subsequent extinctions are also faster than the first extinction [11]. Other phenomena, including conditioned inhibition [3], counter conditioning [3] and feature positive discrimination [16] shed further light on the nature of the associative changes during extinction.

In this paper, we show that the fundamental principles of neural computation, embodied in the Leabra modeling framework [8], spontaneously capture these phenomena of extinction. In particular, we show how synaptic plasticity, bidirectional excitation between cortical regions, and lateral inhibition between cortical regions interact, allowing the effects of previous acquisition and extinction to be maintained side by side. Of particular importance are processes of lateral inhibition, which introduce competition between neurons involved in the encoding of stimuli. Our model encodes the acquisition and extinction using a separate pool of neurons that compete with each other via the lateral inhibitory mechanism. Hence, much of the associational knowledge embedded in the synapses of the acquisition neurons is retained even after extinction. Similarly, many of the changes in extinction neuron synapses wrought during extinction training are retained after reacquisition training. Through this retention of synaptic strengths, our model is able to capture many of the behavioral results described above.

The paper is organized as follows. In the background section, we present a brief description of the Leabra modeling framework. The details of our model are described in the subsequent section. That is followed by a detailed description of the different behavioral results along with the results of our simulation experiments. We conclude the paper with a general discussion of some relevant issues.

2   Related Models

Temporal Difference Reinforcement Learning (TDRL) is unable to reproduce data confirming faster reacquisition after extinction due to the fact that it unlearns acquisition related state values during extinction. Redish et. al [10] proposed an insightful variant of the TDRL model to address this issue. Their model introduced a state-classification process that determines the subject's current state and creates new states and state spaces when observation statistics change. However, the model leaves open issues such as the biological implementation of how the lack of expected reward would be signaled, and how cues would be categorized into situations within the brain. The proposed TDRL model captures the phenomenon of cued renewal by generating a new state on receiving consistently low reward during extinction. The renewal occurs because the associations formed during acquisition are not forgotten, and new extinction associations are formed in a different state. The Leabra model proposed in this paper also captures renewal with context units biasing the response of the subject in different contexts. This allows the network to generate different representations of an association for different contexts.

The idea that acquisition and extinction involve learning in separate pathways has been proposed as an explanation to behavioral data by behavioral scientists [11] and has also been used in some highly successful neural models. Grossberg et. al. [5] assume that extinction of conditioning happens not because of the reversal of learning in the on-pathway, but due to an active process of counter conditioning in the off-pathway. In contrast our model does not 'assume' that lateral inhibition is sufficient for the emergence of independent pathways, rather it is a central finding. In contrast use of seperate pathways for learning and unlearning was an assumption in Grossberg's work, implemented via dedicated learning rules which ensured that seperate sets of weights change during learning and unlearning.

We posit that the biologically grounded property of lateral inhibition is sufficient for the emergence of separate independent pathways for learning and unlearning and offers explanation for a wide variety of behavioral data pertaining to animal conditioning. Our model works on the assumption that the emergence of these phenomena does not necessarily require any specific brain architectures. This is not to suggest that specific brain areas do not participate in the emergence of these behaviors in specific animals, however our model is consistent with findings that suggest that similar behavior should be exhibited in even the most primitive of brains [14]. We have thus kept our network architecture generic i.e. a standard 3 layer network with an input layer, a hidden layer and an output layer.

3   Background

It is imperative that the modeling framework utilized for neurocomputational experiments be realistically grounded in biology. Our proposed model utilizes the well known Leabra cognitive modeling framework [8] for developing a connectionist model that adheres to known neurobiological principles. This section highlights the relevant properties of the Leabra framework that enable the development of biologically plausible connectionist models. Subsequently, the model proposed for demonstrating extnction phenomenon is described in detail.

3.1   Leabra Modeling Framework

The Leabra framework offers a collection of integrated formalisms that are grounded in known properties of cortical circuits but are sufficiently abstract to support the simulation of behaviors arising from large neural systems. Leabra includes dendritic integration using a point-neuron approximation, a firing rate model of neural coding, bidirectional excitation between cortical regions, fast feedforward and feedback inhibition, and a mechanism for synaptic plasticity that incorporates both error-driven and Hebbian learning whilst allowing for lateral inhibition.

The effects of inhibitory interneurons tend to be strong and fast in the cortex. This allows inhibition to act in a regulatory role, mediating the positive feedback of bidirectional excitatory connections between brain regions. Simulation studies have shown that a combination of fast feedforward and feedback inhibition can produce a kind of "set-point dynamics", where the mean firing rate of cells in a given region remains relatively constant in the face of moderate changes to the mean strength of inputs. As inputs become stronger, they drive inhibitory interneurons as well as excitatory pyramidal cells, producing a dynamic balance between excitation and inhibition. Leabra implements this dynamic using a k-Winners-Take-All (kWTA) inhibition function that quickly modulates the amount of pooled inhibition presented to a layer of simulated cortical neural units, based on the layer's level of input activity. This results in a roughly constant number of units surpassing their firing threshold.

In our model, acquisition-related and extinction-related learning occurs in two distinct sets of neurons that compete with each other via this lateral inhibition mechanism. Indeed, it is lateral inhibition, in conjunction with Leabra's synaptic learning mechanism that enables the retention of acquisition knowledge in the face of extinction training and the retention of extinction knowledge in the face of re-acquisition training.

4   The Model

The learning performance of a simple multi-layer Leabra network, as shown in Figure 1, was examined. This model is an extension of the traditional Rescorla-Wagner [13] model of animal conditioning.

In our model, each CS was encoded as a single input unit. The stimulus was recoded over the firing rates of 40 units grouped into a hidden layer. This hidden layer incorporated strong lateral inhibition, using a kWTA parameter of $k = 5$, encouraging only 5 of the 40 units to be active at any one time. The hidden layer had a bidirectional excitatory projection to the output layer. The output layer contained 7 units, with $k = 5$. For the simulation experiments that only used a positive reward, the first 5 units were interpreted as encouraging a positive response in the face of the stimulus, the aggregate activation over these units determining the strength of the response. The final 2 units in the output layer encoded the "null response" generated after extinction training. These units offered a means to suppress the activity in the first 5 units via lateral inhibition. For fear conditioning, the last 2 units encoded the positive response, while the first 5 units encoded the animal's "freezing" response in the presence of the CS. In our model,

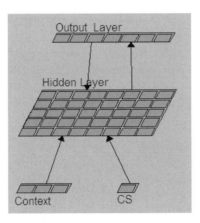

Fig. 1. The Leabra network. Each gray box corresponds to a neural processing unit. Each arrow represents complete interconnectivity between the units in two layers.

the null response was encoded using fewer neurons than the conditioned response in accordance with evidence suggesting greater neural activity during CS+ rather than CS- [15].

Context was encoded by weakly activating a single unit (setting to a value of 0.25, maximum possible value being 1.0) from a pool of contextual input units. These units were connected to the hidden layer units via a random pattern of connectivity, with a 90% probability of each connection being formed. A switching of context in the simulation experiments involved the switching of the active unit in the context layer. For the experiments in which the context was never switched, a single unit in the context layer remained active throughout the simulation.

Leabra's default parameters were used in these simulations, with only a few exceptions. To accommodate the relatively small size of this network, the range of initial random synaptic weights was reduced ($[0.0; 0.1]$ rather than the default range of $[0.25; 0.75]$) and learning rate for synaptic modification was set to a smaller value (0.005, half of the default of 0.01). Also, individual neuron bias weights were removed. The strength of the backward projection from the output layer to the hidden layer was weakened (by setting the wt_scale.rel parameter associated with that layer to a value of 0.05 instead of the default value of 1). This decrease was required because with the default strength, the backward projections strongly activated all the hidden layers in the network, even those for which the corresponding input stimulus was not provided. Modifications of these kinds are common in smaller Leabra networks.

A randomly initialized network was used for each training trial. Each training session was terminated when the sum squared error (SSE) between the network's output and the expected output patterns fell below a criterion value of 1. All simulation experiments were repeated 25 times, and mean results across these runs are reported.

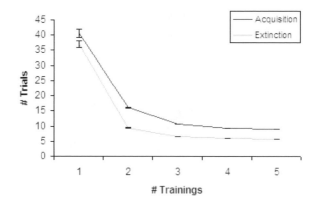

Fig. 2. Simulation 1A. The number of training trials required to reach criterion (Y axis) decreases as number of prior acquisition and extinction training sessions (X axis) increases. Error bars report standard errors of the mean.

5   Experiments

5.1   Simulation 1: Savings

Our first simulation experiment was designed to uncover the degree to which our model exhibits savings. Recall that animals are faster to reacquire an extinguished behavior, as compared to initial acquisition, and they are faster to extinguish a reacquired behavior, as compared to initial extinction. A randomly initialized network was trained to respond upon the presentation of the CS (A+). Once this training reached criterion, the network was trained to not-respond upon the presentation of the CS (A-). This pattern was repeated 5 times. Figure 2 shows the number of trials required for successive acquisition and extinction trainings. Note that the required time quickly decreases. The model predicts that the required number of trials will asymptote to a small value after just a few acquisition-extinction iterations.

The network starts with small initial synaptic weights. Hence, a large change in weights is required for success during the first acquisition training session. During the first extinction training session, the weights to the acquisition neurons start decreasing and the weights to the extinction neurons start increasing. As soon as the extinction neurons win the inhibitory competition, the acquisition neurons tend to fall below their firing threshold. At this stage, the weights to the acquisition neurons stop decreasing, as these neurons are no longer contributing to erroneous outputs. Hence, a significant amount of acquisition related association strength is retained through the extinction process. During the reacquisition training, the weights to the acquisition neurons increase once again and the weights to the extinction neurons decrease. Once again, the weights stop changing as soon as the extinction neurons lose the inhibitory competition. Hence, most of extinction related plasticity is retained through the acquisition process. In this manner, subsequent acquisition and extinction trainings require a very small change in

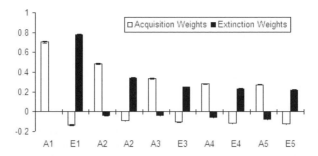

Fig. 3. This graphs plots the change in the summed connection weights in the acquisition pathway and in the extinction pathway (Y axis) during the successive acquisition and extinction trainings (X axis). The change in weights decreases in both the pathways as the number of prior acquisitions and extinctions training sessions increases. There seems to be a slow upward going trend in the weights in both the pathways, which appears to be a quirk of the simulator.

weights(Figure 3). Effectively, acquisition and extinction associations are maintained side by side in the network, allowing for the rapid switching between them based on recent conditioning feedback.

5.2   Simulation 2: Reinstatement

The phenomenon of reinstatement involves the restoration of an extinguished response through the presentation of the US alone. Reinstatement has generally been reported in the domain of fear conditioning. In a typical experiment [12], the animals are first given baseline training for lever pressing. This is followed by fear conditioning, where the animals learn to suppress the lever pressing behavior in the presence of a CS, by pairing that CS with a foot shock. Then, the fear conditioning is extinguished via the omission of the foot shock. At the end of the extinction training, the animals are subject to a non-contingent foot shock without the presentation of the CS. When tested in the presence of the CS, it is observed that the lever pressing behavior is suppressed[1] once again.

Table 1. The four training sessions used in simulation 2. L corresponds to the lever stimulus and T corresponds to a conditioned stimulus. A plus indicates fear conditioning, and a minus indicates extinction training. Note that "LT+" signifies that during the fear conditioning with T stimulus, the lever (L) stimulus was also present, and hence the LT compound was reinforced.

Baseline	Conditioning	Extinction	Reinstatement
L-	L-	L-	L+
	LT+	LT-	

---

[1] Suppression is measured as a ratio; Suppression Ratio = (Responding in the presence of CS) / (Responding in the presence of CS + Baseline responding).

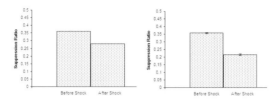

Fig. 4. Simulation 2. Left: Results reproduced from Rescorla's experiment – suppression ratio in the presence of the CS, before and after reinstatement. Right: Simulation results – suppression ratio in the presence of the CS before and after the presentation of the non-contingent shock. Note that there is a greater suppression of the response (smaller suppression ratio) after the shock presentation than before the shock presentation.

This experiment was performed to test if the non-contingent presentation of the US alone results in the reinstatement of an extinguished fear response. The design of this simulation experiment is shown in Table 1. A randomly initialized network was given baseline training, where it learned to press the lever (L-) stimulus. Then, the network was subjected to fear conditioning in the presence of a stimulus (LT+, L-). This was followed by the extinction of the fear conditioning (LT-, L-). Then, the network's responding was tested for in the presence of the CS (LT) and in the absence of the CS (L). This was followed by a single non-contingent presentation of foot shock (L+). Finally, the network's responding for L and LT combination was tested once again. We found that there was negligible change in the responding to L before and after the single shock stimulus presentation. The response magnitude before the shock presentation was $1.881(\pm0.014)$ while after the shock presentation was $1.862(\pm0.005)$ ($t(48) = 1.259$, $p < 0.2128$). However, the responding for the LT combination showed a significant drop ($t(48) = 20.504, p < 0.0421$). See Figure 4.

Most theories of conditioning ignore the fact that the initial baseline training, where the animal learns to press the lever must entail the formation of some associations. In contrast, in our simulations, the lever stimulus (L) forms an integral part of the entire training process. During the baseline training (L-), the lever acquires strong food associations. These food associations survive through the fear conditioning, and L continues to elicit a strong food response. In contrast, after fear conditioning, the fear associations for the LT combination are stronger than the food associations, thereby eliciting a strong fear response. During the extinction training, the fear associations for the LT combination weaken only to the extent required to lose the inhibitory competition. Hence, a large proportion of these associations survive the extinction training. A single shock presentation (L+) results in a sufficient increase in L's associations to cause the fear associations for the LT combination to start winning the inhibitory competition once again. The food associations of L still remain strong enough to win the inhibitory competition when L is presented alone.

5.3   Simulation 3: Renewal

This simulation demonstrates that shifting the animal to a context different from the context of extinction results in a renewal of the conditioned responding. The design of this simulation experiment is shown in Table 2. A randomly initialized network was given baseline training (L-) in three different contexts (A, B and C). After the baseline training, fear conditioning (L-, LT+) was conducted in context A followed by extinction (L-, LT-) in context B. Finally, responding with (LT) and without (L) CS was tested in all three contexts. In accordance with the behavioral results, the network exhibited a greater suppression in context A as compared to context C ($t(48) = 7.712$, $p < 0.0478$). See Figure 5. The suppression in context B was lesser than in context C ($t(48) = 2.21$, $p < 0.0368$). See Figure 6.

Table 2. The three training sessions and a single testing session, used in simulation 3. A, B and C correspond to different contexts, L corresponds to the lever stimulus and T corresponds to a conditioned stimulus. + indicated fear conditioning and - indicates extinction of fear conditioning.

Baseline	Conditioning	Extinction	Test
A: L-	A: L-	B: L-	A: LT
B: L-	A: LT+	B: LT-	B: LT
C: L-			C: LT

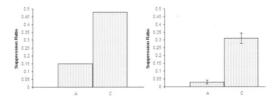

Fig. 5. Simulation 3. Left: Results reproduced from Boutons experiment – suppression ratio in context A and context C. Right: Simulation results – suppression ratio in context A and context C. Note that there is a greater suppression of the response in context A, the context of fear conditioning, as compared to context C, the neutral context.

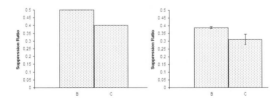

Fig. 6. Simulation 3. Left: Results reproduced from Bouton's experiment – suppression ratio in context B and context C. Right: Simulation results – suppression ratio in context B and context C. Note that there is a lesser suppression of the response in context B, the context where fear conditioning was extinguished, as compared to context C, the neutral context.

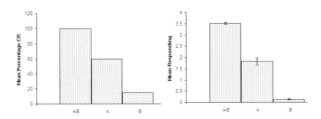

Fig. 7. Simulation 4. Left: Results reproduced from Wagner's experiment – percentage responding for AB compound, A and B. Right: Simulation results – response magnitude for AB, A and B. Note that the response magnitude is greater for AB compound than for A.

Fear conditioning is conducted in context A. Hence, A acquires fear associations. In contrast, due to extinction training, context B acquires extinction related associations. Context C, the neutral context does not acquire either associations. The CS undergoes acquisition training (in context A) and extinction training (in context B). Hence, both the acquisition related and extinction related associations strengthen for it. Hence, when the CS is combined with context A, the response tends to shift in favor of the fear response. When the CS is combined with context B, the response tends to shift in favor of extinction. With context C, the response remains intermediate. Additionally, the presentation of context alone fails to generate any activity in the output layer of our model. Hence, as observed by Bouton, the contexts of training do not acquire any demonstrable associations during this training process.

5.4   Simulation 4: Feature Positive Discrimination

In feature positive discrimination [16], a CS is reinforced in combination with a second CS, and it is not reinforced when presented alone. With this training, the CS combination generates a very strong response. The responding to the first CS alone is negligible. The second CS alone exhibits an intermediate level of responding. The single association theory makes a contradictory prediction for this phenomenon. It predicts that the first CS, starting with a net zero association would acquire a net association that is negative in magnitude. As a result, the association acquired by the second CS will have to be especially strong in order to overcome the subtracting effect of the first CS, when presented in combination. Hence, the responding should be stronger for the second CS alone as compared to the responding for the CS combination.

In this simulation experiment, a randomly initialized network was trained on a stimulus combination (AB+), while one of the stimuli was presented without reinforcement (B-). At the end of this training, the network's performance was measured for A, B and AB compound. In accordance with the behavioral data, the AB compound showed a stronger responding than A ($t(48) = 10.684, p < 0.0262$). See Figure 7.

Stimulus B, which participates in both acquisition and extinction trainings, strengthens its associations in both pathways. Hence, when B is presented alone, the two pathways cancel each other's effects via lateral inhibition, generating very small output. Stimulus A only strengthens its acquisition related associations. As a result, when A is presented alone, it generates a strong response. AB combination generates an even stronger response due to the mutual support of acquisition associations of A and B.

5.5   Simulation 5: Conditioned Inhibition

In conditioned inhibition [3], a CS is reinforced when presented alone, but not reinforced when presented in combination with a second CS. Due to this training, the second CS acquires inhibitory capabilities - it can inhibit the responding, when it is combined with some other reinforced stimulus. The single association theory predicts this result. IT posits that the second stimulus, starting with net zero association would generate a net association that is negative in magnitude. Hence, when combined with the first stimulus, the summed strength of the association becomes smaller, resulting in a diminished responding.

The goal of this experiment was to test if our model predicts conditioned inhibition. A randomly initialized network was reinforced for two different stimuli (A+, B+), while stimulus A was presented without reinforcement in the presence of stimulus C (AC-). At the end of this training, the network was tested for the BC compound. As expected, the BC compound generated a weaker responding as compared to B. The mean responding for the BC compound was $1.459(\pm 0.184)$ while that for B was $4.367(\pm 0.045)$. $(t(48) = 15.374, p < 0.0372)$. This happens because C acquires extinction related associations which inhibit B's responding.

6   Conclusions

What is the biological plausibility of the dual-association hypothesis? The inhibitory circuits that are responsible for reducing the expression of fear are not very well understood. However, there is evidence highlighting the role of the medial prefrontal cortex (mPFC) in memory cicuits for fear extinction [9]. Several distinct nuclei of the amygdala have been shown to be differentially involved in acquisition, extinction and expression of fear responses [6,2,1]. Further evidence shows that extinction does not cause a reversal of plasticity in the acquisition-relevant brain areas [7]. This evidence, when pieced together lends support to the notion that extinction is new learning, rather than erasure of conditioning.

We have proposed a neurocomputational model for the extinction of animal conditioning. This model supports the notion that extinction is not merely a reversal in previously acquired synaptic associations, positing, instead, the existence of a separate pathway that interacts with the acquisition related pathway through the interaction of foundational neural processes, including error-driven synaptic plasticity, bidirectional excitation, and strong lateral inhibition. We have shown that our model captures the relevant patterns of performance exhibited by animals. Another strength of our model is that it does not depend on the specific properties of particular brain systems, such as the hippocampus or the cerebellum. Hence, it helps in explaining why vastly different brains produce similar patterns of learning.

References

1. Amano, T., Unal, C.T., Pare, D.: Synaptic correlates of fear extinction in the amygdala. Nature Neuroscience 13, 1097–6256 (2010)
2. Bellgowan, P.S., Helmstetter, F.J.: Neural systems for the expression of hypoalgesia during nonassociative fear. Behavioral Neuroscience 110, 727–736 (1996)
3. Bouton, M.E.: Context and behavioral processes in extinction. Learning and Memory 11, 485–494 (2004)
4. Bouton, M.E., Brooks, D.C.: Time and context effects on performance in a pavlovian discrimination reversal. Journal of Experimental Psychology: Animal Behavior Processes 19(2), 165–179 (1993)
5. Grossberg, S., Schmajuk, N.A.: Neural dynamics of attentionally-modulated pavlonian conditioning: Conditioned reinforcement, inhibition, and opponent processing. Psychology 15, 195–240 (1987)
6. Maren, S.: Long-term potentiation in the amygdala: A mechanism for emotional learning and memory. Trends in Neuroscience 23, 345–346 (1999)
7. Medina, J.F., Garcia, K.S., Mauk, M.D.: A mechanism for savings in the cerebellum. The Journal of Neuroscience 21(11), 4081–4089 (2001)
8. O'Reilly, R.C., Munakata, Y.: Computational Explorations in Cognitive Neuroscience: Understanding the Mind by Simulating the Brain. MIT Press, Cambridge (2000)
9. Quirk, G.J., Garcia, R., Gonzalez-Lima, F.: Prefrontal mechanisms in extinction of conditioned fear. Biological Psychiatry 60, 337–343 (2006)
10. Redish, A.D., Jensen, S., Johnson, A., Kurth-Nelson, Z.: Reconciling reinforcement learning models with behavioral extinction and renewal: Implications for addiction, relapse, and problem gambling. Psychological Review 114(3), 784–805 (2007)
11. Rescorla, R.A.: Retraining of extinguished pavlovian stimuli. Journal of Experimental Psychology: Animal Behavior Processes 27(2), 115–124 (2001)
12. Rescorla, R.A., Heth, C.D.: Reinstatement of fear to an extinguished conditioned stimulus. Journal of Experimental Psychology: Animal Behavior Processes 104(1), 88–96 (1975)
13. Rescorla, R.A., Wagner, A.R.: A theory of pavlovian conditioning: Variations in the effectiveness of reinforcement and nonreinforcement. In: Classical Conditioning II: Current Research and Theory, pp. 64–99. Appleton-Century-Crofts, New York (1972)
14. Rose, J.K., Rankin, C.H.: Analysis of habituation in c. elegans. Learning Memory 8, 63–69 (2001)
15. Shabel, S.J., Janak, P.H.: Substantial similarity in amygdala neuronal activity during conditioned appetitive and aversive emotional arousal. Proceedings of the National Academy of Sciences 106, 15031–15036 (2009)
16. Wagner, A.R., Brandon, S.E.: A componential theory of pavlovian conditioning. In: Mowrer, R.R., Klein, S.B. (eds.) Handbook of Contemporary Learning Theories, pp. 23–64. Erlbaum, Hillsdale (2001)

Intelligent Supervision for
Robust Plan Execution

Roberto Micalizio, Enrico Scala, and Pietro Torasso

Università di Torino corso Svizzera 185, 10149 Torino
{micalizio,scala,torasso}@di.unito.it

Abstract. The paper addresses the problem of supervising the execution of a plan with durative actions in a just partially known world, where discrepancies between the expected conditions and the ones actually found may arise. The paper advocates a control architecture which exploits additional knowledge to prevent (when possible) action failures by changing the execution modality of actions while these are still in progress. Preliminary experimental results, obtained in a simulated space exploration scenario, are reported.

Keywords: Plan Execution, Intelligent Supervision, Robotic Agents, Control Architecture.

1 Introduction

In the last years a significant amount of interest has been devoted to the area currently known as "planning in the real world", in order to weaken some of the assumptions made in classical planning. While relevant results have been obtained for innovative planning techniques such as conditional and contingent planning, it is worth noting that these forms of planning may be quite expensive from a computational point of view and in many cases it is very hard to anticipate all possible contingencies, since the execution of an action can be perturbed by *unexpected* events which may cause the failure of the action itself. To handle these issues, some methodologies [1,2,3] propose to monitor the execution of a plan and to invoke a repair strategy, typically based on a re-planning step, as soon as action failures are detected. These methodologies, however, are unable to intervene during the execution of an action, as the repair is invoked just after the occurrence of an action failure, that is, when the plan execution has been interrupted. In principle, this problem could be mitigated by anticipating which actions in the plan will not be executable in the future (i.e., threatened actions, see for example [4]), so that the repair strategy can be invoked earlier. Unfortunately, it is not always possible to anticipate the set of threatened actions as the plan executor (i.e., the agent) may have just a partial knowledge of the world where it is operating.

In this paper we propose a control architecture for robust plan execution whose aim is to avoid (at least in some cases) the occurrence of action failures. To reach this goal, the proposed architecture exploits a temporal interpretation

R. Pirrone and F. Sorbello (Eds.): AI*IA 2011, LNAI 6934, pp. 151–163, 2011.

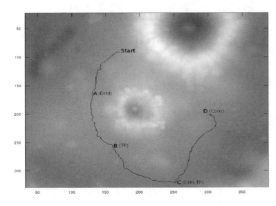

Capital letters A, B, C, D denote the sites the rover has to visit. Each site is tagged with the actions to be performed when the site has been reached: DRILL refers to the drill action, TP to take picture, and COM to data transmission. More actions can be done at the same site, see for instance target C. The black line connecting two targets is the route, predicted during a path planning phase, the rover should follow during a navigate action.

Fig. 1. An example of daily mission plan

module for detecting agent's behavioral patterns which, over a temporal window, describe deviations from the nominal expected behavior. When such potentially hazardous situations are detected, another module, the Active Controller, can decide to change the modality of execution of the current action by taking into account the capability of alternative modalities in alleviating the discrepancy between the actual behavior and the nominal one.

The paper is organized as follows: section 2 describes a space exploration scenario used to exemplify the proposed approach; section 3 presents a basic control architecture which just reacts to action failures, an improved architecture which tries to prevent failures is discussed in section 4; section 5 reports some preliminary experimental results; finally, in section 6, the conclusions.

2   A Motivating Example

This section introduces a space exploration scenario, where a mobile robot (i.e., a planetary rover) is in charge of accomplishing explorative tasks. This scenario presents some interesting and challenging characteristics which made it particularly interesting for the plan execution problem. The rover, in fact, has to operate in a hazardous and not fully observable environment where a number of unpredictable events may occur.

In our discussion, we assume that the rover has been provided with a mission plan covering a number of scientifically interesting sites: the plan includes navigation actions as well as exploratory actions that the rover has to complete once a target has been reached; for instance the rover can:
- drill the surface of rocks;
- collect soil samples and complete experiments in search for organic traces;
- take pictures of the environment.
All these actions produce a certain amount of data which are stored in an on-board memory of the rover until a communication window towards Earth becomes available. In that moment the data can be uploaded; see [5] for a possible

solution tackling the communication problem in a space scenario. For example, a possible daily plan involves: `navigate(Start,A)`; `drill(A)`; `navigate(A,B)`; `tp(B)`; `navigate(B,C)`; `drill(C)`; `tp(C)`; `navigate(C,D)`; `com(D)`. This plan is graphically represented in Figure 1 where a map of a portion of the Martian soil is showed.[1]

It is easy to see that some of these actions can be considered atomic (e.g., take picture), some others, instead, will take time to be completed. For instance, a navigate action will take several minutes (or hours), and during its execution the rover moves over a rough terrain with holes, rocks, slopes. The safeness of the rover could be threatened by too deep holes or too steep slopes since some physical limits of the rover cannot be exceeded. In case such a situation occurs, the rover is unable to complete the action. Of course, the rover's physical limits are taken into account during the synthesis of the mission plan, and regions presenting potential threats are excluded *a priori*.

However, the safeness of the rover could also be threatened by terrain characteristics which can hardly be anticipated. For instance, a terrain full of shallow holes may cause high-frequency vibrations on the rover, and if these vibrations last for a while they may endanger some of the rover's devices. This kind of threat is difficult to anticipate from Earth both because satellite maps cannot capture all terrain details, and because this threat depends on the rover's contextual conditions, such as its speed.

In the following of the paper we propose a control architecture which recognizes potential threats while actions are still under way, and reacts to them by tuning the *execution modality*. For instance, the navigation action can be associated with two execution modalities: high-speed and reduced-speed; slowing down the rover's speed can mitigate the harmful effects of disconnected terrains. As we will see, the solution we propose is sufficiently flexible to change the execution modality not only when threats have been detected, but also when threats terminate and nominal execution modalities can be restored.

3   Basic Control Architecture

As said above, plan execution monitoring becomes a critical activity when a given plan is executed in the real world; differences between the (abstract) world assumed during the planning phase and the actual world may lead, in some cases, to a failure in plan execution. Monitoring the plan execution is hence necessary but it is just the first step (detecting plan failures): a plan repair mechanism should be subsequently activated in order to restore (if possible) nominal conditions. Many plan repair techniques rely on a (re)planning phase to overcome the failure of an action. In some cases, however, this technique may be difficult to apply and too costly, so it should be limited as far as possible. For this reason, in the following we propose a control architecture which tries to limit the necessity of replanning by preventing, as far as possible, the occurrence of plan failures.

---

[1] In the picture, different altitudes are represented in a grey scale where white corresponds to the highest altitude, and black to the lowest.

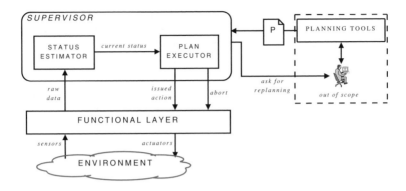

Fig. 2. A basic control architecture

(:durative-action navigate
parameters : (?r - rover ?y - site ?z - site)
duration : (= ?duration navigation-time ?z ?p)
condition : (and(at start(at ?r ?y)
    (over all (and(≤ (pitch-derivative ?r) 5)
        (≤ (roll-derivative ?r) 5)
        (≤ (pitch ?r) 30)
        (≤ (roll ?r) 30))))
effect :(and(at start(not (at ?r ?y)))
    (at end(at ?r ?z)))

```
01 while there are actions in P to be performed
02    rdata_t ← getRawDataFromFL(t)
03    state_t ← StateEstimator(rdata_t)
04    if (checkInvariants(inv_a, state_t) = violation
         ∨ actualDuration(a) > duration(a))
05        send abort to FL
06        ask for replanning
07    if checkOutcome(eff_a, state_t) = succeeded
08        a ←getNextAction(P)
09        if checkPreconditions(pre_a, state_t) =
                         not satisfied
10            ask for replanning
11        else submit a to FL
12    t ← t + 1
```

Fig. 3. An example of durative action   **Fig. 4.** The basic control strategy

For the sake of exposition, before presenting the complete control architecture, we introduce a basic version which just detects failures and reacts to them by aborting the current execution (see Figure 2).

We assume that the given mission plan P is a totally ordered sequence of action instances, which are modeled in PDDL 2.1 [6] (this formalism, in fact, allows to deal with atomic as well as durative actions). Note that, besides preconditions and effects, PDDL 2.1 allows the definition of invariant conditions which the planner must guarantee to maintain during the synthesis of the mission plan. These invariant conditions are exploited in our approach not only in the planning phase, but also to check whether the rover's safeness conditions are maintained during the plan execution. For instance, the "over-all" construct of the navigate action shown in Figure 3 specifies which conditions on the rover's attitude (i.e., the combination of pitch and roll) are to be considered safe, and hence must hold during the whole execution of the action.

The basic architecture of Figure 2, includes two main levels: the Supervisor and the Functional Level (FL). The Supervisor is in charge of managing the execution of the plan P and its control strategy is reported in Figure 4. At each time instant t, the State Estimator gets the raw data provided by the FL and produces an internal representation of the current rover's state possibly by making qualitative abstractions on the raw data. The Plan Executor matches the estimated state $state_t$ against the invariant conditions of the action a currently in execution: in case such conditions are not satisfied or the execution of the action a lasts over the duration indicated in the plan, the action is considered failed. In this case, the Plan Executor aborts the execution (by sending an appropriate abort command to the FL) and asks for a new repair plan P'. In case $state_t$ is consistent with the action model, the Plan Executor establishes whether a has been completed with success (of course, the outcome of a is *success* when the expected effects eff_a hold in $state_t$); in the positive case, the next action in P becomes the new current action to be performed. The new action is actually submitted for execution only after the validation of its preconditions against the current state; in case the action is not executable, the Plan Executor asks for a recovery plan.

The second level of the architecture is the FL, which, from our point of view, is an abstraction of the rover's hardware able to match the actions issued by the Supervisor into lower level commands for the rover's actuators. In doing so, the FL may exploit services such as localization, obstacle avoidance, short range path planning, path following (see [7,8]).

4   Improving the Control

To be more effective, the Supervisor must be able to anticipate plan failures and actively intervene during the execution, not only for aborting the current action, but also for changing the way in which that action is going to be performed. Unfortunately, the pieces of information contained in the mission plan are not sufficient for this purpose and the Supervisor needs additional sources of information complementing the ones in the plan.

4.1   Knowledge for the Active Control

Execution Trajectories. The first extension we introduce is closely related to the actual execution of an action. In the PDDL2.1 model, in fact, one just specifies (propositional) preconditions and effects, but there may be different ways to achieve the expected effects from the given preconditions. For instance, the action $navigate(A, B)$ just specifies that: 1) the rover must be initially located in A and 2) the rover, after the completion of the navigate action, will be eventually located in B; but nothing is specified about the intermediate rover positions between A and B. This lack of knowledge is an issue when we consider the problem of plan execution monitoring. For the monitoring purpose, in fact, it becomes important to detect erroneous behaviors while the action is still under execution. For this reason, we associate each durative action instance a with a parameter trj_a, that specifies a trajectory of nominal rover states. More formally, $trj_a = \{s_0, \ldots, s_n\}$, where s_i $(i : 0..n)$ are, possibly partial, rover states at different steps of execution of a. We just require that both $s_0 \vdash pre_a$ and $s_n \vdash eff_a$ must hold. Therefore, trj_a represents how the rover state should evolve over time while it is performing a. For example, let a be *navigation(A,B)*, trj_a maintains a sequence of waypoints which sketches the route the rover has to follow. Of course, the actual execution of the navigation action may deviate from the given trajectory for a number of reasons (e.g., unexpected obstacles may be encountered along the way). In principle, a deviation from the nominal trajectory does not necessarily represent an issue; in our extended approach, the Supervisor takes the responsibility for tracking these deviations and deciding when they signal anomalies to be faced.

Temporal Patterns. The trajectory associated with an action instance traces a preferable execution path, but it is not sufficiently informative to detect potentially dangerous situations. For example, even though the robot is accurately following the trajectory associated with a navigation action, the safeness of the rover could be endangered by a terrain that can be rougher than expected. Taking into account just the invariant conditions associated with the navigation may not prevent action failures; these conditions, in fact, represent the physical limits the rover should never violate, and when they are violated any reaction may arrive too late. To avoid this situation, the Supervisor must be able to anticipate anomalous conditions before they become so dangerous to trigger an abort. In our approach we associate each action type with a set temporal patterns that describe how the rover should, or should not, behave while it is performing a specific action. Differently from a trajectory, the temporal patterns are defined on sequences of events which abstract relevant changes in the rover state. In the paper we propose the adoption of the chronicles formalism [9] for encoding these temporal patterns. Intuitively, a chronicle is a set of events, linked together by time constraints modeling possible behaviors of a dynamic system over time. The occurrence of events may depend both on the activities carried on by the system itself and on the contextual conditions of the environment where the system is operating.

Execution Modalities. The last extension we introduce consists in associating each action type with a set of *execution modalities*. An execution modality does not interfere with the expected effects of the action; it just represents an alternative way for reaching the same effects. The basic idea is that, while the temporal patterns can be used to anticipate dangerous conditions, the execution modalities could be used to reduce the risk of falling in one of them. For example, a navigate action is associated with the set of execution modalities *mods(navigate)*={*nominal-speed, reduced-speed*}. It is easy to see that in both cases the rover reaches the expected position, but the two modalities affect the navigation in different ways.

4.2   Improved Control Architecture

Relying on the additional pieces of knowledge discussed above, we propose the improved control architecture depicted in Figure 5; three new modules have been added: the State Interpreter (SI), the Temporal Reasoner (TR), and the Active

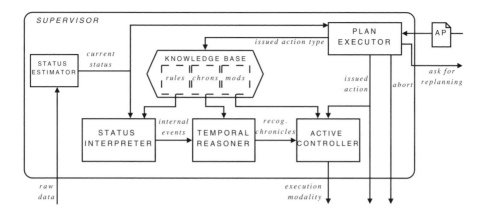

Fig. 5. The improved control architecture

Controller (AC); moreover, a Knowledge Base (KB) is also added to provide the modules with the knowledge associated with a specific action type.

The Supervisor receives in input a plan AP (i.e., an augmented plan where each action is provided with the trj_a parameter discussed above). The actual execution of AP is under the control of the Plan Executor (PE), as in the basic architecture (see Figure 4). The first improvement to the PE strategy is the check that the current $state_t$ is consistent with the constraints imposed by the trajectory trj_a. This improvement is implemented by changing the line 04 in Figure 4 as follows:

if (checkInvariants(inv_a, $state_t$) = $violation$) $\vee$ (actualDuration(a) > duration(a)) $\vee$
   (trajectoryDeviations(trj_a, $state_t$)=$relevant$)

In this way, the PE emits an abort also when the execution of a deviates significantly from the expected trajectory trj_a.[2]

A second and more relevant improvement is about the exploitation of the temporal patterns associated with action a. Since the evaluation of the current execution w.r.t. relevant temporal patterns is a complex activity which requires the coordination of different modules and the decision to changing execution modality (when required), we summarize this process in the macro function *ActiveMonitoring* (depicted in Figure 6). The PE, responsible for the coordination of the internal modules of the Supervisor, invokes *ActiveMonitoring* just before the assessment of an action outcome (see the algorithm in Figure 4, line 07). In the following, we first describe the idea at the basis of the *ActiveMonitoring* and then we sketch how each involved module actually operates.

As said above, *ActiveMonitoring* is aimed at emitting an execution modality relying on the set of chronicles that have been recognized at a given time instant. Since chronicles capture events, it is up to the *StateInterpreter* module to look at the history of the rover state for generating internal events which highlight

---
[2] Since in this paper we are more interested in the problem of correcting the execution by means of the selection of an appropriate execution modality, we do not provide further details about *trajectoryDeviations*.

ActiveMonitoring$(a, t, state_t)$
   $H \leftarrow$ append$(H, state_t)$
   $rules_a \leftarrow$ get-interpretative-rules$(KB, acttype(a))$
   $events_t \leftarrow$ **StateInterpreter**$(H, rules_a)$
   $RC \leftarrow \emptyset$
   $chronicles_a \leftarrow$ get-chronicles$(KB, acttype(a))$
   for each event $e_t \in events_t$
      $chr_a \leftarrow$ get-relevant-chronicle$(e_t, chronicles_a)$
      if **TemporalReasoner**(chr_a, e_t) **emits** *recognized*
         $RC \leftarrow RC \cup \{chr_a\}$
   if $RC \neq \emptyset$
      $mods_a \leftarrow$ get-execution-modalities$(KB, acttype(a), RC)$
      ActiveControl$(RC, mods_a)$ **emits** *mod* to FL

Fig. 6. The strategy for the active monitoring

relevant changes in the rover state. This process is performed in the first three lines of *ActiveMonitoring*: *StateInterpreter* module generates at each time t the set of internal events $events_t$. Each event $e_t \in events_t$ is subsequently sent to the *TemporalReasoner* (i.e., a *CRS*), which consumes the event and possibly recognizes a chronicle chr_a. All the chronicles recognized at time t are collected into the set RC, which becomes the input for the *ActiveController*. This last module has the responsibility for selecting, among a set of possible execution modalities $mods_a$, a specific modality to be sent to the FL.

The State Interpreter generates the internal events by exploiting a set of interpretative rules in $rules(atype)$ (where $atype=acttype(a)$). These interpretative rules have the form *Boolean condition* $\rightarrow$ *internal event*. The Boolean condition is build upon three basic types of atoms: state variables x_i, state variable derivates $\delta(x_i)$, and abstraction operators $qAbs(x_i, [t_l, t_u]) \rightarrow qVals$ which map the array of values assumed by x_i over the time interval $[t_l, t_u]$ into a set of qualitative values $qVals = \{qval_1, \ldots, qval_m\}$. For example, the following interpretative rules:
$(\delta(roll) > limits_{roll} \vee \delta(pitch) > limits_{pitch}) \rightarrow severe\text{-}hazard(roll, pitch)$
is used to generate a *severe-hazard* event whenever the derivate value of either roll or pitch exceeds predefined thresholds in the current rover state.
Another example is the rule:
   $attitude(roll, [t_{current} - \Delta, t_{current}]) = $nominal $\wedge$
   $attitude(pitch, [t_{current} - \Delta, t_{current}]) = $nominal $\rightarrow safe(roll, pitch)$
Where *attitude* is an operator which abstracts the last Δ values of either roll or pitch (the only two variables for which this operator is defined) over the set {nominal, border, non-nominal}.

Note that set of internal events can be partitioned according to the apparatus they refer to; for instance, the *attitude* and *severe-hazard* refer to the rover's mobility; *low-power* instead refers to the rover's battery. We assume that at each time t, $events_t$ can maintain at most one event referring to a specific device.

```
chronicle plain-terrain {
  occurs( (N, +oo), plain-conditions[pitch,
      roll], (t, t+W) )
  when recognized {
   emit event(plain-terrain[pitch,roll],t);
  }
}
```

```
chronicle hazardous-terrain {
  event(medium-hazard[pitch, roll], t1 )
  event(medium-hazard[pitch, roll], t2 )
  event(severe-hazard[pitch, roll], t3)
  t1<t2<t3 ; t2-t1<W1 ; t3-t2<W1
  when recognized { emit
     event(hazardous-terrain[pitch,roll],t)} }
```

Fig. 7. Two chronicle examples in the space exploration scenario

The Temporal Reasoner is essentially a Chronicle Recognition System (CRS) similar to the one proposed by Dusson in [9]. For simplicity, in our approach we assume that each event e_t can be consumed by exactly one active chronicle chr; the function *get-relevant-chronicle* in Figure 6 selects such a chronicle from $chronicles_a$ so that the TR receives in input just the event e_t which can be analyzed by the appropriate chronicle.

A chronicle example associated with the navigation action is given in Figure 7; it allows the Supervisor to indentify a potentially hazardous terrain. This chronicle is recognized when at least N *severe-hazard* events (regarding the parameters pitch and roll) have been detected within an interval of W time instant. The basic idea is that the safeness of the rover may be endangered when it moves at a high speed along a too rough terrain; this kind of threat can be captured by detecting hazardous variation of the roll and pitch parameters in a short time window. Indeed, the event *hazard*, resulting from an interpretation process over the rover's state variables, denotes that, although the rover's state has not violated the physical constraints (and hence it is still nominal), it may become anomalous in the near future.

The Active Controller accomplishes two important activities. First, it selects an execution modality to be issued towards the FL. In principle, such a selection should correct the current robot's behavior smoothly; that is, on one side, the AC's strategy should not be too reactive in order to avoid abrupt changes in the robot's behavior which may be as dangerous as the threat to face; and on the other side, the AC should be able to restore the nominal execution modalities when it is reasonable to presume that no menace is expected in the near future. In our current and preliminary solution, however, the AC is still purely reactive matching a recognized chronicle with a specific execution modality. Second, the AC updates some parameters of the current action according the execution modalities it emits. For instance, when a navigation action is slowed down, it will take more time to be completed, this extra time must be taken into account by the PE during its job. Due to lack of space, we cannot provide further details on this point.

Running example Let us consider the action navigate(A,B), and let us assume that the actual terrain is rougher than expected and causes repeated vibrations while the rover is moving from A to B . The basic architecture would handle this situation by aborting the navigate action, this would have a dramatic effect on the mission plan as the following actions could not be performed. On the contrary, the improved architecture is able to anticipate the threat and to intervene by slowing down the rover, this change of modality reduce the abrupt changes in pitch and roll, so that the action can be completed with success. It is worth noting that the execution modality is changed again (returning to nominal) when the pitch and roll parameters are nominal for a while (see chronicle `plain-terrain` which takes care of this temporal pattern).

5   Experimental Results

The experimental scenario. The approach described in the paper has undergone to a first validation by using as test bed the space exploration scenario previously introduced. The planetary environment has been represented as a Digital Elevation Model (DEM); we assumed that an initial DEM D_{init}, presumably computed from satellites images, is available, and we used it for synthesizing a set of rover's missions. In particular, by taking into account the terrain's characteristics, we have subdivided the rover's missions into two classes: *easy* and *difficult*. Note that the planning phase verifies the feasibility of each navigate action by invoking a path planner that, relying on D_{init}, assesses the validity of the invariant conditions associated with this action type (see Figure 3) and provides also a trajectory in terms of way points.

Obviously, D_{init} is just an approximation of the real terrain, therefore the actual execution of a mission plan may be affected by unexpected environmental conditions. For simulating the discrepancies between D_{init} and the real terrain, we have altered the original DEM by adding a random noise on the altitude of each cell. In our experiments, we have considered 6 noise degrees: from 10 cm to 15 cm, and for each of them we have generated 320 cases: 160 for the *easy* class and 160 for the *difficult* one.

Altogether, in our experiments we have considered up to 1920 navigate actions differing with one another for their starting and ending points, and their length.

To prove the effectiveness of our control architecture, we have simulated the execution of both *easy* and *difficult* cases in each noisy DEM comparing the responses of the two architectures, the basic and the improved, presented above.

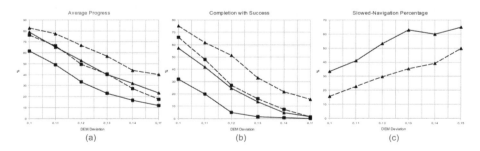

Fig. 8. Experimental results

A simplified simulator of the FL has been implemented in order to generate with a frequency of 1Hz the set of raw data the Supervisor (either basic or improved) has to interpret. For measuring the robustness of the plan execution and for providing some insights in the ability of the Supervsior in tolerating variations in the DEM, we are reporting data about three main parameters concerning the execution of the *navigate* actions:

1) the percentage of navigate actions that were completed successfully.
2) the percentage of progress actually done by the rover with respect to the whole trajectory, computed taking into account both the navigations that were actually completed and the aborted ones.
3) The percentage of steps the navigation has been performed in the slowdown modality w.r.t. the whole trajectory. Of course this datum is relevant just for the improved architecture.

Figure 8 summarizes the results of the tests. The graphs show the average values for the class of difficult cases (solid line), and for the class of the easy ones (dashed line). Each bullet corresponds to the average value of 160 navigations; squares denote the responses of the basic architecture, triangles denote the responses the improved architecture. It is easy to see that the improved architecture always provides better results than the basic one as concerns both the percentage of success and the progress. As expected, in the *difficult* cases, the gains are significant even for small DEM deviations, whereas in the *easy* ones, the gain becomes relevant for larger deviations. The results also show that the mechanism of active control is quite powerful but cannot avoid failures when the noise degree grows too much. A final remark concerns the cost of the intelligent monitoring: while the computational cost is negligible, there is an impact on the actual execution that we estimate as the percentage of steps performed in reduced-speed modality, showed in Figure 8.c. It is easy to see that this percentage is proportional to the noise degree and to hardness of the navigation.

6   Discussion and Conclusions

This paper addresses the problem of robust plan execution, when the environment may (slightly) differ from the one known (or assumed) during the planning phase and unexpected contingencies may arise. Previous works in literature have faced this problem by endowing the plan executor (e.g., a mobile robot) with some form of autonomous behavior. For instance, the control architectures discussed in [7,8,10] support the robot's autonomy by means of three layers of control: the highest one is devoted to the decisional aspects and it is typically based on one (or even more) (re)planning module(s).

Recent works on planning have faced the problem of recovering from an action failure by synthesizing a repairing plan on the fly (e.g., see [3,2,1]). These approaches, however, have been designed to intervene only after a failure has occurred (and therefore when the plan execution has been interrupted). In this paper we propose a control architecture aimed at reducing the necessity of invoking a replanner by preventing, when possible, the occurrence of an action failure. In particular, we have shown how the formalism of chronicles [9] can be used to model patterns of the robot's behavior over a temporal window, and how these patterns are subsequently exploited for anticipating threats. The chronicles formalism represents a viable and efficient solution to the problem of interpreting online the raw data coming from the environment, and hence reasoning about the environment in more abstract terms.

In order to keep potential threats under control, we propose to correct the current robot's behavior through the selection of execution modalities, whose effect is to change the way in which the current action is actually carried on while an action is still in execution. The proposed methodology is therefore a way to enhance the robot's autonomy as it can flexibly switch among modalities according to its contextual conditions. This idea is not completely new, also in [8] the authors suggest a methodology for adjusting the way in which an action is carried on depending on the rover's context. For instance, their navigation has three modalities related to the rover's speed: low, medium, and high. However, their context is just a snapshot (i.e., a set of variables) of the current conditions; conversely, we propose to maintain a "temporal" context by means of chronicles. This allows us to predict how the context will evolve, and hence to anticipate a change of modality. With regard to this point, a central role is played by the Active Controller; in its current preliminary implementation, the AC selects just one execution modality at each time instant. As future work we intend to extend the functionality of this module by allowing the selection of multiple modalities; to reach this result, however, the AC needs to know the dependencies existing between different modalities in order to estimate how their effects could interfere with each other. We are currently investigating the adoption of (probabilistic) causal networks as a possible way to face this challenge.

References

1. Bouguerra, A., Karlsson, L., Saffiotti, A.: Monitoring the execution of robot plans using semantic knowledge. Robotics and Autonomous Systems 56, 942–954 (2008)
2. Bozzano, M., Cimatti, A., Roveri, M., Tchaltsev, A.: A comprehensive approach to on-board autonomy verification and validation. In: ICAPS 2009 Workshop on Verification and Validation of Planning and Scheduling Systems (2009)
3. Micalizio, R.: A distributed control loop for autonomous recovery in a multi-agent plan. In: Proc. IJCAI 2009, pp. 1760–1765 (2009)
4. Micalizio, R., Torasso, P.: Monitoring the execution of a multi-agent plan: Dealing with partial observability. In: Proc. ECAI 2008, pp. 408–412 (2008)
5. Musso, I., Micalizio, R., Scala, E., et al.: Communication scheduling and plans revision for planetary rovers. In: Proc. of i-SAIRAS 2010 (2010)
6. Fox, M., Long, D.: Pddl2.1: An extension to pddl for expressing temporal planning domains. Journal of Artificial Intelligence Research 20, 61–124 (2003)
7. Alami, R., Chatila, R., Fleury, S., Ghallab, M., Ingrand, F.: An architecture for autonomy. International Journal of Robotics Research 17(4), 315–337 (1998)

8. Calisi, D., Iocchi, L., Nardi, D., Scalzo, C., Ziparo, V.A.: Context-based design of robotic systems. Robotics and Autonomous Systems (RAS) - Special Issue on Semantic Knowledge in Robotics 56(11), 992–1003 (2008)
9. Dousson, C., Le Maigat, P.: Chronicle recognition improvement using temporal focusing and hierarchization. In: IJCAI 2007, pp. 324–329 (2007)
10. Nesnas, I.A.: Claraty: A collaborative software for advancing robotic technologies. In: Proc. of NASA Science and Technology Conference (2007)

A Tableau Calculus for a Nonmonotonic Extension of the Description Logic *DL-Lite*_{core}

Laura Giordano[1], Valentina Gliozzi[2], Nicola Olivetti[3], and Gian Luca Pozzato[2]

[1] Dip. di Informatica, U. Piemonte O., Alessandria, Italy
laura@mfn.unipmn.it
[2] Dip. Informatica, Univ. di Torino, Italy
{gliozzi,pozzato}@di.unito.it
[3] LSIS-UMR CNRS 6168, Marseille, France
nicola.olivetti@univ-cezanne.fr

Abstract. In this paper we introduce a tableau calculus for a nonmonotonic extension of the low complexity Description Logic *DL-Lite*_{core} of the *DL-Lite* family. The extension, called $DL\text{-}Lite_c \mathbf{T}_{min}$, can be used to reason about typicality and defeasible properties. The calculus performs a two-phase computation to check whether a query is minimally entailed from the initial knowledge base. It is sound, complete and terminating. Furthermore, it is a decision procedure for $DL\text{-}Lite_c \mathbf{T}_{min}$ knowledge bases, whose complexity matches the known results for the logic, namely that entailment is in Π_2^p.

1 Introduction

The interest for efficient reasoning in Description Logics (DLs) has greatly increased in the last years. This is motivated by the fact that several applications are supposed to handle ontologies comprising thousands of elements, as it happens for instance in the Semantic Web, where ontologies are used to formalize concepts of very large Web repositories. For these applications, it is crucial to dispose of efficient reasoning algorithms. The current reasoning systems (capable of handling sometimes very expressive DLs) are inadequate, as they offer good performances in practice only for relatively small ontologies, but they cannot handle very large data sets. This has led to the study of *low complexity* DLs, such as the logics of the $\mathcal{EL}$ family and the ones of the *DL-Lite* family. The logics of the $\mathcal{EL}$ family [1] are relevant in the bio-medical domain: for instance, medical terminologies, such as the GALEN Medical Knowledge Base, the Systemized Nomenclature of Medicine (SNOMED), and the Gene Ontology used in bioinformatics, can be formalized in small extensions of $\mathcal{EL}$. The logics of the *DL-Lite* family [7] are specifically tailored for effective query answering over DL knowledge bases containing a large amount of data.

Since DLs are used to represent classes and their properties, a nonmonotonic mechanism is wished to express defeasible inheritance of prototypical properties. Nonmonotonic extensions of DLs have been actively investigated since the early 90s, [15, 5, 2, 3, 9, 13, 10, 8]. A simple but powerful nonmonotonic extension of DL is proposed in [13, 12, 10]: in this approach "typical" or "normal" properties can be directly specified by means of a "typicality" operator $\mathbf{T}$ enriching the underlying DL.

R. Pirrone and F. Sorbello (Eds.): AI*IA 2011, LNAI 6934, pp. 164–176, 2011.

In *DL-Lite*$_c$**T**$_{min}$ [12], one can consistently express defeasible inclusions and exceptions such as: typically elephants live in the Savannah, but elephants with a teacher normally do not live in the Savannah. In *DL-Lite*$_c$**T**$_{min}$ the previous inclusions can be formalized as follows: **T**(*Elephant*) $\sqsubseteq$ *LiveInTheSavannah*, *TrainedElephant* $\sqsubseteq$ *Elephant*, *TrainedElephant* $\sqsubseteq$ $\exists$*HasTeacher*.$\top$, **T**(*TrainedElephant*) $\sqsubseteq$ $\neg$*LiveInTheSavannah*. The operator **T** is nonmonotonic, in the sense that from *TrainedElephant* $\sqsubseteq$ *Elephant* it does not follow that **T**(*TrainedElephant*) $\sqsubseteq$ **T**(*Elephant*), and indeed **T**(*TrainedElephant*) and **T**(*Elephant*) can have different properties. Notice that without the **T** operator we would not be able to say that *Elephant* $\sqsubseteq$ *LiveInTheSavannah* and *TrainedElephant* $\sqsubseteq$ $\neg$*LiveInTheSavannah* without saying that *TrainedElephant* is empty.

The typicality operator **T** is characterised by the core properties of nonmonotonic reasoning axiomatized by *preferential logic* [14]. **T** allows to consistently express inclusions as the ones above, but it is not sufficient to perform nonmonotonic inferences. This is why, *DL-Lite*$_c$**T**$_{min}$ comprises a nonmonotonic mechanism based on a minimal model semantics. In the absence of information to the contrary, this mechanism allows to infer that a given individual is a typical instance of the most specific concept it belongs to. For instance, from the KB above, if we only know that *dumbo* is an elephant, then we would conclude that it is a typical elephant and lives in the Savannah, whereas if we knew that it is a trained elephant, then we would conclude that he is a typical trained elephant, and it does not live in the Savannah. Notice that we would not be able to do these inferences without the nonmonotonic mechanism that has been added.

While it has been shown that adding the typicality operator with its minimal-model semantics to other standard DLs, such as $\mathcal{ALC}$ and $\mathcal{EL}^{\perp}$, leads to a very high complexity (in particular, query entailment in $\mathcal{EL}^{\perp}$**T**$_{min}$ is ExpTime hard [12]), it has been recently shown that query entailment in *DL-Lite*$_c$**T**$_{min}$ is in Π_2^p [12]. This result is analogous to the one for *circumscribed DL-Lite*$_{core}$ KBs [3].

Proof methods for nonomonotonic extensions of $\mathcal{ALC}$ and $\mathcal{EL}^{\perp}$, called $\mathcal{ALC}$ + **T**$_{min}$ and $\mathcal{EL}^{\perp}$**T**$_{min}$, have been introduced in [10] and [11], respectively, whereas no proof methods for the logic *DL-Lite*$_c$**T**$_{min}$ have been proposed. In order to fill this gap, in this paper we present a tableau calculus for deciding minimal entailment in *DL-Lite*$_c$**T**$_{min}$. Similarly to $\mathcal{ALC}$ + **T**$_{min}$ and $\mathcal{EL}^{\perp}$**T**$_{min}$, we present a two-phase calculus: in the first phase, candidate models (complete open branches) falsifying the given query are generated, in the second phase the minimality of candidate models is checked by means of an auxiliary tableau construction. The latter tries to build a model which is "more preferred" than the candidate one: if it fails (being closed) the candidate model is minimal, otherwise it is not. As we will discuss in Section 3, there are several differences between the calculus for *DL-Lite*$_c$**T**$_{min}$ and the ones for $\mathcal{ALC}$ + **T**$_{min}$ and $\mathcal{EL}^{\perp}$**T**$_{min}$. As a result, the calculus that we propose here is very simple and does not require any blocking machinery in order to achieve termination. Furthermore, our calculus provides a constructive proof of the upper bound for minimal entailment in *DL-Lite*$_c$**T**$_{min}$.

2   The Typicality Operator T and the Logic *DL-Lite*$_c$T$_{min}$

The language of *DL-Lite*$_c$**T**$_{min}$ is obtained by adding to *DL-Lite*$_{core}$ the typicality operator **T**. The intuitive idea is that **T**(C) selects the *typical* instances of a concept C.

In *DL-Lite$_c$*$\mathbf{T}_{min}$ we can therefore distinguish between the properties that hold for all instances of concept C ($C \sqsubseteq D$), and those that only hold for the normal or typical instances of C ($\mathbf{T}(C) \sqsubseteq D$). Formally, the *DL-Lite$_c$*$\mathbf{T}_{min}$ language is defined as follows.

Definition 1. *We consider an alphabet of concept names $\mathcal{C}$, of role names $\mathcal{R}$, and of individuals $\mathcal{O}$. Given $A \in \mathcal{C}$ and $r \in \mathcal{R}$, we define*

$$C_L := A \mid \exists R.\top \mid \mathbf{T}(A) \qquad C_R := A \mid \neg A \mid \exists R.\top \mid \neg \exists R.\top \qquad R := r \mid r^-$$

A DL-Lite$_c$$\mathbf{T}_{min}$ KB is a pair (TBox, ABox). TBox contains a finite set of concept inclusions of the form $C_L \sqsubseteq C_R$. ABox contains assertions of the form $C(a)$ and $r(a, b)$, where C is a C_L or C_R concept, $r \in \mathcal{R}$, and $a, b \in \mathcal{O}$.

The semantics of *DL-Lite$_c$*$\mathbf{T}_{min}$ is defined by enriching ordinary models of *DL-Lite$_{core}$* by a *preference relation* $<$ on the domain, whose intuitive meaning is to compare the "typicality" of individuals: $x < y$, means that x is more typical than y. Typical members of a concept C, i.e., members of $\mathbf{T}(C)$, are the members x of C that are minimal with respect to this preference relation.

Definition 2 (Semantics of T). *A model $\mathcal{M}$ is any structure $\langle \Delta, <, I \rangle$ where Δ is the domain; $<$ is an irreflexive and transitive relation over Δ that satisfies the following Smoothness Condition: for all $S \subseteq \Delta$, for all $x \in S$, either $x \in Min_<(S)$ or $\exists y \in Min_<(S)$ such that $y < x$, where $Min_<(S) = \{u : u \in S \text{ and } \nexists z \in S \text{ s.t. } z < u\}$. Furthermore, $<$ is multilinear: if $u < z$ and $v < z$, then either $u = v$ or $u < v$ or $v < u$. I is the extension function that maps each concept C to $C^I \subseteq \Delta$, and each role r to $r^I \subseteq \Delta^I \times \Delta^I$. For concepts of DL-Lite$_{core}$, C^I is defined in the usual way. For the $\mathbf{T}$ operator: $(\mathbf{T}(C))^I = Min_<(C^I)$.*

Definition 3 (Model satisfying a Knowledge Base). *Given a model $\mathcal{M}$, I can be extended so that it assigns to each individual a of $\mathcal{O}$ a distinct element a^I of the domain Δ. $\mathcal{M}$ satisfies a KB (TBox,ABox), if it satisfies both its TBox and its ABox, where:*

- *$\mathcal{M}$ satisfies TBox if $\mathcal{M}$ satisifes all its inclusions. An inclusion $C \sqsubseteq D$ is satisifed in $\mathcal{M}$ if $C^I \subseteq D^I$.*
- *$\mathcal{M}$ satisfies ABox if $\mathcal{M}$ satisifes all its formulas $C(a)$ and $r(a, b)$. $\mathcal{M}$ satisifes $C(a)$ if $a^I \in C^I$. $\mathcal{M}$ satisfies $r(a, b)$ if $(a^I, b^I) \in r^I$.*

We assume the unique name assumption.

The operator $\mathbf{T}$ [13] is characterized by a set of postulates that are essentially a reformulation of KLM [14] axioms of *preferential logic* $\mathbf{P}$. $\mathbf{T}$ has therefore all the "core" properties of nonmonotonic reasoning as it is axiomatised by $\mathbf{P}$.

The semantics of the typicality operator can be specified by modal logic. The interpretation of $\mathbf{T}$ can be split into two parts: for any x of the domain Δ, $x \in (\mathbf{T}(C))^I$ just in case (i) $x \in C^I$, and (ii) there is no $y \in C^I$ such that $y < x$. Condition (ii) can be represented by means of an additional modality $\square$, whose semantics is given by the preference relation $<$ interpreted as an accessibility relation. Observe that by the Smoothness Condition, $\square$ has the properties of Gödel-Löb modal logic of provability $\mathbf{G}$. The interpretation of $\square$ in $\mathcal{M}$ is as follows:

$$(\square C)^I = \{x \in \Delta \mid \text{for every } y \in \Delta, \text{ if } y < x \text{ then } y \in C^I\}$$

We immediately get that $x \in (\mathbf{T}(C))^I$ iff $x \in (C \sqcap \square \neg C)^I$. From now on, we consider $\mathbf{T}(C)$ as an abbreviation for $C \sqcap \square \neg C$.

In order to perform nonmonotonic inferences, the semantics of $DL\text{-}Lite_c\mathbf{T}_{min}$ is strenghtened by restricting entailment to a class of minimal (or preferred) models. Intuitively, the idea is to restrict our consideration to models that *minimize the non typical instances of a concept.*

Given a KB, we consider a finite set $\mathcal{L}_\mathbf{T}$ of concepts: these are the concepts whose non typical instances we want to minimize. We assume that the set $\mathcal{L}_\mathbf{T}$ contains at least all concepts C such that $\mathbf{T}(C)$ occurs in the KB or in the query F, where a *query* F is either an assertion $C(a)$, where C is a C_L or C_R, or an inclusion relation $C \sqsubseteq D$, where C is a C_L and D is a C_R. As we have just said, $x \in C^I$ is typical if $x \in (\square \neg C)^I$. Minimizing the non typical instances of C therefore means to minimize the objects not satisfying $\square \neg C$ for $C \in \mathcal{L}_\mathbf{T}$. Hence, for a given model $\mathcal{M} = \langle \Delta, <, I \rangle$, we define:

$$\mathcal{M}_{\mathcal{L}_\mathbf{T}}^{\square^-} = \{(x, \neg \square \neg C) \mid x \notin (\square \neg C)^I, \text{ with } x \in \Delta, C \in \mathcal{L}_\mathbf{T}\}.$$

Definition 4 (Preferred and minimal models). *Given a model* $\mathcal{M} = \langle \Delta <, I \rangle$ *of a knowledge base KB, and a model* $\mathcal{M}' = \langle \Delta', <', I' \rangle$ *of KB, we say that* $\mathcal{M}$ *is preferred to* $\mathcal{M}'$ *with respect to* $\mathcal{L}_\mathbf{T}$, *and we write* $\mathcal{M} <_{\mathcal{L}_\mathbf{T}} \mathcal{M}'$, *if (i)* $\Delta = \Delta'$, *(ii)* $a^I = a^{I'}$ *for all* $a \in \mathcal{O}$, *and (iii)* $\mathcal{M}_{\mathcal{L}_\mathbf{T}}^{\square^-} \subset \mathcal{M}_{\mathcal{L}_\mathbf{T}}'^{\square^-}$. *A model* $\mathcal{M}$ *is a* minimal model *for KB (with respect to* $\mathcal{L}_\mathbf{T}$) *if it is a model of KB and there is no other model* $\mathcal{M}'$ *of KB such that* $\mathcal{M}' <_{\mathcal{L}_\mathbf{T}} \mathcal{M}$.

Definition 5 (Minimal Entailment in $DL\text{-}Lite_c\mathbf{T}_{min}$**).** *A query* F *is minimally entailed in* DL-Lite$_c\mathbf{T}_{min}$ *by KB with respect to* $\mathcal{L}_\mathbf{T}$ *if all models of KB, that are minimal with respect to* $\mathcal{L}_\mathbf{T}$, *satisfy* F. *We write* $KB \models_{\text{DL-Lite}_c\mathbf{T}_{min}} F$.

In [12], a small model construction allowed us to prove the following complexity result:

Theorem 1 (Complexity of entailment in $DL\text{-}Lite_c\mathbf{T}_{min}$**, Theorem 4.6 in [12]).** *The problem of deciding whether* $KB \models_{\text{DL-Lite}_c\mathbf{T}_{min}} F$ *is in* Π_2^p.

3   A Tableau Calculus for $DL\text{-}Lite_c\mathbf{T}_{min}$

In this section we present a tableau calculus $\mathcal{TAB}_{min}^{Lite_c\mathbf{T}}$ for deciding whether a query F is minimally entailed from a KB in the logic $DL\text{-}Lite_c\mathbf{T}_{min}$. As mentioned in the Introduction, the calculus is inspired to the calculi for $\mathcal{ALC} + \mathbf{T}_{min}$ and $\mathcal{EL}^\perp\mathbf{T}_{min}$ introduced in [10] and [11] but it contains a few significant differences in order to obtain a calculus whose complexity matches the result of Theorem 1 above.

The calculus $\mathcal{TAB}_{min}^{Lite_c\mathbf{T}}$ performs a two-phase computation. In the first phase, a tableau calculus, called $\mathcal{TAB}_{PH1}^{Lite_c\mathbf{T}}$, simply verifies whether KB $\cup \{\neg F\}$ is satisfiable in a model, building candidate models. In the second phase another tableau calculus, called $\mathcal{TAB}_{PH2}^{Lite_c\mathbf{T}}$, checks whether the candidate models found in the first phase are *minimal* models of KB, i.e. for each open saturated branch of the first phase, $\mathcal{TAB}_{PH2}^{Lite_c\mathbf{T}}$ tries to build a model of KB which is preferred to the candidate model w.r.t. Definition 4. The whole procedure $\mathcal{TAB}_{min}^{Lite_c\mathbf{T}}$ is formally defined in Definition 8.

Before examining the calculi in detail, we provide some additional formal definitions. First, the negation of a query $\neg F$ is defined as follows: (i) if $F \equiv C(a)$, then $\neg F \equiv (\neg C)(a)$; (ii) if $F \equiv C \sqsubseteq D$, then $\neg F \equiv C(x), (\neg D)(x)$, where x does not occur in KB.

$\mathcal{TAB}_{min}^{Lite_c\mathbf{T}}$ makes use of labels, which are denoted with $x, y, z, \ldots$. Labels represent either a variable or an individual of the ABox. These labels occur in *constraints*. A *constraint* (or *labelled* formula) is a syntactic entity of the form either $x \xrightarrow{r} y$ or $x : C$, where x, y are labels, r is a role and C is either a concept or the negation of a concept of $DL\text{-}Lite_c\mathbf{T}_{min}$ or it has the form $\Box\neg D$ or $\neg\Box\neg D$, where D is a concept. Finally, given a set of constraints S and a role $r \in \mathcal{R}$, we define $r(S) = \{x \xrightarrow{r} y \mid x \xrightarrow{r} y \in S\}$.

3.1 First Phase: The Tableaux Calculus $\mathcal{TAB}_{PH1}^{Lite_c\mathbf{T}}$

Let us first define the basic notions of a tableau system in $\mathcal{TAB}_{PH1}^{Lite_c\mathbf{T}}$. A tableau of $\mathcal{TAB}_{PH1}^{Lite_c\mathbf{T}}$ is a tree whose nodes are pairs $\langle S \mid U \rangle$. S is a set of constraints, whereas U contains formulas of the form $C \sqsubseteq D^L$, representing inclusions $C \sqsubseteq D$ of the TBox. L is a list of labels, used in order to ensure the termination of the tableau calculus. A branch is a sequence of nodes $\langle S_1 \mid U_1 \rangle, \langle S_2 \mid U_2 \rangle, \ldots, \langle S_n \mid U_n \rangle \ldots$, where each node $\langle S_i \mid U_i \rangle$ is obtained from its immediate predecessor $\langle S_{i-1} \mid U_{i-1} \rangle$ by applying a rule of $\mathcal{TAB}_{PH1}^{Lite_c\mathbf{T}}$, having $\langle S_{i-1} \mid U_{i-1} \rangle$ as the premise and $\langle S_i \mid U_i \rangle$ as one of its conclusions. A branch is closed if one of its nodes is an instance of a (Clash) axiom, otherwise it is open. A tableau is closed if all its branches are closed.

In order to check the satisfiability of a KB, we build its *corresponding constraint system* $\langle S \mid U \rangle$, and we check its satisfiability.

Definition 6 (Corresponding constraint system). *Given a knowledge base KB=(TBox,ABox), we define its* corresponding constraint system $\langle S \mid U \rangle$ *as follows:*

- $S = \{a : C \mid C(a) \in ABox\} \cup \{a \xrightarrow{r} b \mid r(a, b) \in ABox\}$
- $U = \{C \sqsubseteq D^\emptyset \mid C \sqsubseteq D \in TBox\}$

Definition 7 (Model satisfying a constraint system). *Let* $\mathcal{M} = \langle \Delta, I, < \rangle$ *be a model as defined in Definition 2. We define a function* α *which assigns to each variable of* $\mathcal{V}$ *an element of* Δ, *and assigns every individual* $a \in \mathcal{O}$ *to* $a^I \in \Delta$. $\mathcal{M}$ *satisfies a constraint F under* α, *written* $\mathcal{M} \models_\alpha F$, *as follows: (i)* $\mathcal{M} \models_\alpha x : C$ *iff* $\alpha(x) \in C^I$; *(ii)* $\mathcal{M} \models_\alpha x \xrightarrow{r} y$ *iff* $(\alpha(x), \alpha(y)) \in r^I$. *A constraint system* $\langle S \mid U \rangle$ *is satisfiable if there is a model* $\mathcal{M}$ *and a function* α *such that* $\mathcal{M}$ *satisfies every constraint in S under* α *and that, for all* $C \sqsubseteq D^L \in U$, *we have that* $C^I \subseteq D^I$.

Proposition 1. *Given a KB=(TBox,ABox), it is satisfiable if and only if its corresponding constraint system* $\langle S \mid U \rangle$ *is satisfiable.*

To verify the satisfiability of KB $\cup \{\neg F\}$, we use $\mathcal{TAB}_{PH1}^{Lite_c\mathbf{T}}$ to check the satisfiability of the constraint system $\langle S \mid U \rangle$ obtained by adding the constraint corresponding to $\neg F$ to S', where $\langle S' \mid U \rangle$ is the corresponding constraint system of KB. To this purpose, the rules of the calculus $\mathcal{TAB}_{PH1}^{Lite_c\mathbf{T}}$ are applied until either a contradiction is generated (Clash) or no other rule is applicable. Given a node $\langle S \mid U \rangle$, for each inclusion $C \sqsubseteq D^L \in U$ and for each label x occurring in the tableau, we add to S the constraint $x : \neg C \sqcup D$: we refer to this mechanism as *unfolding*. As mentioned, each inclusion $C \sqsubseteq D$ is equipped with a list L of labels in which it has been unfolded in the current branch. This is needed to avoid multiple unfolding of the same inclusion by using the same label, generating infinite branches.

The calculus $\mathcal{TAB}_{PH1}^{Lite_c\mathbf{T}}$ is significantly different in four respects from the calculi for $\mathcal{ALC} + \mathbf{T}_{min}$ and $\mathcal{EL}^\perp \mathbf{T}_{min}$. We try to explain such differences in detail.

1. The rule $(\exists^+)$ is split in the following two rules:

$$\frac{\langle S, x : \exists r.\top \mid U \rangle}{\langle S, x \xrightarrow{r} y \mid U \rangle \quad \langle S, x \xrightarrow{r} y_1 \mid U \rangle \cdots \langle S, x \xrightarrow{r} y_m \mid U \rangle} (\exists^+)_1^r$$
$$\text{if } r(S) = \emptyset$$
$$y \text{ new}$$
$$\text{if } y_1, \ldots, y_m \text{ are all the labels occurring in } S$$

$$\frac{\langle S, x : \exists r.\top \mid U \rangle}{\langle S, x \xrightarrow{r} y_1 \mid U \rangle \quad \cdots \quad \langle S, x \xrightarrow{r} y_m \mid U \rangle} (\exists^+)_2^r$$
$$\text{if } r(S) \neq \emptyset$$
$$\text{if } y_1, \ldots, y_m \text{ are all the labels occurring in } S$$

The split of the $(\exists^+)$ in the two rules above reflects the main idea of the construction of a small model at the base of Theorem 1. Such small model theorem essentially shows that $DL\text{-}Lite_c\mathbf{T}_{min}$ KBs can have small models in which all existentials $\exists r.\top$ occurring in KB are satisfied in the model by a single witness y. In the calculus we use the same idea: when the rule $(\exists^+)_1^r$ is applied to a formula $x : \exists r.\top$, it introduces a new label y and the constraint $x \xrightarrow{r} y$ only when there is no other previous constraint $u \xrightarrow{r} v$ in S. Otherwise, rule $(\exists^+)_2^r$ is applied and it introduces $x \xrightarrow{r} y$, with y already occurring in the branch. As a consequence, $(\exists^+)_2^r$ does not introduce any new label in the branch whereas $(\exists^+)_1^r$ only introduces a new label y for each role r occurring in the initial KB in some $\exists r.\top$ or $\exists r^-.\top$, and no blocking machinery is needed to ensure termination.

2. In order to keep into account inverse roles, two further rules for existential formulas are introduced:

$$\frac{\langle S, x : \exists r^-.\top \mid U \rangle}{\langle S, y \xrightarrow{r} x \mid U \rangle \quad \langle S, y_1 \xrightarrow{r} x \mid U \rangle \cdots \langle S, y_m \xrightarrow{r} x \mid U \rangle} (\exists^+)_1^{r^-}$$
$$\text{if } r(S) = \emptyset$$
$$y \text{ new}$$
$$\text{if } y_1, \ldots, y_m \text{ are all the labels occurring in } S$$

$$\frac{\langle S, x : \exists r^-.\top \mid U \rangle}{\langle S, y_1 \xrightarrow{r} x \mid U \rangle \quad \cdots \quad \langle S, y_m \xrightarrow{r} x \mid U \rangle} (\exists^+)_2^{r^-}$$
$$\text{if } r(S) \neq \emptyset$$
$$\text{if } y_1, \ldots, y_m \text{ are all the labels occurring in } S$$

These rules work similarly to $(\exists^+)_1^r$ and $(\exists^+)_2^r$ in order to build a branch representing a small model: when the rule $(\exists^+)_1^{r^-}$ is applied to a formula $x : \exists r^-.\top$, it introduces a new label y and the constraint $y \xrightarrow{r} x$ only when there is no other constraint $u \xrightarrow{r} v$ in S. Otherwise, since a constraint $y \xrightarrow{r} u$ has been already introduced in that branch, $y \xrightarrow{r} x$ is added to the conclusion of the rule.

3. The calculus $\mathcal{TAB}_{PH1}^{Lite_c\mathbf{T}}$ does not need the rule $(\exists^-)$ of $\mathcal{TAB}_{min}^{\mathcal{ALC}+\mathbf{T}}$. Indeed, the only negated existential formulas that can occur in a branch have the form (i) $x : \neg \exists r.\top$ or (ii) $x : \neg \exists r^-.\top$. (i) means that x has no relationships with other individuals via the role r, i.e. we need to detect a contradiction if both (i) and, for some y, $x \xrightarrow{r} y$ belong to the same branch, in order to mark the branch as closed. The clash condition $(\text{Clash})_r$ is added to the calculus $\mathcal{TAB}_{PH1}^{Lite_c\mathbf{T}}$ in order to detect such a situation. Analogously, (ii) means that there is no y such that y is related to x by means of r, then $(\text{Clash})_{r^-}$ is

introduced in order to close a branch containing both (ii) and, for some y, a constraint $y \xrightarrow{r} x$. The clash conditions $(\text{Clash})_r$ and $(\text{Clash})_{r^-}$ are as follows:

$$\langle S, x \xrightarrow{r} y, x : \neg \exists r. \top \mid U \rangle \ (\text{Clash})_r \qquad \langle S, y \xrightarrow{r} x, x : \neg \exists r^-. \top \mid U \rangle \ (\text{Clash})_{r^-}$$

4. In order to build multilinear models of Definition 2, the calculus adopts a strengthened version of the rule $(\square^-)$ used in $\mathcal{TAB}_{min}^{\mathcal{ALC}+\mathbf{T}}$ [10]. We write $\overline{S}$ as an abbreviation for $S, x : \neg\square\neg C_1, \ldots, x : \neg\square\neg C_n$. Moreover, we define $S_{x \to y}^M = \{y : \neg D, y : \square\neg D \mid x : \square\neg D \in S\}$ and, for $k = 1, 2, \ldots, n$, we define $\overline{S}_{x \to y}^{\square^{-k}} = \{y : \neg\square\neg C_j \sqcup C_j \mid x : \neg\square\neg C_j \in \overline{S} \wedge j \neq k\}$. The strengthened rule $(\square^-)$ is also adopted in the calculus for $\mathcal{EL}^\perp \mathbf{T}_{min}$ in [11] and is as follows:

$$\frac{\langle S, x : \neg\square\neg C_1, \ldots, \neg\square\neg C_n \mid U \rangle}{\langle S, y : C_k, y : \square\neg C_k, S_{x \to y}^M, \overline{S}_{x \to y}^{\square^{-k}} \mid U \rangle \ \langle S, y_1 : C_k, y_1 : \square\neg C_k, S_{x \to y_1}^M, \overline{S}_{x \to y_1}^{\square^{-k}} \mid U \rangle \cdots \langle S, y_m : C_k, y_m : \square\neg C_k, S_{x \to y_m}^M, \overline{S}_{x \to y_m}^{\square^{-k}} \mid U \rangle} \ (\square^-)$$

$$\begin{array}{c} y \text{ new} \\ \text{if } y_1, \ldots, y_m \text{ are all the labels occurring in } S, y_1 \neq x, \ldots, y_m \neq x \\ \forall k = 1, 2, \ldots, n \end{array}$$

Rule $(\square^-)$ contains: - n branches, one for each $x : \neg\square\neg C_k$ in $\overline{S}$; in each branch a *new* typical C_k individual y is introduced (i.e. $y : C_k$ and $y : \square\neg C_k$ are added), and for all other $x : \neg\square\neg C_j$, either $y : C_j$ holds or the formula $y : \neg\square\neg C_j$ is recorded; - other $n \times m$ branches, where m is the number of labels occurring in S, one for each label y_i and for each $x : \neg\square\neg C_k$ in $\overline{S}$; in these branches, a given y_i is chosen as a typical instance of C_k, that is to say $y_i : C_k$ and $y_i : \square\neg C_k$ are added, and for all other $x : \neg\square\neg C_j$, either $y_i : C_j$ holds or the formula $y_i : \neg\square\neg C_j$ is recorded. This rule is sound with respect to multilinear models. The advantage of this rule over the $(\square^-)$ rule in the calculus $\mathcal{TAB}_{min}^{\mathcal{ALC}+\mathbf{T}}$ for $\mathcal{ALC} + \mathbf{T}_{min}$ is that all the negated box formulas labelled by x are treated in one step, introducing at most one new label y in

$$\langle S, x : C, x : \neg C \mid U \rangle \ (\text{Clash}) \qquad \langle S, x \xrightarrow{r} y, x : \neg \exists r. \top \mid U \rangle \ (\text{Clash})_r \qquad \langle S, y \xrightarrow{r} x, x : \neg \exists r^-. \top \mid U \rangle \ (\text{Clash})_{r^-}$$

$$\frac{\langle S, x : \exists r. \top \mid U \rangle}{\langle S, x \xrightarrow{r} y \mid U \rangle \ \langle S, x \xrightarrow{r} y_1 \mid U \rangle \cdots \langle S, x \xrightarrow{r} y_m \mid U \rangle} (\exists^+)_1^r \qquad \frac{\langle S, x : \exists r. \top \mid U \rangle}{\langle S, x \xrightarrow{r} y_1 \mid U \rangle \cdots \langle S, x \xrightarrow{r} y_m \mid U \rangle} (\exists^+)_2^r \qquad \frac{\langle S, x : \mathbf{T}(C) \mid U \rangle}{\langle S, x : C, x : \square \neg C \mid U \rangle} (\mathbf{T}^+)$$

$$\begin{array}{c} \text{if } r(S) = \emptyset \\ y \text{ new} \\ \text{if } y_1, \ldots, y_m \text{ are all the labels occurring in } S \end{array} \qquad \begin{array}{c} \text{if } r(S) \neq \emptyset \\ \text{if } y_1, \ldots, y_m \text{ are all the labels occurring in } S \end{array}$$

$$\frac{\langle S, x : \neg\mathbf{T}(C) \mid U \rangle}{\langle S, x : \neg C \mid U \rangle \ \langle S, x : \neg\square\neg C \mid U \rangle} (\mathbf{T}^-) \qquad \frac{\langle S \mid U \rangle}{\langle S, x : \square\neg C \mid U \rangle \ \langle S, x : \neg\square\neg C \mid U \rangle} (cut) \qquad \frac{\langle S \mid U, C \sqsubseteq D^L \rangle}{\langle S, x : \neg C \sqcup D \mid U, C \sqsubseteq D^{L,x} \rangle} (\text{Unfold})$$

$$\begin{array}{c} \text{if } x : \neg\square\neg C \notin S \text{ and } x : \square\neg C \notin S \\ C \in \mathcal{L}_\mathbf{T} \\ x \text{ occurs in } S \end{array} \qquad \begin{array}{c} \text{if } x \text{ occurs in } S \text{ and } x \notin L \end{array}$$

$$\frac{\langle S, x : \exists r^-. \top \mid U \rangle}{\langle S, y \xrightarrow{r} x \mid U \rangle \ \langle S, y_1 \xrightarrow{r} x \mid U \rangle \cdots \langle S, y_m \xrightarrow{r} x \mid U \rangle} (\exists^+)_1^{r^-} \qquad \frac{\langle S, x : \exists r^-. \top \mid U \rangle}{\langle S, y_1 \xrightarrow{r} x \mid U \rangle \cdots \langle S, y_m \xrightarrow{r} x \mid U \rangle} (\exists^+)_2^{r^-} \qquad \frac{\langle S, x : C \sqcup D \mid U \rangle}{\langle S, x : C \mid U \rangle \ \langle S, x : D \mid U \rangle} (\sqcup^+)$$

$$\begin{array}{c} \text{if } r(S) = \emptyset \\ y \text{ new} \\ \text{if } y_1, \ldots, y_m \text{ are all the labels occurring in } S \end{array} \qquad \begin{array}{c} \text{if } r(S) \neq \emptyset \\ \text{if } y_1, \ldots, y_m \text{ are all the labels occurring in } S \end{array}$$

$$\frac{\langle S, x : \neg\square\neg C_1, \ldots, \neg\square\neg C_n \mid U \rangle}{\langle S, y : C_k, y : \square\neg C_k, S_{x \to y}^M, \overline{S}_{x \to y}^{\square^{-k}} \mid U \rangle \ \langle S, y_1 : C_k, y_1 : \square\neg C_k, S_{x \to y_1}^M, \overline{S}_{x \to y_1}^{\square^{-k}} \mid U \rangle \cdots \langle S, y_m : C_k, y_m : \square\neg C_k, S_{x \to y_m}^M, \overline{S}_{x \to y_m}^{\square^{-k}} \mid U \rangle} (\square^-)$$

$$\begin{array}{c} y \text{ new} \\ \text{if } y_1, \ldots, y_m \text{ are all the labels occurring in } S, y_1 \neq x, \ldots, y_m \neq x \\ \forall k = 1, 2, \ldots, n \end{array}$$

Fig. 1. The calculus $\mathcal{TAB}_{PH1}^{Lite_c \mathbf{T}}$

the conclusions. Notice that in order to keep $\overline{S}$ readable, we have used $\sqcup$. This is the reason why our calculi contain the rule $(\sqcup^+)$, even if this constructor does not belong to $DL\text{-}Lite_c\mathbf{T}_{min}$.

The rules of $\mathcal{TAB}_{PH1}^{Lite_c\mathbf{T}}$ are presented in Figure 1. Rules $(\exists^+)_1^r$, $(\exists^+)_1^{r^-}$ and $(\Box^-)$ are called *dynamic* since they can introduce a new variable in their conclusions. The other rules are called *static*. We do not need any extra rule for the positive occurrences of the $\Box$ operator, since these are taken into account by the computation of $S_{x\rightarrow y}^M$ of $(\Box^-)$. The (cut) rule ensures that, given any concept $C \in \mathcal{L}_\mathbf{T}$, an open saturated branch built by $\mathcal{TAB}_{PH1}^{Lite_c\mathbf{T}}$ contains either $x : \Box\neg C$ or $x : \neg\Box\neg C$ for each label x: this is needed in order to allow $\mathcal{TAB}_{PH2}^{Lite_c\mathbf{T}}$ to check the minimality of the model corresponding to the open branch.

The rules of $\mathcal{TAB}_{PH1}^{Lite_c\mathbf{T}}$ are applied with the following standard strategy that takes into account the order $\prec$ of insertion of the labels in the branch: as in [6], if y is introduced in the tableau, then $x \prec y$ for all labels x that are already in the tableau.

Standard strategy: 1. apply a rule to a label x only if no rule is applicable to a label y such that $y \prec x$; 2. apply dynamic rules only if no static rule is applicable.

The above strategy imposes that the static rule (cut) is applied to a label x before $(\Box^-)$ is applied to the same x. This has two benefits. First, it guarantees that $(\Box^-)$ can be applied only once to each x. As it will become clear from the proof of Theorem 5, this has an impact on the overall complexity of the calculus. Indeed, no other $x : \neg\Box\neg C$ can be introduced after the application of $(\Box^-)$ to x, and no further application of $(\Box^-)$ is needed to take into account the newly introduced negated box formula. This is a consequence of the fact that each $x : \neg\Box\neg C$ that will ever occur on the branch has been introduced by (cut) (indeed if $x : \neg\Box\neg C$ has not been introduced by (cut), then $x : \Box\neg C$ has been introduced, which prevents $x : \neg\Box\neg C$ to be introduced at a later stage). Second, since (cut) introduces all possible positive boxed $x : \Box\neg C$ that will ever appear on the branch, the strategy guarantees that for all these formulas $y : \neg C$ and $y : \Box\neg C$ hold for each y introduced by an application of $(\Box^-)$ to x. In this way, we do not need an extra rule $(\Box^+)$.

The calculus $\mathcal{TAB}_{PH1}^{Lite_c\mathbf{T}}$ is sound and complete.

Theorem 2 (Soundness of $\mathcal{TAB}_{PH1}^{Lite_c\mathbf{T}}$). *If KB $\not\models_{DL\text{-}Lite_c\mathbf{T}_{min}}$ F, then the tableau for the constraint system corresponding to KB $\cup$ $\{\neg F\}$ contains an open saturated branch, which is satisfiable (via an injective assignment from labels to domain elements) in a minimal model of KB.*

Theorem 3 (Completeness of $\mathcal{TAB}_{PH1}^{Lite_c\mathbf{T}}$). *Given a constraint system $\langle S \mid U \rangle$, if it is unsatisfiable, then it has a closed tableau in $\mathcal{TAB}_{PH1}^{Lite_c\mathbf{T}}$.*

Let us now analyze termination of $\mathcal{TAB}_{PH1}^{Lite_c\mathbf{T}}$.

Theorem 4 (Termination of $\mathcal{TAB}_{PH1}^{Lite_c\mathbf{T}}$). *Let $\langle S \mid U \rangle$ be the corresponding constraint system of a KB. Any tableau generated by $\mathcal{TAB}_{PH1}^{Lite_c\mathbf{T}}$ for $\langle S \mid U \rangle$ is finite.*

Proof. (Sketch) The following facts allow us to prove termination:
- Rules cannot be reapplied over the same formula without any control. Indeed, the only rule copying its principal formula in its conclusion is (Unfold), but this rule can be applied to $\langle S \mid U, C \sqsubseteq D^L \rangle$ by using the label x only if it has not yet been applied to x in the current branch (i.e., x does not belong to L).

– Only finitely-many labels can be introduced on a branch. Roughly speaking, the $(\exists^+)_1^r$ rule introduces at most one new label y for each role r belonging to the initial KB. The same holds for the rule $(\exists^+)_1^{r^-}$. Moreover, thanks to the properties of $\Box$, it can be shown that the interplay between rules $(\mathbf{T}^-)$ and $(\Box^-)$ does not generate branches containing infinitely-many labels. Intuitively, the application of $(\Box^-)$ to $x : \neg\Box\neg C, x : \neg\Box\neg C_1, \ldots, x : \neg\Box\neg C_k$ adds $y : \Box\neg C$ to the conclusion, so that $(\mathbf{T}^-)$ can no longer consistently introduce $y : \neg\Box\neg C$.

– The (cut) rule does not affect termination, since it is applied only to the finitely many formulas belonging to $\mathcal{L}_\mathbf{T}$.   ∎

Let us conclude this section by estimating the complexity of $\mathcal{TAB}_{PH1}^{Lite_c\mathbf{T}}$. Let n be the size of the initial KB, i.e. the length of the string representing KB, and let $\langle S \mid U \rangle$ its corresponding constraint system. We assume that the size of F and $\mathcal{L}_\mathbf{T}$ is $O(n)$.

Theorem 5 (Complexity of Phase 1). *Given a KB and a query F, the problem of checking whether KB $\cup \{\neg F\}$ is satisfiable is in NP.*

Proof. (Sketch) The calculus builds a tableau for $\langle S \mid U \rangle$ whose branches's size is $O(n)$. This immediately follows from the fact that dynamic rules generate at most $O(n)$ labels in a branch. This is obvious for rules $(\exists^+)_1^r$ and $(\exists^+)_1^{r^-}$. Concerning $(\Box^-)$, consider a branch generated by its application to a constraint system $\langle S, x : \neg\Box\neg C_1 \ldots, x : \neg\Box\neg C_n \mid U \rangle$. In the worst case, a new label y_1 is introduced. Suppose also that the branch under consideration is the one containing $y_1 : C_1$ and $y_1 : \Box\neg C_1$. The $(\Box^-)$ rule can then be applied to formulas $y_1 : \neg\Box\neg C_k$, introducing also a further new label y_2. However, by the presence of $y_1 : \Box\neg C_1$, the rule $(\Box^-)$ can no longer consistently introduce $y_2 : \neg\Box\neg C_1$, since $y_2 : \Box\neg C_1 \in S_{y_1 \to y_2}^M$. Therefore, once $(\Box^-)$ is applied to $\neg\Box\neg C_1 \ldots \neg\Box\neg C_n$ in x, this application generates (at most) one new world y_1 that labels (at most) $n-1$ negated boxed formulas. A further application of $(\Box^-)$ to $\neg\Box\neg C_1 \ldots \neg\Box\neg C_{n-1}$ in y_1 generates (at most) one new world y_2 that labels (at most) $n-2$ negated boxed formulas, and so on. Overall, at most $O(n)$ new labels are introduced by $(\Box^-)$ in each branch. For each of these labels, static rules apply at most $O(n)$ times. Therefore, the length of the tableau branch built by the strategy is $O(n^2)$. Last, to test that a node is an instance of a (Clash) axiom has at most complexity polynomial in n. The same for detecting (Clash)$_r$ and (Clash)$_{r^-}$. Indeed, a node contains a polynomial number of labels, $O(n)$ roles, (hence) a polynomial number of formulas $x : \neg\exists r.\top$ (or $x : \neg\exists r^-.\top$), as well as a polynomial number of formulas $x \xrightarrow{r} y$.   ∎

3.2   The Tableaux Calculus $\mathcal{TAB}_{PH2}^{Lite_c\mathbf{T}}$

Let us now introduce the calculus $\mathcal{TAB}_{PH2}^{Lite_c\mathbf{T}}$ which, for each open saturated branch **B** built by $\mathcal{TAB}_{PH1}^{Lite_c\mathbf{T}}$, verifies whether it represents a minimal model of the KB. First, given an open saturated branch **B** of a tableau built from $\mathcal{TAB}_{PH1}^{Lite_c\mathbf{T}}$, we define $\mathcal{D}(\mathbf{B})$ as the set of labels occurring on **B** and $\mathbf{B}^{\Box^-}$ as the set of formulas $x : \neg\Box\neg C$ occurring in **B**, i.e. $\mathbf{B}^{\Box^-} = \{x : \neg\Box\neg C \mid x : \neg\Box\neg C \text{ occurs in } \mathbf{B}\}$.

A tableau of $\mathcal{TAB}_{PH2}^{\mathit{Lite}_c\mathbf{T}}$ is a tree whose nodes are tuples of the form $\langle S \mid U \mid K\rangle$, where S and U are defined as in $\mathcal{TAB}_{PH1}^{\mathit{Lite}_c\mathbf{T}}$, whereas K contains formulas of the form $x : \neg\Box\neg C$, with $C \in \mathcal{L}_{\mathbf{T}}$. The basic idea of $\mathcal{TAB}_{PH2}^{\mathit{Lite}_c\mathbf{T}}$ is as follows. Given an open saturated branch **B** built by $\mathcal{TAB}_{PH1}^{\mathit{Lite}_c\mathbf{T}}$ and corresponding to a model $\mathcal{M}^{\mathbf{B}}$ of KB $\cup \{\neg F\}$, $\mathcal{TAB}_{PH2}^{\mathit{Lite}_c\mathbf{T}}$ checks whether $\mathcal{M}^{\mathbf{B}}$ is a minimal model of KB by trying to build a model of KB which is preferred to $\mathcal{M}^{\mathbf{B}}$. To this purpose, it keeps track (in K) of the negated box used in **B** ($\mathbf{B}^{\Box^-}$) in order to check whether it is possible to build a model of KB containing less negated box formulas. The tableau built by $\mathcal{TAB}_{PH2}^{\mathit{Lite}_c\mathbf{T}}$ closes if it is not possible to build a model smaller than $\mathcal{M}^{B}$, it remains open otherwise. Since by Definition 4 two models can be compared only if they have the same domain, $\mathcal{TAB}_{PH2}^{\mathit{Lite}_c\mathbf{T}}$ tries to build an open saturated branch containing all the labels appearing on **B**, i.e. those in $\mathcal{D}(\mathbf{B})$. To this aim, the dynamic rules use labels in $\mathcal{D}(\mathbf{B})$ instead of introducing new ones in their conclusions.

The rules of $\mathcal{TAB}_{PH2}^{\mathit{Lite}_c\mathbf{T}}$ are shown in Figure 2. The rules $(\exists^+)^r$ and $(\exists^+)^{r^-}$ introduce $x \xrightarrow{r} y$ and $y \xrightarrow{r} x$, respectively, where $y \in \mathcal{D}(\mathbf{B})$, instead of y being a new label. The choice of the label y introduces a branching in the tableau construction. The rule (Unfold) is applied to *all the labels of* $\mathcal{D}(\mathbf{B})$ (and not only to those appearing in the branch). The rule $(\Box^-)$ is applied to a node $\langle S, x : \neg\Box\neg C_1, \ldots, x : \neg\Box\neg C_n \mid U \mid K\rangle$, when $\{x : \neg\Box\neg C_1, \ldots, x : \neg\Box\neg C_n\} \subseteq K$, i.e. when the negated box formulas $x : \neg\Box\neg C_i$ also belong to the open branch **B**. Even in this case, the rule introduces a branch on the choice of the individual $y_i \in \mathcal{D}(\mathbf{B})$ to be used in the conclusion. In case a tableau node has the form $\langle S, x : \neg\Box\neg C \mid U \mid K\rangle$, and $x : \neg\Box\neg C \notin K$, then

$$\langle S, x : C, x : \neg C \mid U \mid K\rangle \ (\text{Clash}) \qquad \langle S, x \xrightarrow{r} y, x : \neg\exists r.\top \mid U \mid K\rangle \ (\text{Clash})_r \qquad \langle S, y \xrightarrow{r} x, x : \neg\exists r^-.\top \mid U \mid K\rangle \ (\text{Clash})_{r^-}$$

$$\langle S \mid U \mid \emptyset\rangle \ (\text{Clash})_\emptyset \qquad \begin{array}{r}\langle S, x : \neg\Box\neg C \mid U \mid K\rangle \ (\text{Clash})_{\Box^-}\\ \text{if } x : \neg\Box\neg C \notin K\end{array} \qquad \frac{\langle S \mid U \mid K\rangle}{\langle S, x : \Box\neg C \mid U \mid K\rangle \quad \langle S, x : \neg\Box\neg C \mid U \mid K\rangle}\,(\mathit{cut}) \ \begin{array}{l}\text{if } x : \neg\Box\neg C \notin S \text{ and } x : \Box\neg C \notin S\\ x \in \mathcal{D}(\mathbf{B}) \quad C \in \mathcal{L}_{\mathbf{T}}\end{array}$$

$$\frac{\langle S, x : \mathbf{T}(C) \mid U \mid K\rangle}{\langle S, x : C, x : \Box\neg C \mid U \mid K\rangle}\,(\mathbf{T}^+) \qquad \frac{\langle S, x : \neg\mathbf{T}(C) \mid U \mid K\rangle}{\langle S, x : \neg C \mid U \mid K\rangle \quad \langle S, x : \neg\Box\neg C \mid U \mid K\rangle}\,(\mathbf{T}^-)$$

$$\frac{\langle S, x : \neg\Box\neg C_1, \ldots, x : \neg\Box\neg C_n \mid U \mid K, x : \neg\Box\neg C_1, \ldots, x : \neg\Box\neg C_n\rangle}{\langle S, y_1 : C_k, y_1 : \Box\neg C_k, S^M_{x\rightarrow y_1}, \overline{S}^{\Box^{-k}}_{x\rightarrow y_1} \mid U \mid K\rangle \ \cdots \ \langle S, y_m : C_k, y_m : \Box\neg C_k, S^M_{x\rightarrow y_m}, \overline{S}^{\Box^{-k}}_{x\rightarrow y_m} \mid U \mid K\rangle}\,(\Box^-) \ \begin{array}{l}\text{if } \mathcal{D}(\mathbf{B}) = \{y_1, \ldots, y_m\} \text{ and } y_1 \neq x, \ldots, y_m \neq x\\ \forall k = 1, 2, \ldots, n\end{array}$$

$$\frac{\langle S \mid U, C \sqsubseteq D^L \mid K\rangle}{\langle S, x : \neg C \sqcup D \mid U, C \sqsubseteq D^{L,x} \mid K\rangle}\,(\text{Unfold}) \ x \in \mathcal{D}(\mathbf{B}) \text{ and } x \notin L \qquad \frac{\langle S, x : C \sqcup D \mid U \mid K\rangle}{\langle S, x : C \mid U \mid K\rangle \ \langle S, x : D \mid U \mid K\rangle}\,(\sqcup^+)$$

$$\frac{\langle S, x : \exists r.\top \mid U \mid K\rangle}{\langle S, x \xrightarrow{r} y_1 \mid U \mid K\rangle \ldots \langle S, x \xrightarrow{r} y_m \mid U \mid K\rangle}\,(\exists^+)^r \ \text{if } \mathcal{D}(\mathbf{B}) = \{y_1, \ldots, y_m\} \qquad \frac{\langle S, x : \exists r^-.\top \mid U \mid K\rangle}{\langle S, y_1 \xrightarrow{r} x \mid U \mid K\rangle \ldots \langle S, y_m \xrightarrow{r} x \mid U \mid K\rangle}\,(\exists^+)^{r^-} \ \text{if } \mathcal{D}(\mathbf{B}) = \{y_1, \ldots, y_m\}$$

Fig. 2. The calculus $\mathcal{TAB}_{PH2}^{\mathit{Lite}_c\mathbf{T}}$

$\mathcal{TAB}_{PH2}^{Lite_cT}$ detects a clash, called (Clash)$_{\square-}$: this corresponds to the situation in which $x : \neg\square\neg C$ does not belong to **B**, while the model corresponding to the branch being built contains $x : \neg\square\neg C$, and hence is *not* preferred to the model represented by **B**.

The calculus $\mathcal{TAB}_{PH2}^{Lite_cT}$ contains also the clash condition (Clash)$_\emptyset$. Since each application of ($\square^-$) removes the negated box formulas $x : \neg\square\neg C_i$ from the set K, when K is empty all the negated boxed formulas occurring in **B** also belong to the current branch. In this case, the model built by $\mathcal{TAB}_{PH2}^{Lite_cT}$ satisfies the same set of $x : \neg\square\neg C_i$ (for all individuals) as **B** and, thus, it is not preferred to the one represented by **B**. $\mathcal{TAB}_{PH2}^{Lite_cT}$ is sound and complete:

Theorem 6 (Soundness and completeness of $\mathcal{TAB}_{PH2}^{\boldsymbol{Lite_c}T}$). *Given a KB and a query F, let $\langle S' \mid U \rangle$ be the corresponding constraint system of KB, and $\langle S \mid U \rangle$ the corresponding constraint system of KB $\cup \{\neg F\}$. An open saturated branch **B** built by $\mathcal{TAB}_{PH1}^{Lite_cT}$ for $\langle S \mid U \rangle$ is satisfiable by an injective mapping in a minimal model of KB iff the tableau in $\mathcal{TAB}_{PH2}^{Lite_cT}$ for $\langle S' \mid U \mid \mathbf{B}^{\square^-} \rangle$ is closed.*

$\mathcal{TAB}_{PH2}^{Lite_cT}$ always terminates. Termination is ensured by the fact that dynamic rules make use of labels belonging to $\mathcal{D}(\mathbf{B})$, which is finite, rather than introducing "new" labels in the tableau.

Theorem 7 (Termination of $\mathcal{TAB}_{PH2}^{\boldsymbol{Lite_c}T}$). *Let $\langle S' \mid U \mid \mathbf{B}^{\square^-} \rangle$ be a constraint system starting from an open saturated branch **B** built by $\mathcal{TAB}_{PH1}^{Lite_cT}$, then any tableau generated by $\mathcal{TAB}_{PH2}^{Lite_cT}$ is finite.*

It is possible to show that the problem of verifying that a branch **B** represents a minimal model for KB in $\mathcal{TAB}_{PH2}^{Lite_cT}$ is in NP in the size of **B**.

The overall procedure $\mathcal{TAB}_{min}^{Lite_cT}$ is defined as follows:

Definition 8. *Let KB be a knowledge base whose corresponding constraint system is $\langle S \mid U \rangle$. Let F be a query and let S' be the set of constraints obtained by adding to S the constraint corresponding to $\neg F$. The calculus $\mathcal{TAB}_{min}^{Lite_cT}$ says that KB $\models_{DL\text{-}Lite_cT_{min}} F$ iff for each branch **B** built by $\mathcal{TAB}_{PH1}^{Lite_cT}$, either (i) **B** is closed or (ii) the tableau built in (phase 2) by the calculus $\mathcal{TAB}_{PH2}^{Lite_cT}$ for $\langle S \mid U \mid \mathbf{B}^{\square^-} \rangle$ is open.*

Theorem 8 (Soundness and completeness of $\mathcal{TAB}_{min}^{\boldsymbol{Lite_c}T}$). $\mathcal{TAB}_{min}^{Lite_cT}$ *is a sound and complete decision procedure for verifying whether KB $\models_{DL\text{-}Lite_cT_{min}} F$.*

Proof. (Soundness) If for all branches **B** (i) holds, then by Theorem 2, KB $\models_{DL\text{-}Lite_cT_{min}} F$. Consider an open saturated branch **B** for which (ii) holds, by Theorem 6, **B** is not satisfiable via an injective mapping in a minimal model of KB, hence also in this case by Theorem 2, KB $\models_{DL\text{-}Lite_cT_{min}} F$.

(Completeness) Let KB $\models_{DL\text{-}Lite_cT_{min}} F$. For contraposition, let **B** be an open saturated branch (if any) generated by $\mathcal{TAB}_{PH1}^{Lite_cT}$. If this branch was satisfiable by an injective mapping in a minimal model of KB, then by Proposition 1, also KB $\cup \{\neg F\}$ would be, against the hypothesis that KB $\models_{DL\text{-}Lite_c \mathbf{T}_{min}} F$. Hence, **B** is not satisfiable by an injective mapping in a minimal model of KB, and by Theorem 6 the tableau in $\mathcal{TAB}_{PH2}^{Lite_c \mathbf{T}}$ for $\langle S \mid U \mid \mathbf{B}^{\square^-} \rangle$ is open. Hence (i) or (ii) hold. ∎

We can also prove that the complexity of $\mathcal{TAB}_{min}^{Lite_c \mathbf{T}}$ matches the known results for minimal entailment in $DL\text{-}Lite_c \mathbf{T}_{min}$ of Theorem 1:

Theorem 9 (Complexity of $\mathcal{TAB}_{min}^{Lite_c \mathbf{T}}$). *The problem of deciding whether* $KB \models_{DL\text{-}Lite_c \mathbf{T}_{min}} F$ *by means of* $\mathcal{TAB}_{min}^{Lite_c \mathbf{T}}$ *is in* Π_2^p.

Proof. We first consider the complementary problem: KB $\not\models_{min}^{\mathcal{L}_{\mathbf{T}}} F$. This problem can be solved according to the procedure in Definition 8: by nondeterministically generating an open saturated branch of polynomial length in the size of KB in $\mathcal{TAB}_{PH1}^{Lite_c \mathbf{T}}$ (a model $\mathcal{M}^{\mathbf{B}}$ of KB $\cup \{\neg F\}$), and then by calling an NP oracle which verifies that $\mathcal{M}^{\mathbf{B}}$ is a minimal model of KB. In fact, the verification that $\mathcal{M}^{\mathbf{B}}$ is not a minimal model of the KB can be done by an NP algorithm which nondeterministically generates a branch in $\mathcal{TAB}_{PH2}^{Lite_c \mathbf{T}}$ of polynomial size in the size of $\mathcal{M}^{\mathbf{B}}$ (and of KB), representing a model $\mathcal{M}^{\mathbf{B}'}$ of KB preferred to $\mathcal{M}^{\mathbf{B}}$. Hence, the problem of verifying that KB $\not\models_{min}^{\mathcal{L}_{\mathbf{T}}} F$ is in NP$^{\mathbf{NP}}$, that is to say in Σ_2^p, and the problem of deciding whether KB $\models_{DL\text{-}Lite_c \mathbf{T}_{min}} F$ is in CO-NP$^{\mathbf{NP}}$, that is to say in Π_2^p. ∎

4   Related Works and Conclusions

Several nonmonotonic extensions of DLs have been proposed in the literature [15, 5, 2, 3, 9, 13, 10, 8]. Recently, much attention has been devoted to nonmonotonic extensions of low complexity DLs. The complexity of *circumscribed* fragments of the $\mathcal{EL}^\perp$ and DL-lite families have been studied in [3]. A fragment of $\mathcal{EL}^\perp$ for which the complexity of circumscribed KBs is polynomial has been identified in [4].

In this work we have provided a two-phase tableau calculus $\mathcal{TAB}_{min}^{Lite_c \mathbf{T}}$ for checking minimal entailment in a nonmonotonic extension of the Description Logic $DL\text{-}Lite_{core}$, described in [12]. This fills the gap due to the lack of a calculus for this logic. The proposed calculus matches the known complexity results for $DL\text{-}Lite_c \mathbf{T}_{min}$, namely that entailment is in Π_2^p [12].

Acknowledgements. The work has been partially supported by the project "MIUR PRIN08 LoDeN: Logiche Descrittive Nonmonotone: Complessitá e implementazioni".

References

1. Baader, F., Brandt, S., Lutz, C.: Pushing the $\mathcal{EL}$ envelope. In: IJCAI, pp. 364–369 (2005)
2. Baader, F., Hollunder, B.: Priorities on defaults with prerequisites, and their application in treating specificity in terminological default logic. J. of Automated Reasoning (JAR) 15(1), 41–68 (1995)

3. Bonatti, P., Faella, M., Sauro, L.: Defeasible inclusions in low-complexity dls: Preliminary notes. In: IJCAI, pp. 696–701 (2009)

4. Bonatti, P., Faella, M., Sauro, L.: $\mathcal{EL}$ with default attributes and overriding. In: Patel-Schneider, P.F., Pan, Y., Hitzler, P., Mika, P., Zhang, L., Pan, J.Z., Horrocks, I., Glimm, B. (eds.) ISWC 2010, Part I. LNCS, vol. 6496, pp. 64–79. Springer, Heidelberg (2010)

5. Bonatti, P., Lutz, C., Wolter, F.: Description logics with circumscription. In: KR, pp. 400–410 (2006)

6. Buchheit, M., Donini, F.M., Schaerf, A.: Decidable reasoning in terminological knowledge representation systems. J. Artif. Int. Research (JAIR) 1, 109–138 (1993)

7. Calvanese, D., De Giacomo, G., Lembo, D., Lenzerini, M., Rosati, R.: Tractable Reasoning and Efficient Query Answering in Description Logics: The *DL-Lite* Family. J. Autom. Reasoning (JAR) 39(3), 385–429 (2007)

8. Casini, G., Straccia, U.: Rational closure for defeasible description logics. In: Janhunen, T., Niemelä, I. (eds.) JELIA 2010. LNCS, vol. 6341, pp. 77–90. Springer, Heidelberg (2010)

9. Donini, F.M., Nardi, D., Rosati, R.: Description logics of minimal knowledge and negation as failure. ACM Trans. Comput. Log. 3(2), 177–225 (2002)

10. Giordano, L., Gliozzi, V., Olivetti, N., Pozzato, G.L.: Reasoning About Typicality in Preferential Description Logics. In: Hölldobler, S., Lutz, C., Wansing, H. (eds.) JELIA 2008. LNCS (LNAI), vol. 5293, pp. 192–205. Springer, Heidelberg (2008)

11. Giordano, L., Gliozzi, V., Olivetti, N., Pozzato, G.L.: A tableau calculus for a nonmonotonic extension of $\mathcal{EL}^{\perp}$. In: Brünnler, K., Metcalfe, G. (eds.) TABLEAUX 2011. LNCS (LNAI), vol. 6793, pp. 180–195. Springer, Heidelberg (2011)

12. Giordano, L., Gliozzi, V., Olivetti, N., Pozzato, G.L.: Reasoning about typicality in low complexity DLs: the logics $\mathcal{EL}^{\perp}\mathbf{T}_{min}$ and *DL-Lite*$_c\mathbf{T}_{min}$. In: IJCAI, pp. 894–899 (2011)

13. Giordano, L., Gliozzi, V., Olivetti, N., Pozzato, G.L.: $\mathcal{ALC} + \mathbf{T}_{min}$: a preferential extension of description logics. Fundamenta Informaticae 96, 1–32 (2009)

14. Kraus, S., Lehmann, D., Magidor, M.: Nonmonotonic reasoning, preferential models and cumulative logics. Artificial Intelligence 44(1-2), 167–207 (1990)

15. Straccia, U.: Default inheritance reasoning in hybrid kl-one-style logics. In: IJCAI, pp. 676–681 (1993)

Monte-Carlo Style UCT Search for Boolean Satisfiability

Alessandro Previti[1], Raghuram Ramanujan[2],
Marco Schaerf[1], and Bart Selman[2]

[1] Dipartimento di Informatica e Sistemistica Antonio Ruberti,
Sapienza, Università di Roma,
Roma, Italy
[2] Department of Computer Science,
Cornell University,
Ithaca, New York

Abstract. In this paper, we investigate the feasibility of applying algorithms based on the Uniform Confidence bounds applied to Trees [12] to the satisfiability of CNF formulas. We develop a new family of algorithms based on the idea of balancing exploitation (depth-first search) and exploration (breadth-first search), that can be combined with two different techniques to generate random playouts or with a heuristics-based evaluation function. We compare our algorithms with a DPLL-based algorithm and with WalkSAT, using the size of the tree and the number of flips as the performance measure. While our algorithms perform on par with DPLL on instances with little structure, they do quite well on structured instances where they can effectively reuse information gathered from one iteration on the next. We also discuss the pros and cons of our different algorithms and we conclude with a discussion of a number of avenues for future work.

1 Introduction

The Upper Confidence bounds applied to Trees (from now on UCT) algorithm, introduced by Kocsis and Szepesvári in [12], is an (increasingly popular and successful) adaptation of the work on Upper Confidence Bounds (UCB) by Auer, Cesa-Bianchi and others [2,3,4] on the multi-armed bandit problem to tree search. It has been successfully used in many game playing programs, the most notable being MoGo which is one of the strongest computer Go players [10,16]. In this paper we perform a preliminary investigation into the application of UCT-style search algorithms to satisfiability testing of propositional formulas in Conjunctive Normal Form (CNF).

Rather than explore the search space in a depth-first fashion, in the style of DPLL [9,8], UCT repeatedly starts from the root node and incrementally builds a tree based on estimates of node utilities and node visit frequencies computed from previous iterations. In most implementations of UCT, the estimated utility of a new node is computed using Monte-Carlo methods, i.e., by generating

R. Pirrone and F. Sorbello (Eds.): AI*IA 2011, LNAI 6934, pp. 177–188, 2011.

random completions of the search (termed "playouts") and averaging their outcomes. This utility is revised each time the search revisits the node using the estimated values of the children. This technique is especially effective when no adequate heuristics is available to perform this value estimation task.

Here we present in detail a family of algorithms called UCTSAT that employ the UCT search control mechanism but use different mechanisms to estimate the utility of a node. In the first version, called UCTSAT$_h$, a heuristic is used to estimate the initial utility of a node, more precisely, the heuristic used is the fraction of the total set of clauses that are satisfied by the partial assignment associated with the node. While the results have been promising, especially when applied to structured instances, we also experiment with two more variants, called UCTSAT$_{cp}$ and UCTSAT$_{sbs}$, that use search strategies that are closer to the more traditional usage of UCT algorithms, that is using random tryouts in a MonteCarlo style. A very short description of the UCTSAT$_h$ algorithm has been described, but not presented, in [14].

While we do not expect UCTSAT to outperform the state of the art SAT solvers (especially with respect to CPU time), we believe that the development of an algorithm based on a radically different search technique is important for at least two reasons: (a) the hardness of SAT instances is related to the algorithm used [1,7], and hence UCTSAT, which uses different search strategies, can provide useful and new insights into the complexity of SAT instances; and (b) because such algorithms can be useful when included in a portfolio of algorithms (see, for example, [17]) where very different solution techniques can help expand the range of applicability of the portfolio.

The remainder of this paper is organized as follows. In Section 2 we briefly describe the UCT algorithm. Section 3 presents all three versions of the UCTSAT algorithm for satisfiability testing. In Section 4 we present preliminary experimental results from applying UCTSAT to a variety of benchmark problems, and compare it to our own DPLL implementation and to WalkSAT. Section 5 concludes with a discussion of our results, and outlines a few topics that deserve further investigation.

2   Upper Confidence Bounds Applied to Trees (UCT)

Monte-Carlo tree search algorithms such as UCT [12] have recently received a great deal of attention from the planning and game-playing community, in particular due to their success in the domain of Go [10,16]. UCT builds on the UCB1 algorithm for multi-armed bandits [2], which is used to guide the search tree construction process. Exploration of under-sampled actions is balanced against exploitation of known good actions to generate asymmetric trees that are deeper in more promising regions of the search space and shallower elsewhere.

Algorithm 1 describes the recursive procedure UCT uses to build the search tree. $T(s, a)$ is the domain transition function that returns the state s' reached from taking action a in state s. The algorithm maintains two lookup tables — $n(s)$ tracks the number of times state s has been visited and $Q(s)$ tracks the current estimated utility of the state s. The action selection operator $\pi(s)$ is repeatedly applied to descend down the tree until a previously unvisited (or terminal) node k is reached. k is added to the tree and an estimate of its utility is computed which is used to update $Q(s)$ and $n(s)$ for all nodes s on the path from the root node to k, according to lines 11 and 12. Under this scheme, the size of the tree grows by one node on every iteration.

We describe $\pi(s)$ and the utility estimation step in greater detail below:

- **Action Selection:** Given a state s, the action selection operator $\pi(s)$ returns the action a that maximizes an upper confidence bound on the utility of the resulting state $s' = T(s, a)$:

$$\pi(s) = \operatorname*{argmax}_{a} \left(Q(T(s,a)) + c \cdot \sqrt{\frac{\log n(s)}{n(T(s,a))}} \right)$$

  If $n(T(s, a)) = 0$ for an action a, then it is selected first, before any actions are re-sampled. Ties are broken randomly. The constant c is tuned empirically and controls the extent to which exploration or exploitation is favored.
- **Value Estimation:** For terminal nodes, the true utility of the state is returned. For non-terminal nodes, an estimate of the true utility is returned. This estimate can be computed using a domain-independent approach, such as a random playout (which is also the traditional solution), or using a domain-specific heuristic. Notice that the function can return either a definite answer (SAT/UNSAT) or a reward r for the node. To be more precise, both SAT and UNSAT are represented as integer constants in all of the algorithms.

A UCT search consists of repeatedly calling the function given in Algorithm 1 on the root node for as long as time allows. At that point, the action that leads to

---

Algorithm 1. The UCT Algorithm

1: *int* **Function** $UCTRecurse(s : \text{state})$
2: **if** s is a terminal state **then**
3:     Add s to the search tree if $n(s) = 0$
4:     $r \leftarrow$ true utility of s
5: **else if** $n(s) = 0$ **then**
6:     Add s to the search tree
7:     $r \leftarrow$ estimated utility of s
8: **else**
9:     $r \leftarrow UCTRecurse(T(s, \pi(s)))$
10: **end if**
11: $n(s) \leftarrow n(s) + 1$
12: $Q(s) \leftarrow Q(s) + (r - Q(s))/n(s)$
13: **return** r

the state with the highest average utility is returned. Alternate schemes include returning the action with the most number of visits and returning the action with the highest lower confidence bound. In practice, there is little difference between these approaches.

3   UCTSAT

Typical UCT implementations estimate the utility of a node n on the first visit by sampling the search space subsumed by n, via random or pseudo-random playouts. This idea is very appealing when no good heuristics are available for a domain. The pseudo-code for the recursive tree-building component of our procedure (which we call UCTSAT) is given by Algorithm 2.

Algorithm 2. The UCTSAT Algorithm

```
 1: int Function UCTSATRecurse(s : state)
 2: if n(s) = 0 then
 3:     Add s to the search tree
 4:     r ← estimate(s)
 5:     if r = SAT then
 6:         print  "Formula is satisfiable"
 7:         exit
 8:     else if r = UNSAT then
 9:         Mark s as closed
10:     else
11:         var(s) ← chooseVariable()
12:     end if
13: else
14:     r ← UCTSATRecurse(T(s, π(s)))
15:     if r = UNSAT then
16:         if all the children of s are closed then
17:             Mark s as closed
18:             return  UNSAT
19:         else
20:             r ← 0
21:         end if
22:     end if
23:     n(s) ← n(s) + 1
24:     Q(s) ← Q(s) + (r − Q(S))/n(s)
25: end if
26: return  r
```

Analogously to UCT, a UCTSAT search comprises repeated invocations of Algorithm 1 on the root node. UCTSAT behaves like a cross between a backtracking (DPLL-style) and a randomized algorithm (for example, WalkSAT [15]). It is a complete procedure that explores the search space in a very different fashion to that of DPLL. While DPLL only backtracks when it has finished completely evaluating a branch, UCTSAT repeatedly starts from the root node and only goes one level deeper on each iteration. As in UCT, the UCB1 formula is used to control the descent down the tree, where each step involves making a variable assignment and simplifying the original formula. In the flavor of local search methods, the most promising branch is typically chosen at each step, but occasional deviations to sub-optimal branches (that may still lead to solutions) also occur. The search terminates when either:

1. a satisfying assignment is found (line 5)
2. the formula is determined to be unsatisfiable (line 17, when s is the root)
3. or the specified number of iterations is exceeded

We highlight the key differences between Algorithm 2 and Algorithm 1 below:

- **Variable Assignment Look-up Table**: In addition to the look-up tables $n(s)$ and $Q(s)$ employed by UCT, we use an additional look-up table $var(s)$ that stores the variable that will be assigned at state s. This table is updated when the node s is first created (line 11). The function $chooseVariable()$ we used in line 11 of Algorithm 2 returns the variable with the highest number of occurrences in the simplified formula.
- **Action Selection**: In game-tree search, an action corresponds to a move in the game. Here, an "action" is the process of assigning a value to a variable. As in UCT, an upper confidence bound is used to choose among the possible assignments.
- **Handling Terminal Nodes**: In UCT, the information from terminal and non-terminal nodes is propagated up the tree in an identical fashion. UCT-SAT, on the other hand, handles terminal nodes as a special case. When a satisfying assignment is found, the search promptly terminates. When a contradiction is encountered at a node s, it is marked as "closed" and s is never revisited by the search. A negative signal (value of 0) is propagated up to penalize this branch of the tree. When all the children of a node s have been closed (line 17), then s is closed as well — this mechanism propagates information unsatisfiable assignments up the tree.
- **Value Estimation**: We will experiment with various estimation functions, from simple heuristics to different forms of playouts. A more detailed analysis of the comparison between UCT and UCTSAT on this issue will be presented after the definition of the UCTSAT variants.

In the first version of UCTSAT, called UCTSAT$_h$, we replace the playouts with a simple heuristic while retaining the multi-armed bandit approach of balancing exploitation (i.e., DPLL-style depth-first expansion of the search tree) and exploration (i.e., breadth-first expansion). In fact, line 4 is replaced by $r \leftarrow h(s)$, where $h(s)$ is the fraction of clauses that have been satisfied having reached it.

While UCTSAT$_h$ seems a very promising alternative to DPLL-like algorithms, at least on structured instances, here we also want to experiment with two different mechanism to generate playouts. The first such algorithm we developed, called UCTSAT$_{cp}$ (UCTSAT with complete playouts), uses n (complete) random playouts to estimate r. More precisely, we generate n random playouts, each generating a complete assignment to all of the (unassigned) variables of s. For each playout, if it satisfies the formula we are done, otherwise we compute its heuristic value $h(s, s')$ and then compute the average over all the playouts. The pseudo-code for the function is:

Algorithm 3. Complete playout

1: *int* **Function** *estimate*(s : state)
2: *value* $\leftarrow 0$
3: **for** $i = 1$ to n **do**
4:     **for all** $p \in unassignedLiterals(s)$ **do**
5:        $p \leftarrow$ choose a Random value
6:        $s' = update(s, p)$
7:     **end for**
8:     $r \leftarrow h(s, s')$
9:     **if** $r = $ **SAT then**
10:        **print** "Formula is satisfiable"
11:        **exit**
12:     **end if**
13:     *value* $\leftarrow$ *value* $+ r$
14: **end for**
15: **return** *value*$/n$

To fully specify the behavior of UCTSAT$_{cp}$ we still need to clearly define the functions $h(s, s')$, used in line 8 of Algorithm 3. We experimented with various choices, however here we report the results obtained using

$$h(s, s') = \Sigma_{c \in clauses(s)} SatVars(c, s')/sizeof(c, s)$$

where, for each clause of the simplified formula associated to state (partial assignment) s, we compute the number of literals satisfied by s' ($SatVars(c, s')$) and divide it by the size of the clause in s. This metric tries to capture the probability that a solution exists in the close proximity of the current (falsifying) assignment.

While UCTSAT$_{cp}$ performs quite well it can be improved if we allow for a step-by-step choice of the variables in the playouts. More precisely, we define UCTSAT$_{sbs}$ where we choose one (unassigned) variable at the time, assign to it a random variable and then check whether the formula is already falsified in order avoid generating useless complete assignments. Moreover, after each assignment we can perform unit propagation so that forced assignments are immediately performed and we do not choose random values for forced variables. The idea of generating either complete assignments or step-by-step ones is not novel and has already been used in the literature, see, for example, the work of

Lombardi et. al. [13]. These changes are incorporated into this new version of the function *estimate* (Algorithm 4), where $update(s, p)$ not only assigns a value to the variable p but also performs unit-propagation.

Algorithm 4. Step by step playout

 1: *int* **Function** $estimate(s : \text{state})$
 2: $value \leftarrow 0$
 3: **for** $i = 1$ to n **do**
 4:     **while** $\neg contradictory(s)$ **do**
 5:         $p \leftarrow$ random variable in $unassignedLiterals(s)$
 6:         $p \leftarrow$ choose a Random value
 7:         $s' = s$
 8:         $s = update(s, p)$
 9:         **if** $satisfied(s)$ **then**
10:             **return** SAT
11:         **end if**
12:     **end while**
13:     $r \leftarrow h(s, s')$
14:     $value \leftarrow value + r$
15: **end for**
16: **return** $value/n$

We can now summarize the main differences between the 3 variants of UCT-SAT presented in the paper and compare them with the UCT basic algorithm:

- **Value Estimation**: In the style of UCT, both UCTSAT$_{cp}$ and UCTSAT$_{sbs}$ employ a number of random playouts to estimate the utility of a node. Each playout will either find a satisfying assignment or will return a heuristic assessment of the solution found $h(s, s')$ (lines 13 and 8). On the other hand, UCTSAT$_h$ uses a heuristics that computes, for each state s, the fraction of clauses satisfied in s with respect to the total number of clauses. This heuristics makes this variant of UCTSAT more similar to the methodologies used in most variants of DPLL, where the maximization of the number of satisfied clauses is used as a heuristics to choose the branching literal.
- **Using playouts**: In UCT, each playout will report the value of the terminal position reached, these values usually are one of +1 (win), 0 (draw), -1(loss). In our case, if a playout reports 1 (SAT) we are done, so in general, all playouts will report 0 (UNSAT) and, thus, it makes little sense to average these values. To overcome this problem, we assign to each playout a heuristic value $h(s, s')$ that should amount to how far away the playout is from a solution. Since there is clearly no metrics that can exactly compute this value, we tried several such functions. The main difference between UCTSAT$_{cp}$ and UCTSAT$_{sbs}$ is the playouts-generation mechanism, since in UCTSAT$_{cp}$ the playouts are fully generated at the beginning, while in UCTSAT$_{sbs}$ the playouts are generated one literal at the time, thus avoiding to generate too many inconsistencies early in the creation process.

We can now briefly compare the properties of all three variants of UCTSAT with both DPLL-like and WALKSAT-like algorithms. An experimental comparison is discussed at length in Section 4.

The advantages and disadvantages of UCTSAT with respect to DPLL-like algorithms can be summarized as follows:

+ Once UCTSAT has visited all the children of a node, their estimated utilities and visit counts can help it make an informed decision about which assignment to focus on at a node.
+ UCTSAT can exit a dead-end branch without the need to completely explore the branch.
+ The above two properties mean that UCTSAT can, in most cases, create more compact trees.
− UCTSAT needs to keep in memory (almost) all of the visited tree, making it a memory-intensive procedure.
− UCTSAT needs to visit each node multiple times.

We summarize the pros and cons of UCTSAT with respect to local search methods such as WALKSAT below:

+ UCTSAT uses the information obtained from each previous iteration to guide the next one, while (standard) WalkSAT restarts each try from scratch. While more advanced versions of WalkSAT use adaptive strategies to guide the restart, we conjecture that UCT makes a more informed decision.
+ UCTSAT is a complete algorithm that can prove the unsatisfiability of a formula by closing all the nodes.
− UCTSAT needs to retain all of the visited tree and the associated utility estimates and visit counts, making it more memory-intensive than WalkSAT.

4   Experimental Analysis

This work is a preliminary assessment of the feasibility of applying UCT-style methods to solve SAT problems. As such, we have focused our efforts on understanding whether the various variants of UCTSAT are capable of solving SAT instances using smaller search trees than DPLL. To simplify the comparisons, we contrast our algorithms against an in-house, no-frills implementation of DPLL, and against WalkSAT (where the number of flips is used as the comparison benchmark). Our DPLL implementation uses the same heuristic for picking the next variable for assignment as UCTSAT, i.e., the variable with the maximum number of occurrences in the simplified formula. The choice of which branch to explore first is made non-deterministically.

Empirical tuning of the exploration bias constant c revealed that on most instances, a value of c very close to 0 yielded the best performance on average; we therefore fix c to 0 in all our experiments. Our WalkSAT runs use the novelty heuristic with a maximum of 200,000 flips allowed per try. To keep the comparison fair and run-times reasonable, DPLL and UCTSAT also time-out once the size of the search tree exceeds 200, 000 nodes. Since the three algorithms are all non-deterministic, we perform 100 independent runs of each algorithm per instance, and report the average tree size over all successful runs.

Our experiments use instances from the SATLIB repository [11]. We focus our attention only on satisfiable instances; for unsatisfiable instances, UCTSAT$_h$ and DPLL construct identical trees since they use the same variable choosing heuristic. Having fixed the variable ordering, proving unsatisfiability requires both algorithms to visit the same set of nodes. It is meaningless to measure Walk-SAT's performance on unsatisfiable instances since it is an incomplete method. Our first experiment uses uniform random 3-SAT instances of various sizes from the phase-transition region. Figure 1 presents a plot of the size of the trees explored by the various algorithms, as a function of the number of variables in the formula. Each data point corresponds to the average number of visited nodes of the algorithm over 100 instances. Notice that we use a logarithmic scale in all of our plots.

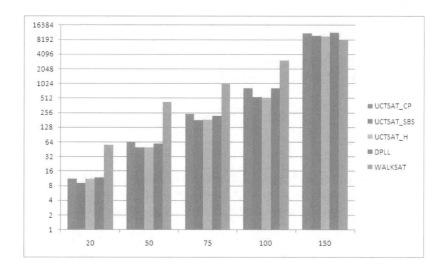

Fig. 1. Average tree sizes for uniform random 3-SAT instances

We observe that the rate of growth is similar for all the algorithms (with the exception of WalkSAT). Moreover, the size of the trees constructed by UCT-SAT and DPLL is generally close, with UCTSAT$_{sbs}$ and UCTSAT$_h$ marginally outperforming DPLL. We believe that this similarity in tree sizes is due to the unstructured nature of these instances. UCTSAT works well when each exploration of the tree yields information that can be successfully used in subsequent ones. Little such information can be gained from unstructured instances and in such settings, UCTSAT only adds overhead to the DPLL machinery.

Our second experiment uses instances from the SAT encoding of graph coloring problems ("flat graph coloring"). These instances are randomly generated but have some underlying structure due to the encoding. These results are presented in Figure 2, and are qualitatively similar to those presented in Figure 1 — UCTSAT$_{sbs}$ slightly outperforms DPLL, while UCTSAT$_{cp}$ has worse performances. UCTSAT$_h$ is comparable with DPLL.

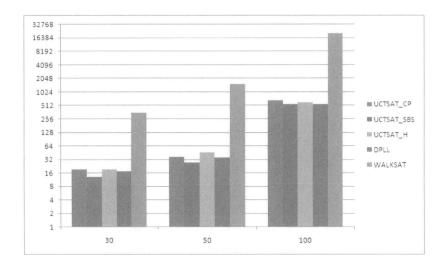

Fig. 2. Average tree sizes for flat graph coloring instances

Finally, we present some results on structured instances drawn from real-world problems, namely instances from circuit fault analysis (single-stuck-at-fault, or SSA).Figure 3 presents the average size of the search tree constructed by DPLL, UCTSAT and WALKSAT on 4 SSA instances. These results show that, when there is an underlying structure in the instances, both the UCTSAT$_{sbs}$ and UCTSAT$_h$ variants of UCTSAT can exploit it very effectively, by building a much smaller tree. It is not very clear why UCTSAT$_{cp}$ performs so poorly on these instances.

5   Discussion and Conclusions

In this paper, we have presented the UCTSAT family of algorithms based on UCT to solve CNF satisfiability problems. This family includes algorithms (UCTSAT$_{cp}$ and UCTSAT$_{sbs}$) using playouts to estimate the utility of nodes as well as an algorithm (UCTSAT$_h$) using a simple heuristic based on the fraction of satisfied clauses. Our initial experimental results show that UCTSAT does not perform well when instances have no underlying structure, but performs very well when it can successfully apply the information it gathers on one iteration on successive visits to the same node in the tree.

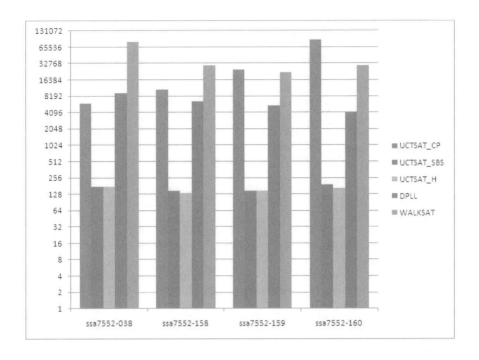

Fig. 3. Average tree sizes for SSA circuit fault analysis instances

UCT-based algorithms have already been successfully used in many applications, mostly in games such as GO [10,16]. All of these applications adopted a playout-based version of UCT, where the estimate is computed by generating random completions of the game. However, our experiments suggest that, at least in settings where a good heuristics to estimate the value of a node is available, using such a heuristics can be competitive (and even outperform) playouts-based algorithms.

There are many interesting avenues for future work. These include:

– Performing a systematic analysis of problem classes to gain insights into the classes of formulas that are better suited to UCTSAT.
– Experimenting with different heuristics to assess the quality of the playouts in both UCTSAT$_{cp}$ and UCTSAT$_{sbs}$.
– Gaining a better understanding of the advantages and disadvantages of using a heuristics to estimate the value of a node w. r. t. the use of playouts.
– Extending UCTSAT to solve Quantified Boolean Formulas (QBF), by extending the algorithms presented in [5,6]. We believe that UCT-style search can be effective in solving QBF instances that have a game-playing structure and is, therefore, closer to algorithms for Go and other games.

References

1. Aguirre, A., Vardi, M.Y.: Random 3-SAT and BDDs: The plot thickens further. In: Walsh, T. (ed.) CP 2001. LNCS, vol. 2239, pp. 121–136. Springer, Heidelberg (2001)
2. Auer, P., Cesa-Bianchi, N., Fischer, P.: Finite-time analysis of the multiarmed bandit problem. Machine Learning 47(2), 235–256 (2002)
3. Auer, P., Cesa-Bianchi, N., Freund, Y., Schapire, R.E.: The nonstochastic multi-armed bandit problem. SIAM Journal on Computing 32(1), 48–77 (2003)
4. Auer, P., Ortner, R.: UCB revisited: Improved regret bounds for the stochastic multi-armed bandit problem. Periodica Mathematica Hungarica 61(1), 55–65 (2010)
5. Cadoli, M., Giovanardi, A., Schaerf, M.: An algorithm to evaluate quantified boolean formulae. In: Proceedings of the Fifteenth National Conference on Artificial Intelligence (AAAI 1998), pp. 263–267 (1998)
6. Cadoli, M., Schaerf, M., Giovanardi, A., Giovanardi, M.: An algorithm to evaluate quantified boolean formulae and its experimental evaluation. Journal of Automated Reasoning 28(2), 101–142 (2002)
7. Coarfa, C., Demopoulos, D.D., San Miguel Aguirre, A., Subramanian, D., Vardi, M.Y.: Random 3-SAT: The plot thickens. Constraints 8(3), 243–261 (2003)
8. Davis, M., Logemann, G., Loveland, D.: A machine program for theorem proving. Communications of the ACM 5(7), 394–397 (1962)
9. Davis, M., Putnam, H.: A computing procedure for quantification theory. Journal of the ACM 7, 201–215 (1960)
10. Gelly, S., Silver, D.: Combining online and offline knowledge in UCT. In: Proceedings of the 24th International Conference on Machine Learning, pp. 273–280. ACM, New York (2007)
11. Hoos, H.H., Stützle, T.: SAT20000: Highlights of Satisfiability Research in the year 2000, chapter SATLIB: An Online Resource for Research on SAT. In: Frontiers in Artificial Intelligence and Applications, pp. 283–292. Kluwer Academic, Dordrecht (2000), Web site available at:
http://www.cs.ubc.ca/~hoos/SATLIB/index-ubc.html
12. Kocsis, L., Szepesvári, C.: Bandit based monte-carlo planning. In: Fürnkranz, J., Scheffer, T., Spiliopoulou, M. (eds.) ECML 2006. LNCS (LNAI), vol. 4212, pp. 282–293. Springer, Heidelberg (2006)
13. Lombardi, M., Milano, M., Roli, A., Zanarini, A.: Deriving information from sampling and diving. In: Serra, R., Cucchiara, R. (eds.) AI*IA 2009. LNCS, vol. 5883, pp. 82–91. Springer, Heidelberg (2009)
14. Previti, A., Ramanujan, R., Schaerf, M., Selman, B.: Applying uct to boolean satisfiability. In: Sakallah, K.A., Simon, L. (eds.) SAT 2011. LNCS, vol. 6695, pp. 373–374. Springer, Heidelberg (2011)
15. Selman, B., Kautz, H.A., Cohen, B.: Local search strategies for satisfiability testing. In: DIMACS Series in Discrete Mathematics and Theoretical Computer Science. Citeseer (1996)
16. Wang, Y., Gelly, S.: Modifications of UCT and sequence-like simulations for Monte-Carlo Go. In: IEEE Symposium on Computational Intelligence and Games, Honolulu, Hawaii, pp. 175–182 (2007)
17. Xu, L., Hutter, F., Hoos, H.H., Leyton-Brown, K.: SATzilla: portfolio-based algorithm selection for SAT. Journal of Artificial Intelligence Research 32(1), 565–606 (2008)

Exploiting Macro-actions and Predicting Plan Length in Planning as Satisfiability

Alfonso Emilio Gerevini, Alessandro Saetti, and Mauro Vallati

Dipartimento di Ingegneria dell'Informazione, Università degli Studi di Brescia
Via Branze 38, 25123 Brescia, Italy
{gerevini,saetti,mauro.vallati}@ing.unibs.it

Abstract. The use of automatically learned knowledge for a planning domain
can significantly improve the performance of a generic planner when solving a
problem in this domain. In this work, we focus on the well-known SAT-based approach to planning and investigate two types of learned knowledge that have not
been studied in this planning framework before: macro-actions and planning horizon. Macro-actions are sequences of actions that typically occur in the solution
plans, while a planning horizon of a problem is the length of a (possibly optimal) plan solving it. We propose a method that uses a machine learning tool for
building a predictive model of the optimal planning horizon, and variants of the
well-known planner SatPlan and solver MiniSat that can exploit macro actions
and learned planning horizons to improve their performance. An experimental
analysis illustrates the effectiveness of the proposed techniques.

Keywords: Machine learning for planning, Planning as satisfiability.

1 Introduction

Learning for planning is an important research field in automated planning research,
that, as demonstrated by the last two planning competitions [6,3], in the recent years
has received considerable attention in the planning community. Starting from the PDDL
formalization of a planning problem, the current learning techniques for deterministic
(classical) planning aim at automatically generating additional knowledge about the
problem, and at effectively using it to improve the performance of a planner.

In this paper, we consider two types of learned knowledge for optimal planners in the
"planning as satisfiability" framework (also called SAT-based planning) [11]: *macro-
actions* and the planning *horizon*. Macro-actions are (usually short) sequences of actions that typically occur in the plans solving the problems of a given planning domain.
The planning horizon of a planning problem is the length of a (possibly optimal) plan
solving the problem, that for classical planning is defined as the number of time steps
in the plan.

Regarding macro-actions, several systems for learning and using them in a planner
have been developed, e.g., [2,16]. However, to the best of our knowledge, the use of
macro-actions in SAT-based planning has never been investigated. Regarding learned
horizons, we are not aware of any existing work about learning this information and
exploiting it in SAT-based planning.

R. Pirrone and F. Sorbello (Eds.): AI*IA 2011, LNAI 6934, pp. 189–200, 2011.
© Springer-Verlag Berlin Heidelberg 2011

We focus our study on SatPlan [11,12,13], one of the most popular and efficient SAT-based optimal planning system. Essentially, first SatPlan uses a preprocessing algorithm to compute a lower (possibly exact) bound k on the optimal planning horizon, and translates the planning problem into a SAT problem, i.e., the satisfiability of a propositional formula in CNF (shortly a CNF) encoding the problem. If the SAT problem is solvable (the CNF is satisfiable), a plan with at most k time steps can be derived from a model of the CNF. If the SAT problem is unsolvable (the CNF is unsatisfiable), SatPlan generates a larger SAT problem using an increased bound, and so on, until it finds a solution or it proves that the original planning problem has no solution.

While SatPlan can use any SAT solver, in this work we concentrate our study on MiniSat [5], a very well-known efficient solver based the DPLL algorithm [4], extended with backtracking by conflict analysis and clause recording [14] and with boolean constraint propagation (BCP) [15].

The paper contains the following contributions in the context of SatPlan: (i) a new variant of MiniSat that can exploit a given set of macro-actions to improve the performance of SAT solving for a CNF encoding a planning problem, (ii) a machine learning technique for constructing a predictive model of the optimal planning horizon for a given problem, and (iii) the use of this model, possibly in combination with macro-actions, to reduce the number of SAT problems during planning. The effectiveness of these techniques is studied in a preliminary experimental analysis, showing that they can lead to significant performance improvements.

The rest of the paper is organized as follows. Section 2 describes a modified version of MiniSat exploiting macro-actions; Section 3 introduces a method for estimating the optimal planning horizon, and proposes another enhanced version of MiniSat for computing shorter plans from a satisfiable CNF; Section 4 presents our experimental results; finally, Sections 5 gives the conclusions.

2   Using Macro-actions during SAT Solving

In order to exploit macro-actions in the SAT-solver of SatPlan, we have modified the well-known solver MiniSAT by using macro-actions to bias the way in which the propositional variables are processed (i.e., selected and instantiated). The intuitive general idea is giving preference to unassigned variables corresponding to actions that would include in the current plan macro actions that are compatible with the current assignment. We shall illustrate the idea with an example after introducing more precisely the notion of macro-action for SatPlan. In the rest of the paper, variable v_i^j of the CNF encoding a planning problem denotes action a_i planned at time step j.

In the context of SatPlan, a macro-action is a sequence of propositional variables representing actions executed at certain time steps. For example, consider a planning problem in the well-known BlocksWorld domain, and assume that $\langle a_1, a_2 \rangle$ is a macro-action for this problem, where a_1 and a_2 abbreviate actions (pick-up A) and (stack A B), respectively. Moreover, suppose that the planning horizon (plan steps) in the SAT problem encoding the planning problem under considerations is 5, and the earliest time steps where a_1 and a_2 can be planned are 3 and 4, respectively. Then the CNF encoding the planning problem contains 3 variables v_1^3, v_1^4 and v_1^5 representing action (pick-up A) planned at time steps 3, 4 and 5, respectively, and 2 variables v_2^4 and

MacroMiniSAT($\mathcal{F}$, M)

Input: The CNF $\mathcal{F}$ encoding a planning problem, a set M of macro actions.

Output: A solution variable assignment W or *failure*.

1. $W \leftarrow \emptyset$;
2. **while** $\exists$ variable v of $\mathcal{F}$ not assigned in W **do**
3. $v \leftarrow$ SelectVariableFromMacros($\mathcal{F}$, W, M);
4.   **if** $v = nil$ **then** $v \leftarrow$ SelectVariable($\mathcal{F}$, W);
5. $W \leftarrow W \cup (v = \text{TRUE})$;
6.   *Propagate value of v*;
7.   **if** the propagation has generated conflicts **then**
8.   **if** the propagation has generated a top level conflict **then return** *failure*;
9.   **else** *Perform backtracking*;
10. **return** W.

Fig. 1. MiniSAT modified for solving the SAT encoding of a planning problem using macro-actions

v_2^5 representing action (stack A B) planned at time steps 4 and 5, respectively. For this CNF, two possible macro-actions are $\{v_1^3, v_2^4\}$ and $\{v_1^4, v_2^5\}$. If, for instance, in the current assignment of the SAT-solver v_2^5 is true, v_2^4 is false, and the other variables are unassigned, then v_1^4 is preferred to v_1^3.

Figure 1 gives a high-level description of a variant of MiniSAT, called MacroMiniSAT, for solving the SAT encoding $\mathcal{F}$ of a given planning problem using a set M of macro-actions. Initially, the current set of assigned variables (W) is empty (step 1). At each iteration of the loop 2–9, an unassigned variable v of the input CNF $\mathcal{F}$ is selected by either procedure SelectVariableFromMacros or procedure SelectVariable, and the value of the selected variable is set to true (steps 3–5). Each variable selection and instantiation is called a (search) *decision*. Then, the effects of the last decision are propagated by unit propagation (steps 6–9): when a clause becomes unary under the current assignment, the remaining literal in the clause is set to true and this decision is propagated, possibly reducing other clauses to unary clauses and repeating the propagation. The propagation process continues until no more information can be propagated. If a conflict is encountered (all literals of a clause are false), a conflict clause is constructed and added to the SAT problem. The decisions made are canceled by backtracking, until the conflict clause becomes unary. This unary clause is propagated and the search process continues.

The main difference w.r.t. MiniSAT concerns SelectVariableFromMacros($\mathcal{F}$, W, M), which uses a macro m selected from a set $M_A \subseteq M$ of macros to determine the next variable to instantiate. A macro m' is in M_A if the three following conditions hold:

1. At least one variable of m' is unassigned;
2. All the assigned variable of m' are true according to the current variable assignment W;
3. m' contains no variable belonging to another macro m'' formed by only variables that are true according to W.

The rationale of condition 2 is to avoid preferring variables representing actions that, given the current variable assignment, will not appear in the solution plan. The motivation of condition 3 is based on the empirical observation that, for many domains, often two macro-actions sharing one or more actions do not appear simultaneously in a solution plan.

If set M_A contains more than one macro, SelectVariableFromMacros prefers the macro m' in M_A according to the actions in a given *relaxed plan* π (see, e.g., [9,10,7,17]) for the planning problem under consideration. Plan π is relaxed in the sense that it does not consider the possible negative interference between planned actions, and it can be quickly computed by a polynomial algorithm (see, e.g., [10]). Essentially, SelectVariableFromMacros chooses a variable from the macro m' in M_A formed by the highest percentage of variables of m' representing actions in π.

For example, consider a BlocksWorld problem in which A and B are on the table, C is on B, and the goal is moving A on B. The following sequence of actions is a possible relaxed plan for our running example: (unstack C B), (pick-up A), (stack A B). In the original planning problem, action (unstack C B) would make the robot arm occupied, but, since this is represented through a negative effect of the action, in the relaxed plan the arm remains free after the execution of (unstack C B), and so action (pick up A) can be planned.

Assume that sequences $\langle a_1, a_2 \rangle$ and $\langle a_3, a_4 \rangle$ are two macro-actions for this BlocksWorld problem, where a_1, a_2, a_3 and a_4 abbreviate (pick-up A), (stack A B), (pick-up C), (stack C B). Moreover, assume that v_2^5 is false in the current variable assignment. Then, M_A is formed by $\{v_1^3, v_2^4\}$, $\{v_3^3, v_4^4\}$ and $\{v_3^4, v_4^5\}$; the percentage of variables in these macros representing actions in the relaxed plan is 100, 0 and 0, respectively. Therefore, SelectVariableFromMacros chooses macro $\{v_1^3, v_2^4\}$.

If the number of macro-actions with the highest percentage of variables representing actions in the relaxed plan is greater than one, then SelectVariableFromMacros uses some secondary criteria to select the most promising macro. These criteria include the ratio between the number of variables assigned as true and the cardinality of the macro, the sum of the variable *activity values* (as defined in [5]), and the time step of the first action in the macro. If none of the these criteria returns a single macro, SelectVariableFromMacros randomly chooses a macro from the set of the best macros. Finally, it returns the earliest (time-step wise) unassigned variable from the best macro.

If set M_A contains no macro, SelectVariableFromMacros returns *nil* and, subsequently, the algorithm uses the standard MiniSAT procedure SelectVariable (as defined in [5]) for choosing an unassigned variable of the CNF.

3   Predicting and Using Learned Horizons in SATPLAN

Typical SAT-based planners like SatPlan generate several unsatisfiable CNF encodings of the given planning problem with different (increasing) plan length bounds before finding a solvable CNF (from which an optimal plan is obtained). Unfortunately, we have observed that, while for a solvable CNF the use of macro-actions can speed up the SAT-solver, often for an unsolvable CNF this is not the case. Hence, in order to better

Table 1. The set of features used to define a predictive model of the planning horizon

Name	Description
$\#G$	number of problem goals
$\#O$	number of problem objects
$\#F$	number of facts in the initial state
$\#A$	number of actions grounded by planner LPG
$\#LM$	number of landmarks computed by planner Lama
$\#ME$	number of mutex exclusive relations computed by LPG
π_r^{FD}	number of actions in the relaxed plan constructed by planner FastDownward
π_r^{FF}	number of actions in the relaxed plan constructed by planner FF
π_r^{Lama}	number of actions in the relaxed plan constructed by Lama
π_r^{LPG}	number of actions in the relaxed plan constructed by LPG

exploit the macro-actions in the whole planning process, we have developed a method for predicting the optimal planning horizon of a problem in a given domain. It should be noted that such a predictive model is an independent technique that can be used without the macro-actions.

The predictive model is constructed using a set of features, given in Table 1, concerning the planning problem, the mutex relations and landmarks in its state space (e.g., [7,17]), and the relaxed solutions used in some heuristics of state-of-the-art satisficing planners. The values of features $\#G$, $\#O$, $\#F$ are derived from the "grounded" description of the planning problem, while the values of the other features are derived using the planning techniques implemented in FastDownward [9], FF [10], Lama [17] and LPG [7]. Essentially, for each training problem Π, the length (number of plan steps) l_π of an optimal solution π for Π is computed using an existing optimal planner; the values of the learning features for Π and l_π provide the data from which the predictive model is generated using a machine-learning tool. In our implementation, we used the well-known tool WEKA [18] with technique M5Rules [8].

The experimental results in Table 2 indicate that, for problems with short optimal plans, the estimated optimal horizon computed by WEKA is sometimes better than the first horizon computed by SatPlan, which is the initial length of Graphplan's planning graph; while for problems with middle-size and long optimal plans, the estimated optimal horizons are always better. Moreover, these results indicate that the predicted plan length can be higher than the actual optimal plan length, while SatPlan's initial horizon is a lower or exact bound. Therefore, in order to use a learned horizon in SatPlan, we have modified its standard behaviour as follows.

If the initial CNF $\mathcal{F}$ encoding the planning problem with a predicted horizon t is solvable, then the process is repeated using a CNF with horizon $r - 1$, where r is the number of time steps in the solution plan computed from $\mathcal{F}$, and so on, until a horizon $q < t$ for which the CNF is unsolvable has been identified (the optimal solution plan is the one generated from the CNF with horizon $q + 1$). Otherwise, the process is repeated with horizon $t + 1$, and the planner stops when a solvable CNF is generated (the optimal solution is the plan derived from the solution of this last CNF).

Table 2. Empirical evaluation of the accuracy of the predicted planning horizon using some problems with different size from 7 known domains: optimal plan length (2nd column), gap between the first plan length bound of SatPlan (defined as the length of the initial Graphplan's planning graph) and the optimal plan length (3rd column), and gap between the predicted planning horizon and the optimal plan length. Smaller Δ-values indicate better estimates.

Problem	Opt. Length	Δ_G	Δ_W
Short plans			
BlocksWorld – 6	8	-2	2
Depots – 3	9	-4	1
Ferry – 3	7	-3	3
Goldminer – 5	8	0	5
Gripper – 4	7	-4	1
Matching-BW – 5	7	-3	4
Sokoban – 6	8	0	2
Middle-size plans			
BlocksWorld – 8	18	-8	-1
Depots – 5	12	-3	2
Ferry – 4	13	-9	1
Goldminer – 7	16	-3	0
Gripper – 8	15	-12	1
Matching-BW – 8	15	-8	0
Sokoban – 7	19	-6	-5
Long plans			
BlocksWorld – 28	81	-47	3
Depots – 10	20	-6	4
Ferry – 11	40	-36	0
Goldminer – 15	92	-18	16
Gripper – 11	23	-20	-1
Matching-BW – 30	46	-28	2
Sokoban – 12	51	-31	4

Each generated CNF can be solved using the modified version of the MacroMiniSAT procedure shown in Figure 2. The gray steps indicate the differences with respect to the version using only macro actions given in Figure 1. At each iteration of the loop 2–13, if the selected variable v represents an action at level (time step) l, procedure PropagateGoalNoop is called to possibly assign true to the unassigned variables encoding "no-ops" at every time step $i > l$ and representing problem goals. As observed below, this is useful when the solution plan has a length that is lower that the current plan horizon.

The loop 4–11 of PropagateGoalNoop assigns true to each unassigned variable v representing a problem goal at level l, and propagates this decision. If a conflict is

MacroMiniSAT$(\mathcal{F}, M)$

Input: The CNF $\mathcal{F}$ encoding a planning problem, a set M of macro actions.

Output: A solution variable assignment W or *failure*.

1. $W \leftarrow \emptyset$;
2. **while** $\exists$ variable v of $\mathcal{F}$ not assigned in W **do**
3. $v \leftarrow$ SelectVariableFromMacros$(\mathcal{F}, W, M)$;
4.   **if** $v = nil$ **then** $v \leftarrow$ SelectVariable$(\mathcal{F}, W)$;
5. $W \leftarrow W \cup (v = \text{TRUE})$;
6.   *Propagate value of v*;
7.   **if** the propagation has generated no conflict **then**
8.   **if** v represents an action **then**
9. $W \leftarrow$ PropagateGoalNoop$(\mathcal{F}, W, Level(v) + 1)$;
10.   **if** $W = failure$ **then return** *failure*;
11.   **else**
12.   **if** the propagation has generated a top level conflict **then return** *failure*;
13.   **else** *Perform backtracking*;
14. **return** W.

PropagateGoalNoop$(\mathcal{F}, W, m)$

Input: The CNF $\mathcal{F}$ encoding a planning problem, a variable assignment W, and a time step m.

Output: A (partial) variable assignment W or *failure*.

1. **if** $m > EndLevel$ **then return** W;
2. **for** $l = m$ **to** $EndLevel$ **do**
3. $G(l) \leftarrow$ set of variables of $\mathcal{F}$ encoding no-ops at level l and representing goals;
4.   **foreach** unassigned variable v in $G(l)$ **do**
5. $W \leftarrow W \cup (v = \text{TRUE})$;
6.   *Propagate value of v*;
7.   **if** the propagation has generated no conflict **then**
8.   **if** all variables of $\mathcal{F}$ are assigned **then return** W;
9.   **else**
10.   **if** the propagation has generated a top level conflict **then return** *failure*;
11.   **else** *Perform backtracking*;

Fig. 2. Algorithms for solving the SAT encoding of a planning problem using a set of macro-actions when the horizon can be higher than the optimal one. Function *Level(v)* returns the time step of the action encoded by variable v. *EndLevel* represents the latest time step of an action encoded in the input CNF $\mathcal{F}$.

generated, and backtracking is performed. The outer loop (steps 2–11) repeats these variable assignments for each time step from the input time step l to the latest time step of an action encoded in the input CNF $\mathcal{F}$.

We experimentally observed that this modified version of MacroMiniSAT generates plans shorter than those generated by the version in Figure 1. Hence, when the predicted horizon is greater than the optimal one, SatPlan generates (and solves through MacroMiniSAT) fewer CNFs.

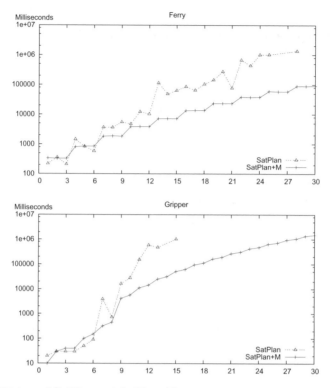

Fig. 3. CPU time of SatPlan and SatPlan+M to solve the SAT problems encoded from the planning problems of domains Ferry and Gripper using the optimal horizon. On the x-axis we have the problem names simplified by numbers.

4   Experimental Results

In this section, we give some results from an experimental analysis aimed at

- understanding the effectiveness of using macro-actions for SAT-based planning;
- evaluating the impact of using learned horizons instead of those computed by SatPlan.

All experimental tests were conducted using an Intel Xeon(tm) 3 GHz machine, with 2 Gbytes of RAM. Unless otherwise specified, the CPU-time limit for each run was 30 minutes, after which termination was forced. (The CPU time used for computing the values of the considered problem features and estimating the plan length is generally negligible w.r.t. the time used for planning, and in our analysis it is ignored.) The CPU time used for computing the values of the problem features and estimating the plan length is generally negligible w.r.t. the time used for planning, and is ignored in our analysis.) In our experiments, macro-actions were computed using the techniques incorporated into Macro-FF [2], a well-known available planning system. However, any other system for computing macro-actions could be used.

Overall, the experiments consider 7 known planning domains: BlocksWorld, Depots, Ferry, Goldminer, Gripper, Matching-BW and Sokoban. However, for

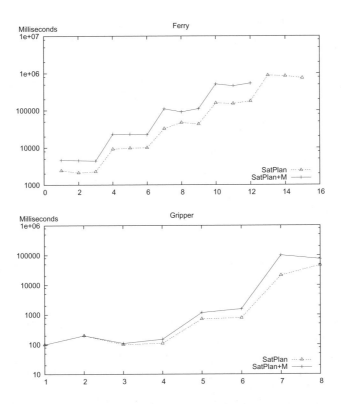

Fig. 4. CPU times of SatPlan and SatPlan+M in domains Ferry and Gripper. On the x-axis we have the problem names simplified by numbers.

testing the use of macro-actions we focus only on domains Ferry and Gripper. Domain Ferry concerns transporting cars between locations using a ferry. Each location is directly connected to every other location; cars can be debarked and boarded; the available ferry can carry at most one car at a time. The only macro-action used in our experiments for this domain is formed by three domain actions (no other macro-action is learned): boarding a car x from a location l_1 to the ferry, moving the ferry from l_1 to another location l_2, and debarking x in l_2. Domain Gripper concerns transporting some balls between two rooms using a robot with two gripper hands. The macro-action used in our experiments for this domain is formed by five actions (again this is the only learned macro-action): picking a ball x from a room r_1 by the right hand, picking a ball y from room r_1 by the left hand, moving the robot from r_1 to another room r_2, dropping x in r_2, and dropping y to r_2.

For the other domains considered in our experiments, the macro-actions computed by Macro-FF are not suitable for SatPlan because they are long and can appear only in sub-optimal plans, or there is a large number of similar macro-actions, involving many variables of the CNF. In these situations, the variant of MiniSat that tries to use macro-actions is not more efficient than the original version.

Table 3. Percentage of solved problems (columns 3-4), average CPU time (columns 5-6) and IPC score (columns 7-8) of SatPlan and SatPlan using the learned horizon estimated by WEKA for 7 known domains

Domain	#Prob	% Solved		Mean CPU Time		Speed score	
		SatPlan	SatPlan+H	SatPlan	SatPlan+H	SatPlan	SatPlan+H
BlocksWorld	60	80	98.33	330.2	321.5	20.6	58.2
Depots	43	100	100	78.6	70.6	30.8	40.7
Ferry	18	83.3	94.4	223.5	176.5	4.9	17.0
Goldminer	44	97.3	95.5	464.5	374.9	31.0	41.3
Gripper	10	80.0	80.0	15.8	10.4	5.8	7.8
Matching-BW	30	96.7	96.67	228.5	146.9	16.9	28.5
Sokoban	140	99.0	100	232.3	93.2	61.0	138.5
Total	345	94.5	98.3	251.1	178.9	171.0	332.1

Figure 3 shows the CPU times required by the original version of SatPlan and the version using macros (abbreviated with SatPlan+M) for domains Ferry and Gripper when the optimal planning horizon is given (and the first generated SAT problem is solvable).

SatPlan+M is almost always faster than SatPlan and solves much larger problems. However, as shown in Figure 4, when the optimal planning horizon is not given, SatPlan+M is usually slower than SatPlan (up to about two times). The main reason why macro-actions slow down the planning process is that they are not useful when a SAT problem is unsolvable, and usually SatPlan generates several unsolvable SAT problems before the solvable one.

In order to better exploit macro-actions, it is important to have accurate bounds on the optimal horizon and minimize the number of the generated unsolvable SAT problems. In the last part of this section, we will show that using macro-actions can be useful when combined with the use of learned horizons (which often are upper bounds, rather than lower bounds as in SatPlan, on the optimal horizons).

Table 3 gives the percentage of solved problems, the average CPU time and the *speed score* of the original version of SatPlan and the proposed version using learned planning horizons (abbreviated with SatPlan+H). The speed score was first introduced and used by the organizers of the 6th International Planning Competition [6] for evaluating the relative performance of the competing planners, and since then it has become a standard method for comparing planning systems. The speed score of a system s is defined as the sum of the speed scores assigned to s over all the considered problems. The speed score assigned to s for a planning problem P is 0 if P is unsolved and $T_P^*/T(s)_P$ otherwise, where T_P^* is the lowest measured CPU time to solve problem P and $T(s)_P$ denotes the CPU time required by s to solve problem P. Higher values of the speed score indicate better performance. The results in Table 3 indicate that the number of problems solved by SatPlan+H is greater than or equal to the number of those solved by SatPlan, and that SatPlan+H is almost always faster than SatPlan, because the speed score of SatPlan+H is very close to the number of considered problems.

Figure 5 shows the CPU time of the original version of SatPlan (which generates the first SAT problem using a lower bound on the optimal horizon through Graphplan) and our version using macros *and* the horizon learned by the proposed method (abbreviated

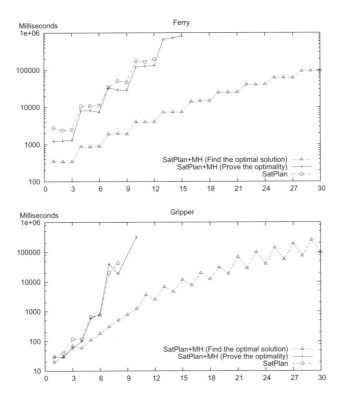

Fig. 5. CPU time of SatPlan and SatPlan+MH to find the optimal solution and to prove that such a solution is optimal for domains Ferry and Gripper. On the x-axis we have the problem names simplified by numbers.

with SatPlan+MH). The results in Figure 5 show that, for the considered domains, the combination of using learned horizons and macro-actions speeds up the planning process of SatPlan. Finally, the CPU time of SatPlan+MH for finding the optimal plan is up to about two orders of magnitude lower than the time of SatPlan, and, moreover, SatPlan+MH solves many more problems. SatPlan+MH is often still faster than SatPlan in proving that a computed solution is optimal, but the performance gap is reduced. The reason why the performance gap is smaller is that, in order to prove optimality, one *un*solvable SAT problem (with horizon equal to the optimal horizon minus one) is generated and, as previously observed, for domains Ferry and Gripper the use of macro-actions for unsolvable SAT problems do not usually increase the performance.

5   Conclusions

We have investigated the use of macro-actions in SAT-based planning based on a variant of a well-known SAT solver that can exploit this information to speed up planning. Moreover, we have presented a new method for predicting the optimal planning horizon through a model generated using standard machine-learning techniques, and shown how to use it in combination with macro actions. A preliminary experimental study indicates that: (i) the use of macro actions can speed up the SAT solver when the CNF encoding of the problem is satisfiable; (ii) the estimate of the optimal plan length computed by our method is more accurate than the bound computed by SatPlan through Graphplan's planning graph and it can be used to improve SatPlan speed; and (iii) the use of macro-actions combined with the learned planning horizon can speed up SatPlan. Future work includes studying the use of macro-actions generated by other tools, e.g., WIZARD [16], and running additional experiments about the impact of macro-actions and learned horizons on the performance of SAT-based planning.

References

1. Blum, A., Furst, M.L.: Fast planning through planning graph analysis. Artificial Intelligence 90, 281–300 (1997)
2. Botea, A., Müller, M., Schaeffer, J.: Learning partial-order macros from solutions. In: Proc. of ICAPS 2005 (2005)
3. Celorrio, S.J., Coles, A., Coles, A.: 7th Int. Planning Competition – Learning Track (2011), http://www.plg.inf.uc3m.es/ipc2011-learning
4. Davis, M., Logemann, G., Loveland, D.: A machine program for theorem-proving. Communications of the ACM 5, 394–397 (1962)
5. Een, N., Sörensson, N.: An extensible SAT-solver. In: Giunchiglia, E., Tacchella, A. (eds.) SAT 2003. LNCS, vol. 2919, pp. 502–518. Springer, Heidelberg (2004)
6. Fern, A., Khardon, R., Tadepalli, P.: 6th Int. Planning Competition – Learning Track (2008), http://eecs.oregonstate.edu/ipc-learn/
7. Gerevini, A., Saetti, A., Serina, I.: Planning through stochastic local search and temporal action graphs. Journal of Artificial Intelligence Research 20, 239–290 (2003)
8. Hall, M., Holmes, G., Frank, E.: Generating Rule Sets from Model Trees. In: Foo, N.Y. (ed.) AI 1999. LNCS, vol. 1747. Springer, Heidelberg (1999)
9. Helmert, M.: The Fast downward planning system. Journal of Artificial Intelligence Research 26, 191–246 (2006)
10. Hoffmann, J., Nebel, B.: The FF planning system: Fast plan generation through heuristic search. Journal of Artificial Intelligence Research 14, 253–302 (2001)
11. Kautz, H., Selman, B.: Planning as satisfiability. In: Proc. of ECAI 1992 (1992)
12. Kautz, H., Selman, B.: Unifying SAT-based and graph-based planning. In: Proc. of IJCAI 1999 (1999)
13. Kautz, H., Selman, B., Hoffmann, J.: SatPlan: Planning as satisfiability. In: Abstract Booklet of the 5th Int. Planning Competition (2006)
14. Marques, S.J.P., Karem, S.A.: GRASP a new search algorithm for satisfiability. In: Proc. of ICCD 1996 (1996)
15. Moskewicz, M.W., Madigan, C.F., Zhao, Y., Zhang, L., Malik, S.: Chaff: Engineering an Efficient SAT Solver. In: Proc. of DAC 2001 (2001)
16. Newton, M., Levine, J., Fox, M., Long, D.: Learning macro-actions for arbitrary planners and domains. In: Proc. of ICAPS 2007 (2007)
17. Richter, S., Westphal, M.: The LAMA planner: Guiding cost-based anytime planning with landmarks. Journal of Artificial Intelligence Research 39, 127–177 (2010)
18. Witten, I.H., Frank, E.: Data mining: Practical machine learning tools and techniques. Morgan Kaufmann, San Francisco (2005)

Clustering Web Search Results with Maximum Spanning Trees

Antonio Di Marco and Roberto Navigli

Dipartimento di Informatica,
Sapienza Università di Roma,
Via Salaria, 113 - 00198 Roma Italy
{dimarco,navigli}@di.uniroma1.it
http://lcl.uniroma1.it

Abstract. We present a novel method for clustering Web search results based on Word Sense Induction. First, we acquire the meanings of a query by means of a graph-based clustering algorithm that calculates the maximum spanning tree of the co-occurrence graph of the query. Then we cluster the search results based on their semantic similarity to the induced word senses. We show that our approach improves classical search result clustering methods in terms of both clustering quality and degree of diversification.

1 Introduction

The huge amount of text nowadays available on the Web makes language-related tasks, such as Information Retrieval, Information Extraction and Question Answering, increasingly difficult. Popular search engines such as Yahoo! and Google in general do a good job at finding a needle in a haystack, i.e., retrieving a small number of relevant results from such an enormous collection of Web pages. However, the current generation of search engines still lacks an effective way to address the issue of lexical ambiguity. In a recent study [34] – conducted using WordNet [25] and Wikipedia as sources of ambiguous words – it was reported that around 3% of Web queries and 23% of the most frequent queries are ambiguous. Such ambiguity is often due to the low average number of query words used by Web users [16]. While the average query length is increasing (now estimated at around 3 words per query) many search engines are addressing the query ambiguity issue by reranking and diversifying their results, so as to return Web pages that are not too similar to each other.

In recent years, Web clustering engines [7] have been proposed as a solution to the issue of lexical ambiguity in Web Information Retrieval. These systems group search results, by providing a cluster for each specific aspect (i.e., meaning) of the input query. Users can then select the cluster(s) and the pages therein that best answer their information needs. However, many Web clustering engines group search results on the basis of their lexical similarity, and therefore suffer from synonymy (same query expressed with different words) and polysemy (different user needs expressed with the same word).

In this paper we present a novel approach to Web search result clustering which is based on the automatic discovery of word senses from raw text – a task we refer to as

R. Pirrone and F. Sorbello (Eds.): AI*IA 2011, LNAI 6934, pp. 201–212, 2011.

Word Sense Induction (WSI). At the core of our approach is the identification of the user query's meaning using a graph-based algorithm which calculates a maximum spanning tree of the co-occurrence graph of the input query. Our experiments on two datasets of ambiguous queries show that our WSI approach boosts search result clustering in terms of both clustering quality and degree of diversification.

2 Related Work

Web directories such as the Open Directory Project are a first solution to query ambiguity. They provide taxonomies for the categorization of Web pages. Given a query, search results are organized by category. This approach has three main weaknesses: first, it is static, thus it needs manual updates to cover new pages; second, it covers only a small portion of the Web; third, it classifies Web pages based on coarse categories. This latter feature of Web directories makes it difficult to distinguish between instances of the same kind (e.g., pages about musicians with the same surname). While methods for the automatic classification of Web documents have been proposed and some problems have been tackled effectively [2], these approaches are usually supervised and still suffer from reliance on a predefined taxonomy of categories.

A different direction consists of associating explicit semantics (i.e., word senses or concepts) with queries and documents, that is, performing Word Sense Disambiguation (WSD, see [26] for a survey). SIR is performed by indexing and/or searching concepts rather than terms, thus potentially coping with both synonymy and polysemy. Over the years, different methods for SIR have been proposed [18,40,24,23, inter alia]. However, contrasting results have been reported on the benefits of these techniques: it was shown that WSD has to be very accurate to benefit Information Retrieval [33] – a result that was later debated [37].

SIR performs WSD using a reference knowledge resource (such as WordNet) and thus suffers from the static nature of the dictionary sense inventory and its inherent paucity of most proper nouns. This latter problem is particularly important for Web searches, as users tend to retrieve more information about named entities (e.g., singers, artists, cities) than concepts (e.g., abstract information about singers or artists).

A third approach to query ambiguity is search result clustering. Given a query, a flat list of text snippets returned from one or more commonly-available search engines is clustered using some notion of textual similarity. At the root of the clustering approach lies van Rijsbergen's [32] cluster hypothesis: "closely associated documents tend to be relevant to the same requests", whereas documents concerning different meanings of the input query are expected to belong to different clusters. Approaches to search result clustering can be classified as data-centric or description-centric [7]. The former focus more on the problem of data clustering than on presenting the results to the user. A pioneering example is Scatter/Gather [13], which divides the dataset into a small number of clusters and, after the selection of a group, performs clustering again and proceeds iteratively. Developments of this approach have been proposed which improve on cluster quality and retrieval performance [17]. Other data-centric approaches use agglomerative hierarchical clustering (e.g., LASSI [42]), rough sets [28] or exploit link information [47]. Description-centric approaches are, instead, more focused on the description to produce for each cluster of search results. Among the most popular and successful approaches are those based on suffix trees [44]. Other methods in the literature are based on formal concept analysis [8], singular value decomposition [30], spectral clustering [11] and graph connectivity measures [14].

Diversification is another research topic dealing with the issue of query ambiguity. Its aim is to reorder top search results using criteria that maximize their diversity. Similarity functions have been used to measure the diversity among documents and between document and query [5]. Other techniques include the use of conditional probabilities to determine which document is most different from higher-ranking ones [9] or use affinity ranking [46], based on topic variance and coverage. More recently, an algorithm called Essential Pages [38] has been proposed to reduce information redundancy and return Web pages that maximize coverage with respect to the input query.

In our work we perform WSI to dynamically acquire an inventory of senses of the input query. Instead of clustering on the basis of the surface similarity of Web snippets, we use our induced word senses to group snippets. This framework was proposed for the first time in [27], where an effective graph algorithm based on triangles and squares was presented. In this paper we use the same framework to introduce maximum spanning trees for WSI-driven search result clustering. Very little further work on this topic has been done: vector-based WSI was successfully shown to improve bag-of-words ad-hoc Information Retrieval [36] and experimental studies [10] have provided interesting, though preliminary, insights into the use of WSI for Web search result clustering. More recently the use of hidden topics has been proposed to identify query meanings [29]. However, topics – estimated from a universal dataset – are query-independent and thus their number needs to be found beforehand. In contrast, we cluster snippets according to a dynamic and finer-grained notion of sense.

3   Approach

Web search result clustering is typically performed in three steps:

1. Given a query q, a search engine (e.g., Yahoo!) is used to retrieve a list of results $R = (r_1, \ldots, r_n)$;
2. A clustering $\mathcal{C} = (C_0, C_1, \ldots, C_m)$ of the results in R is obtained by means of a clustering algorithm;
3. The clusters in $\mathcal{C}$ are optionally labeled with an appropriate algorithm (e.g., see [6] and [43]) for visualization purposes.

In this paper we aim at improving step 2 by means of a graph-based Word Sense Induction algorithm: given a query q, we first use a text corpus to automatically induce the word senses of q (Section 3.1); then we cluster the Web results using the previously-acquired word senses (Section 3.2).

3.1   Word Sense Induction

Word Sense Induction is a task aimed at dynamically identifying the set of senses denoted by a word. These methods acquire word senses from text by grouping word occurrences exploiting the idea that a given word, when used in a specific sense, tends to co-occur with the same neighbouring words [15]. Several approaches to WSI have been proposed in the literature (see [26] for a survey), including context-vector clustering [35], word clustering [22] and co-occurrence graphs [41,27].

A successful approach to WSI is HyperLex [39], a graph algorithm based on the identification of hubs in co-occurrence graphs. However, HyperLex has to cope with a high number of parameters to be tuned [1]. To overcome this issue we propose a simple yet effective graph algorithm for WSI, which we describe hereafter. The algorithm consists of two steps: graph construction and identification of word senses.

Graph construction. Given a target query q, we build a co-occurrence graph $G_q = (V, E)$ such that V is a set of context words related to q and E is the set of undirected edges, each denoting a co-occurrence between pairs of words in V. To determine the set of co-occurring words V, we use the Google Web1T corpus [4], a large collection of n-grams ($n = 1, \ldots, 5$) – i.e., windows of n consecutive tokens – occurring in one terabyte of Web documents. First, for each content word w we collect the total number $c(w)$ of its occurrences and the number of times $c(w, w')$ that words w and w' occur together in any 5-gram (we include inflected forms in the count); second, we use the Dice coefficient to determine the strength of co-occurrence between w and w':

$$Dice(w, w') = \frac{2c(w, w')}{c(w) + c(w')}. \tag{1}$$

The graph $G_q = (V, E)$ is built as follows:

– Our initial vertex set $V^{(0)}$ contains all the content words from the snippet results of query q (excluding stopwords); then, we add to $V^{(0)}$ the highest-ranking words co-occurring with q in the Web1T corpus, i.e., those words w for which $c(q, w) \geq \delta$ and $Dice(q, w) \geq \delta'$ (the thresholds are established experimentally, see Section 4.1). We set $V := V^{(0)}$ and $E := \emptyset$.
– Given a pair of words $\{w, w'\} \in V \times V$, if $max\{\frac{c(w,w')}{c(w)}, \frac{c(w,w')}{c(w')}\} \geq \sigma$, we add edge $\{w, w'\}$ to E with weight $Dice(w, w')$.
– Finally, we remove disconnected vertices.

Identification of word senses. Given the co-occurrence graph G_q for query q, we perform the following steps:

1. Eliminate from G_q all nodes whose degree is 1.
2. Calculate the maximum spanning tree (MST) T_{G_q} of the graph.
3. Work on T_{G_q} by iteratively eliminating the minimum-weight edge $e \in T_{G_q}$ such that its endpoints each have degree ≥ 2 until we obtain N connected components (i.e., word clusters) or there are no more edges to eliminate.

We provide an example of co-occurrence graph for the query *beagle* in Figure 1(a). The maximum spanning tree (step 2 above) is shown in bold, whereas the result of step 3 above, i.e., the final meaning components or word senses, is shown in Figure 1(b).

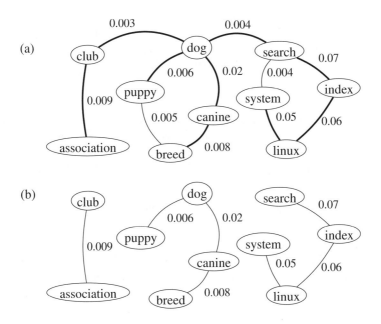

Fig. 1. The *MST* example ($N = 3$): (a) co-occurrence graph G and maximum spanning tree T_G (edges in bold); (b) the word senses induced after edge removal.

3.2 Clustering Web Results

We submit our input query q to a search engine, which returns a list of relevant search results $R = (r_1, \ldots, r_n)$. We transform the text snippet corresponding to each result r_i into a bag of words b_i. To this end, we apply tokenization, stopwords and target word removal, and lemmatization[1]). For instance, given the snippet:

"the *beagle* is a breed of medium-sized dog",

we produce the following bag of words:

{ *breed, medium, size, dog* }.

As a result of the above processing steps, we obtain a list of bags of words $B = (b_1, \ldots, b_n)$. Now, our aim is to cluster our Web results R, i.e., the corresponding bags of words B. To this end, rather than considering the interrelationships between them (as is done in traditional search result clustering), we intersect each bag of words $b_i \in B$ with the sense clusters $\{S_1, \ldots, S_m\}$ acquired as a result of our Word Sense Induction algorithm (cf. Section 3.1). The sense cluster with the largest intersection with b_i is selected as the most likely meaning of r_i. Formally:

$$Sense(r_i) = \begin{cases} \underset{j=1,\ldots,m}{\operatorname{argmax}} |b_i \cap S_j| & \text{if } \underset{j}{\max} |b_i \cap S_j| > 0 \\ 0 & \text{else} \end{cases} \quad (2)$$

---

[1] We use the WordNet lemmatizer.

where 0 denotes that no sense is assigned to result r_i, as the intersection is empty for all senses S_j. Otherwise the function returns the index of the sense having the largest overlap with b_i – the bag of words associated with the search result r_i. As a result of sense assignment for each $r_i \in R$, we obtain a clustering $\mathcal{C} = (C_0, C_1, \ldots, C_m)$ such that:

$$C_j = \{r_i \in R : Sense(r_i) = j\}, \tag{3}$$

that is, C_j contains the search results classified with the j-th sense of query q (C_0 includes unassigned results). Finally, we sort the clusters in our clustering $\mathcal{C}$ based on their "quality". For each cluster $C_j \in \mathcal{C} \setminus \{C_0\}$, we determine its similarity with the corresponding meaning S_j by calculating the following formula:

$$avgsim(C_j, S_j) = \frac{\sum_{r_i \in C_j} sim(r_i, S_j)}{|C_j|}. \tag{4}$$

The formula determines the average similarity between the search results in cluster C_j and the corresponding sense cluster S_j. The similarity between a search result r_i and S_j is determined as the normalized overlap between its bag of words b_i and S_j:

$$sim(r_i, S_j) = sim(b_i, S_j) = \frac{|b_i \cap S_j|}{|b_i|}. \tag{5}$$

Finally, we rank the elements r_i within each cluster C_j by their similarity $sim(r_i, S_j)$. We note that the ranking and optimality of clusters can be improved with more sophisticated techniques [12,19,20,21, inter alia]. However, this is outside the scope of this paper.

4   Experiments

4.1   Experimental Setup

Test Sets. We conducted our experiments on two datasets:

– AMBIENT (AMBIguous ENTries), a dataset which contains 44 ambiguous queries[2]. The sense inventory for the senses (i.e., subtopics)[3] of queries is given by Wikipedia disambiguation pages. For instance, given the *beagle* query, its disambiguation page in Wikipedia provides the senses of dog, Mars lander, computer search service, etc. The most appropriate Wikipedia senses were associated with the top 100 Web results of each query returned by the Yahoo! search engine (overall, 4400 search results were sense tagged).
– MORESQUE (MORE Sense-tagged QUEry results), a new dataset of 114 ambiguous queries which we developed as a complement to AMBIENT following the guidelines provided by the authors of the latter. In fact, our aim was to study the behaviour of Web search algorithms on queries of different lengths, ranging from

---

[2] http://credo.fub.it/ambient
[3] In the following, we use the terms *subtopic* and *word sense* interchangeably.

1 to 4 words. However, the AMBIENT dataset is composed mostly of single-word queries. MORESQUE provides dozens of queries of length 2, 3 and 4, together with the 100 top results from Yahoo! for each query annotated as in the AMBIENT dataset (overall, we tagged 11,400 snippets).

Parameters. Our graph-based algorithm has just one parameter: the maximum number N of clusters. We experimentally set this value to 5 using a small development set of queries and snippets. We used the same development set to learn the three parameters (δ, δ' and σ) needed for the construction of the co-occurrence graphs.

Systems. We compared MST against the best systems reported by [3] (cf. Section 2):

- **Lingo** [30]: a Web clustering engine implemented in the Carrot[2] open-source frame-work[4] that clusters the most frequent phrases extracted using suffix arrays.
- **Suffix Tree Clustering (STC)** [43]: the original Web search clustering approach based on suffix trees.
- **KeySRC** [3]: a state-of-the-art Web clustering engine built on top of STC with part-of-speech pruning and dynamic selection of the cut-off level of the clustering dendrogram.
- **Essential Pages (EP)** [38]: a recent diversification algorithm that selects fundamental pages which maximize the amount of information covered for a given query.
- **Yahoo!:** the original search results returned by the Yahoo! search engine.

The first three of the above are Web search result clustering approaches, whereas the last two produce lists of possibly diversified results (cf. Section 2).

4.2 Experiment 1: Clustering Quality

Measure. While assessing the quality of clustering is a notably hard problem, given a gold standard $\mathcal{G}$ we can calculate the **Rand index** (RI) of a clustering $\mathcal{C}$, a common quality measure in the literature, determined as follows [31]:

$$\text{RI}(\mathcal{C}) = \frac{a}{\binom{|\mathcal{W}|}{2}} \qquad (6)$$

where $\mathcal{W}$ is the union set of all the snippets in $\mathcal{C}$ and a is the number of snippet pairs put into the same (or different) cluster in both $\mathcal{C}$ and $\mathcal{G}$. For the gold standard $\mathcal{G}$ we use the clustering induced by the sense annotations provided in our datasets for each snippet. Similarly to what was done in Section 3.2, untagged results are grouped together in a special cluster of $\mathcal{G}$.

Results. The results of all systems on the AMBIENT and MORESQUE datasets according to the average Rand index are shown in Table 1[5]. In accordance with previous results in the literature, KeySRC performed generally better than the other search result

---

[4] http://project.carrot2.org
[5] For reference systems we used the implementations of [3] and [30].

Table 1. Results by Rand index (percentages)

System	AMBIENT	MORESQUE	All
MST	81.53	86.67	85.24
Lingo	62.75	52.68	55.49
STC	61.48	51.52	54.29
KeySRC	66.49	55.82	58.78

clustering systems, especially on smaller queries. Our Word Sense Induction system, MST, outperformed all other systems by a large margin, thus showing a higher clustering quality. Interestingly, all clustering systems perform more poorly on longer queries (i.e., on the MORESQUE dataset) whereas our WSI system overturns this trend performing better with longer queries.

4.3   Experiment 2: Diversification

Measure. We performed a second experiment to assess the ability of our clustering algorithms to diversify the top results returned by a search engine. For each query q, one natural way of measuring a system's performance is to calculate the **subtopic recall-at-K** [45] given by the number of different subtopics retrieved for q in the top K results returned:

$$\text{S-recall@K} = \frac{|\bigcup_{i=1}^{K} subtopics(r_i)|}{M} \tag{7}$$

where $subtopics(r_i)$ is the set of subtopics manually assigned to the search result r_i and M is the number of subtopics for query q (note that in our experiments M is the number of subtopics occurring in the 100 results retrieved for q, so S-recall@100 = 1). However, this measure is only suitable for systems returning ranked lists (such as Yahoo! and EP). Given a clustering $\mathcal{C} = (C_0, C_1, \ldots, C_m)$, we flatten it to a list as follows: we add to the initially empty list the first element of each cluster C_j ($j = 1, \ldots, m$); then we iterate the process by selecting the second element of each cluster C_j such that $|C_j| \geq 2$, and so on. The remaining elements returned by the search engine, but not included in any cluster of $\mathcal{C} \setminus \{C_0\}$, are appended to the bottom of the list in their original order. Note that the elements are selected from each cluster according to their internal ranking (e.g., for our algorithms we use Formula 5 introduced in Section 3.2).

Results. We compared the output of our system with the original snippet list returned by Yahoo! and the output of the EP diversification algorithm (cf. Section 4.1).

The S-recall@K (with $K = 3, 5, 10, 15, 20$) calculated on AMBIENT + MORES-QUE is reported in Table 2. MST performs best, with a subtopic recall greater than all other systems. We observe that KeySRC and EP perform worse than Yahoo! with low values of K and generally better with higher values of K.

Given that the two datasets complement each other in terms of query lengths (with AMBIENT having queries of length ≤ 2 and MORESQUE with many queries of length ≥ 3), we studied the S-recall@K trend for the two datasets. The results are

Table 2. S-recall@K on all queries (percentages)

System	K=3	K=5	K=10	K=15	K=20
MST	54.7	65.6	79.2	86.7	90.7
Yahoo!	49.2	60.0	72.9	78.5	82.7
EP	40.6	53.2	68.6	77.2	83.3
KeySRC	44.3	55.8	72.0	79.1	83.2

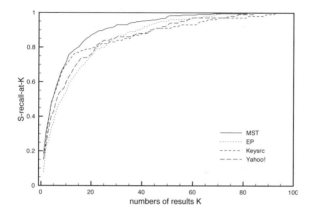

Fig. 2. Results by S-recall@K on AMBIENT

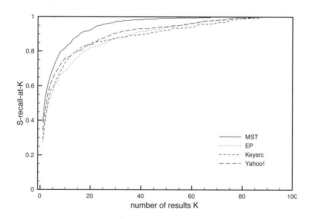

Fig. 3. S-recall@K on MORESQUE

shown in Figures 2 and 3. While KeySRC does not show large differences in the presence of short and long ambiguous queries, our graph-based algorithm does. For instance, as soon as $K = 3$ the MST algorithm obtains S-recall values of 38.72% and 60.46% on AMBIENT and MORESQUE, respectively. The difference decreases as K increases, but is still significant when $K = 15$. We hypothesize that, because they are less ambiguous, longer queries are easier to diversify with the aid of WSI. However, we note that even with low values of K MST obtains higher S-recall than the other systems (with KeySRC competing on AMBIENT when $K \leq 10$).

5   Conclusions

We have presented a new approach to Web search result clustering. Key to our approach is the idea of inducing senses for the target query automatically by means of a simple, yet effective algorithm based on the maximum spanning tree of the cooccurrence graph. The results of a Web search engine are then mapped to the query senses and clustered accordingly.

The paper provides two contributions. First we corroborate our previous finding [27] that WSI greatly improves the quality of search result clustering as well as the diversi- fication of the snippets returned as a flat list. We provide a clear indication on the use- fulness of a loose notion of sense to cope with ambiguous queries. This is in contrast to research on Semantic Information Retrieval, which has obtained contradictory and often inconclusive results. The main advantage of WSI lies in its dynamic production of word senses that cover both concepts (e.g., *beagle* as a breed of dog) and instances (e.g., *beagle* as a specific instance of a space lander). In contrast, static dictionaries such as WordNet – typically used in Word Sense Disambiguation – by their very nature encode mainly concepts. Second, we propose a simple graph algorithm that induces the senses of our queries. Our algorithm has only a single parameter for the sense induction step.

Given the lack of ambiguous query datasets available [34], we hope our new dataset will be useful in future comparative experiments. Moreover, its requirement of a Web corpus of n-grams is not an onerous one, as such corpora are available for several lan- guages and can be produced for any language of interest.

Acknowledgments. We thank Google for providing the Web1T corpus for research purposes; Massimiliano D'Amico for producing the output of KeySRC and EP; Stanis- law Osinski and Dawid Weiss for their help with Lingo and STC; Jim McManus for his useful comments on the original manuscript. The second author gratefully acknowl- edges the support of the ERC Starting Grant MultiJEDI No. 259234.

References

1. Agirre, E., Martínez, D., de Lacalle, O.L., Soroa, A.: Evaluating and optimizing the param- eters of an unsupervised graph-based WSD algorithm. In: Proc. of TextGraphs 2006, New York, USA, pp. 89–96 (2006)
2. Bennett, P.N., Nguyen, N.: Refined experts: improving classification in large taxonomies. In: Proc. of SIGIR 2009, Boston, MA, USA, pp. 11–18 (2009)
3. Bernardini, A., Carpineto, C., D'Amico, M.: Full-subtopic retrieval with keyphrase-based search results clustering. In: Proc. of WI 2009, Milan, Italy, pp. 206–213 (2009)
4. Brants, T., Franz, A.: Web 1t 5-gram, ver. 1, ldc2006t13. In: LDC, PA, USA (2006)
5. Carbonell, J., Goldstein, J.: The use of mmr, diversity-based reranking for reordering docu- ments and producing summaries. In: Proc. of SIGIR 1998, Melbourne, Australia, pp. 335– 336 (1998)

6. Carmel, D., Roitman, H., Zwerdling, N.: Enhancing cluster labeling using Wikipedia. In: Proc. of SIGIR 2009, MA, USA, pp. 139–146 (2009)
7. Carpineto, C., Osiński, S., Romano, G., Weiss, D.: A survey of web clustering engines. ACM Computing Surveys 41(3), 1–38 (2009)
8. Carpineto, C., Romano, G.: Exploiting the potential of concept lattices for information retrieval with CREDO. Journal of Universal Computer Science 10(8), 985–1013 (2004)
9. Chen, H., Karger, D.R.: Less is more: probabilistic models for retrieving fewer relevant documents. In: Proc. of SIGIR 2006, Seattle, WA, USA, pp. 429–436 (2006)
10. Chen, J., Zaïane, O.R., Goebel, R.: An unsupervised approach to cluster web search results based on word sense communities. In: Proc. of WI-IAT 2008, Sydney, Australia, pp. 725–729 (2008)
11. Cheng, D., Vempala, S., Kannan, R., Wang, G.: A divide-and-merge methodology for clustering. In: Proc. of PODS 2005, New York, NY, USA, pp. 196–205 (2005)
12. Crabtree, D., Gao, X., Andreae, P.: Improving web clustering by cluster selection. In: Proc. of WI 2005, Compiègne, France, pp. 172–178 (2005)
13. Cutting, D.R., Karger, D.R., Pedersen, J.O., Tukey, J.W.: Scatter/gather: A cluster-based approach to browsing large document collections. In: Proc. of SIGIR 1992, Copenhagen, Denmark, pp. 318–329 (1992)
14. Di Giacomo, E., Didimo, W., Grilli, L., Liotta, G.: Graph visualization techniques for web clustering engines. IEEE Transactions on Visualization and Computer Graphics 13(2), 294–304 (2007)
15. Harris, Z.: Distributional structure. Word 10, 146–162 (1954)
16. Kamvar, M., Baluja, S.: A large scale study of wireless search behavior: Google mobile search. In: Proc. of CHI 2006, New York, NY, USA, pp. 701–709 (2006)
17. Ke, W., Sugimoto, C.R., Mostafa, J.: Dynamicity vs. effectiveness: studying online clustering for scatter/gather. In: Proc. of SIGIR 2009, MA, USA, pp. 19–26 (2009)
18. Krovetz, R., Croft, W.B.: Lexical ambiguity and Information Retrieval. ACM Transactions on Information Systems 10(2), 115–141 (1992)
19. Kurland, O.: The opposite of smoothing: a language model approach to ranking query-specific document clusters. In: Proc. of SIGIR 2008, Singapore, pp. 171–178 (2008)
20. Kurland, O., Domshlak, C.: A rank-aggregation approach to searching for optimal query-specific clusters. In: Proc. of SIGIR 2008, Singapore, pp. 547–554 (2008)
21. Lee, K.S., Croft, W.B., Allan, J.: A cluster-based resampling method for pseudo-relevance feedback. In: Proc. of SIGIR 2008, Singapore, pp. 235–242 (2008)
22. Lin, D.: Automatic retrieval and clustering of similar words. In: Proc. of the 17th COLING, Montreal, Canada, pp. 768–774 (1998)
23. Liu, S., Yu, C., Meng, W.: Word Sense Disambiguation in queries. In: Proc. of CIKM 2005, Bremen, Germany, pp. 525–532 (2005)
24. Mandala, R., Tokunaga, T., Tanaka, H.: The use of WordNet in Information Retrieval. In: Proc. of the COLING-ACL Workshop on Usage of Wordnet in Natural Language Processing, Montreal, Canada, pp. 31–37 (1998)
25. Miller, G.A., Beckwith, R.T., Fellbaum, C.D., Gross, D., Miller, K.: WordNet: an online lexical database. International Journal of Lexicography 3(4), 235–244 (1990)
26. Navigli, R.: Word Sense Disambiguation: a survey. ACM Computing Surveys 41(2), 1–69 (2009)
27. Navigli, R., Crisafulli, G.: Inducing word senses to improve web search result clustering. In: Proceedings of the 2010 Conference on Empirical Methods in Natural Language Processing (EMNLP), Boston, USA, pp. 116–126 (2010)
28. Ngo, C.L., Nguyen, H.S.: A method of web search result clustering based on rough sets. In: Proc. of WI 2005, Compiègne, France, pp. 673–679 (2005)

29. Nguyen, C.-T., Phan, X.-H., Horiguchi, S., Nguyen, T.-T., Ha, Q.-T.: Web search clustering and labeling with hidden topics. ACM Transactions on Asian Language Information Processing 8(3), 1–40 (2009)
30. Osinski, S., Weiss, D.: A concept-driven algorithm for clustering search results. IEEE Intelligent Systems 20(3), 48–54 (2005)
31. Rand, W.M.: Objective criteria for the evaluation of clustering methods. Journal of the American Statistical Association 66(336), 846–850 (1971)
32. van Rijsbergen, C.J.: Information Retrieval, 2nd edn. Butterworths (1979)
33. Sanderson, M.: Word Sense Disambiguation and Information Retrieval. In: Proc. of SIGIR 1994, Dublin, Ireland, pp. 142–151 (1994)
34. Sanderson, M.: Ambiguous queries: test collections need more sense. In: Proc. of SIGIR 2008, Singapore, pp. 499–506 (2008)
35. Schütze, H.: Automatic word sense discrimination. Computational Linguistics 24(1), 97–124 (1998)
36. Schütze, H., Pedersen, J.: Information Retrieval based on word senses. In: Proceedings of SDAIR 1995, Las Vegas, Nevada, USA, pp. 161–175 (1995)
37. Stokoe, C., Oakes, M.J., Tait, J.I.: Word Sense Disambiguation in Information Retrieval revisited. In: Proc. of SIGIR 2003, Canada, pp. 159–166 (2003)
38. Swaminathan, A., Mathew, C.V., Kirovski, D.: Essential pages. In: Proc. of WI 2009, Milan, Italy, pp. 173–182 (2009)
39. Véronis, J.: HyperLex: lexical cartography for Information Retrieval. Computer Speech and Language 18(3), 223–252 (2004)
40. Voorhees, E.M.: Using WordNet to disambiguate word senses for text retrieval. In: Proc. of SIGIR 1993, Pittsburgh, PA, USA, pp. 171–180 (1993)
41. Widdows, D., Dorow, B.: A graph model for unsupervised lexical acquisition. In: Proc. of the 19th COLING, Taipei, Taiwan, pp. 1–7 (2002)
42. Maarek, Y., Ron Fagin, I.B.S., Pelleg, D.: Ephemeral document clustering for web applications. IBM Research Report RJ 10186 (2000)
43. Zamir, O., Etzioni, O.: Web document clustering: a feasibility demonstration. In: Proc. of SIGIR 1998, Melbourne, Australia, pp. 46–54 (1998)
44. Zamir, O., Etzioni, O., Madani, O., Karp, R.M.: Fast and intuitive clustering of web documents. In: Proc. of KDD 1997, Newport Beach, California, pp. 287–290 (1997)
45. Zhai, C., Cohen, W.W., Lafferty, J.: Beyond independent relevance: Methods and evaluation metrics for subtopic retrieval. In: Proc. of SIGIR 2003, Toronto, Canada, pp. 10–17 (2003)
46. Zhang, B., Li, H., Liu, Y., Ji, L., Xi, W., Fan, W., Chen, Z., Ma, W.-Y.: Improving web search results using affinity graph. In: Proc. of SIGIR 2005, Salvador, Brazil, pp. 504–511 (2005)
47. Zhang, X., Hu, X., Zhou, X.: A comparative evaluation of different link types on enhancing document clustering. In: Proc. of SIGIR 2008, Singapore, pp. 555–562 (2008)

Interdisciplinary Contributions to Flame Modeling

Maria Teresa Pazienza[1] and Alexandra Gabriela Tudorache[1,2]

[1] University of Rome Tor Vergata
pazienza@info.uniroma2.it
[2] Academy of Economic Studies of Bucharest
tudorache.alexandra@gmail.com
http://art.uniroma2.it

Abstract. The world-wide emerging e-society generates new ways to communicate among people with different cultures and backgrounds. Communication systems as forums, blogs, and comments are widely used being easily accessible to end users. Studying and interpreting user generated data/text available on the Internet is a complex and time consuming duty for any human analyst. This study proposes an interdisciplinary approach to modeling the flaming phenomenon (hot, aggressive discussions) in on-line Italian forums. The model is based on the analysis of psycho/cognitive/linguistic interaction modalities among participants to web communities and on state-of-the art machine learning techniques and natural language processing technology. This research gives the opportunity to better understand and model the dynamics of web forums, including the language involved, the interaction between users, the relation between topic and users, language intensity and differences in behavior by age and gender.

Keywords: flaming, flame wars, web forums, flames identification, opinion mining, natural language processing, machine learning.

1 Introduction

Nowadays, web communication is evolving into an important part of people lives. Forums, blogs, and comment systems create a new dimension of communication. User content is generated at a really high pace and in large volumes, often not in a friendly or polite manner. This could easily lead to flame wars. In such dynamic environment it is better to prevent flames than to be forced to close discussions and even ban (exclude) users.

A unique psycho/cognitive/linguistic model of these interactions is still missing, while the demand to process such data for several different applications emerged. This study focuses on modeling in Italian forums the complex flaming phenomena - hot, aggressive discussions, often expressed with a poor language.

In our analysis we considered three types of discussions: *flames, no flames (normal)* and *risky topics*. We identified several specific features and hypothesized that both the inner structure of board language and user behavior could be

R. Pirrone and F. Sorbello (Eds.): AI*IA 2011, LNAI 6934, pp. 213–224, 2011.

good discriminators between flames, risky topics and normal discussions. Each forum is a distinct community with its own users, staff (moderators, administrators) and rules. This closed world model impacts greatly on language, behavior and consequently on flames characteristics: there is not a universal model of flaming! Therefore the analysis should be conducted on data extracted from the same board or forum section.

Based on this model, several experiments were conducted using machine learning methods as detailed in [24]. The aim was to classify flames, risky topics and normal discussions and to observe how each feature included in the model contributes to the general result.

The study could help moderators, forum administration and users to participate to a more friendly, flames free community open to all ages, preventing legal issues correlated to offensive language, privacy and minor abuse. Several applications could benefit from this research as: identify trolls, posts authors on other forums, detect hot topics, email filtering, and identify threats: angry, subversive persons or groups of persons in political or corporate forums and social networks.

2   Related Work

Our research is situated at the confluence of many different disciplines: computational linguistics, psychology and social anthropology. Moreover, it is related to domains as affect analysis and opinion mining.

Ellen Spertus' Smokey [31] was the first attempt to classify aggressive messages in forums. The system was based on Quinlan's C4.5 decision-tree generator to determine (syntactic and semantic) feature based rules. Recently the work of Paltoglou et al. [20] focused on analyzing the emotionality involved in social media responses. The authors demonstrated that for detecting subjectivity on social media the lexicon-based classifier outperforms supervised approaches, while for classic sentiment analysis (on product reviews) machine-learning approaches typically outperform lexicon-based ones.

In the computational area para-language was studied mostly for the Intelligent Conversation Agents. One of the firsts emotional conversation characters was Parry [8]. Cassel et al. [6] proposed a conversation system, between multiple animated human-like agents including synchronized speech, intonation, facial expressions, and hand gestures. In 2009, Pelachaud [27] developed a model of behavior expressivity based on a set of six parameters. In 2010, Gobron et al. [12] introduced a graphical representation of human emotion extracted from text sentences. In the field of opinion mining and sentiment analysis Turney [33] proposed an unsupervised method based on the semantic orientation of the phrases that contain adjectives or adverbs computed as the difference of the mutual information between excellent and poor and the given phrase. Pang et al. [23] used machine learning methods (SVMs, Naive Bayes and Maximum Entropy classifiers), while, Dave et al. used methods based on scoring features [11]. Liu [16] and Pang [21] proposed feature-based opinion summary of multiple reviews while [29] used learned patterns and [38] focused on identifying subjectivity and polarity (orientation). Furthermore dictionary based methods typically relay on WordNet's synsets and hierarchies to acquire opinion words [1]. SentiWordNet is a valuable resource for opinion carrying words in English language [2,21].

3   Web Communication Model

To better understand the complex phenomena that lie behind flaming in web communities (why flames occur) we need to study several aspects related to the communication theory, the interaction within human groups and the way people express their ideas including language and specific web symbols (emoticons).

On-line communication takes place asynchronously in forums, blogs, emails, comment systems and mailing lists between persons belonging to heterogeneous cultures, and having different levels of education, and language knowledge, while facial expression and paralinguistic information is missing. According to Mehrabian [19] 55% of message emotionality is represented by facial expression; 38% by para-linguistics as: pitch, loudness, rate, and fluency; and 7% lies in word semantics. In this context new specific communication encoding has emerged including: emoticons, abbreviations and a specialized English vocabulary. Even with the help of such tools and using our imagination we still miss a great part of communication and in particular the emotional climate. It is up to each person to interpret the message depending of his mood, background, imagination and most important context. Mabry [17] showed that one expression written on a friendly topic can have a colloquial meaning but on a risky topic can promote flames. Furthermore, in the World Wide Web people don't just state opinions individually but interact with other people conferring to on-line communities the dimension of social groups. Such dynamics create an unusual situation in which there is a stable nucleus of users that are a true community, with great level of tacit understanding and empathy [26], and at the same time new members arrive while old members leave. This phenomenon affects the stability of the community and generates bias in communication. In this context early mediation has an important role in smoothing the differences between old and new users as Mabry [17] demonstrated. Suler [32] shows that people behave in cyberspace in different ways than in face-to-face situations. This is called the dis-inhibition effect. As a positive effect, people tend to open up and to communicate: they share personal emotions, fears and wishes or they show unusual acts of kindness and generosity. Otherwise, the dis-inhibition effect may conduct to rude language, harsh criticisms, anger, hatred, and even threats.

Another communication barrier is semantics: the meaning that every person assigns to words. McMenamin, Dongdoo Choi and Coulthard [9,18] studied for the first time the personal language to confirm the authorship of written texts in forensics. Their research shows that a person uses a particular set of meanings of words and expressions depending of education, culture and personality called idiolect.

4   The Flaming Phenomena

For this research a distinction was made between risky topics (that could easily lead to flames) and flames that are the actual aggressive interactions. The web communication model helps identifying the most important features that characterize each discussion type. (see for details [25])

Flames are a sequence of "non constructive", aggressive posts, that have no positive contribution to the discussion. In flames users attack each other at a personal level instead of contrasting the discussion partner for his/her approach, ideas, contribution or argumentation. Flames often induce moderators to close discussions. Sometimes the moderator himself generates flames due to his tough policy or off-topic interventions.

A risky discussion may generate flames and contains both flame and no-flame elements. Therefore risky topics could neither be categorized as flames, nor as normal discussion. In real life the distinction between flames and risky discussions is subjective. It depends on moderator skills, his attitude, level of stress and implication in the subject. Risky topics contain mini-flames (two-three flame posts) that often extinguish by themselves or at the correct intervention of a moderator. As in flames, there are two or maximum three actors (users) participating to arguments. In risky discussions more than one mini-flame could be found and moderators usually don't make hush interventions. Also in such threads users tend to get stormy about ideas and not persons as in flames. When a risky topic turns personal flaming occurs. Moreover, in risky discussions often a good number of off-topic posts are found and the evolution of the thread depends directly on moderator skills and in some measure of users' willingness to argue.

A no-flame ("normal") topic is a discussion that continues without flaming or the risk of it. It is usually a discussion in which moderators do not need to interrupt discussion or recall users. At the most they act as normal users participating in the discussion.

Flames and risky topics share many features and mostly is a subjective classification. In one discussion board one intervention may be considered as flame while in other forum the same topic could be considered at risk or even normal confrontation. This adds difficulty to their correct identification. To provide a flavor of what a flame is we will show in Appendix 1 some examples of the English translation of the Italian corpus. The translation was made trying to preserve as much as possible of the original sense and style to better illustrate emotion involved. We maintained phrase structure and grammatical non concordances, too. All the names of people, companies and forum users were changed to generic names for privacy reasons.

5   Flames and Risky Topics Features

There are different sources that determine normal discussions to become flames. In our model these features are grouped in two main categories: expression and user profiling. Expression is a meta-feature showing how concepts and sentiments are formulated while user profiling captures the behavior of dialog actors.

5.1   Expression Features

Expression includes features as: *discussion topic, general language, personal language, cites,* and *emoticons.* The preliminary analysis of several forums shows that **topics** about politics, sport, social integration, brands and working philosophies (E.g. open source) usually degenerate in flames earlier or later. It is the phenomena of appartenance at a group or ideology that makes people fight for. Topics like small chat or general discussions are less prone to become flames, unless a troll makes a target of that forum - attacking different topics.

Language is a complex and difficult to model phenomenon. Dialog in open communities is even harder to model and analyze. There are differences in language and users background. Furthermore, web communities' dynamics are neither constant nor even linear (see before). Bucci and Maskit [5] demonstrated, in their psychological clinical studies, that ambiguous phrasal constructs, mostly verbal, missing of argumentation are a sign of tension. Also the avoidance of taking responsibility over the expressed ideas introduces flames. E.g. "un" against "il"; "bene", "niente" – Italian, "the" against "this", "the idea" against "my idea"- English. Moreover, Italian forums are characterized by a high context communication. Users address directly to each other at a personal level ("cut and thrust" - "botta e risposta" in Italian). In Italian flames often phrases miss the subject and express a general disagreement. (E.g. " una cosa stupida; non vero" - "it is a stupid idea; it is not true"). Also the use of hypocritical politeness in expressions as: "perdonami se...", "scusa se..." ("excuse me if..."). This kind of expression is present in all the flames to which take part women and in almost 30% of men flames. Culpeper [10] shows that most of the time offensive language generate flames and is also part of flaming. In the analyzed corpus 40% of men use offensive language, while women prefer irony.

Specific symbols and marks also express emotionality as: repeated question or exclamations marks, question marks followed by exclamation marks or the written expression of emoticons like ghghg or lol. Uppercase is used to get the attention or to underline an idea. But in the analyzed forum uppercase is seldom used and we excluded this feature form analysis. A special category of expression is represented by **emoticons**. They integrate the language and act as visual cues for expressing emotions as: anger, happiness, irony.

Often, when arguing, people tend to repeat parts of the discourse of the contender. In forums such phenomena is even more present in the form of **cites** and **cross-cites**. Cites have a double role: they integrate the language used by the user citing and also represent a behavioral feature. Furthermore, unknown **external causes** not related to the discussion topic or interactions can lead to flames as: personal problems or just a user having a bad day.

5.2   User Profiling

User profiling meta-feature includes several individual features as: actors (historic enemies, newbies) personal background and moderation. Forums act as communities where specific individuals (**actors**) interact and compete with each other. Frequently flames are provoked by a small group of users (two or maximum three) in a certain period of time, involving a specific topic or even different topics. When such groups of contenders consolidate they became *historic enemies* [5]. Newly registered persons (**Newbies**) may cause flames simply because they don't know the rules of that specific community or fail to communicate with previously unknown persons. Each user influences directly the development of a flame. **Background** and **education** are directly reflected in the way a person expresses and therefore in the likelihood of provoking flames. Flaming occurs more frequently in forums where the administration uses a tough or a nonlinear **moderation policy**. Moderators can generate flames with nonsense and off topic interventions in no flame or risky discussions. Also discussions where moderators make most interventions usually are hot discussions.

6   Our Approach to Flames and Risky Topics Identification

In this section our approach will be presented starting with information about the corpus structure. Furthermore, the proposed model will be widely discussed including algorithm, software, parametrization and experimental setup.

6.1   Corpus

The corpus used for this experiment is a real life Italian corpus directly extracted from the politics section of a generic forum that hosts also social, sport and general topics. The corpus is composed of a 1540 posts, from which 170 flames, 330 risky posts 1040 no flames. All discussions were selected in the order of creation. The posts were extracted from 195 topics and were written by 73 different users, from which 11 females and 62 males.

Posts were extracted individually, together with several features as: topic, post author, user type (mod, admin or user), date and time of posting, quotes and other users' citations. Each emoticon and special punctuation mark was substituted by its class in the corpus.

Each post was manually annotated by two different annotators and assigned to one of the three classes: flame, risky or no-flame. The agreement rate was: 68 (Cohen's [7] kappa coefficient with 95% confidence interval). This is considered a strong agreement as shown by Gwet [14]. A third annotator selected the final class from the initial annotations.

6.2   The Framework

To model the flaming phenomena, two main meta-categories of features will be analyzed: expression and user profiling (see before). The architecture consists of two main steps: corpus preprocessing and classification (performed for each experiment). In a real life situation the classifier should be retrained from time to time to ensure that the new users and dynamics are included into analysis.

As flames and risky situations develop over several posts, we selected as analysis unit or document (as in information retrieval) a window of 5 consecutive posts. This is the average span of flames in the analyzed forum. E.g. Document 1 will include posts 1-5 from discussion 1, document 2 posts 2-6 from discussion 1. Each analysis unit (document d) will be represented as a vector of features including expression (E_k) and user profiling (UP_u) (see Table 1).

Table 1. Document representation matrix

Documents	Expression Vector	User Profiling Vector
Doc 1	$E_1, \ldots, E_k, E_n$	$UP_1, \ldots, UP_k, UP_u$
...	...	...
Doc 2	$E_1, \ldots, E_k, E_n$	$UP_1, \ldots, UP_k, UP_u$

Feature Selection: To optimize the classification algorithm for each feature category several feature selection strategies were adopted.

Expression Feature Selection: In the model we will include the following expression features: *post contents, punctuation marks, emoticons* and *quotes*. As expression is considered as an extended specific vocabulary it will be represented by the word vector model and the features will be represented as *Tf*idf*, where term frequency is the normalized number of occurrences of a term and inverse document frequency is a measure of the general importance of the term.

Several strategies are employed to normalize expression features. *Post contents* were preprocessed to eliminate the majority of sparse and frequent words reducing the presence of misspelled words, very frequent in forums and common words that don't carry relevant semantic information. (E.g misspelled words as: "inutilite" instead of "inutile" - useless or "avrebe" instead of "avrebbe" - could have, and common words as: "ora", now, "perche" - why). Basili and Moschitti [4] suggest eliminating words having less than 3 characters that do not carry semantics as: articles or prepositions. An Snowball Italian stemmer [28] was used to collapse words to their root. Words sharing the same root have similar semantics even if derive from different parts of speech. (E.g. "elettorale" – electoral, "elettore" - voting person, "eletto" - voted). *Emoticons* were extracted and assigned to 8 different classes: *positive, negative, angry, sad, bitter, holiday, fear* and *ironic* as were categorized on the analyzed forum. *Specific symbols* were also substituted by conventional classes as: *irony* (?!/!?), *multiplequestion* (???), *singlequestion* (?), *multipleexclamation* (!!!), *singleexclamation* (!), *laugh* (ghgh, lol, ahah).

User Profiling Feature Selection: For each analyzed post window user profiling is represented as a vector of features showing users' contribution to each class computed as the general user activity combined with the local user activity (see Eq. 1). *General impact of user activity* (see Eq. 2) expresses the contribution of user u to the entire forum activity and it is computed as a normalized weighted measure of the activity for each class in the training corpus.

It shows both positive and negative contribution to the board. E.g positive: user u has a good presence on forum (many posts) from which the majority no flames; negative: the majority of posts written by user u are flames or that user u has a very low presence). To minimize errors we considered the general user activity to be 0 if the user has not a clear tendency towards that class (it is less than two times the average of the entire forum). Flaming index is computed as the estimate of likelihood for a given user to participate to each class, while user interventions weight is a measure of user impact on the general activity of the analyzed forum. *Local user activity* (see Eq. 4) is the number of posts written by user u in the analysis unit (5 posts window) and represents the contribution of user u to the analyzed document (d). To give the chance of a fair analysis for newbies, inactive and unregistered users, all measures are averaged by the number of days in which the analyzed user is active.

$$UserProfiling = GeneralImpactOfUserActivity * LocalUserActivity. \quad (1)$$

$$GeneralImpactOfUserActivity = UserInterv.Weight * log\frac{1}{ClassIndex}. \quad (2)$$

$$ClassIndex_u = \frac{PostsNoWrittenByUser_u \in Class}{PostsNoWrittenByUser_u}. \quad (3)$$

$$LocalUserActivity = \frac{NoOfPostsWrittenByUser_u \in d}{NoOfAnalysedPosts}. \quad (4)$$

Feature Vectors Builder: For each 5 posts window, feature vectors including expression and user profiling were built. The vectors were ordered first by topic and afterwards by post order following the natural order in the analyzed forum. For this step we used the "Waikato Environment for Knowledge Analysis", Weka [34,35] with an extension of Word Vector Tool [36] and customized software for building the feature vectors. To each window, a class was empirically assigned by trial and error. It will be further considered as a heuristic. The aim was to simulate the choice of a human moderator. A flame post was marked with a 2 score, while risky posts with 1.8 and normal posts with a score of 1. Afterwards, category scores are summed and the biggest one defines the window category. The idea is that a flame post is two times more dangerous than a no-flame post and a risky topic is 1.8 more dangerous than a no-flame.

Classification Algorithm and Evaluation Criteria: As described before the aim is to classify flames, risky topics and a mixed class that contains both flames and risky topics. The no-flame class is considered neutral. The actual classification was performed with parametrized Support Vector Machines using the stratified 10 folds cross validation model [4]. SVM [15] was chosen as currently is considered a state-of-the-art classification algorithm; it is best suited for classification of complex mixed data models [4]. For this experiment was adopted SVMlight [3,15] with a linear kernel. To optimize and balance both precision and recall, we parametrized the lowest cost parameter of SVM. Since topic analysis is subjective and also prone to errors the goal was to avoid overfitting. For evaluation purposes the F1 measure and accuracy will be used (see Eq. 5, 6) [4].

$$F1 = \frac{2 * Precision * Recall}{Precision + Recall} \cdot \tag{5}$$

$$Accuracy = \frac{tp + tn}{tp + fp + tn + fn} \cdot \tag{6}$$

Where: tp (true positives) - topics correctly identified as positives; tn (true negatives) - topics correctly identified as negatives; fn (false negatives) = not found correct topics; fp (false positives) = incorrect topics marked as positives.

7   Results

In this section we show the results of flame, risky topics and mixed (flame and risky topics) identification against no-flames that acts as neutral class. A preliminary experiment on this corpus has been carried on in the context of a paper accepted at the ECSR Journal [24]. In the experiment presented in this paper, as a specific improvement, our aim was to analyze each feature influence in classification. Features were added one by one and average F1 Score and accuracy were computed for each analyzed fold. C parameter was computed in 0.1 steps. For most folds Precision and Recall converged for C=1.2. Language is considered as baseline. The best results for each experiment are reported in bold.

Table 2. Flames, Risky Topics and Mixed Class Recognition average F1 score for 10-fold cross validation with C=1.2

Experiment	Flames / No-flames		Risk / No-flames		Mixed Class	
Feature / Measure	AVG F1	AVG Acc.	AVG F1	AVG Acc.	AVG F1	AVG Acc.
Language	*91.02*	*93.77%*	*83.31*	*84.55%*	*87.73*	*85.31%*
+ *Punctuation*	89.17	92.60%	82.51	83.69%	87.74	85.14%
+ *Emoticons*	90.27	93.26%	**83.53**	**84.75%**	**88.21**	**85.72%**
+ *Quotes*	90.08	93.01%	82.38	83.39%	87.92	85.39%
+ *Topic*	91.21	93.68%	81.17	82.08%	87.40	84.73%
+ *User Profiling*	**91.25**	**93.80%**	81.75	82.72%	87.82	85.29%

8   Discussion and Conclusions

While focusing on a specific problem – flame modeling, this study involved different aspects as: psychology of web communication (forums as social groups, communication barriers, the lack of non verbal for expressing emotion), language correlated problems (mother language, personal language, misspelled words, improper grammar usage), user behavior (determined by the interaction between users, the presence of new users, historic enemies or friends, background and gender).

Experiments confirmed our initial hypothesis: it is possible to recognize flames and risky topics using a supervised machine learning algorithm based on complex features (expression and user profiling). As expected results showed that risky topics are difficult to classify, having features common to both flames and no-flames classes.

We obtained results comparable with the state of the art and a bit more. Our overall average accuracy (all features included) is 83% for risky topics recognition, 85% for mixed class recognition and 94% for flames recognition, while for example the best accuracy obtained by Pang and Lee [22] for sentiment classification is 83%. Shi, L. [30] obtained 87% accuracy for topic based web forum sentiment analysis, while Gupta et al. [13] obtained 83% accuracy for customer care emotional emails recognition. Xu and Zhu [37] obtained 90.94% accuracy for filtering offensive language grammatical relations. Ellen Spertus' Smokey [31] (based on C 4.5 algorithm) obtained a accuracy of 39% for flames recognition.

In our case **expression features** had a major influence in flames, risky topics and mixed class recognition, while user behavior had only a minimal impact on results. Results show that different expression features have different influences on each class (positive or negative), *language* being considered as baseline. *Punctuation* feature has a negative influence on all classes and should be further investigated. Probably this is due to the heterogeneous interpretation of punctuation marks that act as a limited vocabulary, to which users assign different orientations. *Emoticons* have a positive influence on risky topics and mixed class recognition, but are not so relevant to flames. This shows that arguing people mostly write their ideas without the need for an alternative representation, while emoticons are used to smooth tones. *Quotes* have a negative influence on all classes. This is mostly due to the mixing of different writing styles belonging to the original and the quoted author. A better representation of quotes could help refine the model. Furthermore, *topic title* is very important to flames while it is not so important for the analysis of risky topics or mixed class.

Expression analysis could be improved using latent semantic analysis to discover associations of concepts that determine flames, as well as word sense disambiguation and named entity recognition to identify e.g. nicknames of different persons referred in discussions. Moreover, other classifiers (lexicon-based) could be tested as suggested by Paltoglou et al. [20] that focused on analyzing the emotionality involved in social media responses. We will proceed in such direction in the near future.

User profiling contributes positively to each experiment results. This shows that the flaming phenomena depend both on the way users express their ideas and on virtual communities' dynamics. Further research should be necessary to refine user behavior model. (E.g. the analysis of the variance of behavior for each user during flames, risky topics and normal discussions).

References

1. Andreevskaia, A., Bergler, S.: CLaC and CLaC-NB: knowledge-based and corpus-based approaches to sentiment tagging. In: 4th International Workshop on Semantic Evaluations, pp. 117–120 (2007)
2. Baccianella, S., Esuli, A., Sebastiani, F.: SENTIWORDNET 3.0: An Enhanced Lexical Resource for Sentiment Analysis and Opinion Mining. In: LREC 2010 (2010)
3. Basili, R.: Review of Learning to Classify Text Using Support Vector Machines by Thorsten Joachims. Computational Linguistics 29, 655–661 (2003)
4. Basili, R., Moschitti, A.: Automatic Text Categorization: From Information Retrieval to Support Vector Learning. Aracne Editrice, Informatica (2005)
5. Bucci, W., Maskit, B.: A weighted dictionary for Referential Activity. Computing Attitude and Affect in Text (2005)
6. Cassell, J., Badler, N., Steedman, M., Achorn, B., Becket, T., Prevost, S., Stone, M.: Animated conversation: rule-based generation of facial expression, gesture & spoken intonation for multiple conversational agents. In: SIGGRAPH 1994, pp. 413–420 (1994)
7. Cohen, J.: Weighted kappa: Nominal scale agreement provision for scaled disagreement or partial credit. Psychological Bulletin 70, 213–220 (1968)
8. Colby, K.: Artificial paranoia. Artificial Intelligence 2(1) (1971)
9. Coulthard, M.: Author identification, idiolect, and linguistic uniqueness: Forensic linguistics, pp. 431–447. Oxford University Press, Oxford (2004)
10. Culpeper, J.: Impoliteness: Using Language to Cause Offence. Cambridge University Press, Cambridge (2011)
11. Dave, K., Lawrence, S., Pennock, D.M.: Mining the peanut gallery: Opinion extraction and semantic classification of product reviews (2003)
12. Gobron, S., Ahn, J., Paltoglou, G., Thelwall, M., Thalmann, D.: From sentence to emotion: a real-time three-dimensional graphics metaphor of emotions extracted from text. The Visual Computer: IJCG 26, 505–519 (2010)
13. Gupta, N., Gilbert, M., Di Fabbrizio, G.: Emotion Detection in Email Customer Care. In: ACL 2010, pp. 10–16 (2010)
14. Gwet, K.: Handbook of Inter-Rater Reliability. STATAXIS Pub. Company (2010)
15. Joachims, T.: Learning to Classify Text Using Support Vector Machines. Kluwer Academic Publishers, Dordrecht (2002)
16. Liu, B.: Sentiment Analysis and Subjectivity. In: Indurkhya, N., Damerau, F.J. (eds.) Handbook of Natural Language Processing (2010)
17. Mabry, E.A.: Framing Flames: The Structure of Argumentative Messages on the Net. Computer-Mediated Communication 2 (1997)
18. McMenamin, G.R., Choi, D.: Forensic Linguistics: Advances in Forensic Stylistics. CRC Press, Boca Raton (2002)
19. Mehrabian, A.: Silent Messages. Wadsworth Publishing Company, Belmont (1971)
20. Paltoglou, G., Gobron, S., Skowron, M., Thelwall, M., Thalmann, D.: Sentiment analysis of informal textual communication in cyberspace. In: Engage 2010 (2010)
21. Pang, B., Lee, L.: A Sentimental Education: Sentiment Analysis Using Subjectivity Summarization Based on Minimum Cuts. In: ACL 2004, pp. 271–278 (2004)
22. Pang, B., Lee, L.: Opinion Mining and Sentiment Analysis, pp. 1–135 (2008)
23. Pang, B., Lee, L., Vaithyanathan, S.: Thumbs up?: sentiment classification using machine learning techniques. In: ACL 2002, pp. 79–86 (2002)
24. Pazienza, M.T., Lungu, I., Tudorache, A.G.: Flames Recognition for Opinion Mining. ECECSR Journal 3 (to be published, 2011)

25. Pazienza, M.T., Stellato, A., Tudorache, A.G.: Flame, risky discussions, no flames recognition in forums. In: EMOT 2008, Marrakesh, Morocco (2008)
26. Peck, M.S.: The Different Drum: Community Making and Peace. Simon & Shuster, New York (1987)
27. Pelachaud, C.: Studies on gesture expressivity for a virtual agent. Speech Communication Special Issue, 630–639 (2009)
28. Porter, M.F.: Snowball: A language for stemming algorithms (2001)
29. Riloff, E., Wiebe, J.: Learning Extraction Patterns for Subjective Expressions. In: EMNLP 2003 (2003)
30. Shi, L., Sun, B., Kong, L., Zhang, Y.: Web Forum Sentiment Analysis Based on Topics. In: Ninth IEEE CIT 2009. IEEE Computer Society, Washington, DC (2009)
31. Spertus, E.: Smokey: Automatic Recognition of Hostile Messages. In: IAAI 1997, pp. 1058–1065 (1997)
32. Suler, J.: The basic psychological features of cyberspace (2002), http://www-usr.rider.edu/~suler/psycyber/psycyber.html
33. Turney, P.D.: Thumbs Up or Thumbs Down? Semantic Orientation Applied to Unsupervised Classification of Reviews. In: ACL 2002, pp. 417–424 (2002)
34. Weka 3: Data Mining Software in Java, http://www.cs.waikato.ac.nz/ml/weka/
35. Witten, I.H., Frank, E.: Data Mining: Practical Machine Learning Tools and Techniques, 2nd edn. Morgan Kaufmann, San Francisco (2005)
36. The Word & Web Vector Tool, http://nemoz.org/joomla/content/view/43/83/
37. Xu, Z., Zhu, S.: Filtering Offensive Language in Online Communities using Grammatical Relations. In: CEAS 2010 (2010)
38. Yu, H., Hatzivassiloglou, V.: Towards answering opinion questions: separating facts from opinions and identifying the polarity of opinion sentences. In: EMNLP 2003, pp. 129–136 (2003)

Appendix 1 - Textual Data Examples

Flame 381 English translation: *"User1 – my answers are on the other general topic which has been closed... but is correct to open another topic similar with the closed one??' Moderator1 – the posts have been moved. And for certain questions there are the pm. And because you know how to use them.. just use them! User1 – GOOD If I'm the one creating problems I'm going to autoban myself for a few days so everybody should be happy to externalize your legal thoughts.... Thank you all..."*

Risky topic 705 English translation: *"User2 - .... Just heard on the news.... The dwarf there was no Bulgarian edict; mine was just an appeal to the new leaders coming into Company1... that certain things shouldn't happen any more I wonder if there is a limit to stupidity and impudence of that little man!"*

No-flame 754 English translation example: *"User3 - Person2, the wife of the mafia head Person3, has cited for damages the authors of Company2 fiction 'Film1'. How the hell can one tell such stupid things? Person1 that takes the part of Person4 and Person5, the other one that we should accept the mafia. In what country are we living? User4 - In the pulcinella country. User1 - It is a collateral negative effect of the democracy, absolutely shame...".*

Latent Topic Models of Surface Syntactic Information

Roberto Basili[1], C. Giannone[1], Danilo Croce[1], and C. Domeniconi[2]

[1] Dept. of Enterprise Engineering,
University of Roma Tor Vergata, Roma, Italy
{basili,giannone,croce}@info.uniroma2.it
[2] Dept. of Computer Science, George Mason University, USA
carlotta@cs.gmu.edu

Abstract. Topic Models like Latent Dirichlet Allocation have been widely used for their robustness in estimating text models through mixtures of latent topics. Although LDA has been mostly used as a strictly lexicalized approach, it can be effectively applicable to a much richer set of linguistic structures. A novel application of LDA is here presented that acquires suitable grammatical generalizations for semantic tasks tightly dependent on NL syntax. We show how the resulting topics represent suitable generalizations over syntactic structures and lexical information as well. The evaluation on two different classification tasks, such as predicate recognition and question classification, shows that state of the art results are obtained.

1 Introduction

The development of annotated resources for statistical Natural Language Processing (NLP) still represents a bottleneck, limiting the impact on existing technologies. This is particularly true for fine-grained semantic interpretation tasks (such as semantic role labeling, [12]) that usually act over sentences rather than on entire documents. In fact, short texts provide small amounts of lexical information, more exposed to data sparseness and prone to misinterpretation. Generalizations over grammatical structures have been shown useful to learn syntactic disambiguation rules (e.g., parse tree re-ranking in [8]), to classify syntactic arguments in SRL (e.g. [24,22]), as well as to detect the class of a question in QA systems (e.g., [23]). However, the quality and size of available grammatically annotated data is crucial to achieve good accuracy levels. Moreover, portability of the required grammatical information across domains is problematic.

Semi-supervised paradigms [17,4] aim at alleviating this problem, i.e., limit the annotation costs. The idea is to increase the generalization capabilities over limited annotated resources, by first generalizing from large scale unlabeled resources, and then make available the acquired information within a supervised classification settings. Probabilistic topic models are well suited within a semi-supervised perspective. They are generative models of texts used to reveal the underlying topic structure of large corpora. Topic models such as pLSA or LDA [4] have been recently adopted as generative models able to explain word distributions over texts, representing a robust and principled approach to estimate probabilities over unlabeled data.

While topic models have been successfully applied to document classification [4], their application to fine grained linguistic tasks is problematic. Current research on

R. Pirrone and F. Sorbello (Eds.): AI*IA 2011, LNAI 6934, pp. 225–237, 2011.
© Springer-Verlag Berlin Heidelberg 2011

topic-based document classification involves data objects characterized by *a significant amount of lexical information*: document sizes are in the range of 40-1000 words. In this range, the necessary lexical evidence needed to achieve accurate estimation seems guaranteed. In NLP, a number of different tasks are mostly *sentence-driven*, whereas only short texts (i.e., individual sentences) are available. For instance in a target semantic phenomenon, e.g., the category of a question in a QA scenario, is observable only through limited evidence, i.e., one sentence, the estimation of its probability as a distribution over topics can be too coarse-grained. In this context, the open research questions are: which representation is suitable to support a generative topic-based model for fine-grained linguistic phenomena, such as sentences or questions? To what extent can LDA be helpful for increasing the applicability of supervised learning methods in poor training conditions? Which kind of generalization can be achieved for fine-grained topics, when applied at a sentence rather than at a document level?

In this work, we show how embedding of shallow grammatical information in a document representation, as a special case of lexical information, can be used to produce useful generalizations (i.e., latent topics) in a standard LDA setting. The advantage of the proposed approach is a more flexible learning model for fine-grained NLP tasks, with straightforward applications even in poorer training conditions. Moreover, the approach suggested here would be extremely simpler to be applied, and with much weaker requirements in terms of annotated information. Empirical findings in support of this thesis are discussed against two sentence-based semantic tasks, such as predicate detection and question classification.

2   Topic Models for Semantic Interpretation Tasks

A largely used paradigm in unsupervised and semi-supervised text-based learning task is Latent Dirichlet Allocation (LDA) [4]. LDA is a three-level hierarchical Bayesian network that defines a probabilistic generative model for a corpus of documents. The basic idea is that documents are represented as a mixture of latent topics, where each topic is a multinomial distribution over words.

Figure 1 shows the graphical model of LDA. The relationship between documents and words is governed by latent variables z, introduced to capture the responsibility of a particular topic in using a specific word in a document. LDA places Dirichlet priors on the document distributions over topics (α) and on the topic distributions over words (β), thus enabling the model to generalize to unseen documents. The structure of the hidden variables of the model can be learned by inferring their posterior probability distribution, i.e., the topical structure of the collection, given the observed documents.

The generative process of LDA works as follows:

1. Draw $\phi_k \sim Dir(\beta)$, for each topic $k = 1, \ldots, T$;
2. Draw $\theta_d \sim Dir(\alpha)$, for each document $d = 1, \ldots, D$;
3. For each document d, and for each word N_d
   - Draw a topic $z \sim Multinomial(\theta_d)$;
   - Draw a word $w \sim Multinomial(\phi_z)$;

Inverting the generative process, i.e., fitting the hidden variables to the observed data, enables one to learn the distributions of the underlying topics. Although LDA is a fairly simple model, exact inference of the posterior distribution is intractable, and a number of approximate inference techniques have been used, including Gibbs sampling [14] and variational methods [4,20]. Since the introduction of the basic LDA model [4], a variety of topic models have been introduced for different applications, including supervised and semi-supervised approaches [25,3,27,6,7].

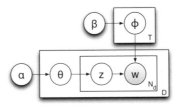

Fig. 1. LDA Topic Model

Recently, several approaches attempted to account for the syntactic structure in the text [13,5]. The model presented in [13] aims at capturing both grammatical constraints as well as semantic generalizations expressed through topic distributions. It extends an HMM model of grammatical dependencies through topic models. The model introduces two distributions that capture short-range syntactic dependencies (e.g., within a sentence) and long-range semantic dependencies (e.g., within a document) between words, respectively. A component is drawn from one of the two distributions depending on the context. Although the model in [13] is shown not to be very effective for document classification tasks, it provides an explicit representation of the syntax: it generates an effective separation of classes of functional words (i.e., closed word classes whose role is eminently syntagmatic, such as determiners or prepositions) from semantically oriented classes, i.e., topics for most of the open word classes (such as nouns). However, the parameter estimation required by this approach is challenging and requires large training data sets, that is a critical issue for most semantic tasks. A similar attempt of modeling syntagmatic as well as lexical generalizations is carried out in [5]. The fundamental difference in [5] is the definition of a single distribution that combines both topics and syntax. Thus the model generates words that agree with both syntactic and semantic structures. Even if the kind of generalization proposed here is less demanding, this method is crucially tight to the supposed availability of a CFG and to the underlying parse trees. A major problem is the quality of the achievable syntactic input (i.e., the parse tree) whose accuracy is a limitation for the syntactic topic models build on top.

An important limitation of most of the above papers is that they discuss experimental evaluation only over limited data sets, i.e. small or artificial benchmarks. Improvement are thus reported in terms of the perplexity index. In this way it is not clear if useful generalizations of lexical and grammatical properties and accurate linguistic predictions are obtained. In contrast, in this paper we focus on relatively stable benchmarks, so that the linguistic generalization ability of a model can be studied in depth. The key objective here is to adopt a very simple (i.e. reusable) generative model that exploits shallow syntactic and lexical information in a flexible and cost-effective manner. Even a standard LDA applied through Gibbs sampling seems to fit the above objectives, as the empirical evaluation of Section 4 shows. Before discussing the technical details of our approach in Section 3, we first introduce hereafter the targeted NLP tasks.

2.1   Semantic Intepretation Tasks

Traditionally, NLP has focused on interpretation tasks acting over specific linguistic units, such as individual sentences or short questions. An example of the former is the recognition of the frame, or predicate, in a sentence for semantic role labeling. The mapping of a question to a semantic class in a Question Answering scenario is an example of the latter. Both tasks can be casted as classification problems as discussed hereafter.

Predicate Recognition as a Classification Task. Semantic role labeling systems based on the *frame semantics* [11] foresee as a preliminary step, the *frame recognition task*. A *frame* is a conceptual structure modeling a *prototypical situation* that is evoked by a sentence, through the occurrence of one of its lexical units (lu) [11]. Due to its conceptual nature, a frame implies an inner relational structure: each frame has a set of prototypical semantic roles, also named semantic arguments, or *Frame Elements* (FE). FEs characterize the participants to the underlying event. For example, in sentences (1) and (2), the lexical unit *fire* evokes two different frames, FIRING[1] and USE_FIREARM[2].

> [1]EMPLOYER *'ll fire [him]*EMPLOYEE [after the speedboat race is over]TIME. (FIRING)   (1)

> [He]AGENT *fired [at Whitlock]*GOAL as he turned into 2nd Avenue. (USE_FIREARM)   (2)

Moreover, in (2), the fragment [He] expresses the AGENT of the (USE_FIREARM) frame, while [at Whitlock] is the GOAL. The semantic difference is basically expressed at the syntactic level, whereas the same lexical unit supports two different constructions, i.e., transitive vs. intransitive form. This in turn implies the different evoked frame elements. The SRL task since since the seminal work of [12] was largely studied. Several SRL systems, (i.e., Shalmaneser [10]) include a module responsible for the detection of individual frames in an input sentence. This predicate recognition task consists in the proper assignment of the frame to the sentence, even when the targeted predicate word is ambiguous (like *fire* in 1 and 2). This corresponds to a complex sentence-based classification task crucially dependent on the underlying NL syntax.

Question Classification. The typical architecture of a QA system foresees three main phases: question processing, document retrieval and answer extraction [16]. Question processing is usually centered around the so called *question classification (QC)* task. It maps a question into one of k predefined answer classes, thus posing constraints on the search space of possible answers. Most accurate QC systems apply supervised machine learning techniques, e.g., Support Vector Machines (SVMs) [23,26] or the SNoW model [18], where questions are encoded using a variety of lexical, syntactic and semantic features. In [18], it has been shown that the questions' syntactic structure contributes

---

[1] In FrameNet, FIRING is defined as the event for which "An EMPLOYER ends an employment relationship with an EMPLOYEE".

[2] USE_FIREARM: "An AGENT causes a FIREARM to discharge, usually directing the projectile from the barrel of the FIREARM (the SOURCE), along a PATH, and to a GOAL."

remarkably to the classification accuracy and this makes the question classification task quite similar to predicate recognition in FrameNet. QC is a sentence-driven task for which a complexity similar to a predicate recognition task arises. In fact questions correspond to short sentences usually exposed to data sparseness. In [26], a supervised approach with composite kernels is applied to incorporate semantic information and to extend a bag-of-words representation. In particular, the authors employ latent semantic kernels [9] to obtain a generalized similarity function between questions from Wikipedia, that improved previous work. One of the best models for Question Classification is presented in [23] where the benefits of grammatical information are proofed by experimenting several syntagmatic kernels (reaching 91.8% of accuracy on the UIUC dataset). High accuracy (about 92% on the TREC 2002 question set) is also reached by [19] that discuss the use of a complex set of syntactic and semantic information from external resources (including Wordnet).

3 Encoding Syntactic Information into a Standard Topic Model

LDA is generally applied to induce latent topics from large text corpora, and use them in supporting predictions over new documents. It allows each document to exhibit multiple topics, that capture the rich inner structure of document categories. LDA traditionally derives class-based lexicalized models in an unsupervised fashion, thus providing a more robust representation against lacks in the training data. In essence, as latent topics capture the distributional evidence embodied by large sets of unlabeled data, LDA works by mapping them into lower dimensional spaces, represented by the T topic distributions (with $T \ll N$, where N is the number of words in the corpus). A possibly small amount of labeled data can be thus expressed in this reduced space where a more effective training of supervised classifier (e.g., an SVM) can be applied. Obviously rare words in the labeled data set and sparse data benefit from the topic generalization of the original feature space.

3.1 Limitations of Purely Lexical LDA for NLP

While LDA is usually employed as a structured model of purely *lexical* information, *word order as well other grammatical information* expressed by a sentence is usually neglected. This can be a significant limitation against strictly syntax-dependent sentence classification tasks. In general, lexical information alone is not sufficient, as other linguistic properties must be observed for enabling the suitable semantic inferences. Syntactic features provide crucial information to estimate the similarity between the question and the candidate answers, as in general explored in tree-kernel-based approaches to Question Classification [23].

While the topic models oriented to syntactic information as discussed in Section 2 have been tested on simple tasks (e.g., POS tagging), their impact on sentence (or question) classification scenarios has not yet been studied in depth. These applications are characterized by very sparse data and the applicability of topic models (e.g. [13]) has not been demonstrated yet.

3.2 A Simple Topic Model for Shallow Syntactic Information

The main idea of this work is to exploit the LDA capability to discover recurrent grammatical patterns over input data. As suggested in Figure 2, the parse tree corresponding
to a simple sentence represents:

- *lexical information* through terminal nodes (e.g., words as $Cognac$, is, ...)
- *coarse-grained grammatical information* through the POS tag characterizing (e.g.,
  NNP, VBZ, ...) the pre-terminal nodes
- *fine-grained grammatical information* according to the subtrees expressing production rules among the non-terminals of the underlying *context free grammar* (CFG).

Examples of the CFG rules shown in Figure 2 are: $S \rightarrow NP\ VP$, $NP \rightarrow NPP$ or
$NP \rightarrow DT\ NN$. Our aim here is to acquire these rules implicitly, as patterns included
in the targeted topic model. Accordingly, we map subtrees into flat features and add
them as novel pseudo-words to the sentence bag-of-word representation. For example,
the partial tree expressed by $VP \rightarrow VBN\ PP$ in Fig. 2 can be represented through a
pseudo token given by the POS sequence *VBN-IN-NNP*.

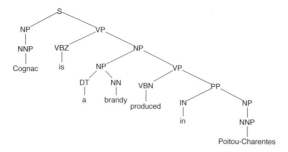

Fig. 2. Parse tree associated to the sentence *Cognac is a brandy produced in Poitou-Charentes*

In this way, lexicalized features (i.e., words) are combined with shallow syntactic
information (i.e., POS sequences as pseudo tokens), by still keeping limited the com-
plexity of the LDA estimation. A document d of length $|d|$ is thus represented as a
set of words (in a bag-of-word fashion) extended by the pseudo token expressing all
the sequences of POS tags whose length is below a fixed threshold n (i.e. n-POS tag
grams). Given the word sequence $d = \{w_1, \ldots, w_{|d|}\}$ whose corresponding POSs
are $\{pos_1, \ldots, pos_{|d|}\}$, the representation of the pseudo tokens is the set of pairs
$\{(w_1.pos_1), \ldots, (w_{|d|}.pos_{|d|})\}$, where each lemmatized word is coupled with its POS.

Sequences of POS tags are also modeled by adding other types of pseudo-tokens as
sequences (i.e., n-grams) of POS tags. Given n as the maximal size of the extracted
sequences, every subsequence of length at most n is mapped into a pseudo-tokens.
These novel "grammatical" tokens of length Δ are expressed as $\{p_j, \ldots, p_{j+\Delta}\}$ where
$\Delta = 1, ..., n$. In these patterns the representation of prepositions (POS tag *IN*) is made
explicit. Every position j for which $pos_j = IN$ is represented through the word w_j, so
that *at-PN* or *of-DT-NN* are obtained as pseudo-tokens for fragments such as "*at Whit-
lock*" or "*of the vineyard*". The representation of the sentence in Fig. 2 is shown in

Table 1. Representation of lexical and grammatical information for the sentence in Fig. 2

BOW	*cognac*.NNP *be*.VBZ *a*.DT *brandy*.NN *produce*.VBN *in*.IN *poitou-charentes*.NNP ...
2-grams	NNP-VBZ VBZ-DT DT-NN NN-VBN VBN-*in* *in*-NNP NNP-.
3-grams	NNP-VBZ-DT VBZ-DT-NN DT-NN-VBN NN-VBN-*in* VBN-*in*-NNP *in*-NNP-.
4-grams	NNP-VBZ-DT-NN VBZ-DT-NN-VBN DT-NN-VBN-*in* NN-VBN-*in*-NNP VBN-*in*-NNP-.
5-grams	NNP-VBZ-DT-NN-VBN VBZ-DT-NN-VBN-*in* DT-NN-VBN-*in*-NNP NN-VBN-*in*-NNP-.

Table 1, here the set of words $(w_i.pos_i)$ and the different n-gram tokens are shown. The *generative model* resulting from the application of LDA to such pseudo-documents captures underlying short-term syntactic relations, through sequences of POS tags.

Using Labeled Information in LDA. The introduction of "grammatical" pseudo-tokens has an impact on the corpus statistics: syntactic regularities should emerge as latent topics, through Gibbs sampling. This representation is expected to enrich the representation for short documents, such as sentences, and improve the statistical robustness against sparse phenomena. Table 3 shows three different topics generated from the representation discussed in this section through LDA in a predicate classification task. Topic 2 is typically related to drinkable liquids or products that are kind of FOOD clearly connected with events of eating or producing (e.g., the INGESTION or MANUFACTURING frames). This clearly results from the generalization of lexical information, as traditionally carried out in topic models. However, Topics 1 and 3 in Table 3, are more syntax-driven and focused around prepositional phrases (i.e., *from-NP* and *with-NP*, in Topic 1 and 3 respectively) that are strongly related to semantic roles: *from-NP* typically expresses spatial relations (i.e., *change of place* events) while *with-NP* introduces company (*Mary played with Ann*) or instrumental (*carved with iron*) relations. In Topics 1 and 3 of Table 3 two special pseudo-words appear: they correspond respectively to the names of the frame F_QUITTING_A_PLACE and to the frame element FE_INSTRUMENT. These are just two examples of how labeled data can be used to augment the linguistic generalization of LDA.

In FrameNet, an annotated sentence explicits *at least* the information about the frame and about the entire set of the predicate's roles. Although this information is available only for the annotated examples, it may play as a strong set of constraints in the topic estimation process. Topic directly inspired by Framenet labels straightforwardly reflect semantic properties (e.g., spatial relations). Training data are thus enriched also with the gold information regarding the *frame* and the *roles*, i.e. frame elements, of the underlying annotated sentence, as a bias for the LDA estimation. Table 2 reports the semantic pseudo-words derivable from the labeled version of the example sentence, according to its MANUFACTURING frame, i.e., *[Cognac]*PRODUCT *is a brandy* ***produced***MANUFACTURING [in Poitou-Charentes region]LOCUS.

Table 2. Representation of semantic information for the labeled version of sentence in Fig. 2

sem_info	MANUFACTURING, PRODUCT, LOCUS.

In the following the representation of unlabeled sentences (as well as test sentences) will follow the example reported in Table 1, while the training sentences will be represented according to the union of the information of Tables 1 and 2.

4   Experiments

In order to prove the effectiveness of the proposed model, we tested it on the predicate recognition and question classification tasks, discussed in the previous sessions. In the following, experimental results of the application of SVMs trained over feature vectors modeled as topic probabilities are reported.

4.1   Task1: Predicate Recognition

In this experiment two large annotated data sets are employed. The first is the FrameNet corpus itself [2], as a significant portion of the BNC corpus: it consists of 150,000 tagged sentences, where frame, lexical unit and frame elements are explicitly annotated for each sentence. The second annotated corpus is the NTI corpus, made available by the SemEval 2007 Task 19[3] [1]: the Nuclear Threat Initiative (NTI). We used the FrameNet corpus as the train set, where 80% is used both to acquire the LDA model and to train the SVM classifier and 20% is held-out for parameter tuning. The NTI corpus is then used as test set; it consists of 8,422 sentences. It is worth noticing that 2,587 sentences are ambiguous, i.e., they contain lus that evoke multiple frames in FrameNet (as the verb *fire*). In this way, we evaluate the generalization capability of the proposed topic model in an out-of-domain test: notice that frames are domain independent linguistic generalizations, while syntactic phenomena are highly corpus specific.

Topic Model Acquisition. We derived the topic mixture model from the training dataset estimating probabilities through the Gibbs-Sampling by fixing the number of latent topics to 50. The sentence representation used to build the model has been enriched through the POS tags (as discussed in Section 3) produced by the LTH dependency parser [15]. n-gram sequences are then derived for all sentences, where $n = 5$ is used. In order to use FrameNet annotations in form of semantic pseudo-words (as in Table 2) every sentence in the training set is also enriched with the frame and the frame element labels, respectively.

In Table 3 first three columns show topics derived from the proposed model that are semantically and syntactically coherent are shown. As already noticed, the event of motion *from a place* clearly characterize the POS sequences (e.g., n-grams from-DT-NN or VBD-from) of Topic 1. Notice that the frame label QUITTING_A_PLACE and its frame element (i.e., SOURCE) confirm this interpretation. The second column is more explicitly oriented to lexically model the notion of FOOD and DRUGS. Topics more oriented to semantic phenomena implied by the frame labels are exemplified by last two columns in Table 3. Topic 4 refers to the COMMERCE events while Topic 5

---

[3] The NTI annotated collection here used represents the train corpus of the SemEval competition. Downloadable at:
nlp.cs.swarthmore.edu/semeval/tasks/task19/data/train.tar.gz.
The test set is not publicly available.

involves more cognitive aspects (e.g., the frame AWARENESS or the FE COGNIZER). Notice that prior knowledge about the sentence frames is only used during the training phase, and does not directly affect inferences over unseen sentences. However, Topic 1 and 3 (in Table 3) seem to suggest that it constraints the parameter estimation, resulting in meaningful topics, i.e., better generalizations of the paradigmatic notion of frame.

Table 3. Topics acquired through LDA (at 50 topics) exhibiting syntactic and lexical information (Topic 1, 2 and 3) and prior knowledge about frames (Topics 4 and 5)

Topic 1	Topic 2	Topic 3	Topic 4	Topic 5
from.IN	wine.NN	with.IN	FE_GOODS	FE_COGNIZER
FE_SOURCE	beer.NN	NN-with	PRP-VB	FE_TOPIC
NN-from	drink.VBD	with-DT-NN	FE_SELLER	PRP-VBP
from-DT-NN	cocaine.NN	with-NN	F_COMMERCE_SCENARIO	about.IN
NN-from-DT	sip.VBD	with-JJ	FE_MONEY	F_AWARENESS
F_QUITTING_A_PLACE	toast.NN	NN-with-DT	MD-PRP-VB	be.VBP
DT-NN-from	dine.VBD	with-DT-JJ	F_COMMERCE_SELL	PRP-VBP-RB
from-NN	champagne.NN	with-DT-JJ-NN	FE_PAYER	about-DT

Predicate recognition. For predicate recognition, sentences are first mapped into their corresponding topic probability distribution. Feature vectors are also augmented through the lexical unit information (e.g., the verb *fire* in 1), in order to obtain a more expressive feature space. An SVM classifier is then trained over the resulting low-dimensional feature space. Thus, each frame f is modeled by a dedicated SVM, whose positive (or negative) examples are sentences for which a predicate is annotated (or is not annotated) with f. Notice that every test sentence is presented together with a specified lu. Every valid frame defined for the lu in FrameNet is a candidate and the multi-classification is obtained through a one-vs-all scheme: frames with the highest margin will be selected.

Here the quality of the overall classification is computed in terms of *accuracy*, i.e., the percentage of sentences labeled with the correct frame. As in the NTI corpus, 1.43 frames per sentence are evoked on average, random selection among possible frames results in a simple baseline of about 69%. If only ambiguous sentences are retained, 2.42 frames per sentence can be evoked in FrameNet, and the baseline amounts to 41% accuracy. The results on the predicate recognition task can be compared with Shalmaneser [10], a toolkit for shallow semantic parsing based on FrameNet. For the frame recognition task Shalmaneser uses a Naive Bayes classifier trained over a rich set of features, e.g., BOW, grammatical functions of the lu, and verb voices. Since we do not know which part of the FrameNet dataset was used to train Shalmaneser, we used the NTI corpus as test set, to do a fair comparison. The syntactic topic model based on n-grams is compared with a topic model built using only the token information, omitting the syntactic information obtained through the n-grams.

Table 4 shows the results obtained with lexical information only (i.e., the pseudo tokens in the row BOW of Table 1), as well as those including all syntactic n-grams ($synt$) and the semantic labels (lbd) from the labeled portion of the dataset. The SVM trained over the LDA model that exploits bag of word (BOW), syntactic information ($synt$) and prior knowledge (lbd) achieves an accuracy of 88.79%, outperforming the Shalmaneser system (79.01%), even if this latter makes explicit use of syntactic features also. In the subset of ambiguous sentences, the robustness is confirmed with an accuracy of 63.51%.

Table 4. Sentence classification accuracy

Approach	Ambiguous predicates	Overall
Random Baseline	41.31%	69.93%
Shalmaneser	51.22%	79.01%
LDA Model (BOW)	60.30%	87.81%
LDA Model (BOW-synt-lbd)	63.51%	88.79%

4.2   Task 2: Question Classification

The second evaluation is carried out on a question classification task. The goal is to classify the questions expressed as the LDA topic mixture. The targeted dataset is the UIUC corpus, largely adopted for benchmarking [18]. UIUC contains a training set of 5,452 questions and a test set of 500 questions, both extracted from TREC. Question classes are organized in two levels of granularity: a first level of 6 coarse-grained classes, like ABBREVIATION, ENTITY, DESCRIPTION and a second level of 50 fine-grained sub-classes, e.g., *Plant* and *Food* are subclasses of the ENTITY category.

Table 5. Topics acquired through LDA (100 topics) for QC

Topic 1	Topic 2	Topic 3	Topic 4
of.IN	who.WP	in-what	world.NNP
kind.NN	?..	die.VB	war.NNP
what-NN-in	be.VBD	BS-in	number.NN
type.NN	the.DT	NNP-NNP-VB	brave.NNP
water.NN	who-VBD	in-what-NN	around.IN
what-NN	BS-who	VB-.	civil.IN
BS-what-NN	BS-who-VBD	BS-in-what	english.NNP
NN-of	VBD-DT	thatcher.NNP	single.JJ

As in the first task, SVM classifiers are used in a one-vs-all fashion computing the quality of classification in terms of accuracy. The training set has been split in 80% for the LDA model estimation and the SVM training, while the remaining 20% is used for the SVM tuning. Sentences are modeled as BOW and sequences of POS tags, while no labeled information (such as the frame labels used in sentence classification) was available for this task. Table 5 reports an excerpt of emerging topics: in line with the previous results, n-grams of POS tends to be predominant in most topics. It is worth noticing that the first three topics capture words and the structures typical of the training sentences (i.e. questions). Moreover, Topic 4 contains mostly lexical information represented by word evoking *war* scenarios.

Again, topic probabilities have been used as features in the SVM algorithm, thus resulting in the so-called LDA kernel ($K(LDA)$). According to the flexibility allowed by kernel-based learning, we also combined the LDA information with a linear kernel based on the traditional bag of words $K(BOW)$ model: the result is a complex kernel hereafter denoted by K(BOW)+K(LDA). Finally, LDA_{BOW} and $LDA_{BOW-ngram}$ denote the LDA model applied only to lexical information and to both lexicals and POS sequences, respectively. Results, compared with the state-of-art achieved by the system discussed in [26] on the same UIUC dataset, are shown in Table 6.

The authors combine a kernel classifier based on BOW with two semantic kernels: one is based on Latent Semantic Indexing applied to Wikipedia, and the other uses semantic information acquired through *manually constructed lists of words*, i.e., a task-specific lexicon related to the answer types. As expected (Section 3), Table 6 shows that syntactic information increases the expressivity of the LDA model, achieving good results even against sparse data. The combination of kernels with LDA as a dimensionality reduction technique (i.e., $K(LDA_{BOW})$) and with lexical information (i.e., $K(BOW) + K(LDA_{BOW-ngram})$) always increases accuracy. The same is observed in [26], where latent semantic kernels $K(LS)$ are introduced. In the coarse-grained test, Table 6 shows how the syntactic generalization supported by the $K(BOW) + K(LDA_{BOW-ngram})$ achieves the best known results on the UIUC dataset, i.e., 91.8% as in [23]. Notice that this improves the best results of [26] (i.e., the $K(BOW) + K(LS) + K(semRel)$) making use of *manually annotated resources that are task specific*. This emerges in the fine-grain task whereas slightly better results depend on the external resources employed. Note how the kernel $K(LS)$ that uses only lexical information gathered by an external corpus like Wikipedia [26] is also weaker than our model, based just on shallow syntax. The impact of such a shallow level of syntactic information is surprisingly good: the increase in performance is +41% (or +47%) between $K(BOW)$ and $K(LDA_{BOW-ngram})$ in the coarse (or fine) grained case.

The results in Table 6 are also remarkable from a computational point of view. The proposed adoption of LDA is straightforward as it only requires the source POS tagged sentences and no parsing is necessary. Moreover, the training time of tree kernel based SVMs on benchmarking data sets are in the order of hours for small data sets and days for large data collections (e.g., Prop Bank, as reported in [21]). The proposed approach was trained over the LDA feature space (i.e., the $K(BOW)+K(LDA_{BOW-ngram}$ kernel) in less than 3 and 10 minutes for the coarse and fine grained task, respectively.

Table 6. Experimental results for the QC task

Kernel	Coarse Accuracy	Fine Accuracy
K(LDA$_{BOW}$)	44.6%	31.6%
K(LDA$_{BOW-ngram}$)	75.2%	59.2%
K(BOW)+K(LDA$_{BOW-ngram}$)	**91.8%**	84.8%
[26]		
K(BOW)	86.4%	80.8%
K(LS)	70.4%	71.2%
K(BOW)+K(LS)	90.0%	83.2%
K(BOW)+K(LS)+K(semRel)	90.8%	**85.6%**
[23]		
Tree Kernels K(BOW)+K(*PartialTrees*)	**91.8%**	-

5   Conclusions

In this paper a simple yet robust topic model that well account for fine-grained linguistic phenomena has been presented. Existing approaches to natural language learning that are sensitive to grammatical information are either topic models based on complex variants of LDA or tree kernel-based discriminative approaches. We show how the basic LDA model can be preserved for its efficiency, and propose sentence expansion as a useful techniques to generalize and expand the standard bags of words representation. The additional pseudo-words, e.g., sequences of POS tags, capture grammatical information (e.g., word syntactic categories) as well as more complex syntactic structures (e.g., POS tag sequences) that extend the original lexical information. The resulting topics provide generalized classes that meaningfully correlate with semantic phenomena. This forms a low-dimensional feature space for the later supervised classification steps. This model achieves very high performances in the predicate recognition and question classification tasks, confirming the state-of-the-art accuracy on standard benchmarking data sets. We showed how the proposed approach is a viable, effective and simpler alternative to complex learning models such as tree kernels. This is particularly true for NL learning task with poor training conditions.

References

1. Baker, C., Ellsworth, M., Erk, K.: Semeval-2007 task 19: Frame semantic structure extraction. In: Proc. of SemEval 2007, Czech Republic, pp. 99–104 (2007)
2. Baker, C.F., Fillmore, C.J., Lowe, J.B.: The berkeley framenet project (1998)
3. Blei, D., McAuliffe, J.: Supervised topic models. In: Proceedings of Advances in Neural Information Processing Systems (2007)
4. Blei, D.M., Ng, A.Y., Jordan, M.I.: Latent dirichlet allocation. Journal of Machine Learning Research 3(4-5), 993–1022 (2003)
5. Boyd-Graber, J., Blei, D.: Syntactic topic models. In: Proceedings of Advances in Neural Information Processing Systems (2008)
6. Boyd-Graber, J., Blei, D., Zhu, X.: A topic model for word sense disambiguation. In: Proc.of the Joint Conference on EMNLP and CoNLL, pp. 1024–1033 (2007)
7. Brody, S., Lapata, M.: Bayesian word sense induction. In: Proceedings of the Conference of the European Chapter of the ACL, pp. 103–111 (2009)
8. Collins, M., Duffy, N.: New ranking algorithms for parsing and tagging: Kernels over discrete structures, and the voted perceptron. In: ACL 2002 (2002)
9. Nello, C., John, S.-T., Huma, L.: Latent semantic kernels. J. Intell. Inf. Syst. 18(2-3), 127–152 (2002)
10. Erk, K., Pado, S.: Shalmaneser - a flexible toolbox for semantic role assignment. In: Proceedings of LREC 2006, Genoa, Italy (2006)
11. Fillmore, C.J.: Frames and the semantics of understanding. Quaderni di semantica 6(2), 222–254 (1985)
12. Gildea, D., Jurafsky, D.: Automatic Labeling of Semantic Roles. Computational Linguistics 28(3), 245–288 (2002)
13. Griffiths, T., Steyvers, M., Blei, D., Tenenbaum, J.: Integrating topics and syntax. In: Proceedings of NIPS 2005, pp. 537–544 (2005)
14. Griffiths, T.L., Steyvers, M.: Finding scientific topics. Proceedings of the National Academy of Sciences, 5228–5235 (2004)
15. Johansson, R., Nugues, P.: Semantic structure extraction using nonprojective dependency trees. In: Proceedings of SemEval 2007, Czech Republic (2007)
16. Kwok, C.C.T., Etzioni, O., Weld, D.S.: Scaling question answering to the web. In: WWW, pp. 150–161 (2001)

17. Landauer, T., Dumais, S.: A solution to plato's problem: The latent semantic analysis theory of acquisition, induction and representation of knowledge. Psychological Review 104(2), 211–240 (1997)
18. Li, X., Roth, D.: Learning question classifiers. In: Proceedings of ACL 2002 (2002)
19. Li, X., Roth, D.: Learning question classifiers: the role of semantic information. Nat. Lang. Eng. 12(3), 229–249 (2006)
20. Minka, T., Lafferty, J.: Expectation-propagation for the generative aspect model. In: Proceedings of the Conference on Uncertainty in Artificial Intelligence, pp. 352–359 (2002)
21. Moschitti, A.: Efficient convolution kernels for dependency and constituent syntactic trees. In: ECML, Berlin, Germany, pp. 318–329 (2006); Machine Learning
22. Moschitti, A., Pighin, D., Basili, R.: Tree Kernels for Semantic Role Labeling. Computational Linguistics Special Issue on Semantic Role Labeling (3), 245–288 (2008)
23. Moschitti, A., Quarteroni, S., Basili, R., Manandhar, S.: Exploiting syntactic and shallow semantic kernels for question answer classification. In: Proceedings of ACL 2007 (2007)
24. Pradhan, S., Hacioglu, K., Krugler, V., Ward, W., Martin, J.H., Jurafsky, D.: Support Vector Learning for Semantic Argument Classification. Machine Learning 60(1-3), 11–39 (2005)
25. Rosen-Zvi, M., Griffiths, T., Steyvers, M., Smyth, P.: The author-topic model for authors and documents. In: Proceedings of Uncertainty in Artificial Intelligence, pp. 487–494 (2004)
26. Tomás, D., Giuliano, C.: A semi-supervised approach to question classification. In: Proc. of the 17th European Symposium on Artificial Neural Networks, Bruges, Belgium (2009)
27. Toutanova, K., Johnson, M.: A bayesian lda-based model for semi-supervised part-of-speech tagging. In: Proceedings of Advances in Neural Information Processing Systems (2008)

Structured Learning for Semantic Role Labeling

Danilo Croce and Roberto Basili

Department of Enterprise Engineering,
University of Roma, Tor Vergata,
Via del Politecnico 1, 00133 Roma
{croce,basili}@info.uniroma2.it

Abstract. The use of complex grammatical features in statistical language learning assumes the availability of large scale training data and good quality parsers, especially for language different from English. In this paper, we show how good quality FrameNet SRL systems can be obtained, without relying on full syntactic parsing, by backing off to surface grammatical representations and structured learning. This model is here shown to achieve state-of-art results in standard benchmarks, while its robustness is confirmed in poor training conditions, for a language different for English, i.e. Italian.

1 Linguistic Features for Inductive Tasks

Language learning systems usually generalize linguistic observations into statistical models of higher level semantic tasks, such as Semantic Role Labeling (SRL). Statistical learning methods assume that lexical or grammatical aspects of training data are the basic features for modeling the different inferences. They are then generalized into predictive patterns composing the final induced model.

Lexical information captures semantic information and fine grained context dependent aspects of the input data. However, it is largely affected by data sparseness as lexical evidence is often poorly represented in training. It is also difficult to be generalized and non scalable, as the development large scale lexical KBs is very expensive. Moreover, other crucial properties, such as word ordering, are neglected by lexical representations, as syntax must be also properly addressed. In semantic role labeling, the role of grammatical features has been outlined since the seminal work by [6]. Symbolic expressions derived from the parse trees denote the position and the relationship between an argument and its predicate, and they are used as features. Parse tree paths are such features, employed in [11] for semantic role labeling. Tree kernels, introduced by [4], model similarity between two training examples as a function of the shared parts of their parse trees. Applied to different tasks, from parsing [4] to semantic role labeling [16], tree kernels determine expressive representations for effective grammatical feature engineering.

However, there is no free lunch in the adoption of lexical and grammatical features in complex NLP tasks. First, lexical information is hard to be properly generalized whenever the amount of training data is small. Large scale general-purpose lexicons are available, but their employment in specific tasks is not satisfactory: coverage in domain (or corpus)-specific tasks is often poor and domain adaptation is difficult. For

R. Pirrone and F. Sorbello (Eds.): AI*IA 2011, LNAI 6934, pp. 238–249, 2011.

example, the lack of lexical information is often claimed as the main responsible for significant performance drops in out-of-domain argument classification [20,11]. Corpus driven methods have been often advocated as a solution according to distributional views on lexical semantics. These are traditionally used to acquire meaning generalizations in an unsupervised fashion (e.g. [18,27]) through the analysis of distributions of word occurrences. In line with previous work, (e.g. [9,5]) we will pursue this line of research, by extending a supervised approach through the adoption of vector based models of lexical meaning, as discussed in Section 2.2.

The adoption of grammatical features and tree kernels is also problematic. First, strict requirements exist in terms of the size of the training data set as high dimensionality spaces are generated, whose data sparseness can be prohibitive. Although specific forms of optimization have been proposed to limit their inherent complexity (e.g. [15]), tree kernels do not scale well over very large training data sets. Moreover, methods for extracting grammatical features from parse trees (e.g. [6]) are strongly biased by the parsing quality. Several studies showed that parsing inaccuracies significantly lower the quality of training data. In [19] experiments over gold parse trees are reported with an accuracy (93%) significantly higher than the ones derived by using automatically derived trees (i.e. vs. 79%). In [14] sequential tagging is applied for SRL and a comparative analysis between two SRL systems is reported. They share the same architecture, but are built on partial vs. full parsing input, respectively. Finally, [11] reports that the adoption of the syntactic parser restricts the correct treatment of FrameNet roles to only the 82% of them, i.e. the only ones that are grammatically recognized. This thus constitutes a strict upper bound for a SRL cascade based on full parsing material.

We want to explore here a possible solution to the above problems through the adoption of shallow grammatical features that avoid the use of a full parser in SRL, combined with distributional models of lexical semantics. While parsing accuracy highly varies across corpora, the adoption of shallower features (e.g. POS n-grams), increases robustness, applicability and minimizes overfitting. At the same time, lexical information is made available in terms of vector spaces derived automatically from large (domain specific) corpora. This aims to increase the quality of the achievable generalization without strict requirements in terms of training data. The expected result is the design of flexible SRL systems, also applicable in poor training conditions, such as for languages where limited resources are available. The open research questions are: which shallow grammatical representation is suitable to support the learning of fine-grained semantic models? Are lexical generalizations provided by distributional analysis of large corpora helpful? Can the grammatical generalizations derived from shallow syntactic representations be profitably augmented through word space models of the involved lexical information?

In the rest of this work, we will show that a structured learning framework, such as SVM^{hmm} [1], benefits from simpler features, and provides competitive performances with respect to richer syntactic representations. Moreover, a distributional method (i.e. Singular Value Decomposition) applied to unlabeled corpora is introduced in order to acquire effective lexical generalizations. The resulting model is then tested and its performances are observed in standard as well as poor training conditions characterizing two languages: English and Italian, respectively.

2   Shallow Parsing and Grammatical Feature engineering

Complex semantic inferences generally require specific feature engineering at the lexical information and grammatical level. For example, sentence (3) is the appropriate answer to question (1), although both sentences (2) and (3) are reasonable candidates.

$$\text{What } \textbf{\textit{French province is Cognac produced in?}} \tag{1}$$

$$\textit{The grapes which } \textbf{produce} \textit{ the } \textbf{Cognac} \textit{ grow } \textbf{in the French province} \text{ ...} \tag{2}$$

$$\textbf{\textit{Cognac is}} \textit{ a brandy } \underline{\textit{made}} \textbf{ }\textit{in Poitou-Charentes.} \tag{3}$$

Suppose we use a lexical overlap rule for solving the above task. Given the large overlap between (1) and (2) (see terms outlined in bold), the wrong answer (2) will be selected. However, when generalizations at the lexical level and syntactic information are made available the correct solution (e.g. 3) can be easily predicted. First, the verb *made* in (3) is a near synonym of *produce*, used in (1) instead. Moreover, its contribution to the syntactic structure of (3) is identical to the prepositional phrase in the question (1). The estimate of the similarity between the question and its candidate answers can be thus made dependent on the lexical generalization of the verb and on the encoding of syntactic features. This latter information is in general explored by tree kernels, in Question Classification [17].

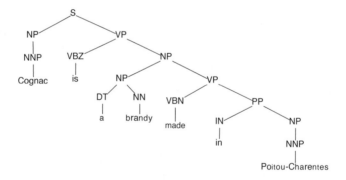

Fig. 1. Parse tree associated to sentence (3)

The parse tree in Figure 1 corresponds to sentence (3) and represents:

– *Lexical information* through its terminal nodes (e.g., $Cognac$, is, $made$, ...)
– *Coarse-grained grammatical information* through the POS tag characterizing pre-terminal nodes (e.g. NNP or VBZ)
– *Fine-grained grammatical information*, through subtrees that correspond to production rules, such as $S \to NP\ VP$, $VP \to VBN\ PP$ or $PP \to IN\ NP$.

In statistical natural language learning, trees are regarded as complex (joint) stochastic events involving their lexical and grammatical components (e.g. [3]). We want to design specific features able to express (i.e. surrogate) the syntactic structures as well as the lexical items in the parse tree, implicitly. Notice how POS tag sequences correspond to subtrees and can be considered their shallow counterpart. They linearly express properties analogous to the Parse Tree Paths in [6]. In other words, subtrees can be artificially surrogated introducing POS tag sequences (or POS n-grams), instead of parse tree paths. For example, the partial tree expressed by VP→VBN PP in Fig. 1 can be represented through the *VBN-IN-NNP* token.

Shallow syntactic information (i.e., the POS n-grams) seemingly to lexicalized information (i.e., simple words) can be thus made available through flat symbolic features. This constrains the capacity of the reference learning machine and is beneficial to the risks of overfitting. A sentence can be thus represented in a bag-of-word fashion, where words as well as the pseudo tokens defining certain POS tag sequences[1], i.e. the POS n-grams, are both made explicit. When a preposition is met, its POS tag (i.e. *IN*) is substituted with the corresponding token, so that *of-DT-NN* is obtained as the pseudo-token for a fragment like "*of the vineyard*". The representation of sentence (3) is shown in Table 1, where words and n-gram are shown.

Table 1. Representation of lexical and grammatical information for sentence (3)

unigrams	cognac.NNP be.VBZ a.DT brandy.NN made.VBN in.IN poitou-charentes.NNP ...
2-grams	NNP-VBZ VBZ-DT DT-NN NN-VBN VBN-in in-NNP NNP-.
3-grams	NNP-VBZ-DT VBZ-DT-NN DT-NN-VBN NN-VBN-in VBN-in-NNP in-NNP-.

Once words and n-gram tokens are made available as individual features, similarities at the lexical and syntagmatic level can be captured. However, near synonymy, as required to capture the analogy between *produce* and *make*, in (1) and (3), respectively, cannot be captured: their discrete matching will provide null contributions during training. We propose to extend the above feature set with vectors representing individual words, according to a distributional word space model (e.g. [27]). While we delay the description of the adopted distributional model to Section 2.2, we assume here that a distributional model as the vector w_i is available for every word w_i. Then the semantic similarity between word pairs (w_i, w_j) (such as *produce* and *make*) is computed during training as the cosine similarity between vectors w_i and w_j. This enhances the features to capture near synonymy properties, required to generalize training evidence (e.g. *produce*) to similar contexts (e.g. *make*).

In the next section we will see how the set of features here introduced is adopted to map the boundary detection and argument classification steps of SRL into a sequence labeling learning tasks.

2.1 Structured Learning for SRL

Semantic role labeling processes have been traditionally partitioned in the two stages of *boundary detection* (BD) and *argument classification* (AC). In BD, given a predicate

---

[1] For examples those POS sequences whose length is smaller than a threshold n

in a sentence, the position and span of its arguments are recognized and annotated. AC deals instead with the assignment of semantic types, i.e. role labels, to the recognized arguments. BD has been often seen as a node labeling task, especially in constituent based approaches [16], whereas one non terminal node strictly corresponds to the span of one argument. However, we want to model the task without relying on parse trees so that only POS tag sequences are available. In analogy with previous work (e.g. [14]), we map BD into a *sequential tagging* task. With respect to BD, each token represents the *beginning* (B), the *inside* (I) or *outside* (i.e. O) of an argument or it can be simply *external* (X) to every argument. For example in the sentence:

$$[Yesterday]_{\text{TIME}} , [a\ robber]_{\text{KILLER}} \textbf{ killed } [a\ guardian]_{\text{VICTIM}} [with\ a\ knife]_{\text{INSTRUMENT}} .$$

the resulting BD can be expressed by labeling each word according to its relative position, i.e.:

$$Yesterday/\text{B} \quad ,/\text{X} \quad a/\text{B} \quad robber/\text{O} \quad killed/\text{X}$$
$$a/\text{B} \quad guardian/\text{O} \quad with/\text{B} \quad a/\text{I} \quad knife/\text{O} \quad ./\text{X}$$

The BD task is here thus a sequence labeling process that determines the individual (correct BIO) class for each token. We can in principle deal with AC in a similar fashion. The resulting sequence labeling process here makes use of role specific labels assigned to individual words. A special X role is reserved to words external to the boundaries, i.e. that are not part of any argument. A voting strategy is applied within each argument and the role label most frequently assigned to the inner members of a boundary is retained as the unique role.

In order to model both problems as sequential tagging tasks ([14]), the SVM^{hmm} model discussed by [1] is proposed[2]. It extends classical SVMs by learning a discriminative model isomorphic to a k-order Hidden Markov Model through the Structural Support Vector Machine (SVM) formulation [26]. In particular, given an observed input sequence $\mathbf{x} = (x_1 \ldots x_l) \in \mathcal{X}$ of feature vectors $x_1 \ldots x_l$, the model predicts a tag sequence $\mathbf{y} = (y_1 \ldots y_l) \in \mathcal{Y}$ after learning a linear discriminant function $F : \mathcal{X} \times \mathcal{Y} \rightarrow \mathbb{R}$ over input/output pairs. The labeling $f(\mathbf{x})$ is thus defined as: $f(\mathbf{x}) = \arg\max_{\mathbf{y} \in \mathcal{Y}} F(\mathbf{x}, \mathbf{y}; \mathbf{w})$.

It is obtained by maximizing F over the response variable, $\mathbf{y}$, for a specific given input $\mathbf{x}$. In these models, F is linear in some combined feature representation of inputs and outputs $\Phi(\mathbf{x}, \mathbf{y})$, i.e. $F(\mathbf{x}, \mathbf{y}; \mathbf{w}) = \langle \mathbf{w}, \Phi(\mathbf{x}, \mathbf{y}) \rangle$.

As Φ extracts meaningful properties from an observation/label sequence pair $(\mathbf{x}, \mathbf{y})$, in SVM^{hmm} it is modeled through two types of features: interactions between attributes of the observation vectors x_i and a specific label y_i (i.e. **emissions** of x_i by y_i) as well as interactions between neighboring labels y_i along the chain (**transitions**). In other words Φ is defined so that the complete labeling $\mathbf{y} = f(\mathbf{x})$ can be computed from F efficiently, i.e. using Viterbi-like decoding algorithm, according to the following linear discriminant function:

$$y = \arg\max_{y} \{ \sum_{i=1\ldots l} [\sum_{j=1\ldots k} (x_i \cdot w_{y_{i-j} \ldots y_i}) + \Phi_{tr}(y_{i-j}, \ldots, y_i) \cdot w_{tr}] \}$$

---

[2] http://www.cs.cornell.edu/People/tj/svm_light/svm_hmm.html

In the training phase, SVM^{hmm} solves the following optimization problem given training examples $(x^1, y^1) \ldots (x^n, y^n)$ of sequences of feature vectors $x^j = (x_1^j, \ldots, x_l^j)$ with their correct tag sequences $y^j = (y_1^j, \ldots, y_l^j)$.

$$min \; \frac{1}{2} \|w\|^2$$

$s.t.$

$$\forall y : \{ \sum_{i=1\ldots l} (x_i^1 \cdot w_{y_i^1}) + \Phi_{tr}(y_{i-1}^1, y_i^1) \cdot w_{tr} \} \geq \{ \sum_{i=1\ldots l} (x_i^1 \cdot w_{y_i}) + \Phi_{tr}(y_{i-1}, y_i) \cdot w_{tr} \} + \Delta(y^1, y)$$

$\ldots$

$$\forall y : \{ \sum_{i=1\ldots l} (x_i^n \cdot w_{y_i^n}) + \Phi_{tr}(y_{i-1}^n, y_i^n) \cdot w_{tr} \} \geq \{ \sum_{i=1\ldots l} (x_i^n \cdot w_{y_i}) + \Phi_{tr}(y_{i-1}, y_i) \cdot w_{tr} \} + \Delta(y^n, y)$$

where $\Delta(y^i, y)$ is the loss function, computed as the number of misclassified tags in the sequence, $(x_i \cdot w_{y_i})$ represents the emissions and $\Phi_{tr}(y_{i-1}, y_i)$ the transitions. The cutting-plane algorithms implemented in SVM^{struct} is applied in order to handle the exponential number of constraints and to solve this problem up to a precision of ϵ in polynomial time [26].

Several benefits are related to the above approach in SRL. First, it implicitly provides a re-ranking ability typical of joint global models of Semantic Role Labeling, [25]. The output labeling is the most likely global solution over a sentence, as the side-effect of Viterbi decoding. Second, SVM^{hmm} naturally expresses a multi-classification process: this means that a single model can be trained for a frame, whose labels refer to all roles. Purely discriminative approaches in a one-vs-all (OVA) scheme require one model per role and are thus much more complex. Finally, in contrast to conventional HMMs, in SVM^{hmm} the observations $x_1 \ldots x_l$ can be naturally expressed in terms of feature vectors. In particular, each word is modeled through a set of lexical and syntactic features, as described in the next section. Section 3 shows that results obtained on the proposed features are competitive with respect to the state-of-the-art.

2.2   Features for SRL Sequence Labeling Tasks

In the discriminative view of SVM^{hmm}, each word is represented by a feature vector, describing its different observable properties. For example, the word *robber* in Table 2 is represented for the BD task through the following features:

- *Position*, i.e. -1 as its relative distance from the target predicate
- *Lexical features*: its lemma (*robber*) and POS tag (*NN*)
- *Semantic features*: the involved predicate (*kill.V*) and the underlying frame (KILLING)
- *Contextual features*: the left and right lexical contexts represented by the 3 words before (*yesterday::RB ,::, a::DT*) and after (*kill::VBD a::DT guardian::NN*); the left and right syntactic contexts as the POS bi-grams and tri-grams occurring *before* (i.e. RB_, , _DT RB_, _DT) and *after* (i.e. PT_DT DT_NN PT_DT_NN) the word. The symbol TP is used to represent the target predicate within any n-gram.

The target classes are specific to the different arguments, and depend on their *phrase type*. In FrameNet, every argument (i.e. frame element) is grammatically realized in a phrase type, so that different argument categories are possible, such as Noun Phrases (NP) or Adverbial Phrases (AVP). Our multiclass problem is thus targeted to label

Table 2. Training data for the BD task: lexical, syntagmatic and task-dependent features

Class						Features		
B-AVP	-4	yesterday	RB		kill.v	KILLING	B3 B2 B1 B3_B2 B2_B1 B3_B2_B1	.::, a::DT robber::NN _DT DT_NN _DT_NN
X	-3	,		,	kill.v	KILLING	B2 B1 yesterday::RB B2_B1 B1_RB B2_B1_RB	a::DT robber::NN kill::VBD DT_NN NN_PT DT_NN_PT
B-NP	-2	a	DT		kill.v	KILLING	B1 yesterday::RB .::, B1_RB RB_, B1_RB_,	robber::NN kill::VBD a::DT NN_PT PT_DT NN_PT_DT
O-NP	-1	robber	NN		kill.v	KILLING	yesterday::RB .::, a::DT RB_, _DT RB_,_DT	kill::VBD a::DT guardian::NN PT_DT DT_NN PT_DT_NN
X	0	kill	VBD	kill.v	KILLING	.::, a::DT robber::NN _DT DT_NN ,_DT_NN	a::DT guardian::NN with ::IN DT_NN NN_with DT_NN_with
...						...		

every token in an argument both with its position (according to the BIO model) as well as the phrase type of the underlying argument. For example, in Table 2, the row for *robber* is a positive example of the label $O - NP$, as *robber* is the ending (O) of a noun phrase NP argument. The phrase type and BIO tags of individual words is then used as a feature in the later *argument classification*: it is first assigned by the BD step, so that it can be used to train the semantic role classifier. While the impact of this information over different corpora and languages will be studied in Section 3, the role of distributional information as an extension of the lexical features, made available to the SVM^{hmm} learner, is discussed hereafter.

Distributional information and Lexical Features. Lexical information is highly affected by data-sparseness, and words as found in test cases (like *made* in (3)) may often result rare or unseen in the training set (whereas *produce* is likely to have been seen instead). Our aim is to increase robustness in the modeling of semantic role recognition by extending lexical information through distributional analysis. For example, we expect that in semantic spaces, e.g. [18], words evoking similar activities to be characterized by similar (i.e. closer) representations.

Semantic spaces have been widely used for representing the meaning of words or other lexical entities ([27]), with successful applications in lexical disambiguation ([22]) or harvesting thesauri [13]. Also lexical features are beneficial in SRL for systems based on Propbank or FrameNet (e.g. [5]). The basic idea is that the meaning of a word can be described by the set of textual contexts in which it appears (*Distributional Hypothesis*, [8]). Here, contexts are words appearing together with target words: such a space models a generic notion of semantic relatedness, i.e. two words close in the space are likely to be either in paradigmatic or syntagmatic relation as in [21]. In a word-based co-occurrence model, contexts are words appearing in a n-window of a target word and n is a parameter that allows the space to capture different aspects of words' behavior.

Latent Semantic Analysis (LSA) [12] is then applied to acquire meaningful generalization of this lexical model. It can be seen as a variant of the Principal Component Analysis idea. LSA finds the best lower dimensional approximation of the original word space, in the sense of minimizing the global reconstruction error, by projecting data along the directions of maximal variance. It captures term (semantic) dependencies by applying a matrix decomposition process called Singular Value Decomposition (SVD) [7]. The original target-by-word matrix M is transformed into the product of three new matrices: U, S, and V so that S is diagonal and $M = USV^T$. Matrix M is approximated by $M_l = U_l S_l V_l^T$ in which only the first l columns of U and V are used, and only the first l greatest singular values are considered. This approximation supplies a way to project term vectors into the l-dimensional space using $Y_{terms} = U_l S_l^{1/2}$. These l newly derived features may be thought of as artificial concepts, each one representing an emerging meaning component as a linear combination of many different words (i.e. contexts). In our model verbs, nouns, adjectives and adverbs are the targets, expressed through a l-dimensional semantic vector. It extends the feature set with l additional real-valued features (i.e. the Distributional Extension shown in Table 3). This source of information is sufficient to imply that *produce* becomes more similar to *make*, than to *grow*, i.e. the generalization needed to solve the QA problem posed in (1-3).

Table 3. Training data for the AC task. The distributional features appear in the last column, where 250 dimensional vectors express the semantics of targeted words (i.e. on the same row)

Class							Features			
TIME	B	AVP	-4	yesterday	RB	kill.v	KILLING	B3 B2 B1 B3_B2 B2_B1 B3_B2_B1	,::, a::DT robber::NN ,_DT DT_NN ,_DT_NN	
X	X	X	-3	,	,	kill.v	KILLING	B2 B1 yesterday::RB B2_B1 B1_RB B2_B1_RB	a::DT robber::NN kill::VBD DT_NN NN_PT DT_NN_PT	
KILLER	B	NP	-2	a	DT	kill.v	KILLING	B1 yesterday::RB ,::, B1_RB RB_, B1_RB_,	robber::NN kill::VBD a::DT NN_PT PT_DT NN_PT_DT	
KILLER	O	NP	-1	robber	NN	kill.v	KILLING	yesterday::RB ,::, a::DT RB_, ,_DT RB_,_DT	kill::VBD a::DT guardian::NN PT_DT DT_NN PT_DT_NN	
X	X	X	0	kill	VBD	kill.v	KILLING	,::, a::DT robber::NN ,_DT DT_NN ,_DT_NN	a::DT guardian::NN with ::IN DT_NN NN_with DT_NN_with	
...							...			

Distr. Extension (label on right side of table)

3   Performance Evaluation

This evaluation aims at understanding the role of surface grammatical information and lexical generalization on the accuracy reachable in SRL, when a sophisticated structured learning algorithm is employed. Two different experiments are discussed. First, we apply this model to the traditional FrameNet benchmark, by comparing our model with state-of-art ones that make direct use of syntactic parse trees. Then, we evaluate the applicability of our method over a different language, i.e. Italian, where much poorer training conditions, i.e. very few annotated examples are available.

3.1   Experiment I: Labeling English Texts

In order to measure the accuracy reachable in SRL by our model we need to employ the POS n-gram features and the structured SVM framework in the boundary detection (BD) and argument classification (AC) tasks. We tested them over the FrameNet annotated corpus, whereas the 90%-10% (i.e 271,560 and 30,173 examples respectively) train-test splitting used in [11] has been adopted. In all the experiments, the version 1.3 of FrameNet is employed. The POS tagging is carried out by the tagger available in the LTH parser [10].

Evaluation of the BD task. In this experiment words are represented through feature vectors such as those described in section 2.1. A lexical and grammatical context of size 3 has been adopted in all tests, as shown in Table 2. Accuracy of this phase is meant as the ratio of the arguments that are *perfectly* detected, i.e. every inner word of an argument receives its correct label. In order to provide a consistent comparison with previous work, recall, precision and F-measure scores are derived by neglecting phrase types, as reported in Table 4. Results for different POS (i.e. verbs (V), nouns (N) and adjectives (ADJ) are reported as predicates in different classes may have very different syntactic behaviors. Summarized results are in column ALL. Column J&N08 allows to compare our model with the richer grammatical representations based on parse trees described in [11]: our F1 score, i.e. 80.22% is even higher than the state-of-art reported in [11], i.e. 79.20% showing that in realistic learning conditions, even shallow grammatical representations are effective. If phrase types are considered and perfect matching implies perfecting PT labeling, the F1 score decreases to 78.76% that it is still a reliable information for the AC stage. A significant number of classification errors are due to the missed long-range dependencies [23], i.e. entire arguments far from the target predicate missed during the sequence labeling. During the feature engineering phase, the *position* feature (see section 2.2) was designed to handle this phenomenon. Even if the error reduction was considerable (about 5%), it can be still considered a non negligible problem.

Table 4. Experimental results for the English BD task

	FN				J&N08
	ADJ	N	V	ALL	ALL
P	82.65%	84.87%	79.44%	81.63%	81.40%
R	77.26%	81.83%	77.41%	78.87%	77.30%
F1	79.87%	83.32%	78.41%	**80.22%**	79.20%

Evaluation of the AC task. In the AC phase, words are labeled through the role labels (as reported in Table 3) and the use of SVM^{hmm} allows to train a single model for each frame. Two different AC methods are employed. In the first (named AC) only the (syntactic) features described in section 2.2 are employed. In the second, called $AC + Dist$, a lexical generalization obtained by the distributional model (i.e. the LSA vector) is used to augment the representation of all verbs, nouns, adjective and adverbs in the target sentence. The underlying word space is derived through the analysis of the ukWak corpus [2], a large scale document collection made by 2 billion tokens. POS tagging is first applied and vectors are computed through the word-based co-occurrence model described in Section 2.2: it is built around a short window of size $n=3$, that better captures the syntactic properties of words. The most frequent 20,000 basic features have been used to build the source M matrix and vector components are weighted according to point-wise mutual information scores. The SVD reduction was applied to M, with a dimensionality cut of $l = 250$. In Table 5, the results for the AC phase are shown, where accuracy corresponds to the percentage of roles that are perfectly labeled. Again results comparable with the state-of-art are obtained. Notice that best results (i.e. 89,49%) are achieved when lexical information is generalized through the distributional model.

Results show that differences between the proposed method and the state-of-the-art SRL system are not relevant. Unfortunately detailed predictions of the latter system are

Table 5. Experimental results for the English AC task

	Accuracy
AC	88.29%
$AC + Dist$	89.49%
J&N08	**89.60%**

not available and a statistical significance test is not applicable. However the outcomes suggest that the accuracy of these SRL systems is achieved by our method even by imposing much lower requirements.

3.2 Experiment II: Labeling Italian Texts

For the evaluation in different training conditions, a gold standard of 987 annotated sentence for Italian, derived from the Europarl corpus and presented in [24], has been adopted. Due to the small size of annotated examples, a 10-cross fold validation has been adopted to provide consistent results. In the BD phase, the arguments involving only verbs have been used, as only these verbal arguments were characterized by a sufficient number of examples. In the AC task only the 28 frames with at least 10 annotated examples were employed, referring to the different POSs. The overall statistics of the Italian datasets are reported in Table 6. The POS tagging is carried out by the TreeTagger[3]. Words are represented through feature vectors such as those described in section 2.1: a lexical and grammatical context of size $n = 3$ has been adopted in all tests.

Evaluation of the BD task. In a such poor training condition, we are not interested in a perfect argument matching of the argument, but also a partial matching is meaningful, on condition that the role head, i.e. the word expressing the semantic of the argument, is correctly labeled. We manually annotated the semantic head in the test examples[4] and results in term of precision, recall and F-measure are shown in Table 7. The F1 score, i.e. 72.83%, suggests that shallow grammatical representations even in this extremely poor training conditions achieve good performance. Notice that the precision score, i.e. 80.92% confirms how this model can be effectively trained on few annotated sentences, also emphasizing that no syntactic parser is necessary.

Table 6. Italian data set (used in 10-fold cross validation)

	Predicates	Arguments
BD (*only verbs*)	807	1,510
AC (>10 *examples*)	638	1,149

---

[3] http://www.ims.uni-stuttgart.de/projekte/corplex/TreeTagger/

[4] In the sentence of section 2.1 the semantic heads are *yesterday, robber, guardian* and *with*.

Evaluation of the AC task. In the AC phase, the same approach described in section 3.1 has been adopted. Here the underlying word space is derived through the analysis of the itWaC corpus [2], a large scale document collection composed by a 2 billion tokens. The performance results for the AC phase are shown in Table 7: accuracy here corresponds to the percentage of roles that are correctly labeled. A baseline for this task can be obtained by assigning the most frequent role (i.e. frame element) of each frame to every argument in a test sentence: this amounts to an accuracy of 55.26%. Even in this experiment, the lexical generalization provides better results, i.e. 70.27% accuracy. Notice that, as a 10-fold cross validation test is applied, each training dataset relies only on 570 annotated sentences. This is challenging as, on average, only 22 example sentences per frame are made available for learning.

Table 7. Experimental results for the Italian BD and AC tasks

	P	80.92%
BD	R	66.21%
	F1	**72.83%**
AC	Acc	69.71%
$AC + Dist$		**70.27%**

4   Conclusions

In this paper grammatical n-grams and distributional lexical semantic information are proposed as flexible and largely applicable features for language learning in SRL scenarios, combined with sophisticated SVM-based learning, i.e. the Structural Support Vector Machine (SVM) [26]. Sequences of POS tags are proposed as an effective model of grammatical information. All experiments show that state-of-the-art results are achieved or closely approximated by our modeling. These simple kernels over POS n-grams are quite effective as our model equals the performance of accurate syntagmatic kernels. The experiment on boundary detection demonstrate that a very robust BD model is obtained, achieving state-of-the-art performances on the FrameNet dataset. Its computational advantages are significant: SVM^{hmm} builds the models for all verbal predicates in less than 100 minutes over the whole FN-BNC training set, while it tags texts at a rate of about 2,000 sentences per second[5]. The resulting SRL system can be also ported to a new languages, as the experiments on Italian demonstrate. In synthesis, the simple approach here proposed is far more applicable as it poses much weaker software requirements. This makes it a quite appealing solution for complex NLP applications on a realistic scale.

References

1. Altun, Y., Tsochantaridis, I., Hofmann, T.: Hidden Markov support vector machines. In: Proceedings of the International Conference on Machine Learning (2003)

---

[5] No preprocessing or I/O overhead is obviously considered here.

2. Baroni, M., Bernardini, S., Ferraresi, A., Zanchetta, E.: The wacky wide web: a collection of very large linguistically processed web-crawled corpora. Language Resources and Evaluation 43(3), 209–226 (2009)
3. Collins, M.: Three generative, lexicalised models for statistical parsing. In: Proceedings of ACL 1997, pp. 16–23 (1997)
4. Collins, M., Duffy, N.: Convolution kernels for natural language. In: Proceedings of Neural Information Processing Systems (NIPS), pp. 625–632 (2001)
5. Croce, D., Giannone, C., Annesi, P., Basili, R.: Towards open-domain semantic role labeling. In: ACL, pp. 237–246 (2010)
6. Gildea, D., Jurafsky, D.: Automatic Labeling of Semantic Roles. Computational Linguistics 28(3), 245–288 (2002)
7. Golub, G., Kahan, W.: Calculating the singular values and pseudo-inverse of a matrix. Journal of the Society for Industrial and Applied Mathematics 2(2) (1965)
8. Harris, Z.: Distributional structure. In: Katz, J.J., Fodor, J.A. (eds.) The Philosophy of Linguistics. Oxford University Press, Oxford (1964)
9. Huang, F., Yates, A.: Distributional representations for handling sparsity in supervised sequence-labeling. In: ACL-IJCNLP 2009, vol. 1, pp. 495–503. ACL, Morristown (2009)
10. Johansson, R., Nugues, P.: Dependency-based syntactic-semantic analysis with propbank and nombank. In: Proceedings of CoNLL 2008, Manchester, UK, August 16-17 (2008)
11. Johansson, R., Nugues, P.: The effect of syntactic representation on semantic role labeling. In: Proceedings of COLING, Manchester, UK, August 18-22 (2008)
12. Landauer, T., Dumais, S.: A solution to plato's problem: The latent semantic analysis theory of acquisition, induction and representation of knowledge. Psychological Review 104(2), 211–240 (1997)
13. Lin, D.: Automatic retrieval and clustering of similar word. In: Proceedings of COLING-ACL, Montreal, Canada (1998)
14. Màrquez, L., Comas, P., Gimènez, J., Català, N.: Semantic role labeling as sequential tagging. In: Proceedings of CoNLL-2005 Shared Task (2005)
15. Moschitti, A.: Efficient convolution kernels for dependency and constituent syntactic trees. In: ECML, Berlin, Germany, pp. 318–329 (2006); Machine Learning
16. Moschitti, A., Pighin, D., Basili, R.: Tree kernels for semantic role labeling. Computational Linguistics 34 (2008)
17. Moschitti, A., Quarteroni, S., Basili, R., Manandhar, S.: Exploiting syntactic and shallow semantic kernels for question answer classification. In: Proc. of ACL 2007, pp. 776–783 (2007)
18. Pado, S., Lapata, M.: Dependency-based construction of semantic space models. Computational Linguistics 33(2) (2007)
19. Pradhan, S., Hacioglu, K., Krugler, V., Ward, W., Martin, J.H., Jurafsky, D.: Support vector learning for semantic argument classification. Machine Learning Journal (2005)
20. Pradhan, S.S., Ward, W., Martin, J.H.: Towards robust semantic role labeling. Comput. Linguist. 34(2), 289–310 (2008)
21. Sahlgren, M.: The Word-Space Model. Ph.D. thesis, Stockholm University (2006)
22. Schutze, H.: Automatic word sense discrimination. Journal of Computational Linguistics 24, 97–123 (1998)
23. Steedman, M.: The syntactic process. MIT Press, Cambridge (2000)
24. Tonelli, S., Pianta, E.: Frame information transfer from english to italian. In: LREC 2008 (2008)
25. Toutanova, K., Haghighi, A., Manning, C.D.: A global joint model for semantic role labeling. Comput. Linguist. 34(2), 161–191 (2008)
26. Tsochantaridis, I., Joachims, T., Hofmann, T., Altun, Y.: Large margin methods for structured and interdependent output variables. J. Machine Learning Reserach 6 (December 2005)
27. Turney, P.D., Pantel, P.: From frequency to meaning: Vector space models of semantics. Journal of Artificial Intelligence Research 37, 141–188 (2010)

Cross-Language Information Filtering: Word Sense Disambiguation vs. Distributional Models

Cataldo Musto, Fedelucio Narducci, Pierpaolo Basile,
Pasquale Lops, Marco de Gemmis, and Giovanni Semeraro

Department of Computer Science,
University of Bari "Aldo Moro", Italy
{cataldomusto,narducci,basilepp,lops,degemmis,semeraro}@di.uniba.it
http://www.di.uniba.it/

Abstract. The exponential growth of the Web is the most influential factor that contributes to the increasing importance of text retrieval and filtering systems. Anyway, since information exists in many languages, users could also consider as relevant documents written in different languages from the one the query is formulated in. In this context, an emerging requirement is to sift through the increasing flood of multilingual text: this poses a renewed challenge for designing effective multilingual Information Filtering systems. *How could we represent user information needs or user preferences in a language-independent way?*

In this paper, we compared two content-based techniques able to provide users with cross-language recommendations: the first one relies on a knowledge-based word sense disambiguation technique that uses Multi-WordNet as sense inventory, while the latter is based on a dimensionality reduction technique called Random Indexing and exploits the so-called *distributional hypothesis* in order to build language-independent user profiles.

Since the experiments conducted in a movie recommendation scenario show the effectiveness of both approaches, we tried also to underline strenghts and weaknesses of each approach in order to identify scenarios in which a specific technique fits better.

Keywords: Cross-language Recommender System, Content-based Recommender System, Word Sense Disambiguation, Random Indexing.

1 Introduction

Nowadays the amount of information we have to deal with is usually greater than the amount of information we can process in an effective way. For this reason, user modeling and personalized information access are becoming essential to propose only (or firstly) the information that appear relevant or someway related to the informative need of the target user.Information Filtering (IF) systems are rapidly emerging in this context since they are helpful for carrying out this task in an effective way. These systems adapt their behavior to individual users by learning their preferences and storing them in a *user profile*. Filtering

R. Pirrone and F. Sorbello (Eds.): AI*IA 2011, LNAI 6934, pp. 250–261, 2011.

algorithms, exploiting the information stored in user profiles, perform a progressive removal of non-relevant content according to information about user interests, preferences or specific needs. Specifically, the content-based filtering approach [18] analyzes a set of documents (usually textual descriptions of items previously rated as relevant by an individual user) and builds a model or profile of user interests based on the features (usually keywords) that describe the target objects. The profile is then exploited to recommend new relevant items. If the profile accurately reflects user preferences, the information access process could be effective, since the profile could be used to filter search results, by deciding whether a user is interested in a specific item/document or not and, in the negative case, preventing it from being displayed. On the other side, these approaches have to deal with at least two kinds of problems: firstly, traditional keyword-based profiles are unable to capture the semantics of user interests because they are primarily driven by string matching operations. If a string, or some morphological variant of it, is found in both the profile and the document, a match is made and the document is considered as relevant. However, string matching suffers from problems of *polysemy*, the presence of multiple meanings for one word, and *synonymy*, multiple words with the same meaning. The result is that, due to synonymy, relevant information can be missed if the profile does not contain the exact keywords in the documents while, due to polysemy, wrong documents could be deemed as relevant. Another relevant problem related to string matching approaches is the strict connection with the user language: an English user, for example, frequently interacts with information written in English, so her (keyword-based) profile of interests mainly contains English terms. In order to receive suggestions of items whose textual description is in a different language, she must explicitly express her preferences on items in that specific language, as well. This means that the information already stored in the user profile cannot be exploited to provide suggestions for items whose description is provided in other languages, although they share some common features (e.g. an Italian and an English movie might share the same features but their plots could be written in two different languages). In this paper we investigated a simple research question: *how could we represent user profiles in order to create a mapping between preferences expressed in different languages and to provide cross-language recommendations with minimum costs?* We addressed this issue by comparing two different approaches: the first one exploits a Word-Sense Disambiguation technique based on MultiWordnet while the second one is based on an assumption typical of the so-called *distributional models*. It assumes that in every language each term often co-occurs with the same other terms (expressed in different languages, of course), thus by representing content-based user profiles in terms of the co-occurences of its terms, user preferences could become inerently independent from the language and this is sufficient to provide the user with cross-language recommendations. In this work we used a dimensionality reduction technique based on distributional hypothesis, called Random Indexing.

The paper is organized as follows. Section 2 analyzes related works in the area of cross-language filtering and retrieval. Recommendation models are presented in section 3 and section 4, while the design of the experimental session carried out in a movie recommendation scenario is described in Section 5. Conclusions and future work are drawn in the last section.

2   Related Work

Up to our knowledge, the topic of Cross-Language and Multilanguage Information Filtering is not yet properly investigated in literature.

An attempt to define an effective multilingual information filtering system is proposed in [21]. The system is based on the fuzzy set theory. More specifically, the semantic content of multilingual documents is represented using a set of universal content-based topic profiles, encapsulating all feature variations among multiple languages. Using the co-occurrence statistics of a set of multilingual terms extracted from a parallel corpus (collection of documents containing identical text written in multiple languages), fuzzy clustering is applied to group semantically-related multilingual terms to form topic profiles.

Recently, the Multilingual Information Filtering task at CLEF 2009[1] has introduced the issues related to the cross-language representation in the area of Information Filtering. Damankesh et al. [5], propose the application of the theory of Human Plausible Reasoning (HPR) in the domain of filtering and cross language information retrieval. The system utilizes plausible inferences to infer new, unknown knowledge from existing knowledge to retrieve not only documents which are indexed by the query terms but also those which are plausibly relevant.

The state of the art in the area of cross-language Information Retrieval is undoubtedly richer, and can certainly help in designing effective cross-language Information Filtering systems. Oard [17] gives a good overview of the approaches for cross-language retrieval. In [14] the authors propose an approach to build a model of the user's interests based on word senses rather that on simply words. The approach relyes on MultiWordNet to perform Word Domain Disambiguation and to create synset-based multilingual user profiles shown effective for news filtering. The most recent approaches to Cross-Language Retrieval mainly rely on the use of large corpora like Wikipedia. Potthast et al. [20] introduce CL-ESA, a new multilingual retrieval model for the analysis of cross-language similarity. The approach is based on Explicit Semantic Analysis (ESA) [8], extending the original model to cross-lingual retrieval settings. Furthermore, Juffinger et al.[1] recently presented the cross language retrieval system developed for the Robust WSD Task at CLEF 2008[2]. Finally, Gonzalo et al. [9] discuss ways in which EuroWordNet (EWN) [25] can be used in multilingual information retrieval activities, focusing on two approaches to Cross-Language Text Retrieval that exploit the EWN database as a large-scale multilingual semantic resource.

The use of techniques for dimensionality reduction, such as Random Indexing [22], in the area of both monolingual and multilingual Information Filtering

---

[1] http://www.clef-campaign.org/2009.html
[2] http://www.clef-campaign.org/2008.html

is a relatively new topic. The effectiveness of this approaches has already been demonstrated in [4] with an application for image and text data. Recently the research about semantic vector space models gained more and more attention: S-Space[3] and Semantic Vectors (SV)[4] are the first packages developed in this area. The SV package was implemented by Widdows [26]: it implements a Random Indexing algorithm and defines a negation operator based on quantum mechanics. Some initial investigations about the effectiveness of the Semantic Vectors for retrieval and filtering tasks are reported in [3] and [16].

3 First Approach: Learning Profiles Based on MultiWordnet

In a classic content-based recommender system the item properties are represented in the form of *textual slots*. For example, a movie can be described by slots *title, genre, actors, summary*. In this approach we can imagine a general architecture composed by the three main components: a *Content Analyzer*, a *Profile Learner*, and a *Recommender*.

The *Content Analyzer*, which relies on META (Multi Language Text Analyzer) [2], a tool for the analysis and the processing of textual documents, allows introducing semantics in the recommendation process by analyzing documents in order to identify relevant concepts representing the content. This process selects, among all the possible meanings (senses) of each (polysemous) word, the correct one according to the context in which the word occurs. In this way, documents are represented using concepts instead of keywords, in an attempt to overcome problems due to natural language ambiguity, and to the diversity of languages. This step requires the identification of a repository for word senses and the design of an automated procedure for defining word-concept associations. For the first requirement we exploited the *MultiWordNet* lexical ontology [19]. Similary to WordNet, the basic building block for MultiWordNet is the synset (SYNonym SET), a structure containing sets of words with synonymous meanings, which represents a specific meaning of a word. Some words have several different meanings, and some meanings can be expressed by several different word forms. Polysemy and synonymy can be viewed as complementary aspects of this mapping. In MultiWordNet the Italian WordNet is strictly aligned with English WordNet 1.6 [15]. For the second requirement we implemented a Word Sense Disambiguation (WSD) procedure that, given some generical textual content represented through the classical bag-of-words (BOW), allows to obtain a richer synset-based vector space representation, called bag-of-synsets (BOS), where each word (or each *set* of words, for bigrams or trigrams) that occurs in the original BOW is mapped on the MultiWordnet concept it refers to. In the BOS model, a synset vector, rather than a word vector, corresponds to a document. Since each concept is represented through an unique *id* that is independent from the language, by shifting the representation from BOW to BOS

---

[3] http://code.google.com/p/airhead-research/
[4] http://code.google.com/p/semanticvectors/

we obtained a new and unified representation that is language-independent for both English and Italian documents.

The generation of the cross-language user profile is performed by the *Profile Learner*, which infers the profile as a binary text classifier [24] since each document has to be classified as interesting or not with respect to the user preferences. Therefore, the set of categories is restricted to c_+, the positive class (*user-likes*), and c_- the negative one (*user-dislikes*). The induced probabilistic model is used to estimate the *a posteriori* probability, $P(c|d_j)$, of document d_j belonging to class c. The algorithm adopted for inferring user profiles is a Naïve Bayes text learning approach, widely used in content-based recommenders, which is not presented here because already described in [7]. The profile learning process for user u starts by selecting all items (disambiguated documents) and corresponding ratings provided by u. Each item falls into either the positive or the negative training set depending on the user rating, in the same way as previously described in this section. Therefore, given a new document (previously disambiguated) d_j, the *recommendation step* consists in computing the a-posteriori classification scores $P(c_+|d_j)$, used to produce a ranked list of potentially interesting items, from which items to be recommended can be selected. Finally the *Recommender* exploits the cross-language user profile to suggest relevant items by matching concepts contained in the semantic profile against those contained in documents to be recommended (previously disambiguated). The user might receive recommendations in her own mother tongue, or in languages she knows. This is a decision of the specific application in which the recommender is integrated.

4   Second Approach: Distributional Models

The second strategy used to represent items content in a semantic space relies on the distributional approach. This approach represents documents as vectors in a high dimensional space, such as WordSpace [23]. The core idea behind WordSpace is that words and concepts are represented by points in a mathematical space, and this representation is learned from text in such a way that concepts with similar or related meanings are near to one another in that space (geometric metaphor of meaning). Replacing words with documents results in a high dimensional space where similar documents are represented close. Therefore, semantic similarity between documents can be represented as proximity in a n-dimensional space. The main characteristic of the geometric metaphor of meaning is not that meanings are represented as locations in a semantic space, but rather that similarity between documents can be expressed in spatial terms, as proximity in a high-dimensional space. One of the great virtues of the distributional approach is that documents space can be built using entirely unsupervised analysis of free text. According to the *distributional hypothesis* [10], the meaning of a word is determined by the rules of its usage in the context of ordinary and concrete language behavior. This means that words are semantically similar to the extent that they share *contexts* (surrounding words). Co-occurrence is defined with respect to a context, for example a document. Hence, words are similar if they have the same contexts, that is to say, they are similar if they occur in the same documents. It is important to underline here that a word is represented by a vector in a high dimensional space. Since these techniques are expected to handle efficiently high dimensional vectors, a common choice is to adopt *dimensionality reduction* that allows for representing high-dimensional data in a lower-dimensional space without losing information. *Latent Semantic Analysis (LSA)* [13] collects the text data in a co-occurrence matrix, which is then decomposed into smaller matrices with singular-value decomposition (SVD), by capturing latent semantic structures in the text data. The main drawback of SVD is scalability. Differently from LSA, *Random Indexing* (RI) [11] targets the problem of dimensionality reduction by removing the need for the matrix decomposition or factorization. RI incrementally accumulates context vectors, which can be later assembled into a new space, thus it offers a novel way of conceptualizing the construction of context vectors.

RI is based on the concept of Random Projection: the idea is that high dimensional vectors chosen randomly are "nearly orthogonal". This yields a result that is comparable to orthogonalization methods, such as Singular Value Decomposition [13], but saving computational resources.

Formally, given a $n \times m$ matrix A (in this scenario it represents the classical term/document matrix) and a $m \times k$ matrix R made up of k m-dimensional random vectors, we define a new $n \times k$ matrix B as follows:

$$A^{n,m} \cdot R^{m,k} = B^{n,k} \quad k << m \tag{1}$$

The new matrix B is a more compact representation of the original matrix A and it has the property to preserve the distance between points known as Johnson-Lindenstrauss lemma, if the distance between two any points of A is d, then the distance d_r between the corresponding points in B will satisfy the property that $d_r = c \cdot d$. A proof of that property is reported in [6].

Specifically, RI builds two spaces, namely WordSpace and DocumentSpace, by following three steps:

1. a context vector is assigned to each document. This vector is sparse, high-dimensional and ternary, which means that its elements can take values in {-1, 0, 1}. A context vector contains a small number of randomly distributed non-zero elements, and the structure of this vector follows the hypothesis behind the concept of Random Projection;
2. context vectors are accumulated by analyzing terms and documents in which terms occur. In particular, the semantic vector for a term is computed as the sum of the context vectors for the documents which contain that term. Context vectors are multiplied by term occurrences.
3. the semantic vector for a document is computed as the sum of the semantic vectors for the terms which occur in that document. Semantic vectors ot temrs are multiplied by term occurrences.

In this approach for each movie we extracted its plot (in English and Italian) and we built a multilingual space. The main difference between a multilingual space and a monolingual one is that in this space each movie has two fields F_{L1} and F_{L2}, which store the same content but in two different languages (the plot in English and Italian). It is important to underline that not necessarily the content of F_{L2} is the perfect translation of F_{L1}. The power of distributional approaches is that two terms, in different languages, are similar because they share the same context. To build multilanguage space we need to generate four spaces: two *WordSpace* SW_{L1} and SW_{L2} and two *DocumentSpace* SD_{L1} and SD_{L2}. These spaces are built as follows:

1. a context vector is assigned to each movie (plot) as described in RI algorithm. We call this space RB (random base);
2. the semantic vector for a term in SW_{L1} is computed as the sum of the context vectors in RB for the movies (plots) which contain that term in the field F_{L1};
3. the semantic vector for a movie (plot) in DW_{L1} is computed as the sum of the semantic vectors for the terms in SW_{L1} which occur in that movie (plot) in the field F_{L1};
4. the semantic vector for a term in SW_{L2} is computed as the sum of the context vectors in RB for the movies (plots) which contain that term in the field F_{L2};
5. the semantic vector for a movie (plot) in DW_{L2} is computed as the sum of the semantic vectors for the terms in SW_{L2} which occur in that movie (plot).

Given a multilingual space built in that way, in order to provide recommendations we have also to build user profiles. In these work we compared two approaches called W-RI and W-SV, thoroughly described in [16]. In the first approach, given the set of the movies that a user liked in the past (namely, whose rating explicitly provided by the user is over a certain threshold) the user profile is computed as the sum of the semantic vectors for all of this movies in DW_{L1} or DW_{L2}. In the latter one the user profile is built in the same way, but the approach gives a bigger weight to the movies that the user liked more. The user profiles can be seen as a new element of the *DocumentSpace* and can be istantiated in the vector space.

Since all the four spaces share the same random base RB, this makes possible to compare elements that belong to different spaces. For example we can compute how a user profile in DW_{L1} (or, respectively, in DW_{L2}) is similar to a movie in DW_{L2} (or, respectively, in DW_{L1}) and we exploited this property in order to calculate items similarity and provide users with cross-lingua recommendation. Specifically, in this approach the user receives as recommendations the items whose similarity is the higher w.r.t. her profile.

5   Experimental Evaluation

The goal of the experimental evaluation was to measure the predictive accuracy of both content-based multilingual recommendation approaches, by comparing the language-independent (cross-language) user profiles represented through

BOSs with the W-SV and W-RI approaches based on distributional hypothesis and Random Indexing. More specifically, we would like to test 1) whether user profiles learned using examples in a specific language can be effectively exploited for recommending items in a different language, 2) whether the accuracy of a cross-language recommender system is comparable to that of a monolingual one and 3) whether a specific approach gets a significative improvement w.r.t. the other ones, becoming preferable in some specific recommendation scenario.

Experiments were carried out in a movie recommendation scenario in which the languages adopted in the evaluation phase are English and Italian.

5.1   Users and Dataset

The experimental work has been performed on a subset of the MovieLens dataset[5], containing 100,000 ratings provided by 943 different users on 1,628 movies. The original dataset does not contain any information about the content of the movies. The content information for each movie was crawled from both the English and Italian version of Wikipedia. In particular the crawler gathers the *Title* of the movie, the name of the *Director*, the *Starring* and the *Plot*. For both approaches the text in each slot has been tokenized, stemmed and the stopwords have been removed. For the model based on the bayesian classifier the POS tag has been identified before running the WSD algorithm.

In order to learn accurate user profiles, we have not performed the evaluation for those users who provided less than 20 ratings. Moreover, we selected all the movies for which both the English and Italian description is available. To sum up, the dataset after this processing contained 40,717 ratings provided by 613 different users on 520 movies.

5.2   Design of the Experiment

User profiles are learned by analyzing the ratings stored in the MovieLens dataset. Each rate was expressed as a numerical vote on a 5-point Likert scale, ranging from 1=strongly dislike, to 5=strongly like. The effectiveness of the rec-ommendation approaches has been evaluated by means of *Precision@n*, where n has been set as 5 and 10. In the experiment, an item is considered *relevant* for a user if the rating is greater than or equal to 3. The dataset used in the experiment is really unbalanced in terms of positive and negative ratings (83% positive, 17% negative).We designed two different experiments, depending on 1) the language of items used for learning profiles, and 2) the language of items to be recommended:

- EXP#1 – *ENG-ITA*: profiles learned on movies with English description and recommendations provided on movies with Italian description;
- EXP#2 – *ITA-ENG*: profiles learned on movies with Italian description and recommendations produced on movies with English description.

---

[5] http://www.grouplens.org

We compared the results against the accuracy of classical monolanguage content-based recommender systems:

- EXP#3 – *ENG-ENG*: profiles learned on movies with English description and recommendations produced on movies with English description;
- EXP#4 – *ITA-ITA*: profiles learned on movies with Italian description and recommendations produced on movies with Italian description.

We executed one experiment for each user in the dataset. The ratings of each specific user and the content of the rated movies have been used for learning the user profile and measuring its predictive accuracy, using the aforementioned measures. Each experiment consisted of:

1. selecting ratings of the user and the description (English or Italian) of the movies rated by that user;
2. splitting the selected data into a training set Tr and a test set Ts;
3. using Tr for learning the corresponding user profile by exploiting the:
   - English movie descriptions (EXP#1) or Italian movie descriptions (EXP#2);
4. evaluating the predictive accuracy of the induced profile on Ts, using the aforementioned measures, by exploiting the:
   - Italian movie descriptions (EXP#1) or English movie descriptions (EXP#2);

In the same way, a single run for each user has been performed for computing the accuracy of monolingual recommender systems, but the process of learning user profiles from Tr and evaluating the predictive accuracy on Ts has been carried out using descriptions of movies in the same language, English or Italian. The methodology adopted for obtaining Tr and Ts was the 5-fold cross validation [12].

5.3   Discussion of Results

Results of the experiments are reported in Table 1 and 2, averaged over all the users.

Table 1. Precision @5

Experiment	W-SV	W-RI	Bayes
EXP#1 – ENG-ITA	84,65	84,65	85,61
EXP#2 – ITA-ENG	84,85	84,63	85,20
EXP#3 – ENG-ENG	85,23	85,29	85,23
EXP#4 – ITA-ITA	85,27	84,84	85,71

By summing up, in this experimental session we tried to compare two very different approaches. The first one, based on a classical Bayes classifier, exploits external linguistic knowledge and relies on the assumption that the BOS can be

Table 2. Precision @10

Experiment	W-SV	W-RI	Bayes
EXP#1 – ENG-ITA	84,73	84,43	84,60
EXP#2 – ITA-ENG	84,77	84,54	84,56
EXP#3 – ENG-ENG	85,10	84,86	84,89
EXP#4 – ITA-ITA	85,11	84,86	84,93

an effective bridge to represent user preferences expressed in different languages. The second one, based on Random Indexing, does not require any linguistic pre-processing and is totally based on the distributional hypothesis. It assumes that the similar distribution of the terms, even in different languages, makes the preferences independent from the language and a simple projection of the user profile built in one language into the space built in another one is sufficient to provide the user with cross-language recommendations. In general, the main outcome of the experimental session is that the strategy implemented for providing cross-language recommendations is quite effective for both approaches. There is no significative difference by comparing the accuracy of the models previously presented. More specifically, user profiles learned using examples in a specific language can be effectively exploited for recommending items in a different language, and the accuracy of the approach is comparable to those in which the learning and recommendation phase are performed on the same language. Specifically, the approach based on the bayesian classifier gained the best results in the *Precision@5*. This means that this model has an higher capacity to rank the best items at the top of the recommendations list. Furthermore, it is worth to note that the comparison of the results of the Exp#1 with the results of the Exp#3 shows that the cross-lingua recommendations based on the profiles learned in English improve the Precision with respect to the monolingual one. This is due to the better accuracy of the WSD process for english contents, for which the disambiguation process introduces less noise when the BOS are built.

However the approach based on the bayesian classifier and MultiWordNet might seem too elaborate, because of the several operations needed to represent documents as bag-of-synsets. On the other side, the absence of a linguistic pre-processing is one of the strongest point of the approaches based on the distributional model and the results gained by the W-SV and W-RI models in the Precision@10 further underlined the effectiveness of this approach. Indeed, in all of the experiments, the Precision of the W-SV model is higher with respect to the W-RI model and the Bayesian one. The first result, that confirms the results already presented in [16], shows the importance of modeling negative user preferences and the goodness of the negation operator based on Quantum logic.

In conclusion, both approaches gained good results. Even in most of the experiments the cross-lingua recommendation approaches get worse results w.r.t. the mono-lingual ones, the difference in the predictive accuracy does not appear statistically significative. In general the bayesian approach fits better in scenarios where the number of items to be represented is not too high, and this can justify the application of the pre-processing steps required for building BOSs, while the distributional models, thanks to their simplicity and effectiveness, fit better in scenarios where real-time recommendations that ensure a good accuracy need to be provided.

6   Conclusions and Future Work

In this paper we presented a comparison between two content-based approaches for providing cross-language recommendations. The key idea behind the first one is to provide a bridge among different languages by exploiting a language-independent representation of documents and user profiles based on word meanings, called bag-of-synsets, while the second one relies on a totally unsupervised learning method based on the distributional hypothesis. Experiments were carried out in a movie recommendation scenario, and the main outcome is that the accuracy of cross-language recommmendations is comparable to that of classical (monolingual) content-based recommendations for both approaches. In the future, we are planning to investigate the effectiveness of both models on different domains and datasets. More specifically, we are working to extract cross-language profiles by gathering information from social networks, such as Facebook, LinkedIn, Twitter, etc., in which information are generally available in different languages.

References

1. Andreas Juffinger, R.K., Granitzer, M.: A Wikipedia-Based Multilingual Retrieval Model. In: Evaluating Systems for Multilingual and Multimodal Information Access, pp. 155–162 (2009)
2. Basile, P., de Gemmis, M., Gentile, A., Iaquinta, L., Lops, P., Semeraro, G.: META - MultilanguagE Text Analyzer. In: Proceedings of the Language and Speech Technnology Conference - LangTech 2008, Rome, Italy, February 28-29, pp. 137–140 (2008)
3. Basile, P., Caputo, A., Semeraro, G.: Semantic vectors: an information retrieval scenario. In: Melucci, M., Mizzaro, S., Pasi, G. (eds.) IIR 2010 - Proceedings of the First Italian Information Retrieval Workshop, Padua, Italy, January 27-28, pp. 1–5 (2010)
4. Bingham, E., Mannila, H.: Random projection in dimensionality reduction: applications to image and text data. In: KDD 2001, pp. 245–250. ACM, New York (2001)
5. Damankesh, A., Singh, J., Jahedpari, F., Shaalan, K., Oroumchian, F.: Using Human Plausible Reasoning as a Framework for Multilingual Information Filtering. In: Peters, C., Di Nunzio, G.M., Kurimo, M., Mostefa, D., Penas, A., Roda, G. (eds.) CLEF 2009. LNCS, vol. 6241. Springer, Heidelberg (2010)
6. Dasgupta, S., Gupta, A.: An elementary proof of the Johnson-Lindenstrauss lemma. Tech. rep., Technical Report TR-99-006, International Computer Science Institute, Berkeley, California, USA (1999)
7. de Gemmis, M., Lops, P., Semeraro, G., Basile, P.: Integrating Tags in a Semantic Content-based Recommender. In: Proc. of the 2008 ACM Conf. on Recommender Systems, RecSys 2008, Lausanne, Switzerland, October 23-25, pp. 163–170 (2008)

8. Gabrilovich, E., Markovitch, S.: Computing Semantic Relatedness Using Wikipedia-based Explicit Semantic Analysis. In: Veloso, M.M. (ed.) IJCAI, pp. 1606–1611 (2007)
9. Gonzalo, J., Verdejo, F., Peters, C., Calzolari, N.: Applying EuroWordNet to Cross-Language Text Retrieval, vol. 32, pp. 185–207. Springer, Netherlands (1998)
10. Harris, Z.: Mathematical Structures of Language. Interscience, New York (1968)
11. Kanerva, P.: Sparse Distributed Memory. MIT Press, Cambridge (1988)
12. Kohavi, R.: A Study of Cross-Validation and Bootstrap for Accuracy Estimation and Model Selection. In: Proc. of IJCAI 1995, pp. 1137–1145 (1995)
13. Landauer, T.K., Dumais, S.T.: A Solution to Plato's Problem: The Latent Semantic Analysis Theory of Acquisition, Induction, and Representation of Knowledge. Psychological Review 104(2), 211–240 (1997)
14. Magnini, B., Strapparava, C.: Improving user modelling with content-based techniques. In: Bauer, M., Gmytrasiewicz, P.J., Vassileva, J. (eds.) UM 2001. LNCS (LNAI), vol. 2109, pp. 74–83. Springer, Heidelberg (2001)
15. Miller, G.: WordNet: An On-Line Lexical Database. International Journal of Lexicography 3(4) (1990) (Special Issue)
16. Musto, C.: Enhanced vector space models for content-based recommender systems. In: Proceedings of the Fourth ACM Conference on Recommender Systems, RecSys 2010, pp. 361–364. ACM, New York (2010), http://doi.acm.org/10.1145/1864708.1864791
17. Oard, D.W.: Alternative Approaches for Cross-Language Text Retrieval. In: AAAI Symposium on Cross-Language Text and Speech Retrieval. American Association for Artificial Intelligence, pp. 154–162 (1997)
18. Pazzani, M.J., Billsus, D.: Content-Based Recommendation Systems. In: Brusilovsky, P., Kobsa, A., Nejdl, W. (eds.) Adaptive Web 2007. LNCS, vol. 4321, pp. 325–341. Springer, Heidelberg (2007) iSBN 978-3-540-72078-2
19. Pianta, E., Bentivogli, L., Girardi, C.: MultiwordNet: developing an aligned multilingual database. In: Proc. of the 1st Int. WordNet Conference, Mysore, India, pp. 293–302 (2002)
20. Potthast, M., Stein, B., Anderka, M.: A wikipedia-based multilingual retrieval model. In: Macdonald, C., Ounis, I., Plachouras, V., Ruthven, I., White, R.W. (eds.) ECIR 2008. LNCS, vol. 4956, pp. 522–530. Springer, Heidelberg (2008)
21. Chau, R., Yeh, C.-H.: Fuzzy multilingual information filtering. In: 12th IEEE International Conference on Fuzzy Systems, FUZZ 2003, pp. 767–771 (2003)
22. Sahlgren, M.: An introduction to random indexing. In: Methods and Applications of Semantic Indexing Workshop, TKE 2005 (2005)
23. Sahlgren, M.: The Word-Space Model: Using distributional analysis to represent syntagmatic and paradigmatic relations between words in high-dimensional vector spaces. Ph.D. thesis, Stockholm University, Department of Linguistics (2006)
24. Sebastiani, F.: Machine Learning in Automated Text Categorization. ACM Computing Surveys 34(1) (2002)
25. Vossen, P.: Introduction to EuroWordNet. Computers and the Humanities 32(2-3), 73–89 (1998)
26. Widdows, D.: Orthogonal negation in vector spaces for modelling word-meanings and document retrieval. In: ACL 2003: Proceedings of the 41st Annual Meeting on Association for Computational Linguistics, pp. 136–143. Association for Computational Linguistics, Morristown (2003)

Double-Sided Recommendations: A Novel Framework for Recommender Systems

Fabiana Vernero

Department of Computer Science, University of Turin,
c.so Svizzera, 185 - 10149 Torino (Italy)
vernerof@di.unito.it

Abstract. Recommender systems actively provide users with suggestions of potentially relevant items. In this paper we introduce *double-sided recommendations*, i.e., recommendations consisting of an item and a group of people with whom such an item could be consumed. We identify four specific instances of the double-sided recommendation problem and propose a general method for solving each of them (social comparison-based, group-priority, item-priority and same-priority methods), thus defining a framework for generating double-sided recommendations.

We present the experimental evaluation we carried out, focusing on the restaurant domain as a use case, with the twofold aim of 1) assessing user liking for double-sided recommendations and 2) comparing the four proposed methods, testing our hypothesis that their perceived usefulness varies according to the specific problem instance users are facing. Our results show that users appreciate double-sided recommendations and that all four methods -and, in particular, the group-priority one- can generate useful suggestions.

Keywords: double sided recommendations, recommender systems, recommendations to groups, recommendations of groups, social networking.

1 Introduction

In a scenario where the available contents on the Web are constantly growing, *recommender systems* emerge as a specific information filtering technique which actively provides users with suggestions of potentially relevant items, thus helping them to deal with the so called "information overload" problem. Different approaches are usually distinguished based on the information which is needed to generate suggestions: content-based systems [15] employ some knowledge about user preferences and needs, on the one hand, and item features, on the other hand, while collaborative filtering systems [17] base on the opinions of a large community of users.

Most recommender systems target single individuals. In recent years, however, systems which provide recommendations for groups have emerged, based on the idea that some types of recommended items are at least as likely to be used by groups as by individuals, for example vacations, movies, restaurants or cultural events (e.g., concerts or exhibitions).

In many situations, however, a group for which a suggestion could be generated is not necessarily predefined: most of us interact with different individuals, and with

R. Pirrone and F. Sorbello (Eds.): AI*IA 2011, LNAI 6934, pp. 262–273, 2011.

different formal and informal groups, from time to time. Also, our social networks usually comprise various (and sometimes overlapping) communities, which relate to different "aspects" of our social lives. For example, we can have friends -and, possibly, different groups of friends-, relatives, school mates, colleagues and sometimes even previous colleagues, as well as occasional acquaintances.

Imagine you are planning to dine out next weekend. Common sense suggests that different restaurants might represent the best choice according to the people you choose to dine. On the other hand, if you are really eager to try a certain (type of) restaurant, some of your contacts may be more willing to accompany you than others. Restaurant and group choices are deeply interconnected. Thus, the most appropriate question you would like a recommender system in the restaurant domain to answer might neither be "Where could I go?", nor "Where could I go with a certain group of friends?", but "Where could I go and with whom?". Similar questions may arise in all other domains where the recommended items are usually enjoyed by groups rather than by single individuals.

To the best of the author's knowledge, no recommender systems exist at present which were explicitly designed to answer questions like these. In this paper, we first introduce the idea of recommendations where both an item and a group of people with whom such an item should be consumed are suggested, and call them *double-sided recommendations*. Then, we propose a framework for generating double-sided recommendations.

According to our framework, different instantiations of the double-sided recommendation problem may exist, depending on contextual and occasional elements or on a personal preference for a certain framing of the problem itself. For example, some users might prefer to be recommended an item they can really enjoy, and see the company of other people as an additional treat or just as a way to adhere to unwritten social rules (for example, some individuals might not feel at ease going to a restaurant or to the movies alone; however, they may be quite flexible as far as a company is concerned, provided that it consists of people they like); on the contrary, other users might be primarily interested in spending some time in good company and be ready to compromise on an item which can suit the group as a whole, even if it is not their preferred option.

In our framework, we identify four possible instances of the double-sided recommendation problem, each of which sets different priorities and thus requires a different approach for generating appropriate recommendations. For each problem instance, we provide a generic solution method and some detail on how we exploited it in order to generate double-sided recommendations in our use case. Notice that, in this paper, the focus is on providing a general description of the framework, rather than on discussing specific computational details. Given the basic ideas expressed in the four proposed methods, we believe that different specific techniques might be used in order to compute double-sided recommendations.

An empirical evaluation was carried out in the restaurant domain with the twofold aim of: 1) assessing user liking for double-sided recommendations, demonstrating that these represent a novel and useful service, and 2) comparatively evaluating the four proposed solution methods, as far as their capability of providing *useful* recommendations is concerned.

The paper is structured as follows: Section 2 presents state of the art literature about recommender systems, with a special focus on recommendations to groups, recommendations of groups and social network-based recommendations; Section 3 describes our framework and Section 4 explains how we evaluated our approach and reports our results. Section 5 concludes the paper, with a discussion of possible future work.

2 Related Work

Double-sided recommendations consist of an item and a group of people with whom such an item could be consumed. Thus, they are based on what we could call "single-sided recommendations", i.e., traditional recommendations where only one element -be it an item or a group- is suggested either to a single individual or to a group.

Most literature on recommender systems has focused on the task of recommending potentially relevant items to single individuals. Traditional recommender systems are usually classified according to the information they use in order to assess item relevance and generate recommendations: in *content-based* systems, items are recommended which are similar to those the target user liked in the past [15], while in *collaborative filtering* systems, items are recommended which were positively evaluated by users with similar tastes and interests with respect to the target user [17]; finally, in systems which adopt *hybrid approaches*, both types of techniques are used in order to compensate for their respective weaknesses and reach better performances [4].

Some recent approaches have started to consider the social network of the target user as a source of information for generating recommendations, based on the observation that friend-provided suggestions can be more appreciated than those offered by an anonymous system [18], and often proceeding from the consideration that, although many social content sites and recommender systems are appearing which integrate social networking features, no specific guidance is usually provided for selecting interesting items among the huge volume of network-generated contents. Guy et al. [8] found that users prefer recommendations generated taking into account their social network with respect to recommendations based on user-user similarity, as in collaborative filtering, especially when explanations are provided which highlight which people are related to each recommended item. Carmagnola et al. [5] claimed that the mere fact of being part of a social network may cause individuals to modify their attitudes and behaviours because of social influence dynamics, and proposed SoNARS, a recommendation algorithm which explicitly targets users as members of their social network.

Specific issues arise when item recommendations are provided to groups rather than to single individuals. According to Jameson [10], group recommenders are characterized according to 1) the way information about group member preferences is acquired, 2) the way recommendations to groups are generated, 3) the way recommendations are explained (either to individual group members, or to subgroups, or to the group as a whole), and 4) the way group members are eventually helped to achieve consensus. Most related work examines the problem of choosing an appropriate aggregation strategy, depending on the system goals, e.g., maximizing average satifaction or minimizing misery. Recent approaches in recommendations to groups focused on issues related to balancing group and individual satisfaction [11], considering interactions among group members with different personalities (e.g., assertive or cooperative) [16], and explicitly handling disagreement [1].

Finally, a few approaches have considered recommending groups. Most of them focus on suggesting users to affiliate to existing, explicitly-defined communities, based either on structural properties of their social networks (e.g., user proximity to a community [20] or the number and relevance of friends who already belong to it [5]) or on content-related features, such as predicted user interest for the topics which are usually associated to such communities [2]. However, groups to recommend could also be generated on-the-fly by taking into account the social networks of the target users, for example in case a well-matched group of friends to invite to a party should be suggested. Many works in the area of complex network analysis actually focus on the task of identifying relevant subgroups, i.e., sets of nodes (corresponding to individuals in social networks) which are densely connected to each other, while only few links exist which connect them to external nodes. Different approaches exist which either operate in an *a-priori* manner, taking a whole social network as their only input, as in hierarchical clustering-based methods (see for example [14]), or aim at identifying local communities for a given node (see for example [6]). In recent work, methods for finding local communities which contain a set of target nodes [19], and for detecting possibly overlapping communities, thus taking into account the fact that each individual may belong to more than a group [13], are also proposed.

3   Double-Sided Recommendation Framework

We call "double-sided" recommendations where both an item and a group of people with whom such an item should be used are suggested, and formally define them as follows:

DEFINITION (**Double-sided recommendation**): "Given a target user t, a set of contacts C of the target user and a set of candidate items I, we call a double-sided recommendation either a pair $< i, G >$ where $i \in I$ and $G \subseteq C$; or an N-tuple $< i, G_1, ..., G_{N-1} >$, where $G_n \subseteq C$ and $G_1, ..., G_{N-1}$ are alternative group options, given a certain recommended item; or an N-tuple $< i_1, ..., i_{N-1}, G >$, where $i_n \in I$ and $i_1, ..., i_{N-1}$ are alternative item options, given a certain recommended group."

This definition implies the following assumptions: first, information about user interests or opinions with respect to domain items should be available; otherwise, no item recommendation would be possible. Second, information about the social network of the target user t is needed; otherwise, we would not be able to generate recommendations of groups. More specifically, we assume that a measure of *relationship strength* can be computed in order to assess how important a certain contact c is to the target user t. In the case of our specific implementation, this measure depends on the type and number of actions performed by t which refer to or have an effect on c, such as sending a message or inviting to join a group. A measure of how relevant a group is to the target user (which we will refer to as *groupScore* in the following Sections) is computed as a mean of relationship strength values with respect to all group members.

The proposed definition of double-sided recommendations is very general by design. In fact, it was formulated so that it can encompass all the different situations (i.e., specific *instances* of the double-sided recommendation problem) that users in need of double-sided recommendations may be facing, depending on contextual and occasional elements or on a personal preference for a certain framing of the situation. In our framework, we identify four possible problem instances:

- **Instance 1.** Users are looking for an item to enjoy with some of their contacts. They are very concerned in making a "socially approved" choice and would like to know what the others would do in their place.
- **Instance 2.** Users are interested in spending some time in good company, and they would like to find an item which can please all the people they will meet.
- **Instance 3.** Users are interested in enjoying a pleasant item, and they would like to know who, among their contacts, could keep them company.
- **Instance 4.** Users are interested in enjoying an item in company, and the choice of both a suitable item and a good company are equally important;

Since each specific problem instance sets different priorities, we propose four different methods for solving them: the Social Comparison-based (instance 1), the Group-priority (instance 2), the Item-priority (instance 3), and the Same-priority (instance 4) recommendation method. The last three methods are referred to as *component-based*, since they all base on the identification of structural subcomponents in the social network of the target user (see Section 3.2) in order to generate recommendations of groups.

3.1 Social Comparison-Based Recommendation Method

Method. Taking inspiration from past work on exploiting social influence dynamics in the recommendation process [5] and, in particular, from social comparison theory[1], this method suggests items that were positively evaluated, on average, by relevant others. User relevance depends on both user similarity and user affiliations (i.e., relationship strength), based on the idea that close contacts are more likely to exert some influence. A group recommendation is generated for each item by selecting only the contacts who expressed a positive opinion about it. Moreover, the list of all the people who evaluated each item is highlighted in order to leverage social influence.

Detail on how we computed recommendations. In our case, items are recommended based on a threshold value for $itemRelevance_i$: $I x \mathcal{P}(C) \rightarrow Q$, which is a weighted mean of the opinions $itemRelevance_{ic}$ expressed by each contact $c \in R \in \mathcal{P}(C)$ of the target user who reviewed item i. In particular, $itemRelevance_{ic}$: $I x C \rightarrow Q$ is computed based on the number, type, and value of actions user c performed on item i. Action *type* is treated as a weight, considering that different types of actions (e.g., rating or bookmarking) may provide different evidence about the strength of user interests for a certain item [12]. As for action *value*, we consider that user actions may have

---

[1] According to social comparison theory, people who are uncertain about what they should be thinking or doing usually seek information about the opinions of relevant others in order to form their own attitudes and behaviours [7].

a different polarity (e.g., a rating may be positive or negative) and intensity (e.g., a rating of 4 is more positive than a rating of 3). In the case of ratings, the action value corresponds to the rating itself. In the case of actions such as tagging or commenting, action values might be determined by means of some language analysis. However, for simplicity, we decided to determine the value of such actions based on the value of other actions for which it is simply determined (such as rating), assuming that the actions of a certain user on a certain item share the same polarity. A default value is used if no other actions were performed. Finally, the opinion $itemRelevance_{ic}$ of each specific user c is weighted according to the relevance of c to the target user, which is obtained as a mean of his or her scores for relationship strength and similarity with respect to the target user. Similarity depends on user preferences for domain items and is computed based on a variation of the formula for the standard deviation.

Social comparison-based recommendations consisting each of an item, a list of contacts who reviewed it and a recommended group are assigned a recommendation score called $totalScore$, which is the sum of $itemRelevance_i$ and $groupScore$, and are ordered based on it for presentation to the target users.

3.2   Component-Based Recommendation Methods

Methods. The following three methods we propose share three main aspects: 1) individual user preferences for recommendable items are predicted according to a content-based approach; 2) group preferences for items are predicted by aggregating individual group member preferences; 3) recommendable groups are generated based on meaningful substructures (in our case, connected components) which can be identified in the social network of the target user and on simple social rules.

The three proposed methods differ for the facet which is prioritized in providing a double-sided recommendation to the target user: either the item, or the group, or both.

In *group-priority method*, the most relevant groups for the target user are selected at first among all the recommendable ones, based on a threshold score, and then the two best item options are identified for each selected group, according to group preferences.

In *item-priority method*, the best items for the target user are selected at first based on a threshold score for individual preferences, recommendable groups are generated for each item by taking into account only the contacts of the target user whose preference for such an item is higher than a threshold, and then the two best group options are identified according to both group preferences for the item and group relevance to the target user.

In *same-priority method*, all recommendable groups for the target user are combined with all available contents. The best options are selected based on a score which depends on group preferences for the recommended item and group relevance to the target user.

Detail on how we computed recommendations. *Individual preferences* for items are predicted according to a content-based approach. More specifically, a score $itemIndividualScore_{iu}: IxU \rightarrow Q$, indicating how interesting item i is expected to be to user u, is computed for every item-user couple, taking into consideration: 1) user u's interests with respect to the domain, 2) overall item interestingness, and 3) specific item interestingness to user u, if available.

User interests are represented in the user model as *<feature, value>* pairs, where *feature* corresponds to a category in the reference domain, e.g., "traditional Piedmontese restaurants" for the restaurant domain, and *value* represents the level of interest of user u for that category. In our case, user model values are derived from user actions (such as rating an item) which can be considered indicators for user interest in a certain category. Since each possible item i can be mapped to a category, the level of interest of user u for item i is derived from their interest for the corresponding category.

Overall item interestingness is considered a property of item i, which can be derived from the actions and evaluations of the whole user community. At the moment, only the average user rating is taken into account.

Specific item interestingness to user u is considered a property of the "item-user" pair, which can be derived from the actions and evaluations user u performed on item i (at the moment, we only deal with user bookmarks). We consider specific item interestingness since we assume that favourite items may be included in double-sided recommendations even if users are already aware of them.

Group preferences for items are predicted by aggregating individual preference scores. More specifically, a score $itemGroupScore_{iG}$: $IxP(C) \rightarrow Q$, indicating how interesting item i is expected to be to group $G \in P(C)$ as a whole, is computed as an average of the individual preference scores of all group members with respect to item i.

Recommendable groups are generated starting from connected components in the social network of the target user. Such a social network is represented as a graph and may contain either all the contacts of the target user, or, in case an item has already been selected for recommendation, only those contacts whose preference for that item is higher than a threshold. Three sets of connected components, considering respectively family, friendship and all relationships, are identified. Groups are generated from the connected components by: a) eliminating duplicates, and b) applying simple social rules: for example, if the target user has a partner and he or she is not included in a certain recommendable group, another group can be built which also includes him or her. The same rule applies for the target user's best friend, i.e., the contact for whom the value of relationship strength is maximum. Moreover, two groups including, respectively, only the target user's partner and only the target user's best friend are added, if they do not result from connected components.

4   Evaluation

We chose the *restaurant domain* as a use case for our evaluation, considering that restaurants represent a typical example of items which can be recommended to groups, and that people can be assumed to dine with different groups on different occasions.

4.1   Evaluation Overview

The evaluation method we adopted consisted of several steps. First, we recruited the experimental subjects among Facebook users, through a snowballing sampling strategy[2].

---

[2] In snowballing sampling, experimental subjects usually tell the researchers about other individuals who possess the desired characteristics to take part into the study. In our case, it allowed us to recruit people who were connected to each other.

We opted for Facebook users for two reasons: on the one hand, they are accustomed to interactive social websites, and can therefore be considered target users for double-sided recommendations; on the other hand, this allowed us to observe real social networks. 172 people (60% female and 40% male, aged 19-65) accepted to take part to the evaluation at that time. Then, we analyzed their social networks in order to gather information about their social relationships, and their strenght and type (based on Facebook data, we distinguished among "friend", "family" and "significant other"). This information was stored for successive use.

Experimental subjects were actively involved in the following step, which we call the *opinion gathering phase*: they were asked to use iFOOD[3], an adaptive recommender system in the restaurant domain, for a twenty-day period. This phase aimed at 1) building user models containing information about user interests with respect to different types of restaurants, and at 2) gathering user evaluations of the system contents. 29 subjects out of 172 accepted to take part to this phase (17 female and 12 male, aged 19-62).

Information from the opinion gathering phase was combined with information about social networks in order to generate four personalized recommendation lists (one with each of the proposed recommendation methods) for each experimental subject.

In the following step, which we call the *main evaluation phase*, experimental subjects evaluated the double-sided recommendations they were presented, and answered a final short survey. Further detail abouth this phase is provided in the next section.

4.2   Main Evaluation Phase

This evaluation phase has two goals: assessing user liking for double-sided recommendations and comparatively evaluating the four proposed recommendation methods.

Hypotheses. We hypothesized that users appreciate double-sided recommendations (H1). We also hypothesized that all four methods can provide useful double-sided recommendations (H2); however, we expect that their performances vary according to i) the type of double-sided recommendation problem the experimental subjects are facing during the evaluation (externally provided problem instance definition, H3), and to ii) the way they usually experience this problem in their real life (personal problem instance definition, H4). In particular, we expect that recommendations generated with a certain method can prove especially useful if users are experiencing the corresponding problem instance (social comparison method is associated to problem instance 1, group-priority method to problem instance 2, item-priority method to problem instance 3 and same-priority method to problem instance 4).

Subjects. All the 29 subjects who took part to the opinion gathering phase were selected as experimental subjects.

Experimental design. Mixed 4 X 5 factorial design, consisting of one within-subject variable (double-sided recommendation method), with four levels (social comparison, group-priority, item-priority and same-priority), and one between-subject variable (externally provided problem instance definition), with five levels (instance 1, 2, 3, 4, and control situation, where no explicit description of the problem instance subjects are

---

[3] http://www.piemonte.di.unito.it/progettoDSR/
DialogManager?page=home

facing is provided). The experimental subjects were randomly assigned to five groups, corresponding to the five levels of the between-subject variable (9 were assigned to the control group and 5 each to the other four groups).

Material. Recommendations were presented by means of simple web pages. A different page was devoted to each one of the four recommendation lists and navigation was devised so that users could access the following page only after they had completed their tasks on the current one. With the aim of avoiding order-effects, recommendation lists were presented in random order to each experimental subject. Moreover, an initial web page was devoted to the explanation of the experimental task, while a short online survey was presented in the end.

Measures. We evaluated recommendation *usefulness*, i.e., how useful each recommendation is to the experimental subjects in solving the specific problem instance they are facing [9]. We measured recommendation usefulness by means of a 5-point Likert scale, where the first position corresponds to "not useful at all" and the last one to "very useful". Each recommendation was accompanied by the scale to use for its evaluation.

In the survey, *question 1* asked users to express their level of liking for double-sided recommendations in the restaurant domain. *Question 2* aimed at assessing the way users experience the double-sided recommendation problem in their real life (personal problem instance definition): they were provided with four sentences describing the four problem instances we identified in our framework (sentence 2a mapped to instance 1, sentence 2b to instance 2, sentence 2c to instance 3, and sentence 2d to instance 4), and were asked to assess how much each sentence described a situation they experience in their everyday life. All answers should be provided by means of 5-point Likert scales.

Experimental task. Subjects assigned to groups other than the control one were asked to imagine they were facing a specific problem instance, which was described to them according to the instance definitions we provided in our framework presentation (Section 3). All subjects were asked to evaluate at least the first ten recommendations (if available) in each list and to complete the final survey.

Results. Only three subjects out of 29 did not complete the evaluation of all four recommendation lists and did not filled in the final survey. On the whole, 347 evaluations were collected for the same-priority method, 41 for the group-priority method, 284 for the item-priority method and 294 for the social comparison-based method.

User liking for double-sided recommendations. User answers to *question 1* tell us that most subjects were positively impressed by double-sided recommendations (H1): their average rating is 4,38 and 88,4% of subjects expressed a definitely positive opinion, choosing a rating of 4 or 5. A further confirmation of user liking comes from their evaluations of recommendation usefulness: the average evaluation, 3,88, is quite satisfactory. In addition, 67,4% users evaluated the recommendations they received as very useful, rating them 4 or 5.

Comparison among recommendation methods. The average usefulness evaluations are definitely positive for all methods (H2, see Figure 1(b)). However, it seems that the group-priority method is able to provide more useful recommendations, both considering the whole data set and with respect to most problem instances. Thus, we can assume

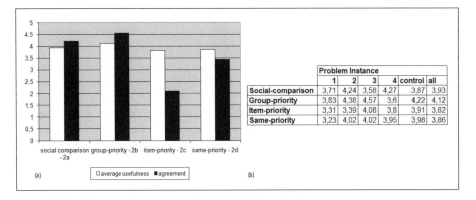

Fig. 1. (a)Comparison between the average usefulness of double-sided recommendation methods and average user agreement with the sentences describing the corresponding problem instances. (b)Average recommendation usefulness evaluations with respect to externally provided problem instance definitions.

that the group is more relevant than the restaurant when planning to dine out, at least to our experimental subjects. This seems to be confirmed by the fact that subjects tended to agree the most with sentence 2b (which maps to problem instance 2 and group priority method) in the final survey (see dark bars in Figure 1(a)).

Effect of problem instance. We first consider the *externally provided problem instance definition* (see Figure 1(b)). Recommendations generated with the group-priority method achieve the highest average usefulness evaluations with respect to three problem instances out of four. However, if we focus on a certain problem instance, it can be observed that the method we expected to have the best performance is always at least second best, as for average recommendation usefulness. Considering the *personal problem instance definition*, we can notice from Figure 1(a) that the values representing the average recommendation usefulness for the four recommendation methods (based on the whole dataset) are quite close to those representing user average agreement with the sentence describing the corresponding problem instance, although user agreement with instance 3 (question 2c) is much lower than the average usefulness of item-priority recommendations. In order to further investigate this issue, we performed a correlational study and found that the average usefulness of recommendations generated with a certain method and for a certain user is positively, although weakly, related to the level of agreement of such a user with the sentence describing the corresponding problem instance ($r=0.226$, with $p = 0,05$). Considering both types of problem instance definitions, our results do not yet allow us to validate our hypotheses; however, they seem to support our idea that the specific instance users are facing influences the perceived usefulness of recommendations generated with different methods, even if its effect is not so clearly defined as expected (H3, H4).

5   Conclusion and Future Work

In this paper we have introduced *double-sided recommendations* and proposed a framework, consisting of four different recommendation methods, for generating them.

Results of the experimental evaluation we carried out, focusing on the restaurant domain as a use case, showed that experimental subjects appreciated the possibility of receiving suggestions consisting of a restaurant and a group of people to dine with. Moreover, all the proposed methods proved effective in generating useful recommendations. We also studied the effect of the specific problem instance users are facing on the perceived usefulness of double-sided recommendations generated with different methods, taking into account both externally provided and personal problem instance definitions. Our results seem to support the idea that some connection between problem instance and perceived recommendation usefulness is actually present, although it is not so clearly defined as it was expected and our data do not yet allow us to statistically confirm our hypothesis.

Although our results refer to a small number of subjects, and should therefore be treated cautiously, we believe that they can be interesting to designers of recommender systems. First, we showed that double-sided recommendations are a novel and useful service: as a consequence, they could be integrated in recommender systems which deal with items which are consumed by groups as often as by individuals (e.g., restaurants, movies or cultural events). As for the specific recommendation methods to use, we found that group-priority method generates particularly useful suggestions, at least in our use case domain. Thus, such a method could be used safely both if it is known that contextual elements and/or personal preferences will determine a specific scenario for group-priority recommendations, and as a default option. The other methods might be used, alone or in conjunction with group-priority method, if it is known that users are facing the specific problem instance they were designed for. However, based on our results, we expect that users will relatively rarely face a situation where they prefer to privilege the item aspect, or where they want to assign the same importance to item and group aspects, in the restaurant domain.

Further research is required in order to investigate the extendability of our results to other domains. For example, it might be possible that methods other than group-priority tend to provide the most useful double-sided recommendations in a different domain. Moreover, an important issue which should be dealt with in future work regards how to determine which type of double-sided recommendation problem users are facing, if the correlation between problem instance and perceived recommendation usefulness is confirmed. For example, this might be inferred from user feedback on recommendations provided with different methods, or from the quality and quantity of their actions on system contents and users (e.g., users who often interact with their social network might be expected to give higher priority to the group aspect).

Acknowledgements. This work has been partially supported by Regione Piemonte, grant ICT Converging Technologies 2007, Piemonte Project.

References

1. Amer-Yahia, S., Roy, S.B., Chawla, A., Das, G., Yu, C.: Group recommendation: Semantics and efficiency. PVLDB 2(1), 754–765 (2009)
2. Baatarjav, E., Phithakkitnukoon, S., Dantu, R.: Group recommendation system for facebook. In: Chung, S. (ed.) OTM 2008, Part II. LNCS, vol. 5332, pp. 211–219. Springer, Heidelberg (2008)

3. Brusilovsky, P., Kobsa, A., Nejdl, W. (eds.): Adaptive Web 2007. LNCS, vol. 4321. Springer, Heidelberg (2007)
4. Burke, R.D.: Hybrid web recommender systems. In: Brusilovsky, et al. (eds.) [3], pp. 377–408
5. Carmagnola, F., Vernero, F., Grillo, P.: Sonars: A social networks-based algorithm for social recommender systems. In: Houben, G.-J., McCalla, G., Pianesi, F., Zancanaro, M. (eds.) UMAP 2009. LNCS, vol. 5535, pp. 223–234. Springer, Heidelberg (2009)
6. Chen, J., Zaïane, O., Goebel, R.: Local community identification in social networks. In: ASONAM 2009: Proceedings of the 2009 International Conference on Advances in Social Network Analysis and Mining, pp. 237–242. IEEE Computer Society, Washington, DC (2009)
7. Festinger, L.: A theory of social comparison process. Human Relations 7, 117–140 (1954)
8. Guy, I., Zwerdling, N., Carmel, D., Ronen, I., Uziel, E., Yogev, S., Ofek-Koifman, S.: Personalized recommendation of social software items based on social relations. In: RecSys 2009: Proceedings of the Third ACM Conference on Recommender Systems, pp. 53–60. ACM, New York (2009)
9. Herlocker, J.L., Konstan, J.A., Terveen, L.G., Riedl, J.T.: Evaluating collaborative filtering recommender systems. ACM Trans. Inf. Syst. 22, 5–53 (2004)
10. Jameson, A., Smyth, B.: Recommendation to groups. In: Brusilovsky, et al. (eds.) [3], pp. 596–627
11. Kim, J.K., Kim, H.K., Oh, H.Y., Ryu, Y.U.: A group recommendation system for online communities. International Journal of Information Management 30(3), 212–219 (2010)
12. Kobsa, A., Koenemann, J., Pohl, W.: Personalised hypermedia presentation techniques for improving online customer relationships. Knowl. Eng. Rev. 16(2), 111–155 (2001)
13. Lancichinetti, A., Fortunato, S., Kertész, J.: Detecting the overlapping and hierarchical community structure in complex networks. New Journal of Physics 11(3), 033015 (2009)
14. Newman, M.E.: Detecting community structure in networks. The European Physical Journal B - Condensed Matter and Complex Systems 38(2), 321–330 (2004)
15. Pazzani, M.J., Billsus, D.: Content-based recommendation systems. In: Brusilovsky, et al. (eds.) [3], pp. 325–341
16. Recio-Garcia, J.A., Jimenez-Diaz, G., Sanchez-Ruiz, A.A., Diaz-Agudo, B.: Personality aware recommendations to groups. In: RecSys 2009: Proceedings of the Third ACM Conference on Recommender Systems, pp. 325–328. ACM, New York (2009)
17. Schafer, J.B., Frankowski, D., Herlocker, J.L., Sen, S.: Collaborative filtering recommender systems. In: Brusilovsky, et al. (eds.) [3], pp. 291–324
18. Sinha, R., Swearingen, K.: Comparing recommendations made by online systems and friends. In: Proceedings of the DELOS-NSF Workshop on Personalization and Recommender Systems in Digital Libraries (2001)
19. Sozio, M., Gionis, A.: The community-search problem and how to plan a successful cocktail party. In: KDD 2010: Proceedings of the 16th ACM SIGKDD International Conference on Knowledge Discovery and Data Mining, pp. 939–948. ACM, New York (2010)
20. Vasuki, V., Lu, Z., Natarajan, N., Dhillon, I.: Affiliation recommendation using auxiliary networks. In: RecSys 2010: Proceedings of the Forth ACM Conference on Recommender Systems. ACM, New York (2010)

Intelligent Self-repairable Web Wrappers

Emilio Ferrara[1] and Robert Baumgartner[2]

[1] Dept. of Mathematics, University of Messina,
V. Stagno d'Alcontres 31, 98166 Italy
`emilio.ferrara@unime.it`
[2] Lixto Software GmbH, Favoritenstrasse 16/DG, 1040 Vienna, Austria
`robert.baumgartner@lixto.com`

Abstract. The amount of information available on the Web grows at an incredible high rate. Systems and procedures devised to extract these data from Web sources already exist, and different approaches and techniques have been investigated during the last years. On the one hand, reliable solutions should provide robust algorithms of Web data mining which could automatically face possible malfunctioning or failures. On the other, in literature there is a lack of solutions about the maintenance of these systems. Procedures that extract Web data may be strictly interconnected with the structure of the data source itself; thus, malfunctioning or acquisition of corrupted data could be caused, for example, by structural modifications of data sources brought by their owners. Nowadays, verification of data integrity and maintenance are mostly manually managed, in order to ensure that these systems work correctly and reliably. In this paper we propose a novel approach to create procedures able to extract data from Web sources – the so called *Web wrappers* – which can face possible malfunctioning caused by modifications of the structure of the data source, and can automatically repair themselves.

Keywords: Web data extraction, wrappers, automatic adaptation.

1 Introduction

The actual panorama of distribution of information through the Web depicts a clear situation: there is an incredible amount of data delivered under the form of Web data sources and a corresponding need of capability of mining this information in a reliable and efficient way. Mining information from Web sources is a task which can obviously be useful in several different area of the knowledge. Moreover, this topic interests both the academia and the enterprises. For example, consider the following scenarios: i) a research group which needs to acquire a dataset of information delivered through online services, say for example an online database publishing, day by day, information about the mapping of some genes; ii) a company for which it is essential, for marketing and product placement, to monitor the trends of pricing of services offered by its competitors, provided through the Web. Both the two actors need to extract, possibly, a huge amount of data during an extend period of time (e.g., months), at regular intervals (say, each day). One important aspect in both the cases is the reliability and

R. Pirrone and F. Sorbello (Eds.): AI*IA 2011, LNAI 6934, pp. 274–285, 2011.

the quality of data extracted. It is utterly important that acquired information is correct, because the research group can not accept corrupted data and the comparison with competitors would fail in case of bad product data.

These two examples highlight common requirements in the panorama of Web data mining, and depict different related problems. Although in literature some techniques to design systems for the extraction of data from Web sources have been presented, there is a lack of work in the area of their maintenance. An ample number of questions and problems related to the possibility of automatizing the process of maintenance are still uncovered. This work tries to focus on some aspects related to the maintenance of these systems. We first introduce the theoretical background required to create intelligent procedures of Web data extraction. Then, we explain how to face malfunctioning likely to happen during the extraction process, for example caused by modifications in the structure of the data source. The second point in particular is the main focus of this work. Let us contextualize this problem: essentially there exist two different approaches to extract information from Web sources. The first one relies on machine learning platforms [5]; a system analyzes, possibly, huge amount of positive and negative examples during a training period, and, then, it infers some set of rules that makes it able to perform its tasks in the same domain or Web site. Different approaches rely on logic-based algorithms which analyze the structure of the data source and induct some procedures to extract required information exploiting structural characteristics of the Web source to identify and find required data. The second approach utilizes the knowledge a human can bring in about a particular site or domain. The wrapper is generated in a way that the human creates the rules and navigation paths together with the system in a supervised and interactive fashion. Still, the system can assist the wrapper designer and offer possibilities that make the wrapper execution as robust as possible, even in case of structural changes. From now, in this work we assume that the platform we are going to describe and improve adopts the latter philosophy.

Organization of the paper. We describe related work in Section 2. In Section 3, the algorithmic background is introduced, describing an efficient tree matching technique. Section 4 covers the design of robust and adaptable procedures of Web data extraction, henceforth called *intelligent self-repairable Web wrappers*. Then, in Section 5 we describe the adaptation process during wrapper execution. We explain how these procedures can automatically, in an autonomous way, face malfunctioning, trying to adapt themselves to the modifications that possibly caused problems. A prototype has been implemented on top of a state-of-the-art extraction platform, the Lixto Visual Developer. Performance of this system are shown in Section 6, by means of precision and recall scores. Section 7 concludes summarizing our main achievements and depicting some future work.

2   Background and Related Work

We split related literature in three main topics: i) Web data extraction systems; ii) maintenance and related problems; iii) tree matching algorithms.

Web data extraction systems The work related to systems of Web information extraction is manifold but well depicted by several surveys. Laender et al. [13] provided the first rigorous taxonomical classification of Web data extraction systems. Kushmerick [11] classified several finite-state approaches to generate wrappers, such as the wrapper induction, natural language processing approaches and hidden Markov models. Sarawagi [17] provided the most comprehensive survey on the information extraction panorama. This work covers different existing techniques explaining several approaches. In the last years, first Baumgartner et al. [1] and later Ferrara et al. [8] provided two different surveys on the discipline of Web data extraction. The first is mainly addressed to practitioners, the latter focuses on application fields of this discipline.

Maintenance and related problems. Although some interesting work, we can identify a general lack of solutions provided in the area of the Web wrapper maintenance. Kushmerick [12,10] for first introduced the concept of wrapper maintenance as the process of verifying the correct functioning of the data extraction procedures and manually, automatically or in a semi-automatic way, intervene in case of malfunctioning. Lerman and Minton [14], instead, faced both the problems of verifying the correctness of data extracted by a wrapper and eventually try to repair it. Their approach is a mix of machine learning techniques. Another approach based on machine learning has been provided by Chidlovskii [4]; he described a system which can automatically classify Web pages in order to extract information from those pages which can be handled adopting both conventional extraction rules and ensemble methods of machine learning, such as the content features analysis. Meng et al. [15] developed the SG-WRAM (Schema-Guided WRApper Maintenance) slightly modifying the perspective of Web wrappers generation, observing that changes in Web pages, even substantial, always preserve syntactic features (i.e., syntactic characteristics of data items like data patterns, string lengths, etc.), hyperlinks and annotations (e.g., descriptive information representing the semantic meaning of a piece of information in its context). Finally, another heuristic approach has been presented by Raposo et al. [16]; they adopted a collected sample of positive labeled examples during the normal execution of the wrappers, to be exploited in case of malfunctioning, in order to re-induct the broken wrapper ensuring a good accuracy of the process.

Tree Matching. In general, the process of comparing the structure of two trees is a well-known classic problem. The possibility of transforming a tree into another one, through a sequence of (possibly different) operations, is another well-known algorithmic challenge, namely the *tree editing* problem. The minimum number of elementary transformations, such as adding/removing nodes, relabeling nodes or moving nodes, represents the *distance* between two trees. This value can be used to represent the measure of dissimilarity between two trees. The tree edit distance problem is a well-known NP-hard problem [3]. Several approximate solutions have been advanced during the years; the most appropriate algorithm to face the problem of matching up similar trees, has been suggested by Selkow [18]. This technique relies on the concept of finding isomorphic elements present in both the two compared trees, implementing a light-weight recursive top-down resolution during which the algorithm evaluates the position of nodes to measure the degree of isomorphism between them, analyzing and comparing their sub-trees. Different versions of this algorithm exist; each of them presents some optimizations. Ferrara and Baumgartner [6,7] so as Yang [19] adopt *weights*, obtaining a variant of this algorithm with the capability of discovering clusters of similar sub-trees. An interesting evaluation of the simple tree matching and its weighted version, presented by Kim et al. [9], has been performed exploiting these two algorithms to extract information from HTML Web pages. These optimized algorithms underly the design of our self-repairable Web wrappers.

3   The Tree Matching Algorithm

This work relies on some assumptions: i) Web pages are represented by using DOM trees, as the HTML standard imposes[1]; ii) it is possible to identify elements within a DOM tree by using the XPath language[2]; iii) the logics of XPath underly the functioning of Web wrappers (this is further explained in following sections and in [1,2]). Given these milestones, the main idea of our approach is to compare two trees, one representing the original Web page and another representing the page after that some modifications occurred. This is practical in order to automatize the adaptive process of automatic repairing of our wrappers. To do so, we utilize a variant of the seminal Simple Tree Matching (STM) [18], optimized by Ferrara and Baumgartner [6,7]. Let $d(n)$ be the degree of a node n (i.e., the number of first-level children); let $T(i)$ be the i-*th* sub-tree of the tree rooted at node T; let $t(n)$ be the number of total siblings of a node n including itself. The *Weighted Tree Matching* here described (see Algorithm 1) optimizes the simple tree matching, for our specific domain.

4   Web Wrappers

In supervised and interactive wrapper generation, the application designer is in charge of deciding how to characterize Web objects that are used for traversing the Web and for extracting information. It is one of the most important aspects of a wrapper to be resilient against changes (both changes over time and variations of similarly structured pages), and parts of the robustness of a data extractor depend on how the application designer configures it. However, it is crucial that the wrapper generation system assists the wrapper designer and suggests how to make the identification of Web objects and trails through Web sites as stable as possible.

---

[1] http://www.w3.org/TR/DOM-Level-2-HTML/html.html
[2] http://www.w3.org/TR/xpath/

Algorithm 1. WeightedTreeMatching(T', T'')

1: **if** T' has the same label of T'' **then**
2:    $m \leftarrow d(T')$
3:    $n \leftarrow d(T'')$
4:    **for** $i = 0$ to m **do**
5:       $M[i][0] \leftarrow 0$;
6:    **for** $j = 0$ to n **do**
7:       $M[0][j] \leftarrow 0$;
8:    **for all** i such that $1 \leq i \leq m$ **do**
9:       **for all** j such that $1 \leq j \leq n$ **do**
10:          $M[i][j] \leftarrow \mathrm{Max}(M[i][j-1], M[i-1][j], M[i-1][j-1] + W[i][j])$ where $W[i][j] = \mathrm{WeightedTreeMatching}(T'(i-1), T''(j-1))$
11:    **if** $m > 0$ AND $n > 0$ **then**
12:       return M[m][n] * 1 / $\mathrm{Max}(t(T'), t(T''))$
13:    **else**
14:       return M[m][n] + 1 / $\mathrm{Max}(t(T'), t(T''))$
15: **else**
16:    return 0

4.1   Robust XPath Generation and Fall-Back Strategies

In Lixto Visual Developer (VD) [2], a number of mechanisms are offered to create a resilient wrapper. During recording, one task is to generate a robust XPath or regular expression, interactively and supported by the system. During wrapper generation, in many cases only one labeled example object is available, especially in automatically recorded deep Web navigation sequences. In such cases, efficient heuristics in XPath generation and fallback strategies during replay, are required. Typical heuristics during recording for reliably identifying such single Web objects include:

- Generalization of a chosen XPath by using form properties, element properties, textual properties and formatting properties. During replay, these ingredients are used as input for an algorithm that checks in which constellation to best apply this property information to satisfy the integrity constraints imposed on a rule (e.g., as result a single instance is required).
- DOM Structural Generalization – starting from the full path, several generalized paths are created, using only characteristic elements and characteristic element sequences. A number of stable anchor points are identified and stored, from which relative paths to this object are created. Typical stable anchor points are identified automatically and include, e.g., the outermost table structure and the main content area (being chosen upon factors such as the longest content).
- Positional information is considered if the structurally generalized paths identify more than one element. In this case, during execution, variations of the XPath generated with this "index heuristics" are applied on the active Web page, removing indexes until the integrity constraints of the current rule are satisfied.

- Attributes and properties of elements are taken into account, in particular of the element of choice, but we also consider ancestor attributes if the element attributes are not sufficient.
- Attributes that make an element unique are preferred, i.e., similar elements are checked for distinguishing criteria.
- Attribute Values are considered, if attribute names are not sufficient. Attribute Value Fragments are considered, if attribute values are not sufficient (using regular expressions).
- The ID attributes are used as far as possible. If an ID is unique and meaningful for characterizing an element it is considered in the fallback strategies with a high weight.
- Textual information and label information is used, only if explicitly turned on (since this might fail in case of a language switch).

The output of the heuristic step is a "best XPath" shown to the wrapper designer, and a set of XPath expressions and priorities regarding when to use which fallback strategy, stored in the configuration. Figure 1 illustrates which information is stored by the system during recording. In this case, a drop down was selected by the application designer, and the system decided that the "id" attribute is the most reliable one and is chosen as best XPath. If this evaluation fails, the system will apply heuristics based on the (in this example, three) stored fallback XPaths, which mainly exploit form and index properties. In case one of the heuristics generates results that do not invalidate the defined integrity constraints, these Web objects are considered as result.

During generation of rules (e.g., "extract") and actions (e.g., "click"), the wrapper designer imposes constraints on the results to be obtained, such as:

- Cardinality Constraints: restrictions on the number of results, e.g., exactly one element or at least one element must be matched.
- Data Type Constraints: restrictions on the data type of a result, e.g., a result must be of type integer or match a particular regular expression.

Constraints can be defined individually per rule and action, or defined globally by using a schema on the output data model.

4.2 Configuring Adaptable Wrappers

The procedures described in the previous section do not adapt the wrapper, but address situations in which the initially chosen XPath does no longer match and simply try different ones based on this one. In the configuration of wrapper adaptation, we go one step beyond: on the one hand we exploit tree and string similarity techniques to find the most similar Web object(s) on the new page, and on the other hand, in case the adaptation is triggered, the wrapper is changed on the fly using the new configuration created by the adaptation algorithms.

As before, integrity constraints can be imposed on extraction and navigation rules. Moreover, the application designer can choose whether to use wrapper

Fig. 1. Robust Web Object Detection in Visual Developer

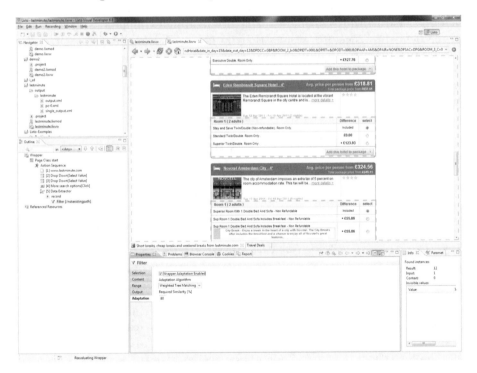

Fig. 2. Configuration of Wrapper Adaptation in Lixto VD

adaptation on a particular rule in case the constraints are violated during runtime. When adaptation is chosen, alternatively to using XPath-based means to identify Web objects we store the actual result subtree. In case of HTML leaf elements, which are usually the elements under consideration for navigation actions, we instead store the tree rooted at the n-th ancestor of the element, and the additional fact where the result element is located within this tree. In this way, tree matching can also be exploited for HTML leaf elements.

Wrapper designers can choose between various similarity measures: this includes in particular the Simple Tree Matching algorithm [18] and the Weighted Tree Matching algorithm described in Section 3. In future, further algorithms will extend the capabilities of the tool, e.g., a bigram-based tree matching that is capable to deal with node permutations in a more favorable fashion. In addition to the similarity function, one can choose certain parameters, e.g., whether to use the HTML element name as node label or instead to use spelling attributes such as *class* and *id* attributes. Figure 2 illustrates the configuration of wrapper adaptation in Visual Developer.

5   Automatic Wrapper Adaptation

5.1   Self-repairing Rules

Figure 3 describes the adaptation process. The wrapper adaptation process is triggered upon violation of defined constraints. In case in the initial wrapper an element is detected with an XPath, the adaptation procedure substitutes this by storing the subtree of a matched element. In case the wrapper definition already stores the example tree, and the similarity computation returns results that violate the defined constraints, the threshold is lowered or raised until a perfect match is generated.

During runtime, the stored tree is compared to the elements on the new page, and the best fitting element(s) are considered as extraction results. During configuration, wrapper designers can choose an algorithm (such as the Weighted Tree Matching), and a similarity threshold. The similarity threshold can be constant, or defined to be within an interval of acceptable thresholds. During execution, various thresholds within the allowed range are considered, and the one generating the best fit with respect to the defined constraints is chosen.

As a next step, the stored tree is refined and generalized so that it maximizes the matching value for both the original subtree and the new trees, reflecting the changes of a Web page over time. This generalization process generates a simple tree grammar, a "tree template" that is allowed to use occurrence indicators (one or more element, at least one element, etc.) and optional depth levels. In further runs, the tree template is compared against the sub trees of an active Web page during execution. First, the algorithm checks which trees on the new page satisfy the tree template. In case the results are within the defined integrity

Wrapper Adaptation Process

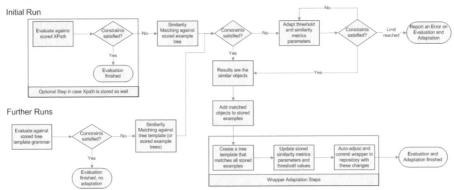

Fig. 3. Wrapper Adaptation Process

constraints, no further action is taken. In case the results are not satisfying, the system searches for most similar trees based on the defined distance metrics; in this case, the wrapper is auto-adapted, the tree template is further refined and the threshold or threshold interval is automatically re-adjusted. At the very end of the process, the corrected wrapper is stored in the wrapper repository and committed to a versioning system to keep track of all changes.

5.2　Wrapper Re-induction

In practice, single adaptation steps of rules and actions are embedded into the whole execution process of a wrapper and the adapted wrapper is stored in the repository after all adaptation steps have been concluded. The need for adapting a particular rule influences the further execution steps.

Usually, wrapper generation in VD is a hierarchical top-down process – e.g., first, a "hotel record" is characterized, and inside the hotel record, entities such as "rating" and "room types". To define a rule to match such entities, the wrapper designer visually selects an example and together with system suggestions generalizes the rule configuration until the desired instances are matched. To support the automatic adaptation process during runtime, as described above, the wrapper designer further specifies what it means that extraction failed. In general, this means wrong or missing data, and with integrity constraints one can give indications how correct results look like. The upper half of Figure 4 summarizes the wrapper generation.

During wrapper creation, the application designer provides a number of configuration settings to this process. This includes:

- Threshold Values.
- Priorities/Order of Adaptation Algorithms used.
- Flags of the chosen algorithm (e.g., using HTML element name as node label, using id/class attributes as node labels, etc.).

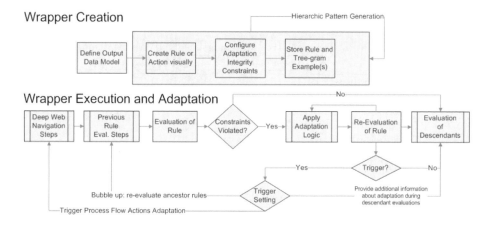

Fig. 4. Diagram of the Web wrapper creation, execution and maintenance flow

- Triggers for bottom-up, top-down and process flow adaptation bubbling.
- Whether stored tree-grams and XPath statements are updated based on adaptation results to be additionally used as inputs in future adaptation procedures (reflecting and addressing regular slight changes of a Web page over time).

Triggers in Adaptation Settings can be used to force adaptation of further fragments of the wrapper as depicted in the lower half of Figure 4.

- Top-down: forcing adaptation of all/some descendant rules (e.g., adapt the "price" rule as well to identify prices within a record if the "record" rule was adapted).
- Bottom-up: forcing adaptation of a parent rule in case adaptation of a particular rule was not successful. Experimental evaluation pointed out that in such cases it is often the problem that the parent rule already provides wrong or missing results (even if matched by the integrity constraints) and has to be adapted first.
- Process flow: it might happen that particular rule matches can no longer detected because the wrapper evaluates on the wrong page. Hence, there is the need to use variations in the deep web navigation actions. In particular, a simple approach explored at this time is to use a switch window or back step action to check if the previous window or another tab/popup provides the required information.

6   Performances Measurement

For our initial performance evaluation we tested the robustness of our Wrappers against real world use-cases. Actual areas of interest for Web data extraction

Table 1. Experimental performance evaluation in real world scenarios

		Simple Tree Matching Precision/Recall			Weighted Tree Matching Precision/Recall		
Scenario	thresh.	tp	fp	fn	tp	fp	fn
Delicious	40%	100	4	-	100	-	-
Ebay	85%	200	12	-	196	-	4
Facebook	65%	240	72	-	240	12	-
Google news	90%	604	-	52	644	-	12
Google.com	80%	100	-	60	136	-	24
Kelkoo	40%	60	4	-	58	-	2
Techcrunch	85%	52	-	28	80	-	-
Total	-	1356	92	140	1454	12	42
Recall	-	90.64%			97.19%		
Precision	-	93.65%			99.18%		
F-Measure	-	92.13%			98.18%		

problems include social networks, retail market and Web communities. We defined a total of 7 scenarios and designed 10 adaptive wrappers each. Results, by means of precision, recall and F1-score, are as shown in Table 1. Column *thresh.* represents the fixed threshold value; *tp*, *fp* and *fn* summarize true and false positive, and false negative, respectively. Performance obtained by using *simple* and *weighted* tree matching are good; these algorithms are definitely viable solutions to our initial purpose and provide high degree of reliability (F-Measure > 90%).

7 Conclusions and Future Work

In literature, several implementations of systems to extract data from Web sources have been presented, but there is a lack of solutions about their maintenance. This paper tries to address this problem, describing adaptive techniques to make Web data extraction systems, based on wrappers, self-maintainable, adopting algorithms optimized to this purpose. So, enhanced Web wrappers become able to recognize structural modifications of Web sources and to adapt their functioning accordingly. Characteristics of our self-repairable solution are discussed in details, providing first experimental results to evaluate its robustness. More experimentation has to come in the next future.

Moreover, as for future work, additional algorithms would be included in order to improve the capabilities of the adaptation feature; in particular, a viable idea could be to generalize a bigram-based tree matching algorithm capable of dealing with node permutations in a more efficient way with respect to Simple Tree Matching based algorithms adopted as to date. Similarly, the Jaro-Winkler distance could be adapted to our tree matching problem in order to better reflect missing or added node levels, so as improving performance of our adaptation process. Finally, the tree-grammar could be extended to classify different topologies of templates (those frequently adopted by Web pages), in order to define several standard protocols of automatic adaptation, to be adopted in specific contexts.

References

1. Baumgartner, R., Gatterbauer, W., Gottlob, G.: Web data extraction system. Encyclopedia of Database Systems, 3465–3471 (2009)
2. Baumgartner, R., Gottlob, G., Herzog, M.: Scalable web data extraction for online market intelligence. Proceedings of the VLDB Endowment 2(2), 1512–1523 (2009)
3. Bille, P.: A survey on tree edit distance and related problems. Theoretical Computer Science 337(1-3), 217–239 (2005)
4. Chidlovskii, B.: Automatic repairing of web wrappers by combining redundant views. In: Proceedings of the 14th International Conference on Tools with Artificial Intelligence, pp. 399–406. IEEE, Los Alamitos (2003)
5. Esposito, F., Malerba, D., Di Pace, L., Leo, P.: A machine learning approach to web mining. In: AI* IA 1999: Advances in Artificial Intelligence, pp. 190–201 (2000)
6. Ferrara, E., Baumgartner, R.: Automatic wrapper adaptation by tree edit distance matching. In: Hatzilygeroudis, I., Prentzas, J. (eds.) Combinations of Intelligent Methods and Applications. Smart Innovation, Systems and Technologies, vol. 8, pp. 41–54. Springer, Heidelberg (2011)
7. Ferrara, E., Baumgartner, R.: Design of automatically adaptable web wrappers. In: Proceedings of the 3rd International Conference on Agents and Artificial Intelligence, pp. 211–217 (2011)
8. Ferrara, E., Fiumara, G., Baumgartner, R.: Web data extraction, application and techniques: A survey. Technical Report (2011)
9. Kim, Y., Park, J., Kim, T., Choi, J.: Web information extraction by HTML tree edit distance matching. In: Proceedings of the International Conference on Convergence Information Technology, pp. 2455–2460. IEEE, Los Alamitos (2008)
10. Kushmerick, N.: Wrapper verification. World Wide Web 3(2), 79–94 (2000)
11. Kushmerick, N.: Finite-state approaches to Web information extraction. Extraction in the Web Era, 77–91 (2003)
12. Kushmerick, N., et al.: Regression testing for wrapper maintenance. In: Proceedings of the National Conference on Artificial Intelligence, pp. 74–284 (1999)
13. Laender, A., Ribeiro-Neto, B., da Silva, A., Teixeira, J.: A brief survey of web data extraction tools. ACM Sigmod Record 31(2), 84–93 (2002)
14. Lerman, K., Minton, S., Knoblock, C.: Wrapper maintenance: A machine learning approach. Journal of Artificial Intelligence Research 18(1), 149–181 (2003)
15. Meng, X., Hu, D., Li, C.: Schema-guided wrapper maintenance for web-data extraction. In: Proceedings of the 5th ACM International Workshop on Web Information and Data Management, pp. 1–8. ACM, New York (2003)
16. Raposo, J., Pan, A., Alvarez, M., Hidalgo, J.: Automatically generating labeled examples for web wrapper maintenance. In: Proceedings of the IEEE/WIC/ACM International Conference on Web Intelligence, pp. 250–256 (2005)
17. Sarawagi, S.: Information extraction. Foundations and Trends in Databases 1(3), 261–377 (2008)
18. Selkow, S.: The tree-to-tree editing problem. Information Processing Letters 6(6), 184–186 (1977)
19. Yang, W.: Identifying syntactic differences between two programs. Software: Practice and Experience 21(7), 739–755 (1991)

Synthesis of Collective Tag-Based Opinions in the Social Web*

Federica Cena, Silvia Likavec, Ilaria Lombardi, and Claudia Picardi

Dipartimento di Informatica, Università di Torino, Italy
{cena,likavec,lombardi,picardi}@di.unito.it

Abstract. This paper presents an approach to personalized synthesis of tag-based users' opinions in a social context. Our approach is based on an enhanced tagging framework, called iT$_A^G$, where tags are enriched with structure and expressivity and can be addressed to different features of a resource and weighed by relevance. Our main contribution is a synthesis of the collective opinions that is *multi-faceted*: it shows different points of view on the same resource, rather than averaging the opposite opinions, or choosing the one with the most supporters. If the social tool provides user modeling and trust mechanisms, our synthesis can also be *personalized*, taking into account both the user's *social network* (considering only the opinions of trusted authors) and her *user model* (considering only the features the user likes). In addition, we propose an innovative visualization modality for iT$_A^G$s, which allows for an at-a-glance impression of all the opinions on a given resource, including significant differences in point of view. We evaluated the iT$_A^G$ framework to test (i) its expressiveness for providing opinions, and (ii) the effectiveness of our synthesis with respect to traditional tag clouds.

Keywords: social web, tagging systems, personalized synthesis.

1 Introduction

In the context of social applications, users often participate in the community life by providing their opinions on the resources the community life revolves around (e.g. books, music, pictures, etc.). To do so, they can *rate, tag* or write *free text* comments on items. Social applications could use such meta-data available on the resources for different purposes: to learn about users' preferences or to provide the other users with the synthesis of such a content. The possibility to do this depends on the typology of the user-generated content: ratings are the simplest one to be aggregated as average values, but they are not very informative on the qualities or shortcomings the ratings are based on. Free text comments are very informative but they are difficult to be effectively processed and synthesized. Tags lie in between ratings and free-text comments for richness of information and computability. Our work moves from the observation that (i) traditional tags are more suited to express facts (e.g. for content classification) rather than opinions, since they do not possess enough richness and structure to allow users

---

* This work has been supported by PIEMONTE Project - People Interaction with Enhanced Multimodal Objects for a New Territory Experience.

R. Pirrone and F. Sorbello (Eds.): AI*IA 2011, LNAI 6934, pp. 286–298, 2011.

to express complex and multi-faceted opinions; (ii) the *tag clouds*, commonly used by most of the social applications to present tags in an aggregated form, are often difficult to browse and are not very informative.

Our goals were to overcome such limitations of tagging. First, our aim was to enable expressing of elaborate opinions using tags, i.e., giving users the possibility to use tags in order to express: *judgement* (liking or disliking a feature), *relevance* (saying that a feature is more important than another one) and *scope* (referring an opinion to only a part of a resource). Second, our goal was to merge opinions given by means of tags by different users into a synthesis representing an overview of the collective opinion. The synthesis should be (i) *multi-faceted*, i.e., present contrasting opinions, and (ii) *personalized*, i.e., take into account both the *social network* of the person it will be shown to (the synthesis will consider only the opinions of the users the person trusts)[1] and her *user model* (the synthesis will show only the features of a resource the user considers relevant). This brings about a need to find an innovative *visualization modality* which allows for an at-a-glance opinion of a large amount of users' tags on a resource, giving the possibility to discover different points of view on it.

The paper presents the iT$_A$G framework, an enhanced tagging framework, where tags are enriched with structure and expressivity, so that they can be addressed to different features of a resource and weighed by relevance, and where an approach to opinion synthesis is provided. We report the results of the evaluation of: (i) the expressiveness of the iT$_A$G framework for communicating opinions, (ii) the effectiveness of our synthesis with respect to traditional tag clouds, applied to a social environment for opinion-sharing on restaurants.

The paper is organized as follows. Section 2 introduces the notion of iT$_A$Gs. Section 3 describes our approach to iT$_A$Gs semantic interpretation, used further in iT$_A$Gs synthesis in Section 4. Section 5 discusses the results of the iT$_A$G framework evaluation, followed by the discussion of related work in Section 6 and conclusions in Section 7.

2   Introducing the iTag Concept

Let us consider an object $O \in \mathbb{O}$ the user wishes to comment upon, where $\mathbb{O}$ is the set of all the objects in the domain. We assume that: (i) the object O is of a distinguished type, $type(O)$; (ii) a hierarchy of *facets* $\mathbb{F}$ is associated with $type(O)$, where each facet $F \in \mathbb{F}(type(O))$ denotes something about the objects of this type the user may want to comment on. $\mathbb{F}$ contains $type(O)$ itself, as the root of the facet hierarchy. Figure 1 shows a possible hierarchy of facets associated with the *restaurant* type.

An iT$_A$G can be assigned to a specific facet of a given object. We represent an iT$_A$G as a labelled circle of a given size, placed above (positive impression) or below (negative impression) the given facet (see Figure 2). In other words, an iT$_A$G can express an *opinion* (iT$_A$G label, typically an adjective), *scope* (choice of a given facet), *judgement* (iT$_A$G placement) and *relevance* (iT$_A$G size). Formally, an iT$_A$G is defined as follows:

---

[1] Users prefer the recommendations generated by the users they trust rather than the suggestions generated by computer programs, see [1,3].

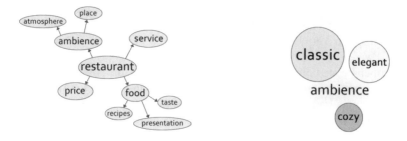

Fig. 1. A hierarchy of facets for type *restaurant*   **Fig. 2.** Representations of iTAGs as spheres

Definition 1 (iT$_A$G). *An iT$_A$G I is a tuple I = ⟨a, O, F, L, p, S⟩ where*

- *a is the iT$_A$G author;*
- *O ∈ 𝕆 is the object the iT$_A$G refers to;*
- *F ∈ 𝔽(type(O)) is the facet of type(O) the iT$_A$G is assigned to;*
- *L is the label (typically an adjective) that refers to facet F of object O;*
- *p is the polarity (positive or negative) which describes the user's judgement of a given facet: iT$_A$G placed above (below) the given facet denotes a positive (negative) opinion on a given facet;*
- *S ∈ [θ, 1] is the size that expresses the relevance the iT$_A$G author gives to L w.r.t. her global impression of O (θ > 0 is the minimum threshold provided by the system).[2]*

For example, in the following iT$_A$G *I*, Jo comments positively on Alfredo's food, which she finds *tasty* in a relevant way (size is 0.8):

$$I = ⟨Jo, Alfredo's, food, tasty, +, 0.8⟩.$$

3   iTags Interpretation

We assume that each facet *F* is associated with a set of properties ℙ(*F*). Properties represent the attributes which are relevant when commenting on the facet; they are known to the system but invisible to the user. For example, the *ambience* facet of type *restaurant* could be associated with the following property set: {*classicism, elegance, comfort, spaciousness, lighting, quietness, cleanliness*}.

Interpreting an iT$_A$G *I* = ⟨a, O, F, L, p, S⟩ means finding out which property *P* of the facet *F* the label *L* is addressing. The label *L* in *I* can relate to the property *P* in two ways: it can confirm the property (e.g. *calm* w.r.t. *quietness*) or oppose it (e.g. *noisy* w.r.t. *quietness*). Therefore, we define an *interpreted* iT$_A$G in the following way:

---

[2] The user may give a high relevance value to a given impression for several reasons. She may consider it relevant because the corresponding facet *F* is very important to her, or because the feature she is considering is very prominent. For example, the label *expensive* may be large because the restaurant is very expensive or because the tagger thinks that it being expensive is important in her judgment. We do not distinguish the reasons behind a given relevance value.

Definition 2 (iT_AG interpretation). *For a given* iT_AG $I = \langle a, O, F, L, p, S \rangle$, *an interpreted* iT_AG $\hat{I}$ *is a tuple* $\hat{I} = \langle a, O, F, L, p, (P, r), S \rangle$ *where*

- a, O, F, L, p, S *are the same as in Definition 1;*
- $P \in \mathbb{P}(F)$ *is the property the label L refers to;*
- $r \in \{0, 1\}$ *is the relationship between L and P; r = 1 means that L confirms P and r = 0 means that L opposes P.*

As an example, consider the following iT_AGs:

$$I_1 = \langle Jo, Alfredo's, ambience, refined, +, 0.7 \rangle \quad I_2 = \langle Meg, Alfredo's, ambience, simple, +, 0.4 \rangle.$$

Both iT_AGs can be related to the property *elegance* of the facet *ambience*, but the labels *refined* and *simple* express opposite meanings. Therefore, the interpretation of these two iT_AGs would result in:

$$\hat{I}_1 = \langle Jo, Alfredo's, ambience, refined, (elegance, 1), +, 0.7 \rangle$$
$$\hat{I}_2 = \langle Meg, Alfredo's, ambience, simple, (elegance, 0), +, 0.4 \rangle.$$

We propose an automated interpretation method based on WordNet [10], which works on iT_AGs whose labels are descriptive adjectives, possibly combined with the negation *not* or with an adverb of degree, such as *very*, *scarcely*, etc. Other iT_AGs are left uninterpreted; that they will not be used in the iT_AG synthesis but will be individually visible to users.

We briefly recall that WordNet organizes adjectives in synset clusters. Each cluster C is characterized by a focal synset $foc(C)$, expressing the "main" adjective, while the other "satellite" synsets express similar, more specialized notions (e.g., if the focal sysnset is represented by *fast*, its satellites are *prompt*, *alacritous*, etc.). The most relevant semantic relation between adjectives is that of *antonymy*. Synset clusters come in pairs $(C, \overline{C})$, where the two focal synsets are direct antonyms. Given C, we can determine its opposite $\overline{C}$, such that $foc(C)$ is the antonym of $foc(\overline{C})$. Satellite synsets are not considered direct antonyms, rather conceptual opposites or, as WordNet puts it, *indirect* antonyms. For example, *slow* is the direct antonym of *fast*, while *sluggish* is conceptually opposite to *alacritous*, but they are not antonyms. Hence, WordNet uses a bipolar adjective structure, the two poles being direct antonyms, each surrounded by satellites representing similar adjectives.

For our purposes, each bipolar structure in the WordNet adjective organization corresponds to a property. One of the two poles is selected as representative; any word in that pole synset or in one of its satellites *confirms* the property, while any word in the opposite pole synset or in one of its satellites *opposes* the property.

Since we also consider adverbs of degree as adjective modifiers, and WordNet does not offer any means to derive the "direction" of the modification, we pre-partition the set of adverbs of degree in two: *positive* adverbs enhance or intensify the meaning of the adjective, while *negative* ones diminish or negate it. The negative set obviously contains *not*. Therefore, our approach can be summarized as follows:

- Given a label L used for a facet F, we search for the words contained in it in WordNet, to find out whether L is indeed an adjective, possibly accompanied by an adverb of degree ad. Any other combination of words is discarded.

- If L is an adjective, we consider the WordNet cluster C it belongs to. If the noun obtained from $foc(C)$ or $foc(\overline{C})$ belongs to $\mathbb{P}(F)$, then the property we seek is represented by the pair $(C, \overline{C})$. If $foc(C) \in \mathbb{P}(F)$, $r = 1$. If $foc(\overline{C}) \in \mathbb{P}(F)$, $r = 0$.
- If neither $foc(C)$ nor $foc(\overline{C})$ belongs to $\mathbb{P}(F)$, then $foc(C)$ is added in as a new property representative, and r is set to 1.
- In case L is accompanied by an adverb of degree ad, if ad is a negative adverb according to our partition, r is reversed (it becomes 1 if it was 0, and vice versa).

Interpretation allows us to understand which property the tag author is addressing, and whether she thinks the property is present or not, but it does not say whether the tag author *likes* the presence or absence of that property. In the above example, I_1 and I_2 express opposite opinions on the *elegance* property. However, two iTₐGs may express the same opinion with opposite judgements: two iTₐG authors may both think that the resource or the facet has a given property, but one of them likes it, the other one does not. This is a difference in *polarity*. If we consider the *relationship r* between label and property, and the *polarity p* of the iTₐG author's impression, we have four different possibilities. Each of these possible combinations is called an *aspect* of the property. For example (see Figure 3), the four aspects of the *elegance* property could be represented by the labels *chic* ("it's elegant, I like this", $p = +$, $r = 1$), *sophisticated* ("it's *too* elegant, I don't like this", $p = -$, $r = 1$), *simple* ("it's not elegant, but I like this", $p = +$, $r = 0$), *shabby* ("it's not elegant at all, I don't like this", $p = -$, $r = 0$).

4   iTags Synthesis

The aim of iTₐG synthesis is to provide users with a comprehensive and immediate aggregation of what people think about a given object, and to offer an effective representation of the overall opinion, which is the most meaningful for the user. In doing this, we take into account:

- the existence of niches of people whose opinions differ from the majority[3];
- the *social network* of the target user, since other people's opinions weigh differently depending on how much the user trusts them on the topic[4];
- the *user model* of the target user, considering only the facets relevant for her.

In order to produce a meaningful synthesis that takes into account the above issues, we partition the set of iTₐGs associated with a given object O first according to the facet F and property P, and then according to the *aspect*, i.e. the (*relationship, polarity*) pair. The rationale behind this lies in our approach to synthesis, which is the following:

---

[3] As an example, suppose that 30 people out of 50 think that a restaurant is *cheap* while the remaining 20 think it is *expensive*. Going with the prevailing opinion would mean showing only the *cheap* fraction. On the other hand, computing an average of the users' impressions, imagining that *cheap and* expensive lie on the same scale of *cheapness* with opposite signs, would lead to showing something like *moderately cheap*. We think that none of these solutions correctly portrays the collective impression on the restaurant.

[4] In the case of a restaurant, one would probably trust more the impression of a well-known enogastronomic journalist, than those of her gym friend who usually goes for the cheapest meal around.

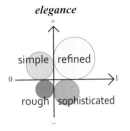

Fig. 3. The four synthesis aspects of the *elegance* property

Fig. 4. The synthesized itag for the *elegance* property

1. We merge all iT₄Gs that refer to the same facet, the same property, and have the same relationship and polarity. These iT₄Gs are essentially stating the same concept, only in different words (labels) and with different relevance (sizes). In order to merge them, we need to select a representative label and find an average size. As we will see, in doing this we will take into account the social network and the trust level.

2. We decide which facets we want to show to a given user, by considering her interest in them as expressed in the user model.

3. For such facets, we show the merged iT₄Gs for all four aspects, provided they are relevant enough (i.e. the resulting merged size is above a given threshold). We wish to show a minority's opinion only if it is a significant.

As an example of a possible result, Figure 3 shows the merged iT₄Gs for the four aspects of the *elegance* property in the *ambience* facet, while Figure 4 shows the visualization in the iT₄G system of the synthesized iT₄G. Where a single iT₄G appears as a circle around its facet, a synthesized iT₄G resembles a flower with at most four petals.

Let us now formalize how we compute a label L and a size S for a merged iT₄G. Recall that we merge a set of iT₄Gs that refer to the same object $\bar{O}$, to the same facet $\bar{F}$ and to the same property $\bar{P}$, and that belong to the same aspect $(\bar{r}, \bar{p})$. We will denote the set of such iT₄Gs as $\text{ITAGS}(\bar{O}, \bar{F}, \bar{P}, \bar{r}, \bar{p}) = \{\hat{I}_1, \dots, \hat{I}_n\}$.

Both label and size depend on the user u asking for the synthesis (*target user*), and on her social relationship with the iT₄Gs *authors*, since the u may value more the opinions of specific people (e.g. official experts or trusted people in her social network).

In what follows we clarify how people associated with the target user u influence the iT₄G synthesis. The people belonging to the model of the target user u are divided into groups $g_1^u, \dots, g_m^u$ called *trust groups*. Each group g_j^u, $j = 1, \dots, m$, has a *weight* $\omega_j^u \in [0, 1]$ associated with it. The user u trusts differently people in each group, e.g. she could have a group of "friends" weighted 0.6 and a group of "experts" weighted 1. The weight assigned by the target user to each group can be initially set to default values and then tuned according to her preferences or her behavior in the social application.[5] We denote by $\{a_1, \dots, a_k\}$ the tagging authors who have tagged the object O.

---

[5] Notice that the way the trust level is computed is out of the scope of this paper. It is understood that the trust values are topic dependent. In case there is no trust value, the approach works by considering a unique group.

Definition 3 (Trust value). *For the tagging author* a_j, $j \in \mathbb{N}$, *its* trust value $\tau_{a_j}^u \in [0, 1]$
w.r.t. the target user u is computed as:

1. *If* $\forall i \in \mathbb{N}$, $a_j \notin g_i^u$, $\tau_{a_j}^u = 0$;
2. *If* $\exists! i \in \mathbb{N}$, $a_j \in g_i^u$, $\tau_{a_j}^u = \omega_i^u$;
3. *If* $\exists k \in \mathbb{N}$, $a_j \in g_1^u \cap \cdots \cap g_k^u$ *and* $\omega_1^u, \ldots, \omega_k^u$, *are the weights for* $g_1^u, \ldots, gu_k$,

$$\tau_{a_j}^u = \omega_1^u + \sum_{j=2}^{k} \omega_j^u \prod_{h=1}^{j-1} (1 - \omega_h^u). \tag{1}$$

This means that: (i) the opinion of an author not belonging to any group of interest
for the target user is discarded; (ii) if the author belongs just to one interest group her
trust value is equal to the weight of that group; and (iii) if the author belongs to several
interest groups for the target user, her trust value is computed according to Formula 1.

The following definition computes a weight for each interpreted iṰG, taking into
account the sizes of the interpreted iṰGs, and the trust values of the tagging authors.[6]
The weight represents the contribution that each iṰG gives to the synthesis.

Definition 4 (Weight of interpreted iṰG). *For a given target user u, the weight of an*
interpreted iṰG $\hat{I}_j \in \text{iTags}(\bar{O}, \bar{F}, \bar{P}, \bar{r}, \bar{p}) = \{\hat{I}_1, \ldots, \hat{I}_n\}$ *is computed as follows:*

$$W(\hat{I}_j) = S_j \tau_{a_j}^u \Big/ \sum_{a \in \text{Auth}(\bar{O}, \bar{F})} \tau_a^u \tag{2}$$

- $\tau_{a_j}^u$ *is the trust value of the author* a_j *w.r.t. the target user u (see Eq. 1);*
- S_j *is the size of the* iṰG *provided by the author* a_j *(we assume* $S_j \in [\theta, 1], \theta > 0$*).*
- $\text{Auth}(\bar{O}, \bar{F})$ *is the set of the authors that tagged the facet* $\bar{F} \in \mathbb{F}(type(\bar{O}))$.

The weight of each iṰG can be computed w.r.t. all tagging authors, w.r.t. all tagging
authors who have tagged F, or w.r.t. the tagging authors who have tagged P. In our
opinion the best option to consider is the second one: a tagging user neglecting a whole
facet probably means she does not find it relevant, while a tagging user not mentioning a
property in a facet she is tagging probably means she has a neutral opinion with respect
to that property.

We use the computed weight to select a label L for the merged iṰG in the set
$\{L_1, \ldots, L_h\}$ of all labels used in $\text{iTags}(\bar{O}, \bar{F}, \bar{P}, \bar{r}, \bar{p})$. Given $\text{iTags}(\bar{O}, \bar{F}, \bar{P}, \bar{r}, \bar{p}) = \{\hat{I}_1, \ldots, \hat{I}_n\}$, then for a given label L_i we can assume without loosing generality that
$\{\hat{I}_1, \ldots, \hat{I}_{h-1}\}$ are the ones using L_i while $\{\hat{I}_i, \ldots, \hat{I}_n\}$ are the ones using some other label.

Definition 5 (Resulting label). *The* weight *associated with* L_i *is given by:*

$$W(L_i) = \sum_{j=1}^{h-1} W(\hat{I}_j) \tag{3}$$

---

[6] The sum of the trust values is not 1, therefore we use it to normalize the weighted size.

Then we select as label L for the merged iT$_A$G the L_i with the highest value for $W(L_i)$.

The size S of a merged iT$_A$G is calculated by adding the weights of the corresponding interpreted iT$_A$Gs. To avoid iT$_A$Gs of extremely small size, we introduce a minimum threshold $\theta \in (0, 1]$.

Definition 6 (Resulting size). *The size S of the merged* iT$_A$G *for* ıTAGS$(\bar{O}, \bar{F}, \bar{P}, \bar{r}, \bar{p}) = \{\hat{I}_1, \ldots, \hat{I}_n\}$ *is:*

$$S = \begin{cases} \sum_{i=1}^{n} W(\hat{I}_i) & if \sum_{i=1}^{n} W(\hat{I}_i) \geq \theta \\ 0 & otherwise \end{cases} \quad (4)$$

5 A Preliminary Evaluation of the Approach

In order to evaluate whether our iT$_A$G framework effectively achieves the goals of (i) allowing the communication of opinion and (ii) providing effective collective opinion's synthesis, we carried out a preliminary evaluations with users, targeted at answering the following research questions:

Research Question 1 (RQ1). Does our tagging framework allow people to express complex opinions? Is it intuitive to use and easily understandable?

Research Question 2 (RQ2). Is the personalized multi-facet synthesis more informative than traditional tag clouds?

We selected 38 users[7] which were divided into two groups. The first one, consisting of 18 participants, used the framework to provide and examine single iT$_A$Gs; their experience allowed us to investigate **RQ1**. The second group, with the remaining 20 participants, had to evaluate synthesized iT$_A$Gs and thus provide information concerning **RQ2**.

We asked the users in the first group to comment on 3 restaurants (tagging at least 2 facets for each restaurant), using iT$_A$Gs. Afterwards, all the users were asked to look at 3 iT$_A$Gs provided by other people, and describe their understanding of the other people's opinion on those restaurants.

The 20 users of the second group were then asked to chose 2 restaurants from a list of 20 restaurants that were tagged in the previous phase. Each restaurant had a detailed description of different facets (price, cuisine, ambient, etc). The tags for the restaurants of their choice were presented both as a traditional tag cloud and as synthesized iT$_A$Gs; the users were asked to compare the two presentations.

Finally, all users answered a questionnaire about their experience with the iT$_A$G system.

The goal of the evaluation was to answer address iT$_A$Gs' expressiveness and immediacy in conveying opinions, the overall usability of the interface, and the correct interpretation of iT$_A$Gs. More specifically we examined:

---

[7] Users were recruited among the contacts and colleagues of the authors, according to an availability sampling strategy. Even if non-random samples are not statistically representative, they are often used in psychology research and usability testing, during early evaluation phases.

1. *Expressiveness of* iT$_A$G *editing interface*. This implies two further questions:
   - Does the iT$_A$G framework allow users to freely express opinions?
   - Is the specific interface we developed usable?

   Regarding both questions, users had to provide a number on a 4 point scale (1 being "absolutely no" and 4 being "definitely yes"). For the first experiment, we obtained a mean of 3.3 and a mode of 4. For the second experiment we obtained a mean of 3.2 with a mode of 3. We can conclude that the system expressiveness is good.
2. *Expressiveness of* iT$_A$G *viewing interface*. This implies two further questions:
   - Does the iT$_A$G framework allow users to correctly understand the opinion the iT$_A$G authors wanted to communicate?
   - Are iT$_A$Gs immediate and do they communicate opinions at a glance?

   Users had the chance to express a first impression looking only briefly at the iT$_A$G, and then to examine in more depth the iT$_A$G structure possibly zooming over the smaller labels. All users correctly understood the taggers' opinions after the first brief examination. We also asked the users what they thought the circle size expressed. Most of the users (87% in the first experiment, 83% in the second one) answered that they interpreted the circle size as the relevance of the iT$_A$G with respect to the overall comment. The remaining people either saw no particular meaning associated to size, or thought it was a quantification of the label. Finally, we asked the users which was in their opinion the major advantage of iT$_A$G system (if any) with respect to traditional tagging systems. For 61% the advantage of iT$_A$G system is immediacy: opinions can be understood at a glance; for 17% the advantage is the possibility to refer the words to different facets of the resource. Hence, we can conclude that iT$_A$G framework is expressive and communicatively rich.
3. *Expressiveness of the synthesis*. 89% of the users preferred our synthesis to the traditional tag cloud, for the following reasons: (i) the overall opinion is clear at a glance (78%); (ii) it is quicker to read (60%); (iii) it is more informative (71%); (iv) it presents also the niche opinions (82%).

6   Related Work

Our work aims at enhancing tagging with capability of express complex opinions, in order to provide a personalized synthesis of tags in social applications. This implies (i) interpreting and synthesizing the tags, (ii) adapting the synthesis to target users and (iii) visualizing the synthesized tags.

Interpretation and synthesis. The interpretation techniques for tags depend on the tags typology (free, structured or facet-based tags). *Free tags* give the highest freedom to users, but they are difficult to process and interpret. Some work uses techniques from machine learning and artificial intelligence [13]; others use clustering methods or similar mathematical approaches [4]. Yet another approach is to map the tags to an existing domain ontology or semantic knowledge base as DBpedia [15], using some similarity measures to compute the distances between words from a syntactic [6] and semantic point of view [5]. *Structured tagging* provides more information, since it forces users to focus on a specific subject and to assign values to a set of predefined metadata fields (see BibSonomy [13] for documents and VRA Core Vr4[8] for multimedia). Although tag interpretation is easier for structured tagging, too much complexity discourages users from providing tags. To solve the processing problem, we adopted a compromise between freedom and structure using *facets* [17], which involve creating a bottom-up classification. There are several proposals for applying facets to social tagging applications mostly with the aim to classify a tag associating it to one or more facets [18]. In this case, facets are tag categories (people, time, place, etc.), possibly organized in a hierarchy, that can help to clarify the meaning of the tag [16]. Even though our facets are organized in a hierarchical structure, they do not serve as tag classifiers. They rather represent different features the user can comment on by expressing her opinion with the iT$_A$Gs.

Personalization. To our knowledge, there are no other proposals to personalize tag clouds according to trust measures. Some works in information retrieval proposed personalized search tools that return only the tags that agree with the user main interests given in her user profile [7]. In social bookmarking systems several authors propose to recommend tags to users, i.e., to propose the tags that better fit a user's needs. Our aim is different: we do not suggest resources to users, we rather provide a synthesis of the opinions of the people they trust. In this sense, our work is similar to trust-based recommender systems, which generate personalized recommendations by aggregating the opinions of users in the trust network. Even if it is not our main goal, another application of our work is to use iT$_A$Gs for recommending resources to users, as it has been recently proposed for example in collaborative filtering approaches, i.e., to recommend an item given the similarity of its tags with the tags used for another item she liked [9], or to compute users' similarity starting from the tags they used [14].

Visualization. In social web, the method used the most for visually representing information are fuzzy aggregations of *tag clouds* [12], where terms are organized in alphabetical order and presented in a compact space. Tag clouds enable visual browsing by showing the list of the most popular tags, alphabetically ordered and weighted by font size and thickness. The selection of tags to be shown in clouds is based on frequency, which results in high semantic density and in a limited number of different topics dominating the whole cloud [4]. Moreover, alphabetical order is convenient only when the user already knows what she is looking for. In fact, tag clouds facilitate neither visual scanning nor representation of semantic relationships among tags.

In [19] the authors present *tag expression*, a tag cloud-based interface that allows users to rate a movie with a tag and an associated feeling (like, dislike, or neutral) which measures the user's opinion about the movie. Similarly to us, the user can express an opinion on one of item features (tags in *tag expression* and facets in iT$_A$G), but the synthesis approaches are different. In fact, they simply flatten the opinions in a unique average value, not considering different points of view.

In order to overcome the limitations of traditional tag clouds, several methods have been proposed. [11] presents a new tag cloud layout that shows tag similarity[9] at a

---

[8] www.vraweb.org/projects/vracore4/VRA_Core4_Intro.pdf
[9] One easy and commonly used technique to evaluate the similarity of two tags is to count their co-occurrences, i.e., how many times they are used to annotate the same resource.

glance. Based on co-occurrence similarity, data clustering techniques are used to aggregate tags into clusters whose members are similar to each other and dissimilar to members of other clusters. The result is a tag cloud where semantically similar tags are grouped horizontally whereas similar clusters are vertical neighbors. Instead of clustering, we propose the four aspects layout as a form of iT$_A$G aggregation, since the aim of our synthesis is to summarize the impressions of different users about a certain property of a facet, by condensing agreeing options and by relating the opposite ones. An alternative approach is proposed by *tagFlakes* [8], a system that helps the user navigate tags in a hierarchical structure, where descendant terms occur within the context defined by the ancestor terms. Similarly, we adopt a form of aggregation, in which we group tags that refer to the same property. However, our aggregation is *multi-aspect*, in order to show and highlight disagreements, as well as similarities.

7   Conclusions and Future Work

We presented an enhanced tagging framework, called iT$_A$G, which allows the users of a social application to share their opinions on resources, and that allows for personalized synthesis of users' opinions, enabling easer understanding of a huge amount of users' tags on a resource. We introduce a method for interpreting tags, which partitions the tags according to the *property* of the resource they describe and their *relationship* with the given property. Next, we propose a personalized synthesis method which takes into account how much the user trusts the tag authors. We assume that the trust measure is provided by the social application using some existing state of the art approach (such as [2]). This is out of the scope of the paper. The main contributions of our work are the following:

(i) A novel tagging modality, which enables expressing complex opinions on a resource, and at the same time enables their interpretation by the system.

(ii) A method for iT$_A$Gs interpretation, which assigns iT$_A$Gs to specific properties, without using Natural Language Processing techniques.

(iii) A personalized synthesis of users' opinions using the above interpretation. The iT$_A$Gs synthesis is (i) *multi-faceted*, it maintains the differences in opinions of different users, and (ii) *personalized*, it is based on the user model and her social network.

(iv) A visualization of the synthesis that allows for an immediate understanding of the collective opinion.

Preliminary evaluation showed the users appreciated the iT$_A$G framework. We intend to perform a more rigorous and comprehensive evaluation in the future.

Another open problem is how to resolve the problem of labels polysemy in the process of mapping labels to properties.

Currently we are working on the use of iT$_A$Gs for recommendation purposes, by defining a notion of *distance* between iT$_A$Gs. The goal is twofold: being able to recommend a resource to a user due to the similarity between the resource reputation and the user's tastes, and being able to find similar users for social recommendation purposes. At the same time, we are working on exploiting such tag-based opinions to enrich the user model, also with dislike values.

As future work, we aim at investigating the possibility to combine our research with the results of the Sentiment Analysis field, which aims to determine the attitude of a person with respect to some topic. In particular, we plan to exploit in our framework the SentiWordnet system,[10] a lexical resource for opinion mining.

References

1. Andersen, R., Borgs, C., Chayes, J., Feige, U., Flaxman, A., Kalai, A., Mirrokni, V., Tennenholtz, M.: Trust-based recommendation systems: an axiomatic approach. In: 17th Int. Conference on World Wide Web, WWW 2008, pp. 199–208. ACM, New York (2008)
2. Aroyo, L., Meo, P.D., Ursino, D.: Trust and reputation in social internetworking systems. In: Proc. of the Workshop on Adaptation in Social and Semantic Web, at UMAP 2010 (2010)
3. Bar-Ilan, J., Shoham, S., Idan, A., Miller, Y., Shachak, A.: Structured versus unstructured tagging: a case study. Online Information Review 32(5), 635–647 (2008)
4. Begelman, G., Keller, P., Smadja, F.: Automated Tag Clustering: Improving search and exploration in the tag space. In: Collaborative Web Tagging Workshop at WWW 2006 (2006)
5. Cattuto, C., Benz, D., Hotho, A., Stumme, G.: Semantic grounding of tag relatedness in social bookmarking systems. In: Sheth, A.P., Staab, S., Dean, M., Paolucci, M., Maynard, D., Finin, T., Thirunarayan, K. (eds.) ISWC 2008. LNCS, vol. 5318, pp. 615–631. Springer, Heidelberg (2008)
6. Damerau, F.J.: A technique for computer detection and correction of spelling errors. Communications of the ACM 7(3), 171–176 (1964)
7. Daoud, M., Tamine, L., Boughanem, M.: A personalized graph-based document ranking model using a semantic user profile. In: De Bra, P., Kobsa, A., Chin, D. (eds.) UMAP 2010. LNCS, vol. 6075, pp. 171–182. Springer, Heidelberg (2010)
8. Di Caro, L., Candan, K.S., Sapino, M.L.: Using tagFlake for Condensing Navigable Tag Hierarchies from Tag Clouds. In: 14th ACM SIGKDD Int. Conference on Knowledge Discovery and Data Mining, KDD 2008, pp. 1069–1072. ACM, New York (2008)
9. Durao, F., Dolog, P.: A personalized tag-based recommendation in social web systems. In: Houben, G.-J., McCalla, G., Pianesi, F., Zancanaro, M. (eds.) UMAP 2009. LNCS, vol. 5535. Springer, Heidelberg (2009)
10. Fellbaum, C. (ed.): WordNet: An Electronic Lexical Database. MIT Press, Cambridge (1998)
11. Hassan-Montero, Y., Herrero-Solana, V.: Improving Tag-Clouds as Visual Information Retrieval Interfaces. In: Int. Conference on Multidisciplinary Information Sciences and Technologies, InScit 2006 (2006)
12. Hoeber, O., Liu, H.: Comparing tag clouds, term histograms, and term lists for enhancing personalized web search. In: Int. Workshop on Web Information Retrieval Support Systems at IEEE/WIC/ACM Int. Conference on Web Intelligence, pp. 309–313 (2010)
13. Hotho, A., Jäschke, R., Schmitz, C., Stumme, G.: Emergent semantics in bibsonomy. In: Informatik 2006 – Informatik für Menschen. Band 2. Lecture Notes in Informatics, vol. P-94. Gesellschaft für Informatik (2006)
14. Liang, H., Xu, Y., Li, Y., Nayak, R.: Tag based collaborative filtering for recommender systems. In: Wen, P., Li, Y., Polkowski, L., Yao, Y., Tsumoto, S., Wang, G. (eds.) RSKT 2009. LNCS, vol. 5589, pp. 666–673. Springer, Heidelberg (2009)
15. Mirizzi, R., Ragone, A., Di Noia, T., Di Sciascio, E.: Semantic tag cloud generation via dBpedia. In: Buccafurri, F., Semeraro, G. (eds.) EC-Web 2010. LNBIP, vol. 61, pp. 36–48. Springer, Heidelberg (2010)

---

[10] http://sentiwordnet.isti.cnr.it/

16. Quintarelli, E., Resmini, A., Rosati, L.: FaceTag: Integrating Bottom-up and Top-down Classification in a Social Tagging System. In: IA Summit 2007: Enriching Information Architecture (2007)
17. Ranganathan, S.R.: Prolegomena to Library Classification. Asia Publishing House, New York (1967)
18. Spiteri, L.: Incorporating facets into social tagging applications: An analysis of current trends. Cataloguing and Classification Quarterly 48(1), 94–109 (2010)
19. Vig, J., Soukup, M., Sen, S., Riedl, J.: Tag expression: tagging with feeling. In: 23nd ACM Symp. on User Interface Software and Technology, pp. 323–332 (2010)

Propagating User Interests in Ontology-Based User Model*

Federica Cena, Silvia Likavec, and Francesco Osborne

Dipartimento di Informatica, Università di Torino, Italy
{cena,likavec,osborne}@di.unito.it

Abstract. In this paper we address the problem of propagating user interests in ontology-based user models. Our ontology-based user model (OBUM) is devised as an overlay over the domain ontology. Using ontologies as the basis of the user profile allows the initial user behavior to be matched with existing concepts in the domain ontology. Such ontological approach to user profiling has been proven successful in addressing the cold-start problem in recommender systems, since it allows for propagation from a small number of initial concepts to other related domain concepts by exploiting the ontological structure of the domain. The main contribution of the paper is the novel algorithm for propagation of user interests which takes into account i) the ontological structure of the domain and, in particular, the level at which each domain item is found in the ontology; ii) the type of feedback provided by the user, and iii) the amount of past feedback provided for a certain domain object.

Keywords: User model, ontology, propagation of interests.

1 Introduction

In different areas of the Web, personalization and adaptation are crucial concepts nowadays, since they help users find what they really want and need. From e-commerce to e-learning, from tourism and cultural heritage to digital libraries, users benefit from tailoring the content and visualization techniques to their own needs. A simple way to capture the user preferences and/or interests and account for differences in needs of individual users is provided by user modeling. In user-adaptive and recommender systems [7,1], a User Model stores the available information about a user by maintaining user properties such as interests, preferences, knowledge, goals and other facts considered relevant for the application. The information in the user model is then used by adaptive systems, applying some reasoning strategies, to provide personalization services to each user (e.g. adapting the interface or the contents order, or recommending certain items to users).

User models can be constructed in different ways, e.g. by exploiting the information provided by users upon registration, clustering users in stereotypes, obtaining the information from other applications, deriving information from users' behavior etc. The

---

* This work has been supported by PIEMONTE Project - People Interaction with Enhanced Multimodal Objects for a New Territory Experience.

R. Pirrone and F. Sorbello (Eds.): AI*IA 2011, LNAI 6934, pp. 299–311, 2011.

latter case is particularly relevant, since users' interaction with the system can be implicitly monitored and recorded by the system, providing a rich source for further analysis with minimal user intervention.

There are several approaches to representing user models, from simple property-value pairs, to more complex probabilistic approaches, such as Bayesian networks. Very often the user model is conceived as an overlay over the domain model, where the user's current state (such as interest or knowledge) with respect to domain concepts is recorded [3]. For each domain model concept, an individual overlay model stores a value which is an estimation of the user's attitude to this concept. For example, an overlay model of user interests can be represented as a set of pairs ⟨concept, value⟩, with one pair for each domain concept.

Ontologies as explicit specifications of domain concepts and relationships between them are emerging as powerful formalisms for knowledge representation with associated reasoning mechanisms. In the last years, there is a growing tendency to use ontologies to represent the domain models. In this context, one of the promising approaches to represent a user model is by conceiving it as an *overlay over the domain ontology*.

In this paper we address the problem of propagating user interests in ontology-based user models. Our ontology-based user model (OBUM) is devised as an overlay over the domain ontology. Using ontologies as the basis of the user profile allows the initial user behavior to be matched with existing concepts in the domain ontology. Such ontological approach to user profiling has been proven successful in addressing the cold-start problem in recommender systems, since it allows for propagation from a small number of initial concepts to other related domain concepts by exploiting the ontological structure of the domain [9].

The main contribution of the paper is the novel algorithm for propagation of user interests which takes into account:

- the ontological structure of the domain and, in particular, the level of the object receiving the user feedback in the ontology;
- the type of feedback provided by the user;
- the amount of past feedback provided for a certain domain object.

In addition, our approach allows for bi-directional propagation of user interests in the ontology, i.e. both bottom-up and top-down. This approach contributes to resolution of the cold-start problem, thus improving the adaptation of the system at the beginning. It also alleviates the sparsity problem (i.e., the much bigger number of items than the number of items rated by users) which often plagues recommender systems.

The rest of this paper is organized as follows. We begin by presenting our specific algorithm for propagating user interests on an ontology in Sect. 2. Then, we describe how we apply our approach on an existing application in Sect. 3, followed by the results of a preliminary evaluation in Sect. 4. In Sect. 5 we present some related work and background. Finally, we conclude and give some directions for future work in Sect. 6.

2   Our Approach

In this paper, we propose an approach to propagating user interests in a domain ontology, starting from the user behavior in the system. Our approach is based on the following requirements:

- the domain model has to be represented using an *ontology*, where domain items are modeled as instances of the concepts in the ontology;[1]
- the user model has to be represented as an overlay over the domain ontology;
- user interaction with the system has to be recorded in order to infer user interests.

We chose to use ontologies to represent the objects of the domain for several reasons. Ontologies guarantee exact semantics for each statement, and avoid semantic ambiguities. Since they are expressed with standard formats and technologies, they allow for extensibility and re-usability. But most importantly, their structure allows for powerful and rigorous reasoning over data (inheritance, subsumption, classification etc.).

Regarding the user model, we decided to employ an *ontology-based user model*, where user interests are recorded for the classes of the domain ontology. Each ontological user profile is an instance of the reference ontology where every domain object in the ontology has an interest value associated to it. This means that each node N in the domain ontology can be seen as a pair $\langle N, \mathcal{I}(N) \rangle$, where $\mathcal{I}(N)$ is the interest value associated to N denoting the user interest in the object represented with the node N.

As for the last point, we chose to infer user interests indirectly by observing user behavior with the system. When interacting with the system, users provide valuable feedback about their interests that the system records implicitly and can use to incrementally create (and update) the user model by modifying the interest values for certain domain objects. According to Kobsa [7], possible user actions are: selecting, tagging, rating, commenting or bookmarking an item. Each of these actions is assigned a certain weight f, according to its strength as a signal of user interest. For example, clicking on a certain domain object denotes less interest than bookmarking it. All these actions are being registered in the log files in order to permit for their later retrieval and further analysis. Our idea is to use this feedback to:

- infer user interest in an object receiving the feedback;
- calculate the interests in other related domain objects, such as ancestors or descendants.

Each time a user provides a feedback, the following steps take place. First, we calculate the level of interest in the node that received the feedback (*sensed interest*). The sensed interest is added to the initial node interest and used for the subsequent propagation phase. Then, starting from this value, we calculate the *propagated interest* for the nearest nodes. During the propagation, the algorithm traverses vertically the ontology graph, and for each node it meets, the original interest is incremented by the *propagated interest*, which depends on the sensed interest, the distance from the initial node and the amount of the past feedback received.

The user interest $\mathcal{I}$ in a certain item is calculated as follows:

- for the node which receives direct user feedback:

$$\mathcal{I}(N) = \mathcal{I}_{\mathbf{o}}(N) + \mathcal{I}_{\mathbf{s}}(N) \tag{1}$$

- for the node which receives an interest value propagated vertically:

$$\mathcal{I}(M) = \mathcal{I}_{\mathbf{o}}(M) + \mathcal{I}_{\mathbf{p}}(M) \tag{2}$$

---

[1] For the time being, our method is designed for a static domain, where dynamic modification of the domain is not managed.

where
  - $\mathcal{I}_\mathbf{O}$ (**old interest**) is the old value for the user interest (initially equal to zero);
  - $\mathcal{I}_\mathbf{S}$ (**sensed interest**) is the value obtained from the direct feedback of the user;
  - $\mathcal{I}_\mathbf{p}$ (**propagated interest**) is the value obtained by vertical propagation.

We explain below how to calculate sensed interest and propagated interest.

2.1 Inferring User Interest (Sensed Interest)

Sensed interest is the value that shows how much of a direct feedback from the user the given node "senses". Thus, we introduce the concept of a *sensitivity* of a node. The sensitivity of a certain node depends on its position in the ontology: if the user provides feedback about the node lower down in the ontology, the effect is stronger than when the feedback is received for the node higher up, requiring a lower amount of feedback as a signal of strong user interest for a lower node. In fact, since the lowest nodes in the ontology represent specific concepts, they signal more precise interest than interest expressed in more general concepts, represented by upper classes in the ontology. For example, declaring interest in `Sparkly_White_Wine` gives more precise information with respect to declaring interest in generic item `Wine`).

In order to calculate the interest sensed by a given node N, we use the following formula (adaptation of Stevens' power law [18] used to relate the intensity of the physical stimulus and its sensed intensity):

$$\mathcal{I}_\mathbf{S}(N) = \frac{l(N) + 1}{max + 1}\mathsf{f}(N)^b \tag{3}$$

where $l(N)$ is the level of the node that receives the feedback, max is the level of the deepest node in the ontology, $\mathsf{f}(N)$ is the feedback obtained from the user for the node N, $b \in \mathbb{R}$ is a constant ($0 < b < 1$) which controls how strongly the node senses each different type of direct feedback. Thus, it is possible to account for the case where one type of feedback has a stronger impact than the other, but also to keep b constant so as to perceive all user actions equally.

For example, if a leaf node N ($l(N) = max$) receives the first feedback $\mathsf{f} = 5$ its sensed interest is $\mathcal{I}_\mathbf{S}(N) = 5^b$. Since $\mathcal{I}_\mathbf{O}(N) = 0$, the cumulative interest $\mathcal{I}(N) = 5^b$. When the same node receives the second feedback $\mathsf{f} = 10$, $\mathcal{I}_\mathbf{S}(N) = 10^b$ and $\mathcal{I}(N) = 5^b + 10^b$.

2.2 Propagating User Interest (Propagated Interest)

The main idea is that the effect of a feedback for a given node (i.e. the interest in a certain domain object) can be propagated *vertically* in the ontology, upward to the ancestors as well as downward to the descendants of a given node[2]. For example, if a person is interested in red wine `Barbera_d'Asti`, it is safe to assume that the same person might be interested in a specific kind of this wine, such as `Vietti_Barbera_d'Asti_Tre_Vigne_2008`, but also in `Red_Wine` in general. Of course, the interest for an object that can arise from this assumption is less strong than the original and will depend on the

---

[2] This idea is based on the similarity derived from IS-A relations. Concepts connected by IS-A relations are somehow similar, according to taxonomy based approach [13], since descendant concepts are subclasses of a class and the subclasses inherit attributes of the upper classes.

conceptual distance between the two nodes. Therefore we can assume that the interest we can propagate vertically is inversely correlated to the distance from the node that received the feedback. In order to correctly propagate the interests, the contributions of the various nodes to the propagation must be balanced. In particular, the node contribution to the interest propagation process should be decreasing with the amount of feedback already spread by that node. This is needed to prevent the non-proportional propagation of interests when the user concentrates solely on one or a few items. In that case, this input is redundant and does not add any information. On the contrary, we want to reward those cases where an ancestor node receives feedback from a good part of its sub-nodes, showing a consistent interest for that class.

The *propagated interest* is the value of "indirect" interest that a node can receive as a result of vertical propagation in the ontology. It is calculated modulating the sensed interest $\mathcal{I}_s(N)$ of the node N that receives the feedback by the exponential factor which describes the attenuation with each step up or down (simulating attenuation in physics), and a weight inversely correlated with the amount of feedback already received by the node as follows:

$$\mathcal{I}_p(M) = \frac{e^{-kd(N,M)}}{1 + \log(1 + n(M))} \mathcal{I}_s(N) \tag{4}$$

where $d(N, M)$ is the distance between the node N receiving the feedback and the node M receiving the propagated interest, $n(M)$ is the number of actions performed in the past on the node M and $k \in \mathbb{R}$ is a constant. Varying the attenuation coefficient k, it is possible to control how much interest is propagated depending on the type of feedback received: for instance, favoring the feedback resulting from bookmarking rather than the feedback from simply visiting the page.

3 Use Case

We exploited our approach in gastronomic domain, using the *WantEat* application [10] as a use case. *WantEat* is a part of an ongoing project which aims at providing a "Social Web of Entities", where a network of people and intelligent objects is built to enable their mutual interaction. People can interact in a natural way with these objects, accessing information and stories about them. They can navigate social networks of objects and people, annotate, rate, bookmark and comment the objects. The behavior of the system is adaptive, since it personalizes the interaction according to the preferences of individual users. In particular, the order of the objects presented to the users takes into account the user model.

The domain of the project is gastronomy: enhanced objects can be products such as cheeses and wines, as well as places of origin, shops, restaurants, market stalls etc. Such a domain is represented using an upper ontology which imports additional sub-ontologies describing particular areas of the domain (products, recipes, actors, places). The domain objects are modeled as the instances of the classes of the ontology.

Following the requirements of our approach described in Sect. 2, the user model is represented as an overlay over such an ontology, associating, for each user, an interest value for every domain class in the ontology. Such interests are derived by the system

Table 1. Weights associated to user actions

Action	Weight
Bookmarking an object	9
Tagging an object	7
Commenting on an object	5
Rating an object	1*vote

from the feedback the user provides in his interaction with a given object, which can be a certain category (e.g. wine, cheese, etc.) or a specific item (e.g. a specific wine or cheese). There are five possible typologies of user feedback in the system: (i) clicking on an object, (ii) tagging, (iii) commenting, (iv) voting on a scale from 1 to 5, or (v) putting an item into favorites. Each type of feedback has different impact on the user model, based on the strength of interest indication. The lowest interest indicator is obtained from clicking on a certain domain item, followed by commenting, tagging, voting and putting an item into favorites (this is the indicator of the highest interest). All these actions are registered in the log files and analyzed. According to the strength of interest indication, each of these actions is assigned a certain weight f, following the approach developed in [4]. For example, clicking on a certain domain object can have $f = 3$, whereas tagging can have $f = 8$ (see Table 1).

Hence, the users indicate their interests directly for the domain objects. We want to use these values to calculate users' interests in such objects, but also the interests in some other related domain objects.

After having collected the users' feedback, we applied our approach for inferring interests and propagating them, taking into account (as explained in Sect. 2):

1. the type of feedback provided by the user (see Table 1);
2. the amount of feedback (the number of actions the user performed while interacting with the system);
3. the level in the ontology of the object that received the feedback.

As an example, let us consider the following scenario. Tom uses the application to find a good restaurant nearby and he discovers a restaurant famous for its cold cuts. Since he particularly likes ham, he bookmarks Prosciutto_Crudo (raw ham). With this information, the system is able to a) infer a value of interest for this class, and b) infer the level of interest in the related classes.

The bookmarking process assigns a feedback of 9 to the class Prosciutto_Crudo (according to Table 1). Prosciutto_Crudo is a leaf at the maximum level of the ontology ($l(PC) = max$, see Fig. 1) and we assume $b = 0.5$. Thus, $I(PC) = I_S(PC) = 9^{0.5} = 3$. Starting from this sensed interest, we propagate this value according to (4). Since the class is a leaf, the propagation is directed only to the ancestor classes.

The first ancestor class we meet is Salumi_Interi_Crudi (whole raw cold cuts) (see Fig. 1). We assume that the distance between the nodes is equal to the number of edges between them (1 in this case) and $k = 0.8$. Besides, the class Salumi_Interi_Crudi did not receive any feedback in the past. The system will derive the interest of the node $I(SIC) = I_p(SIC) = I_S(PC)e^{-0.8} = 1.35$. The second ancestor class we meet is Salumi_Interi. Let us assume that this class has already received 4 feedback

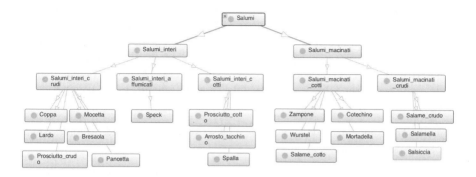

Fig. 1. The portion of domain ontology representing cold cuts

in the past. Hence, it will receive $\mathcal{I}_{\mathbf{p}} = 3e^{-1.6}/1.7 = 0.36$. The propagation phase stops when we get to the root.

The same process occurs for all other feedback. Of course, using other values for k and b we can obtain different propagation behavior. We can see that, as Tom simply expresses his interest for `Prosciutto_Crudo`, the system starts to infer his interest in all the correlated classes. After a little feedback the application will be able to suggest to Tom other objects he probably likes: for example, it will know that he prefers `Salumi_Interi_Crudi` to `Salumi_Interi_Cotti` (whole cooked cold cuts). The propagation process works also the other way around, hence the application will be ready to suggest similar objects like `Speck`, using the feedback Tom may have given on a ancestor class like `Salumi_Interi_Affumicati` (whole smoked cold cuts).

4   Evaluation of the Approach

We used the application described in the previous section to test our approach for propagating interest values in the ontology. In particular, the evaluation was performed on the portion of the ontology described in the above scenario (see Fig. 1). One of the most important personalization operations performed by the application is the ordering of objects returned by a search or displayed during the navigation. Producing a list that mirrors the user's preferences is a challenging task which provided a valuable testing environment for our algorithm.

Note that in our approach, we consider both the direct feedback provided for a certain object and the propagated interest. However, in the evaluation we decided to only focus on the contribution provided by vertical propagation. This may be thought of as a borderline case, in which we have no feedback on the objects to order, thus we have to rely solely on the propagation technique presented above.

Hypothesis. We assumed that our algorithm can be used to predict user's interests in certain objects of a given domain, starting from various classes in the ontology and propagating the interest to the related concepts (ancestors and descendants of a given class). We wanted to compare the lists of domain items generated by our algorithm with the ordered lists provided by the users themselves.

Experimental Design. We designed a questionnaire to collect users' preferences. It was divided into two parts, in order to test both upward and downward propagation. In the first part, the users were asked to rate 8 non-leaf classes of our ontology (e.g. Salumi_Interi_Cotti) on a scale from 1 to 10. In the second part, they had to indicate 4 leaves (e.g. Prosciutto_Crudo, Speck etc.) they "like very much" (simulating the bookmarking action) and four leaves they "like enough" (simulating the tagging action) among the total of 15 leaf classes.

Subjects. The sample included 92 subjects, 19-45 years old, recruited according to an availability sampling strategy[3].

Measures and Material. Users were asked to connect to the web site with written instructions and compile the anonymous questionnaire. Users' ratings were registered in a database. The evaluation followed two phases:

- We tested the *downward* propagation by generating a list of leaves based on the feedback provided for the higher classes and comparing it with the list provided by the user.
- We tested the *upward* propagation by generating a list of classes based on the feedback provided for the leaves and comparing it with the list provided by the user.

To compare the two lists we used the Pearson's correlation coefficient r between ranks, that measures the degree of association between two ordered lists of items[4].

Results. We have collected the total of 1,472 ratings and we calculated Pearson's correlation coefficient for 92 pairs of lists. Note that $-1 \leq r \leq 1$, where $r = 1$ corresponds to perfect association (the two lists follow the same order) and $r = -1$ corresponds to perfect inverse association.

a) Downward propagation. Since the leaves that could be voted were eight, whereas the options to signal the degree of user interest were only two, we have several tied ranks and the degree of association r in the downward propagation test is restricted to 5 values. These values correspond one to one to the number of errors in the generated lists: $r = 1$ corresponds to no errors, $r = 0.5$ to 1 error, $r = 0$ to 2 errors, and so on. Figure 2 shows the distribution of cases for various values of r. Notice that in 73% of the cases we were able to generate the lists with no more than 1 error. In almost one fifth of the cases we were able to generate the list equal to the user list. Only 25% of the cases had no benefit from the technique. The application of the downward propagation has almost never negative effects: only two cases showed a medium inverse association and not one a strong inverse association.

b) Upward propagation. The ordered lists used for testing the upward propagation were constructed by choosing 4 bottom classes the users "liked very much" and 4 bottom classes the users "liked enough". We assigned numerical values on a scale from

---

[3] Even though random sampling is the best way of having a representative sample, these strategies require a great deal of time and money. Therefore much research in social science is based on samples obtained through non-random selection, such as the availability sampling, i.e. a sampling of convenience, based on subjects available to the researcher, often used when the population source is not completely defined.

[4] In this case it is not correct to talk about linear correlation since there are no independent and dependent variables related by a linear relation.

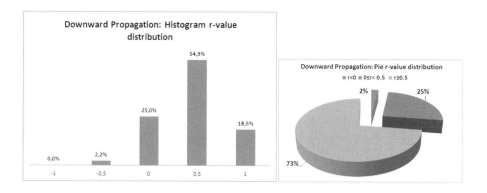

Fig. 2. Upward propagation: the distribution of cases for various values of r

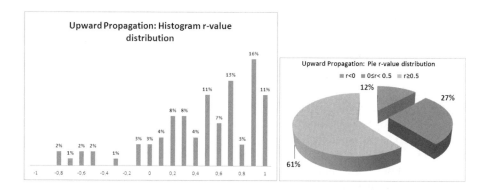

Fig. 3. Upward propagation: the distribution of cases for various values of r

1 to 10 to "like very much" (9) and "like enough" (7) to simulate bookmarking and tagging. Then we generated personalized lists using these 8 feedback. Figure 3 shows the distribution of cases for various association coefficients r, which in this case cover 20 values. We can see that in 88% of the cases our algorithm was able to generate a list with a positive association with the one provided by the user. In 61% of the cases $r \geq 0.5$. More impressive, in 27% of the cases we obtained $r \geq 0.9$, value which is usually considered "very good". Only in 12% of the cases $r < 0$.

Discussion. The preliminary tests described above aimed at evaluating the possibilities of our algorithm, in particular a propagation technique that works both upwards and downwards. Both propagation modalities gave positive results. For a great number of the cases we were able to generate a perfect list or a list with a high degree of association. The cases that could present the user with a misleading list are practically zero for the downward propagation and very low for the upward one.

5   Related Work

The most common way to model the user in recommender systems is to use a vector of items ratings, and to provide recommendations comparing these vectors, in order to find similar users (collaborative-filtering approach) or similar items (content-based approach) [1]. Rather than using simple feature vector models and computing user similarity on this whole set of items, in our work the definition of user profiles is relative to the ontology, giving rise to *ontological user profiles*. This allows for association of user interests with relevant concepts in the ontology, rather than single atomic entities, and for a dynamic update of such user interests in the ontology.

Similar work exploiting ontological user profiles can be found in [16,9].

As in our case, in [16] the ontological user profiles are instances of the pre-existing domain ontology. Similarly to us, their algorithm incrementally updates user profiles, based on the results of user interaction with the system. To update the interest values they exploit a spreading activation approach, thus treating the ontology as a semantic network, and propagate the values only in one direction. Instead, we perform a bi-directional propagation of the values. They do not take into consideration the amount of feedback on each node, the type of feedback received and the level of the node in the ontology.

Similar approach is the one of [9], where the user feedback on research papers is collected and effects the interest in ontological topics in the user model. Relationships between topics in the ontology are exploited to infer other topics of interest. Differently from us, they propagate the interest in a static manner: the interest value for a specific class is spread to the super-class always as 50% of its value. Another difference is that they use a time decay function, which weights more the recently seen papers than the older ones, while we weight more the past interaction with respect to the new one.

Another work which performs a similar value propagation in the ontology is the one of [15], where the authors use spreading activation techniques combined with classical search in order to improve the search process in a certain semantic domain. They apply their technique using a hybrid instances network, where each relation instance is assigned both a semantic label and a numerical weight based on certain properties of the ontology. The initial values for the spreading are obtained using classical search techniques on ontological concepts and used further to find related concepts. This technique makes it possible to find concepts which do not contain any of the words specified in the initial search. A point to notice is that they also use the constant that functions as attenuation factor, diminishing the activation with each processed node.

Other works propose the use of ontological user models in a different sense with respect to us. They create a specific ontology to represent the user features in the user model (demographic features such as age, gender, profession, etc). Examples are the General User Model Ontology (GUMO) [6] and the Unified User Context Model (UUCM) [8]. Differently from them we only use the ontology to model the domain, and represent the user model as an overlay over such a domain ontology. As a mixed approach which uses ontologies for user and domain modeling, we can cite Onto-bUM [12]. In addition to Domain Ontology, authors introduce the User Ontology which describes different features of the users, as well as the Log Ontology which describes the user interaction.

The propagation of values on an ontology can occur only among similar concepts. The similarity of two concepts in an ontology can be measured in different ways: by using the *information content* in an IS-A taxonomy (given by the negative logarithm of the probability of occurrence of the class in a text corpus) [14], by using the ontology graph structure, i.e. calculating the *distance between nodes* [11], by using RDF(S) or OWL DL primitives such as *rdf:id*, *owl:subClassOf* etc. to estimate partial similarities between concepts [2], etc. In this paper, we consider only the IS-A based similarity.

6   Conclusions and Future Work

This paper describes a promising approach to vertical propagation of user interests in an ontology-based user model. Starting from a given node in the ontology it is possible to propagate user interests to its ancestors and descendants. The novelty of our algorithm stems from the fact that we take into account the ontological structure of the domain, letting the level of each node influence the propagated interest values (the nodes lower down in the ontology can sense and propagate more). The high amount of past feedback for certain nodes helps decrease the propagation process for these nodes and prevents the non-proportional propagation of interests throughout the ontology. In addition, there is a possibility to treat various types of feedback differently, hence giving them different levels of importance.

The main contributions of our work are the following:

1. solution of the cold start problem - it is easier to obtain interest values for more domain items even with low amount of the user feedback;
2. alleviation of the sparsity problem - the number of items with associated interest values increases faster;
3. improvement of the recommendation accuracy - the items recommended to users mirror more closely their interests.

We tested our approach in a preliminary evaluation, obtaining satisfactory results for both downward and upward propagation. When producing the lists simulating user interests, 73% of the created lists for downward propagation and 61% for upward propagation had correlation coefficient $r \geq 0.5$. For the downward propagation, $r = 1$ for 18.5% of the cases and for the upward propagation $r \geq 0.9$ for 27% of the cases.

The next step will be the evaluation of our approach in more realistic conditions of usage of the system, in order to take into account other feedback typologies. We intend to test our framework at Bra "Cheese" festival in September 2011. We also plan on evaluating our approach in a different domain, represented by a different ontology, since the way the ontology is built can be the bottleneck of the approach. In fact, the approach is highly dependent on the knowledge representation.

Another possible future direction is to exploit ontological user profile and our propagation of interests to compute users' similarity, as in [17]. Since the propagated interest depends on the distance between the node that receives the direct feedback and the one that receives the propagated feedback, it would be interesting to see how our approach would behave when the notion of exponentially decreasing edge lengths in the ontology is used (see [5]). Another interesting feature that we would like to explore is adding constraints to our propagation algorithm along the lines of [15], such as concept type constraints (no propagation to certain kinds of nodes) or distance constraints (stopping the propagation upon reaching the nodes too far from the initial node). Note that in this version we are not considering compound concepts modeled by "part-whole" relations. We intend to achieve this by considering properties of domain items in the ontology, thus enabling also "horizontal propagation" among concepts. Finally, we plan on introducing some temporal aspects to the propagation mechanism.

References

1. Adomavicius, G., Tuzhilin, A.: Toward the next generation of recommender systems: A survey of the state-of-the-art and possible extensions. IEEE Transactions on Knowledge and Data Engineering 17(6), 734–749 (2005)
2. Bach, T.-L., Dieng-Kuntz, R.: Measuring Similarity of Elements in OWL DL Ontologies. In: Proc. of the AAAI 2005 Workshop on Contexts and Ontologies: Theory, Practice and Applications (2005)
3. Brusilovsky, P.: Methods and techniques of adaptive hypermedia. User Modeling and User Adapted Interaction 6(2-3), 87–129 (1996)
4. Carmagnola, F., Cena, F., Console, L., Cortassa, O., Gena, C., Goy, A., Torre, I., Toso, A., Vernero, F.: Tag-based user modeling for social multi-device adaptive guides. User Modeling and User-Adapted Interaction 18, 497–538 (2008)
5. Chiabrando, E., Likavec, S., Lombardi, I., Picardi, C., Theseider Dupré, D.: Semantic similarity in heterogeneous ontologies. In: Proc. of the 22nd ACM Conference on Hypertext and Hypermedia, Hypertext 2011, pp. 153–160. ACM, New York (2011)
6. Heckmann, D., Schwartz, T., Brandherm, B., Schmitz, M., von Wilamowitz-Moellendorff, M.: GUMO - the General User Model Ontology. In: Ardissono, L., Brna, P., Mitrović, A. (eds.) UM 2005. LNCS (LNAI), vol. 3538, pp. 428–432. Springer, Heidelberg (2005)
7. Kobsa, A., Koenemann, J., Pohl, W.: Personalized hypermedia presentation techniques for improving online customer relationship. The Knowledge Engineering Review 16(2), 111–155 (2001)
8. Mehta, B., Niederée, C., Stewart, A., Degemmis, M., Lops, P., Semeraro, G.: Ontologically-Enriched Unified User Modeling for Cross-System Personalization. In: Ardissono, L., Brna, P., Mitrović, A. (eds.) UM 2005. LNCS (LNAI), vol. 3538, pp. 119–123. Springer, Heidelberg (2005)
9. Middleton, S.E., Shadbolt, N.R., De Roure, D.C.: Ontological user profiling in recommender systems. ACM Transactions on Information Systems 22, 54–88 (2004)
10. PIEMONTE Team. WantEat: interacting with social networks of intelligent things and people in the world of enogastronomy. In: Interacting with Smart Objects Workshop 2011 (2011)
11. Rada, R., Mili, H., Bicknell, E., Blettner, M.: Development and application of a metric on semantic nets. IEEE Transactions on Systems Management and Cybernetics 19(1), 17–30 (1989)
12. Razmerita, L., Angehrn, A., Maedche, A.: Ontology-based user modeling for knowledge management systems. In: Brusilovsky, P., Corbett, A.T., de Rosis, F. (eds.) UM 2003. LNCS, vol. 2702, pp. 213–217. Springer, Heidelberg (2003)
13. Resnik, P.: Using information content to evaluate semantic similarity in a taxonomy. In: Proc. of the 14th Int. Joint Conference on Artificial Intelligence, pp. 448–453 (1995)
14. Resnik, P.: Semantic similarity in a taxonomy: An information-based measure and its application to problems of ambiguity in natural language. Journal of Artificial Intelligence Research 11, 95–130 (1999)

15. Rocha, C., Schwabe, D., Aragao, M.P.: A hybrid approach for searching in the semantic web. In: Proc. of the 13th Int. Conference on World Wide Web, WWW 2004, pp. 374–383. ACM, New York (2004)
16. Sieg, A., Mobasher, B., Burke, R.: Web search personalization with ontological user profiles. In: Proc. of the 16th ACM Conference on Information and Knowledge Management, CIKM 2007, pp. 525–534. ACM, New York (2007)
17. Sieg, A., Mobasher, B., Burke, R.: Improving the effectiveness of collaborative recommendation with ontology-based user profiles. In: Proc. of the 1st Int. Workshop on Information Heterogeneity and Fusion in Recommender Systems, HetRec 2010, pp. 39–46. ACM, New York (2010)
18. Stevens, S.: On the psychophysical law. Psychological Review 64(3), 153–181 (1957)

Integrating Commonsense Knowledge into the Semantic Annotation of Narrative Media Objects

Mario Cataldi[1], Rossana Damiano[1], Vincenzo Lombardo[1],
Antonio Pizzo[2], and Dario Sergi[3]

[1] Dipartimento di Informatica and CIRMA, Università di Torino, Italy
[2] Dipartimento Dams and CIRMA, Università di Torino, Italy
[3] Show.it, Torino, Italy
{cataldi,rossana,vincenzo}@di.unito.it, antonio.pizzo@unito.it,
sergi@show.it

Abstract. In this paper we present an innovative approach for semantic annotation of narrative media objects (video, text, audio, etc.) that integrates vast commonsense ontological knowledge to a novel ontology-based model of narrative, Drammar (focused on the dramatic concepts of 'character' and 'action'), to permit the annotation of their narrative features.

We also describe the annotation workflow and propose a general architecture that guides the annotation process and permits annotation-based reasoning and search operations. We finally illustrate the proposed annotation model through real examples.

Keywords: Media annotation, narrative annotation, ontology.

1 Introduction

Nowadays, in a web 2.0 reality, everyone produces stories by creating new materials or by editing and re-interpreting existing media objects (video, text, audio, etc). A large part of these materials are very likely to have a narrative content of some kind. Consider, for example, home made fiction and personal stories in YouTube videos or narratives in blogs and news sites: with such an explosion of contents, search tools for narrative are required. As a consequence, the issue of efficient annotation of the narrative contents is becoming critical.

The narrative format, then, is intrinsically relevant: according to cognitive psychologists [2], it corresponds to the way people conceptualize stories. In cross-media communication, the narration of stories from the standpoint of characters (so, in a *dramatic* fashion) has become a shared feature of media, as pointed out by [7,23].

Currently, many retrieval systems exploit the freely associated user generated annotations (tags, descriptions and meta-data) to provide access to the contents [13]; unfortunately, these annotations are often lacking or loosely related to the story incidents. For example, consider a movie segment in which a character

R. Pirrone and F. Sorbello (Eds.): AI*IA 2011, LNAI 6934, pp. 312–323, 2011.
© Springer-Verlag Berlin Heidelberg 2011

escapes from a prison; usually, the annotation of this narrative object relies on its perceivable properties: the actor who plays the character, his physical aspect, the location of the prison, the name of the director. The narrative features of the segment (the actions of the character and the purpose of these actions) are not accounted by current approaches, while they are useful for retrieval and editing, in a search and reuse perspective.

In this paper we present a novel semantic annotation system of narrative media objects (video, text, audio, etc.), centered on the two notions of *character* and *action*. This system is part of the CADMOS project, (Character-centred Annotation of Dramatic Media ObjectS)[1]; the ultimate goal of CADMOS is to test the benefits of narrative annotation for the production and reuse of narrative multimedia. In order to limit the arbitrariness of the annotation, we leverage a large–scale ontology-based semantic knowledge. In this way, the resulting annotation can be used for advanced retrieval operations to provide a more flexible and efficient access to the media objects.

The paper is structured as follows: we first survey, in Section 2, the relevant literature on story in computational systems; in Section 3 we describe the architecture for annotation and retrieval of narrative media objects. Then, in Section 4 we describe the narrative ontology, its theoretical background and the implementation, while in Section 5 we analyze how the narrative annotation of media object is constrained to shared semantics by integrating large commonsense ontologies. Finally, in Section 6 we present a real case study based on the annotation of Alfred Hitchcock's movie "North by Northwest" (1959).

2 Related Work

In this section, we survey some relevant research projects that employ some form of annotation for the analysis and production of media object. These projects, mostly targeted on video, encompass some narrative concepts as part of the annotation they rely on. In Section 4, we review the literature on narrative and drama studies that support the basic tenets of the narrative model underlying our approach.

The Advène project [21] addresses the annotation of digital video, and is not specifically targeted to the narrative content. In Advène, the fragments of the video are annotated with free textual description of the content, cross-segment links, transcribed speech, etc. This information can be exploited to provide advanced visualization and navigation tools for the video. For example, as a result of the annotation, the video becomes available in hypertext format. The annotation is independent from the video data and is contained in a separate package that can be exchanged on the net.

Complementary to this effort, the EU-funded ANSWER project[2], aims at defining a formal language for script and scene annotation. However, the

---

[1] Cadmos project is funded by Regione Piemonte, Polo di Innovazione per la Creatività Digitale e la Multimedialità, 2010–2012, POR-FESR 07–13.

[2] http://www.answer-project.org/

perspective of ANSWER is not the annotation of existing material for search and reuse, but the pre-visualization of a media object to be produced, with the aims of helping the author to pursue her/his creative intent and of optimizing the production costs. Again, ANSWER does not address the narrative aspects, but rather the filmic language by which the narrative will be conveyed. This choice is explicit in the project design, since it relies on the semantic layer provided by a Film Production ontology. This ontology constitutes the reference model for the Director notation, the input language for the pre-visualization services.

A media independent project is provided by the OntoMedia ontology, exploited across different projects (such as the Contextus Project [10]) to annotate the narrative content of different media objects, ranging from written literature to comics and tv fiction. The OntoMedia ontology [11] mainly focuses on the representation of events and the order in which they are exposed according to a timeline. In this sense, it lends itself to the comparison of cross-media versions of the same story, for example, a novel and its filmic adaptation, while it does not cover in a detailed way the role of the individual characters and the description of their behavior.

The CADMOS project shares with these approaches the basic assumption that a media object can be segmented into meaningful units and, given some kind of formal or semi-formal description, the units can be accessed and navigated. Moreover, it does not restrict its interest to video, although it recognizes that video constitutes a most suitable test bed for a character-based, narrative annotation: being the most representative example of what Esslin terms 'dramatic media', i.e., media displaying live action [7], the video medium assigns the characters a primary role in the exposition of stories.

Finally, the objectives of the I-Search project (a unIfied framework for multimodal content SEARCH) partially overlap with those of CADMOS. I-SEARCH defines a unified framework, called Rich Unified Content Description (RUCoD) which describes a Content Objects (be it a textual item, a 3D model, or else) in terms of its intrinsic features, social features and user-related, emotional properties [5]. Based on this description, content objects are delivered to users through a innovative graphical interface. Differently from CADMOS, I-SEARCH includes a low level analysis component in its architecture for the automatic acquisition of content information; however, it does not include an explicit representation of this information, partly because it assigns more relevance to emotional and social features

3   The Architecture of CADMOS

The architecture of CADMOS, illustrated in Figure 1, includes six main modules: the User Interface, the Annotation Manager, the Ontology Framework, the Ontology Mashup, NL-to-Onto module and the DMO Repository.

The dramatic media objects (DMO) are stored and indexed within the DMO Repository. A User Interface assists the user in the annotation process. The entire annotation workflow is led by the Annotation Manager which communicates with the DMO Repository (permitting to analyze and visualize the original

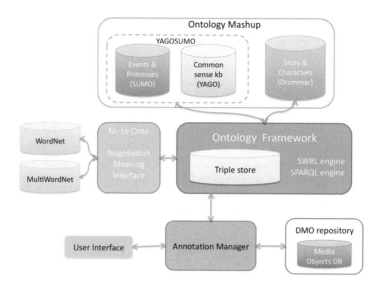

Fig. 1. The architecture of the Cadmos project

media objects) and with the Ontology Framework, which acts as the pivot component of the architecture. The Annotation Manager and the user interface are an extension of the Cinematic system, described in [15].

The Ontology Framework carries out the reasoning services requested by the Annotation Manager; for example bridging the gap between the natural language input from the user and the ontological knowledge (Ontology Mashup). The mediation among these two modules is conducted through the NL-to-Onto module, which exploits the integration of WordNet with Multi-WordNet[3] to help the user disambiguate the intended meaning of the input descriptions, and translate them to the correspondent English terms. In fact, as explained in detail in Section 5, given a user input expressed in her/his native language (currently, the only available languages are English, Italian, Spanish, Portuguese, Hebrew and Romanian), this module first tries to disambiguate the sense of the inserted term (in the user native language) by proposing to the user different possible meanings of the term; then, when the user has selected the most suitable meaning, it translates the correspondent term in English by leveraging the WordNet linguistic knowledge.

Finally, the Ontology Mashup maps each English term to an ontology expression, using the vast knowledge expressed by two well-known ontologies: the Suggested Upper Merged Ontology (SUMO [19][4]) and Yet Another Great Ontology (YAGO [25][5]), merged into YAGOSUMO [16][6]. YAGOSUMO incorporates almost 80 millions of entities from YAGO (which is based on Wikipedia and

---

[3] http://multiwordnet.fbk.eu/english/home.php

[4] http://www.ontologyportal.org/

[5] http://www.mpi-inf.mpg.de/yago-naga/yago/

[6] http://www.mpi-inf.mpg.de/~gdemelo/yagosumo.html

WordNet) into SUMO, a highly axiomatized formal upper ontology. This combined ontology provides a very detailed information about millions of entities, such as people, cities, organizations, and companies and can be positively used not only for annotation purposes, but also for automated knowledge processing and reasoning. This univocal mapping is possible thanks to the integration of WordNet in YAGOSUMO. While YAGOSUMO represents the commonsense knowledge for describing characters, objects and actions, the narrative ontology expresses the knowledge about the dramatic structures of the narrative domain. The role of these two knowledge sources in the annotation process is described in detail in Section4.

It is important to note that the current architecture also permits annotation-based user queries through the User Interface; in this case, the Ontology Framework translates the user request into a SPARQL and performs the requested operation on the triple store (which contains the annotated information). The result is returned to the Annotation Manager that also retrieves the relevant associated media objects and presents them to the user through the User Interface.

4   The Drammar Ontology

According to literary studies [20], from a dramatic point of view, each story develops along two orthogonal axes: *characters* and *plot*; in fact, each story contains a series of incidents, made of characters' actions and, sometimes, unintentional, or naturally occurring, events. The plot can be recursively segmented into units; in cinema, for example, they usually form three layers, respectively called 'scenes', 'sequences' and 'acts' [14].

In drama, the character plays a central role, as it is the medium through which the story is conveyed to the audience. As acknowledged by contemporary aesthetics [4], character is a powerful instrument of identification [3], contributing to the emotional engagement of the audience "in sympathy with the narrative character" [8]. Thus, the typical workflow of linear storytelling relies upon the notions of "character bible" [24], which reports the characters' typical attitudes, personality and behaviors, and the "story", organized into a set of hierarchical *units*, which form the plot tree.

The formal model of agency known as BDI (Belief, Desire, Intentions) has proven effective to model story characters in computational storytelling. According to this model, characters, given their *beliefs* about the world, devise plans to achieve their goals (or *desires*), and form the *intentions* to execute the actions contained in their plans. While beliefs and intentions constitute the mentalistic component of the model, actions constitute the external, perceivable component of the model. In a story, actions have different levels of granularity in the plot tree. At the high level, characters' actions can be described as complex actions, that incapsulate sequences of simpler actions at the lower levels – in the same way as the action "dating" includes "inviting somebody out", "reserving a table", etc. In parallel, at the highest levels, characters' goals tend to persist, while low–level, immediate goals tend to be continuously modified in reaction to the plot incidents.

Being inspired by the notion of bounded rationality [1], the BDI model, by itself, is not sufficient to capture the essentials of characters. In fact, it is important to notice that emotional and moral aspects must be accommodated in a rational background to model actual characters. According to the cognitive framework proposed by Ortony, Clore and Collins [18], emotions stem from the appraisal of one's and others' actions based on a combination of self-interest and moral evaluation. Cognitive studies, then, have pointed out the relation between intentions, i.e., action plans, and emotions and formalized their integration in a computational model [9].

Finally, the sequence of the incidents in a story is represented by the changes in the world state. This component accounts for the narratologists' claim that plot incidents must be causally connected to each other as a necessary condition for story construction [22]. For the story to be consistent, the state of the world that holds after a certain unit must be consistent with the logical preconditions of the unit that follows it in the narration.

The Drammar ontology has been designed with the twofold goal of providing, at the same time, an instrument for the conceptual modeling of drama facts, and a formal tool for the practical task of annotating the narrative properties of dramatic media objects. The ontology is organized to reflect the tripartite structure of plot, characters and units mentioned above. The essential part of the ontology is shown in Figure 2.

The top level of the ontology consists of four disjoint classes. The DramaDy-namics, the DramaEntity, the Structure classes (which contain the definition of proper drama), and the Relation class (which encodes some relevant properties of the expressive means by which drama is exposed to the audience in a specific narrative, such as the visual properties that relates characters and objects on scene).

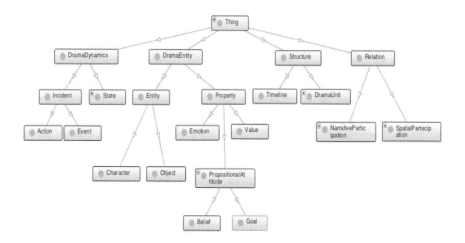

Fig. 2. The class hierarchy of the Drammar ontology

The DramaDynamics class models the evolution of drama though the sequence of incidents (actions or events) that bring the story world from a state to another, affecting the characters' mental states, i.e., their belief, goals, values and emotions. Incidents occur in drama units; states are established as a consequence of the incidents, and constitute the units' preconditions and effects. States can hold in the story world (FactualState class) or in the characters' mind (RepresentedState class), i.e., constitute the content of their beliefs or goals (for example, a character has the goal that a certain state is true or believes it to be true).

The DramaEntity class contains characters (and objects) and their properties. Of paramount importance for drama annotation are the characters' properties that represent their propositional attitudes, i.e., beliefs and goals. Drama entities and drama dynamics are connected through the characters' properties of being agent of actions in order to achieve certain goals. Action class represents a well know design pattern, as, through its properties, it links the execution of actions by the characters with a specific segments of the drama.

The Structure class encompasses the notion of plot tree (a hierarchically organized set of units) and the relation of units with the media objects. Incidents happen inside drama units, with characters doing actions purposely to bring about their goals, under the influence of their values and emotions. The property of units of having preconditions and effects connects the incidents in a unit with the state holding before and after it occurs.

The Relation class models the relations among the drama objects at the expressive level, such as the spatial relations. The latter, together with the scene layout, rely on the filmic codes through which the characters' actions are exposed to the audience.

5   Ontology-Based Meaning Negotiation

Annotation ambiguities are one of the most critical issue that every annotation method should take into account: in fact, ambiguities create a mismatch between what the annotator intended and the community requirements. They are usually caused by unclear (e.g. incomplete, inconsistent, overlapping) definitions of the terminology employed within the annotation. Ontologies are instrumental in facilitating this negotiation process in large scale online communities. In fact, ontologies currently play a central role in the development and deployment of many data applications, especially in media environment. They can be defined as structured information that describes, explains, locates, and makes easier the retrieval, use, and management of information resources. Within the CADMOS approach, we explored ways to develop ontology-guided meaning negotiation [17], with the goal of avoiding annotation ambiguities. However, since the use of ontologies can be difficult in the annotation process, we explored ways to develop ontology-guided meaning negotiation from natural language input.

In CADMOS, the YAGOSUMO ontology allows the use of a shared vocabulary for describing resource content and capabilities, whose semantics is described in a (reasonably) unambiguous and machine-processable form. Ontologies also

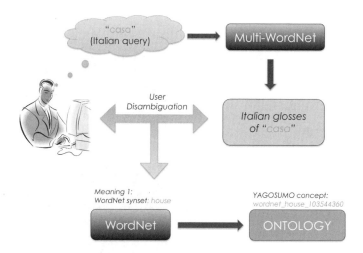

Fig. 3. Ontology-based disambiguation method: the annotator starts the process by searching for some term in her own language. The system retrieves the glosses related to the different meanings of the inserted term and proposes them to the annotator. The annotator selects the most suitable gloss and the system retrieves the related ontological concept by using the associated WordNet synset.

enable effective cooperation between multiple annotators or for establishing consensus where unsupervised (or semi-supervised) systems cooperate with human annotators. Therefore, this domain knowledge helps exclude terminological and conceptual ambiguities, due to ambiguous interpretations. When using YAGO-SUMO for annotation purposes, the annotator can be constrained to use its terminology, through a negotiation step. Moreover, in order to facilitate the work of the annotator, we include in our system, a multi-language tool that permits to initially express the annotation by using the language of the annotator. For example (see Figure 3), the Italian word "casa" can be mapped to different synsets in Multi-Wordnet, ranging from the 'house' and 'home' to 'firm', but only one of them maps to the YAGOSUMO concept that corresponds to the annotator's meaning (here, 'house'). In particular, for each element in the annotation interface (presented in Figure 4), which is associated to a class/predicate in the Drammar Ontology, the system implements the following steps:

- the annotator initially expresses the content as a short natural language term in her/his own language: the term will be forwarded to Multi-Wordnet[7] and the possible meanings of the inserted term are shown by reporting them to the annotator with the different related glosses;
- then, the annotator is able to disambiguate the meaning of the term by selecting the gloss that best match her/his initial thought;
- each gloss is the mapped univocally to the representative English WordNet synset;

---

[7] http://multiwordnet.fbk.eu/english/home.php

Fig. 4. The prototype of the CADMOS Annotation Interface

- as a final step, YAGOSUMO takes as input the synset and returns the internal related concept. Moreover, it is then possible to analyze the assertions related to the concept and the predicates in which it is involved.

In Figure 3 the entire process is shown (with a concise example): the annotator starts the process searching for the meaning of the italian term "casa" (the Italian translation of "house"). Using the NL-to-Onto module (Section 3) the system retrieves five different meanings and reports them to the user by showing the related italian glosses. Using these natural language explanations, the annotator is now able to identify the intended sense of the inserted term. Therefore, the system automatically retrieves the related WordNet synset (by using the WordNet sub-module) and identifies the ontology concept associated to the synset (by exploiting the integration between YAGOSUMO and WordNet store in the Ontology Mashup module).

6   Annotation Example

In order to describe the output of the annotation process we resort to an example, taken from Hitchcock's "North by Northwest". In particular, we describe the annotation of an excerpt of the film, where the protagonist, Roger (Cary Grant), is approaching, unseen, the villain's luxurious house to find out where Eve (Eva Marie Saint) is segregated. The related video segment (a media object) shows Roger in the act of approaching the gate of the house in order to gain a view of the inside.

Therefore, considering the introduced dramatic annotation model, *roger* is an instance of the `Character` class, so it also belongs to the `DramaEntity` class. The existence of this individual however, is not tied to any particular segment of the story: the relation between *roger* and the segment we are considering is given by the fact that he is the agent of an action occurring within the segment. The ontology contains a commitment toward the intentional stance [6], as any occurrence of the action class must be tied to a character's goal that constitutes the purpose for which it is executed. The following triples, represented in Turtle format and stored in the triple store of the Ontology Module (see Section 3), describe Roger's physical aspect, his propositional attitudes and the actions he performs.

```
1 :roger rdf:type <http://www.cadmos-project.it/drammar.owl#Character> ,
2                  owl:NamedIndividual ;
3         :actor "http://www.mpii.de/yago/resource/Cary_Grant"^^xsd:anyURI ;
4         :aspect "http://www.mpii.de/yago/resource/wordnet_chic_104813712"^^xsd:anyURI ;
5         :gender "http://www.mpii.de/yago/resource/wordnet_man_110288763"^^xsd:anyURI ;
6         :isAgentOf :actionRoger ;
7         :hasGoal :goalRoger ;
8         :appearsIn :unit1 .
```

In detail, the properties from line 3 to 5 (actor, aspect, gender) state that the character's role is played by the actor Cary Grant and that he looks like an elegant man. Note that the annotation has been constrained through the use of the YAGO knowledge base (from YAGOSUMO, stored in the Ontology Mashup module, see Section 5). Lines 6 to 8 record its relation to other individuals: the `isAgentOf` property specifies the action it performs, the `hasGoal` property connects it to his goals and the `appearsIn` property links the character instance to unit instances. The relation between a certain action and the goal instance that constitutes its purpose is represented as a property of the action itself, as in the following fragment:

```
1 :actionRoger rdf:type <http://www.cadmos-project.it/drammar.owl#Action> ,
2                        owl:NamedIndividual ;
3               :type "http:sumoOntology/Walking"^^xsd:anyURI ;
4               :hasPurpose :goalRoger ;
5               :hasPosition1 :roger ;
6               :occursIn :unit1 .
```

The action instance represented in these triples is connected to the unit it occurs in (line 6, `occursIn`), to its action type (`type`, line 3), i.e., the ontological definition of the action in the domain knowledge (selected by the annotator through the mechanism described in Section 5), to the goal that motivates the character (line 4, `hasPurpose`) and to the entities that fill the action's thematic roles in its natural language counterpart (expressed according to a positional notation inspired by VerbNet [12], with *roger* as the first argument of the "walk" verb).

This structured annotation permits to properly define not only the basic features of the selected media object (the elements that are shown and their fundamental visual properties) but also the dramatic features and their correlations implied by the narrated story. Thus, the annotation can be also positively used for retrieval purposes by leveraging the drama knowledge imposed by the model and the vast common sense entities definitions reported within the ontologies (stored in the Ontology Mashup module). Moreover, the RDF triple storage also permits, through the Ontology Framework module, reasoning operations that can be used for advanced query operations (for example, the user can now retrieve media objects where some character is performing an action for a specific purpose, i.e. a goal).

7   Conclusions

In this paper we presented a novel semantic annotation system of media objects that relies on an explicit model of narrative, encoded as an ontology, focused on the dramatic concepts of character and action. We described the theoretical motivations for this work as well as a concrete ontology-based proposal for the annotation process. Moreover, we highlighted the general architecture for a narrative-based annotation system and we presented a novel ontology-based terminology disambiguation method that permits to negotiate the meaning of each term inserted by the annotator. This annotation model enables the interoperability among the processes that create new narrative content and the content aggregators, with benefits for the reuse of existing dramatic media objects and for the creation of new (and already annotated) ones.

References

1. Bratman, M.E., Israel, D.J., Pollack, M.E.: Plans and resource-bounded practical reasoning. Computational Intelligence 4, 349–355 (1988)
2. Bruner, J.: The narrative construction of reality. Critical Inquiry 18(1), 1–21 (1991)
3. Carroll, N.: Beyond Esthetics: Philosophical Essays. Cambridge University Press, New York (2001)
4. Carroll, N. (ed.): Special Issue on Narrative, Journal of Aesthetics and Art Criticism, vol. 67. Wiley Blackwell (2009)
5. Daras, P., Semertzidis, T., Makris, L., Strintzis, M.: Similarity content search in content centric networks. In: Proceedings of the International Conference on Multimedia, pp. 775–778. ACM, New York (2010)
6. Dennett, D.C.: The Intentional Stance. The MIT Press, Cambridge (1987)
7. Esslin, M.: The Field of Drama. Methuen, London (1988(1987))
8. Giovannelli, A.: In Sympathy with Narrative Character. Journal of Aesthetics and Art Criticism 67, 83–95 (2009)
9. Gratch, J., Marsella, S.: The architectural role of emotion in cognitive systems. Integrated Models of Cognition Systems, 230 (2007)
10. Jewell, M.O., Lawrence, K.F., Tuffield, M.M., Prugel-Bennett, A., Millard, D.E., Nixon, M.S., Schraefel, M.C., Shadbolt, N.R.: Ontomedia: An ontology for the representation of heterogeneous media. In: Multimedia Information Retrieval Workshop (MMIR 2005) SIGIR. ACM SIGIR (2005)
11. Jewell, M., Lawrence, K., Tuffield, M., Prugel-Bennett, A., Millard, D., Nixon, M., Shadbolt, N.: OntoMedia: An ontology for the representation of heterogeneous media. In: Proceeding of SIGIR Workshop on Mutlimedia Information Retrieval. ACM SIGIR (2005)

12. Kipper, K.: VerbNet: A broad-coverage, comprehensive verb lexicon. Ph.D. thesis, University of Pennsylvania (2005)
13. Kiryakov, A., Popov, B., Terziev, I., Manov, D., Ognyanoff, D.: Semantic annotation, indexing, and retrieval. Web Semantics: Science, Services and Agents on the World Wide Web 2(1), 49–79 (2004)
14. Lavandier, Y.: La dramaturgie. Le clown et l'enfant, Cergy (1994)
15. Lombardo, V., Damiano, R.: Semantic annotation of narrative media objects. Multimedia Tools and Applications, 1–33 (2011), http://dx.doi.org/10.1007/s11042-011-0813-2, doi:10.1007/s11042-011-0813-2
16. Melo, G.D., Suchanek, F., Pease, A.: Integrating yago into the suggested upper merged ontology. In: Proceedings of the 2008 20th IEEE International Conference on Tools with Artificial Intelligence, vol. 01, pp. 190–193. IEEE Computer Society, Washington, DC (2008)
17. Moor, A.D.: Ontology-guided meaning negotiation in communities of practice. In: Proc. of the Workshop on the Design for Large-Scale Digital Communities at the 2nd International Conference on Communities and Technologies, pp. 21–28 (2005)
18. Ortony, A., Clore, G., Collins, A.: The Cognitive Structure of Emotions. Cambridge University Press, Cambridge (1988)
19. Pease, A., Niles, I., Li, J.: The suggested upper merged ontology: A large ontology for the semantic web and its applications. In: Working Notes of the AAAI 2002 Workshop on Ontologies and the Semantic Web (2002)
20. Prince, G.: A Dictionary of Narratology. University of Nebraska Press (2003)
21. Richard, B., Prié, Y., Calabretto, S.: Towards a unified model for audiovisual active reading. In: Tenth IEEE International Symposium on Multimedia, pp. 673–678 (December 2008), http://liris.cnrs.fr/publis/?id=3719
22. Rimmon-Kenan, S.: Narrative Fiction: Contemporary Poetics. Routledge, New York (1983)
23. Ryan, M.L.: Narrative Across Media: The Language of Storytelling. University of Nebraska Press (2004)
24. Seger, L.: Creating Unforgettable Characters. Henry Holt and Company, New York (1990)
25. Suchanek, F.M., Kasneci, G., Weikum, G.: Yago: A Core of Semantic Knowledge. In: 16th International World Wide Web Conference (WWW 2007). ACM Press, New York (2007)

An Application of Fuzzy Logic to Strategic Environmental Assessment

Marco Gavanelli[1], Fabrizio Riguzzi[1], Michela Milano[2],
Davide Sottara[2], Alessandro Cangini[2], and Paolo Cagnoli[3]

[1] ENDIF - Università di Ferrara
{marco.gavanelli,fabrizio.riguzzi}@unife.it
[2] DEIS, Università di Bologna
{michela.milano,davide.sottara}@unibo.it,
alessandro.cangini@studio.unibo.it
[3] ARPA Emilia-Romagna
pcagnoli@arpa.emr.it

Abstract. Strategic Environmental Assessment (SEA) is used to evaluate the environmental effects of regional plans and programs. SEA expresses dependencies between plan activities (infrastructures, plants, resource extractions, buildings, etc.) and environmental pressures, and between these and environmental receptors. In this paper we employ fuzzy logic and many-valued logics together with numeric transformations for performing SEA. In particular, we discuss four models that capture alternative interpretations of the dependencies, combining quantitative and qualitative information. We have tested the four models and presented the results to the expert for validation. The interpretability of the results of the models was appreciated by the expert that liked in particular those models returning a possibility distribution in place of a crisp result.

1 Introduction

Regional planning is the science of the efficient placement of land use activities and infrastructures for the sustainable growth of a region. Regional plans are classified into types, such as agriculture, forest, fishing, energy, industry, transport, waste, water, telecommunication, tourism, urban and environmental plans to name a few. Each plan defines activities that should be carried out during the plan implementation. Regional plans need to be assessed under the Strategic Environmental Assessment (SEA) directive, a legally enforced procedure aimed at introducing systematic evaluation of the environmental effects of plans and programs. This procedure identifies dependencies between plan activities (infrastructures, plants, resource extractions, buildings, etc.) and positive and negative environmental pressures, and dependencies between these pressures and environmental receptors.

[3] proposed two logic based methods to support environmental experts in assessing a regional plan: one based on constraint logic programming and one

R. Pirrone and F. Sorbello (Eds.): AI*IA 2011, LNAI 6934, pp. 324–335, 2011.

based on probabilistic logic programming. Both methods translate qualitative dependencies into quantitative parameters (interpreted in the first model as linear coefficients and in the second as probabilities). However, transforming qualitative elements into numbers without a proper normalization runs the risk of summing non homogeneous terms. In addition, not all impacts should be aggregated in the same way: some pressures may indeed be summed, some receptors present a saturation after a given threshold, while for others a different combination is required in case of positive and negative synergies between activities or between pressures.

To deal with qualitative information, we employ fuzzy logic, a form of multi-valued logic that is robust and approximate rather than brittle and exact. We propose four alternative models. The first model modifies the linear one (i.e., the one implemented via the constraint programming approach) in terms of quantitative fuzzy concepts. The second model is a qualitative interpretation of the dependencies exploiting many-valued logics and gradual rules. The third and the fourth are variants of more traditional fuzzy models that use fuzzy partitions of the domains of variables and provide a semi-declarative definition of the relations using fuzzy rules. All the models are parametric in the definition of the combination operators.

We consider as a case study the assessment of Emilia Romagna regional plans. We describe specific experiments on the regional energy plan explaining the strength and weakness of each model. The models have been extensively tested and the results have been proposed to environmental experts, who appreciated in particular the fourth model, combining linear and fuzzy logic features, guaranteeing high expressiveness and proposing results in a very informative way.

2 Strategic Environmental Assessment

Regions are local authorities that include among their tasks the planning of interventions and infrastructures. Particular emphasis is devoted to energy, industry, environment and land use planning. Before any implementation, regional plans have to be environmentally assessed, under the *Strategic Environmental Assessment Directive*. SEA is a method for incorporating environmental considerations into policies, plans and programs that is prescribed by EU policy.

In the Emilia Romagna region the SEA is performed by applying the so-called *coaxial matrices*, that are a development of the network method [7]. The first matrix defines the dependencies between the activities contained in a plan and positive and negative pressures on the environment. The dependency can be *high, medium, low* or *null*. Examples of negative pressures are energy, water and land consumption, variation of water flows, water and air pollution and so on. Examples of positive pressures are reduction of water/air pollution, reduction of greenhouse gas emission, reduction of noise, natural resource saving, creation of new ecosystems and so on. The second matrix defines how the pressures influence environmental receptors. Again the dependency can be *high, medium, low* or *null*. Examples of environmental receptors are the quality of surface water and groundwater, quality of landscapes, energy availability, wildlife wellness and so on. The matrices currently used in Emilia Romagna contain 93 activities, 29 negative pressures, 19 positive pressures and 23 receptors.

The SEA is now manually performed by environmental experts on a given plan. A plan defines the so-called *magnitude* of each activity: magnitudes are real values that intuitively express *"how much"* of an activity is performed with respect to the quantity available in the region. They are a percentage for each activity.

3   Fuzzy and Many-Valued Logic

After the introduction of the concept of fuzzy set [9] by Zadeh for modeling vague knowledge and partial degrees of truth, much work has been done in various research areas to apply the concept of fuzziness to existing fields, including formal logic. Historically, two possible approaches have been adopted [5]: one, more mathematically oriented, belongs to the family of many-valued logics and is called fuzzy logic "in a narrow sense", while the other, fuzzy logic "in a broader sense", is closer to Zadeh's original definition and uses a softer approach.

"Fuzzy" many-valued logics are a truth-functional generalization of classical logic. Atomic predicates p/n are considered fuzzy relations, whose truth degree is given by their associated membership function μ_P. Thus predicates can have truth values in the range $L = [0, 1]$. In order to construct and evaluate complex formulas, logical connectives, quantifiers and inference rules (e.g. modus ponens) are generalized to combine truth degrees. For example, the conjunction operator can be defined using any *t-norm* $\star$, such as the minimum, the product or Łukasiewicz's norm. Likewise, the disjunctive connective is defined using an *s-norm* and the implication depends on the t-norm definition by residuation [4]. A rule $C \leftarrow_i A$, then, can be used to entail a fact C with a degree of at least c, provided that a fact matching with A exists with degree $a > 0$ and that the implication $\leftarrow$ itself has a degree $i > 0$.

In [9], Zadeh introduced the concept of *fuzzy linguistic variable*, a qualitative construct suitable to describe the value of a quantitative variable X with domain Δ_X. Each linguistic value λ_j belongs to a finite domain Λ and is associated to a fuzzy set A_j. Together, the sets define a fuzzy partition of Δ_X iff $\forall x \in \Delta_X$: $\sum_j \mu_{A_j}(x) = 1$. The membership values of an element x to a set can either be interpreted as the *compatibility* of x with the concept expressed by a linguistic variable, or as the *possibility* that x is the actual value of X, assuming that x is unknown save for the fact that it belongs to A_j. (For a complete discussion on the relation between compatibility and possibility, see [2]).

Fuzzy partitions are usually used in conjunction with fuzzy rules to approximate complex functions $y = f(x)$ by fuzzifying the function's domain and range, then matching the resulting input and output sets using rules [1]. Different types of rules have been proposed: "Mamdani" rules infer fuzzy consequences from fuzzy premises and have the form x *is* $A_j \Rightarrow_\varepsilon y$ *is* B_k[1]; Fuzzy Additive Systems (FAS), instead, entail quantitative values. In the former case, then, it is

---

[1] A_j and B_k are fuzzy sets and *is* is the operator evlauating set membership.

necessary to collect the different sets entailed by the various rules, combine them - usually by set union - into a single possibility distribution and finally, if appropriate, apply a defuzzification process [6] to get a crisp consequence value. In the latter case, instead, the quantitative values are directly available and can be aggregated, e.g., using a linear combination.

4   Models

[3] proposed two logic-based approaches for the assessment of environmental plans. The first, based on Constraint Logic Programming on Real numbers (CLP($\mathcal{R}$)), interprets the coaxial matrix qualitative values as coefficients of linear equations. The values, suggested by an environmental expert, were 0.25 for *low*, 0.5 for *medium*, and 0.75 for *high*. The advantages of such a model are its simplicity, efficiency and scalability, but, due to its linearity, it assumes that positive and negative pressures derived from planned activities can be always summed. While, in general, pressures can indeed be summed, in some cases a mere summation is not the most realistic relation and more sophisticated combinations should be considered.

The second, based on Causal Probabilistic Logic Programming, gave a probabilistic interpretation to the matrices. The same numeric coefficients have been used to define the likelihood of a given pressure (or receptor) being affected by an activity (respectively, a pressure). While probability laws allow for a different combination strategy, the relations used by the experts are vague and they have a gradual nature rather than a stochastic one: how an activity (respectively a pressure), when present, affects a pressure (respectively receptor) is usually a matter of degree of truth/possibility and not of chance.

The purpose of this paper, then, is to provide alternative models of the dependencies, exploiting the concepts and mechanisms of Multi-Valued Logic and Fuzzy Logic. The first step in formalizing the required concepts is to redefine the involved variables (such as activities' magnitude, pressures and environmental receptors) in terms of fuzzy sets and linguistic variables.

There are two main approaches for representing each variable (activity, pressure or receptor) in our model. The first is to define a many-valued predicate, $mag/1$, whose truth value represents the magnitude of that variable, i.e. represents how much the considered variable is "large" in terms of a truth value in the interval $[0, 1]$. Notice that, although the predicate is the same, the membership function is different for each variable. For example, if we have the activity *road construction* and atom $mag(road)$ has truth value 0.7, this means that the plan involves the building of a significant amount (0.7) of roads with respect to the current situation, while a smaller truth value would correspond to a smaller number of kilometers of roads to be built.

In the second approach, a fuzzy linguistic variable is defined for each variable, creating a fuzzy partition on its domain. The partition contains one fuzzy set for each value of the linguistic variable. The sets are used to describe different levels of magnitude: we consider a five-set fuzzy partition of each variable's domain consisting of the sets *VeryLow*, *Low*, *Average*, *High* and *VeryHigh*.

The second degree of freedom we have is the selection of the aggregation method for the results, i.e. the choice of the s-norm used to combine the results of the application of rules with the same consequent. For example, consider two pressures such as energy consumption and odor production; the overall energy consumption is the sum of the consumptions due to the single activities, but the same hardly applies to odor production. An activity that produces a strong odor may "cover" weaker odors, so a good aggregation for this kind of variable should be the maximum (or geometrical sum). The aggregation strategy becomes even more important in the case of environmental receptors.

Pressures can be either "positive" or "negative": translating positive pressures into positive contributions, and negative ones into negative contributions would be an approximation since the former do not always cancel the latter. If a linear model returns a final result of 0 for a given receptor, there is no way of telling whether that value is the combination of large positive and negative contributions canceling each other, or we are simply in the case of absence of significant pressures influencing that receptor. So, we have chosen to split the individual receptor variables into two parts, one considering only positive and one considering only negative pressures affecting that receptor. The strategy for the combination of the two can then be decided on an individual basis.

In the remainder of this section, we will provide a description of the four classes of models which can be designed, according to different combinations of the underlying logic (many-valued vs classical fuzzy) and aggregation style (linear vs non-linear).

4.1   Many-Valued Logic Models

Model I. To begin with, we revised the existing constraint based model in terms of quantitative fuzzy concepts. The original formulation [3] takes as input the activities, in terms of their relative magnitudes, and calculates pressures as $p_j = \sum_{i=1}^{N_a} m_{ij} * a_i$. Each coefficient m_{ij} quantifies the dependency between the activity i and the pressure j according to the qualitative value in the matrix M. The values $a_{i:1..N_a}$, instead, are the magnitudes of each activity: the values represent the increment of an activity A_i as a percentage in relation to the existing A_i^0, in order to make the different activities comparable. For example, a magnitude of 0.1 for activity *"thermoelectric plants"* means increasing the production of electricity through thermoelectric energy by 10% with respect to the current situation.

Likewise, the influence on the environmental receptor r_k is estimated given the vector of environmental pressures $p_{j:1..N_p}$ calculated in the previous step. An alternative formulation of the model, this time in logic terms, is composed by the following Horn clauses:

$$contr(Press_j, Act_i) \Leftarrow_{\beta_{i,j}} mag(Act_i) \wedge impacts(Act_i, Press_j) \qquad (1)$$

$$mag(Press_j) \Leftarrow_1 \exists\, Act_i : contr(Press_j, Act_i) \qquad (2)$$

	Lin.	Non Lin.
MVL	Model I	Model II
Fuzzy	Model IV	Model III

Fig. 1. Model classification by type of logic and aggregation style

where we use the auxiliary predicate *contr* to describe the contribution from a single source, whereas *mag* describes the aggregate contributions.

The value $\beta_{i,j}$ is a normalization coefficient, that makes the maximum possible value of truth for $mag(Press_j)$ equal to 1 when the truth degree of $mag(Act_i)$ for all the impacting activities is equal to 1. Its default value can be changed by the environmental expert to obtain other behaviors.

In order to replicate the behavior of the linear model, we need to (i) configure the $mag/1$ predicate to use a linear membership function, (ii) configure the $impacts/2$ predicate to use a membership function derived directly from the matrix, i.e. $\mu_{impacts(Act_i,Press_j)} = m_{ij}$ where m_{ij} is the real value obtained from the qualitative dependencies as in [3] , (iii) configure the $\wedge$ operator to use the product t-norm, (iv) configure the $\exists$ quantifier to use a linear combination s-norm and (v) configure the reasoner to use gradual implications and the product t-norm to implement modus ponens.

The critical point is that the logic operators do not aggregate values, which have only a quantitative interpretation, but degrees of truth, which have a more qualitative interpretation. If one wants the degree to be proportional to the underlying quantitative value, the use of scaling coefficients might be mandatory since a degree, having an underlying logic semantics, is constrained in $[0, 1]$. Intuitively, the coefficients model the fact that, even if an individual piece of evidence is true, the overall proof may not: the coefficient, then, measures the loss in passing from one concept to the other (which, from a logical point of view, is a gradual implication). If the coefficients are chosen accurately, the aggregate degree becomes fully true only when all the possible contributions are fully true themselves. As a side effect, the normalization function used by the predicate $mag/1$ cannot map the existing amount A_i^0 of a given activity to 1, since that is not the theoretical maximum of a new activity, and the contributions of the individual activities require a scaling by a factor $\beta_{i,j}$ before being aggregated.

Model II. After a more detailed discussion with the expert, however, it turned out that no single model alone — qualitative or quantitative, linear or non-linear — could capture the full complexity of the problem, mainly because the relations between the entities are different depending on the actual entities themselves.

A purely linear model has also other limitations: for example, some public works are already well consolidated in the Emilia Romagna region (e.g. roads), so that even a large scale work would return a (linearly) normalized activity value around 1. Others, instead, are relatively new and not well developed (e.g.

wind plants), so even a small actual amount of work could yield a normalized value of $5 \div 10$, unsuitable for logic modelling as well as being unrealistic given the original intentions of the experts. In order to cope with this problem, we decided to adopt a non-linear mapping between the amount of each activity and its equivalent value, using a sigmoid function:

$$a_i = \frac{1 - e^{-A_i/(k_i A_i^0)}}{1 + e^{-A_i/(k_i A_i^0)}} \qquad (3)$$

This expression behaves like a linear function for small relative magnitudes, while saturates for larger values, not exceeding 1. The relative magnitude can be further scaled using the parameter k_i, provided by the expert, to adjust the behaviour for different types of activities. Moreover, the normalization function (3) is a proper membership function for the fuzzy predicate $mag/1$, defining how large the scale of an activity is with respect to the existing and using the parameter k_i to differentiate the various entities involved.

This membership function, however, also slightly changes the semantics of the linear combination. In the original linear model, we had a sum of quantitative elements, measured in activity-equivalent units and weighted by the coefficients derived from the matrix, and we tried to replicate the same concept in Model I. Now, instead, we have a proper fuzzy count of the number of activities which are, at the same time, "large" and "impacting" on a given pressure. Notice, however, that we still need gradual implications, in order to use the standard, "or"-based existential quantifier to aggregate the different contributions.

The second extension we introduce in this model involves the relation between pressures and receptors. While in Model I it is sufficient to replicate rules (1) and (2), here we keep the positive and negative influences separated:

$$contrPos(Press_j, Rec_k) \Leftarrow_{\gamma_{j,k}} mag(Press_j) \wedge impactsPos(Press_j, Rec_k)$$
$$contrNeg(Press_j, Rec_k) \Leftarrow_{\delta_{j,k}} mag(Press_j) \wedge impactsNeg(Press_j, Rec_k)$$
$$influencePos(Rec_k) \Leftarrow_1 \exists\, Press_j : contrPos(Press_j, Rec_k)$$
$$influenceNeg(Rec_k) \Leftarrow_1 \exists\, Press_j : contrNeg(Press_j, Rec_k)$$

In order to combine the positive and negative influences, their relation has to be expressed explicitly. For example, rule (4) states that positive and negative influences are interactive and affect each other directly; rule (5) defines the concept of beneficial pressures explicitly, while rule (6) stresses those receptor which have been impacted in an absolute way. As usual, the operator definitions can be chosen on a case-by-case basis to better model the relations between the particular pressures and receptors.

$$influencePos(Rec_k) \Leftrightarrow_\varepsilon \neg\, influenceNeg(Rec_k) \qquad (4)$$
$$benefit(Rec_k) \Leftarrow_1 influencePos(Rec_k) \wedge \neg\; influenceNeg(Rec_k) \quad (5)$$
$$hit(Rec_k) \Leftarrow_1 influencePos(Rec_k) \vee influenceNeg(Rec_k) \quad (6)$$

4.2   Fuzzy Models

In models I and II, the elements of the coaxial matrices are converted into simple scaling coefficients. To increase the expressiveness of this mapping, we assumed that each label is actually an indicator for some kind of predefined function, for which we do not provide an analytic expression, but a fuzzy logic approximation [1]. So, we created fuzzy partitions on the domain of each activity, pressure and receptor: in particular, each partition consists of 5 triangular membership functions, not necessarily equally spaced on the domains. These sets have been associated to the linguistic values *VeryLow*, *Low*, *Average*, *High* and *VeryHigh*. Then, we used rules such as $VeryLow(Act) \Rightarrow VeryLow(Pres)$ to map (linguistic) values from one domain onto (linguistic) values of the corresponding range, according to the connections expressed in the matrices.

In both model III and IV we gave the same interpretation to the matrix elements, using linear functions with slope 1, 0.5 and 0.25 for *"high"*, *"medium"* and *"low"* respectively. These functions, then, have been fuzzified as shown schematically in Figure 2. A *VeryHigh* input is mapped onto a *VeryHigh*, *Average* or *Low* output, respectively, when the label in a cell of a coaxial matrix is *high*, *medium* or *low*. The mapping can easily be changed by altering the rules and allows to define non-linear relations as well as linear ones. In fact, the use of a fuzzy approximation gives a higher flexibility to the system, while keeping the evaluation robust.

Model III and IV. Model III is a canonical fuzzy system, where the relations between (i) activities and pressures and (ii) pressures and positive and negative effects on receptors are defined using fuzzy rules. The inputs, the activities' magnitudes, are no longer normalized, but fuzzified: the rules are then evaluated using the min-max composition principle [1] and chained, propagating the inferred fuzzy distributions from the pressures onto the receptors. If needed, the resulting possibility distributions can then be defuzzified to obtain a crisp "impact" value for each receptor. Using this model, it is possible to distinguish pressures which are affected by the different activities at different levels.

According to the experts, however, this model suffers from a drawback: being purely qualitative, the degrees inferred for each fuzzy set tell only whether a pressure/receptor will *possibly* be affected with any level of intensity. Suppose for example that activity a_i generates pressure p_j with *Low* intensity. If the magnitude of a_i is sufficiently large, the system will entail that it is (fully) possible that p_j has a *Low* component. This answer is not quantitative: it would not allow to distinguish this case from one where many different activities, all individually generating pressure p_j with low intensity, are present at the same time. Thus, this model is appropriate when only a qualitative answer is sufficient.

To overcome this limitation, we created Model IV as a minor extension of Model III, exploiting the same concepts used in Model II. We used gradual rules to scale and give an additional quantitative meaning to the consequence degrees: $VeryLow(Act) \Rightarrow_\beta VeryLow(Pres)$. Then, we allowed the norms to be configurable, so the min-max composition principle was replaced by a more general t-s norm composition principle. The min-max model, now a special case, is

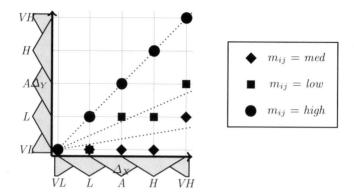

Fig. 2. Relation between an activity a_i and a pressure p_j depending on the value m_{ij} appearing in the coaxial matrix

still admissible and is suitable for situations where the various inputs are not interactive, whereas the probabilistic sum and the algebraic sum s-norms are more suitable when the sources are independent or exclusive.

5   Model Evaluation

We implemented the four models using the Jefis Library [8], and tested the software on a system equipped with a Core 2 Duo T6600 processor and 4GB RAM. Using the full content of the coaxial matrices to derive the rules and a past regional energy plan as a realistic test case, the evaluation required less than 1 second, guaranteeing that computation time is not a critical factor for the proposed system, considering also that the task is not to be performed in hard real time. Instead, we focus on the assessment of the expressiveness of the four models. Here, due to space limitations and for the sake of readability, we will discuss a more focused test case rather than a whole plan.

We assume that the matrices contain only two activities - *Incinerators* (INC) and *Wastewater Treatment Plants* (WTP), two pressures - *Noxious Gases Emission* (NG) and *Odor Emission* (OD), and one environmental receptor - *Landscape Quality* (LQ). In our example, the two activities influence both pressures, albeit in a different way. The first pressure, NG, is assumed to be linear in the causes: different independent sources simply increase the amount of gas released into the atmosphere; the latter, OD, is not linear: since odors cover each other, we assumed the sources to be independent but interactive. Moreover, both pressures affect negatively the considered receptor, with an influence strength shown by the matrix excerpt in Table 1. When computing the effects of the pressures, instead, we assumed them to be independent and non-interactive, so the positive (respectively, negative) impact on the receptor is given by the best (resp. worst) effect induced by a pressure. Notice that only models II and IV are able to capture these differences, since model I is linear by default, and model III

Table 1. Excerpt of the Coaxial Matrices

		NG	OD
Act	INC	H	M
	WTP	M	H
Rec	LQ	L	H

is non-interactive by default. Given this simplified matrix, the models were set up as follows. Regarding activities, we assume that the initial input is already expressed in terms of equivalent units: a planned magnitude of 100 units is equivalent to the currently existing amount of the same activity. In our test, we set $A_{INC} = 90$ and $A_{WTP} = 180$.

To perform a linear normalization in Model I, we assume that plans cannot involve values greater than 200, effectively planning the construction of no more than twice the existing. When the sigmoidal normalization is used in Model II, an upper limit is not necessary; however, for comparison purposes, we chose values for k_{INC} and k_{WTP} such that the result is the same as when the linear normalization is applied, i.e. the resulting normalized values are 0.45 and 0.9 respectively. Models III and IV, instead, do not require an explicit normalization, since it is performed implicitly by the fuzzification of the magnitudes. Since they are already expressed in equivalent units, all activities share the same domain - the range [0..200], which has been partitioned using sets labelled *VL, L, A, H* and *VH*. All sets are uniformly spaced and use triangular membership functions, except *VH*, which uses a "right-shoulder" function, and *VL*, which uses a "left-shoulder" function. As already pointed out, in Model I and II the values of both matrices are mapped onto 0.75, 0.5 and 0.25; Model III and IV, instead, map the values onto different set of rules, as shown in figure 2. When evaluating the pressure NG, Model II and IV also use scaling coefficients $\beta = 0.5$ (i.e. the reciprocal of the number of activities) to avoid saturation.

We now discuss the evaluation of the different models. First, we consider the relation between INC, planned with magnitude 90, and NG, which is *high* according to the matrix.

I Using rule (1) with the product norm, the linearly normalized magnitude, 0.45, is scaled by the coefficient 0.75, yielding a contribution of 0.3375.

II Similarly, the sigmoidal normalization yields 0.45. This time, however, the gradual rule entails a pressure magnitude of 0.16875.

III The fuzzification of the input yields a reshaped partition $\{L/0.75, A/0.25\}$ describing the magnitude of the activity. After the adequate set of rules has been applied, one obtains a contribution for the distribution of the pressure, which incidentally is identical: $\{L/0.75, A/0.25\}$.

IV The result is analogous in Model IV, save for the effect of the gradual rule: $\{L/0.375, A/0.125\}$

Table 2. Intermediate and final results of the evaluation

		I	II	III	IV
Act	INC	0.45	0.45	$\{L/0.75,\ A/0.25\}$	$\{L/0.75,\ A/0.25\}$
	WWTP	0.90	0.90	$\{H/0.5,\ VH/0.5\}$	$\{H/0.5,\ VH/0.5\}$
Press	NG	0.79	0.39	$\{L/0.75,\ A/0.5\}$	$\{L/0.625,\ A/0.375\}$
	OD	0.90	0.72	$\{VL/0.75,\ L/0.25,\ H/0.5,\ VH/0.5\}$	$\{VL/0.75,\ L/0.25,\ H/0.5,\ VH/0.5\}$
Rec	$LQ_\vee^-$	0.87	0.54	$\{VL/0.75,\ L/0.25,\ H/0.5,\ VH/0.5\}$	$\{VL/0.75,\ L/0.25,\ H/0.5,\ VH/0.5\}$
	LQ_+^-	0.87	0.64	$\{VL/0.75,\ L/0.25,\ H/0.5,\ VH/0.5\}$	$\{VL/1.00,\ L/0.25,\ H/0.5,\ VH/0.5\}$
	$LQ_\oplus^-$	0.87	0.58	$\{VL/0.75,\ L/0.25,\ H/0.5,\ VH/0.5\}$	$\{VL/0.95,\ L/0.25,\ H/0.5,\ VH/0.5\}$

In order to compute the overall degrees for the pressure NG, one must also take
into account the contributions due to the WTP activity, planned with magnitude
180: according to the coaxial matrix, the relation between the two is *medium*.

I The second contribution, 0.9, is summed to the previous one, yielding 0.7875.
II This model gives a contribution of 0.225. Depending on the chosen s-norm,
   this value is combined with the other value of 0.16875: since gas emissions
   are additive, we use Łukasiewicz's *or* to get a combined value of 0.39375.
III The fuzzified input, $\{H/0.5,\ VH/0.5\}$, is mapped onto the output as $\{L/0.5,$
   $A/0.5\}$ due to the use of a different set of rules, in turn due to the relation
   between the two being defined as *medium*. The fuzzy union of the previous
   and current contributions gives $\{L/0.75,\ A/0.5\}$.
IV The combination of gradual fuzzy rules, and the use of Łukasiewicz's *or*
   leads to a final result of $\{L/(0.375 + 0.25),\ A/(0.125 + 0.25)\}$

The same procedure is repeated for OD. Notice that, due to the initial modelling
assumptions, a more appropriate s-norm for models II and IV is the "probabilistic sum". Once both pressures have been evaluated, the inference propagation
pattern is applied once more to obtain the final degree/distribution for the receptor LQ. Given the initial assumption of non-interactivity, the "max" s-norm
would be more appropriate in models II and IV, however we performed the
computations also using the bounded and noisy sum norms for comparison. All
the intermediate and final results are reported in table 2. Notice that we only
consider negative effects on the environmental receptor because all pressures
considered in this example were considered as negative pressures.

6   Conclusions

In this paper we proposed a fuzzy logic approach to SEA. We implemented
four models, with different features and informative capabilities. Model I is an
implementation of a linear model in fuzzy logic; it cannot distinguish interactive
from non-interactive effects, and scenarios with many small effects from scenarios
with a few large ones. Model II can cope with the former problem, but not with
the latter, while Model III tackles the latter but not the former. Model IV
combines the two aspects in a single model and was considered by the expert

as the most expressive and informative. Moreover, instead of an unrealistically precise single value, as proposed in [3], Model IV now proposes a possibility distribution over the values that can be expected for environmental receptors.

These models introduce a more qualitative approach than [3] and show that fuzzy logic provides a tool for SEA that is more appealing for the domain experts.

In this work we implemented the simple one-way relation included into the co-axial matrices already used in the Emilia-Romagna region. However, environmental systems are very complex, and seldom relations are only in one direction, but environmental receptors could have an effect on the impacts or raise the need to perform some compensation activity in the future regional plans. In future work, we plan to study such effects.

References

1. Dubois, D., Prade, H.: Fuzzy Sets and Systems: Theory and Applications. Academic Press, London (1980)
2. Dubois, D., Prade, H.: Possibility theory, probability theory and multiple-valued logics: a clarification. Annals of Mathematics and Artificial Intelligence 32(1-4), 35–66 (2001)
3. Gavanelli, M., Riguzzi, F., Milano, M., Cagnoli, P.: Logic-Based Decision Support for Strategic Environmental Assessment. Theory and Practice of Logic Programming, Special Issue 10(4-6), 643–658 (2010); 26th Int'l. Conference on Logic Programming (ICLP 2010)
4. Hájek, P.: Metamathematics of Fuzzy Logic (Trends in Logic), 1st edn. Springer, Heidelberg (2001)
5. Novák, V.: Mathematical fuzzy logic in narrow and broader sense - a unified concept. In: BISCSE 2005. The Berkeley Initiative in Soft Computing, Univ. of California, Berkeley (2005)
6. Saade, J.: A unifying approach to defuzzification and comparison of the outputs of fuzzy controllers. IEEE Transactions on Fuzzy Systems 4(3), 227–237 (1996)
7. Sorensen, J.C., Moss, M.L.: Procedures and programs to assist in the impact statement process. Tech. rep., Univ. of California, Berkeley (1973)
8. Wulff, N., Sottara, D.: Fuzzy reasoning with a Rete-OO rule engine. In: Governatori, G., Hall, J., Paschke, A. (eds.) RuleML 2009. LNCS, vol. 5858, pp. 337–344. Springer, Heidelberg (2009)
9. Zadeh, L.A.: Fuzzy sets. Information and Control 8(3), 338–353 (1965)

Using Planning to Train Crisis Decision Makers

Amedeo Cesta[1], Gabriella Cortellessa[1],
Riccardo De Benedictis[1], and Keith Strickland[2]

[1] CNR, Italian National Research Council, ISTC, Rome, Italy
[2] Cabinet Office, Emergency Planning College, York, UK

Abstract. Training for crisis decision makers poses a number of challenges that range from the necessity to foster creative decision making to the need of creating engaging and realistic scenarios in support of experiential learning. This article describes our effort to build an end-to-end intelligent system, called the PANDORA-BOX, that supports a trainer in populating a 4-5 hours training session with exercises for a class of decision makers to teach them how to achieve joint decisions under stress. The paper gives a comprehensive view of the current system and in particular focuses on how AI planning technology has been customized to serve this purpose. Aspects considered are: (a) the timeline-based representation that acts as the core component for creating training sessions and unifying different concepts of the PANDORA domain; (b) the combination of planning and execution functionalities to maintain and dynamically adapt a "lesson plan" on the basis of both trainees-trainer interactions and individual behavioral features and performance; (c) the importance of keeping the trainer in the loop preserving his/her responsibility in creating content for the class but endowing him/her with a set of new functionalities.

1   Introduction

This paper presents work done for creating an intelligent support system to facilitate training of decision makers in emergency scenarios. Indeed, all too often, shortcomings in the management of the emergency do not stem from the ignorance of procedures but from difficulties resulting from the individual response to the challenge of operating in such a context, particularly when additional unexpected problems arise. Crisis management is of major importance in preventing emergency situations from turning into disasters. In these critical circumstances, there is a tremendous necessity of effective *leaders*. Nevertheless, the ambiguity, urgency and high risk of a crisis situation posit some constraints on the abilities of leaders. For example, given the severe time pressure imposed by the crisis, they have little time to acquire and process information effectively. As a consequence, they are required to assess information and making decisions under tremendous psychological stress and physical demands. Authors are working within an EU project called PANDORA[1] whose goal is to build a training environment for supporting trainers to engage a class of crisis managers to take decisions when confronted with realistic simulations. An important aspect is the decision making

---

[1] http://www.pandoraproject.eu/

R. Pirrone and F. Sorbello (Eds.): AI*IA 2011, LNAI 6934, pp. 336–347, 2011.
© Springer-Verlag Berlin Heidelberg 2011

seen as a collective action of managers from different institutions. This is a key requirement for instantiating a realistic crisis situation. Hence the basic infrastructure requires to connect different trainees and, additionally, in case some basic institutional role is missing among the current trained group, to simulate them by simple artificial characters (called NPCs for *non player characters*). For any of the real trainees in the class a continuous loop of create/update of a trainee model is actuated. Goal of the continuous user modeling is to have a level of individual adaptation of the stimuli during the class. The proactive part of the environment is composed by a *back-end* that uses planning technology for deciding the content for the lesson (*aka* a set of causally connected stimuli) and by a *front-end* that uses multimedia technology for rendering content in an engaging way.

This paper is dedicated to show how planning technology has been molded in order to create a support for continuous animation and adaptation of lesson content. Additionally it describes specific features of the first year demonstrator of the project called the PANDORA-BOX. The writing is organized as follow: Section 2 gives some further context on the training environment, elaborates on its potential users then introduces the system architecture synthesized for this project, Section 3 describes the use of planning technology in the PANDORA-BOX, Section 4 sketches the current interaction environment that allows both trainer and trainees to use the system.

2   Training for Crisis Decision Makers

Usually, when referring to planning connected to crisis management in emergency situations we have in mind the intervention plans for people who directly goes in operational theaters (see for example [1]). Indeed in reality there are different leaders whose decisions are relevant under crisis. Additionally the success of crisis management often depends not only on the ability to apply well established procedures, but also on the effectiveness of high-level strategic choices and the ability of decision makers to anticipate the possible consequences of actions by means of flexible and forward-looking reasoning. To summarize, the three different levels corresponding to different roles of crisis decision makers are the following:

- At the **operational level** we have the so called **bronze commanders**, people directly operating on crisis scenarios and performing practical activities and actions, the results of which are monitored and communicated to higher levels;
- At the **tactical level** we have **silver commanders** responsible for translating high level strategic decisions into actions and related resources allocations. At this level the expected results from the various tasks are identified but details are expected to be filled in by bronze commanders;
- The **strategic level** is the theater for **gold commanders**. At this level the key issues of a critical situation are identified, and priorities are established at a sufficiently high level of abstraction. Strategies for resolving the crisis are also decided which are then communicated to the lower levels for their detailed specification and implementation.

The choices at the strategic level are particularly important and critical for the success of crisis management in particular for preventing dangerous domino effects.

It is worth underscoring how different roles of crisis managers correspond to different level of decision-making: specifically, at strategic level, decision making is mainly unstructured and not describable in term of programmed procedures to apply, being mainly related to novelty and unpredictability of the catastrophic event. In this light it is rather a non-programmed decision making effort.

Most of the state-of-the-art support systems and training simulators are aimed at the operational or tactical levels. On the other hand PANDORA is specifically targeted to the strategic decision makers thus presenting difficult challenges at both modeling and computational level. Additional challenges come from the need to foster quick decisions under stress and from the need to encourage creative decision-making.

Among the main objectives of gold commanders during crisis management we can mention: protect human life and, as far as possible, property; alleviate suffering; support the continuity of everyday activity and the restoration of disrupted services at the earliest opportunity. The speed with which recovery strategies are identified to contain and resolve the crisis also has a great influence on the loss of whatever nature. For this reason the strategic decision maker has to develop an ability to quickly react and decide with the overall goal to obtain a rapid recovery of the normal growth thus minimizing the losses. In this light, training plays a fundamental role. At the strategic level, training aims to teach decision-makers to focus on the possible consequences of their actions, to integrate and test the compatibility of plans, to decide in agreement with other organizations and between different nations, to promote continuity of efforts and have a well defined focus.

2.1 The PANDORA Approach

Goal of PANDORA is to build an intelligent training environment able to generate a spectrum of realistic simulations of crisis scenarios. Important points are: (1) the reproduction of stressful factors of the real world crisis; (2) the personalization of the planned stimuli according to different trainees and (3) the support for the dynamic adaptation of "lesson plans" during the training time-horizon.

The system design has followed a user-centered approach, based on a close cooperation with the training experts who have profoundly influenced the shaping of the system. Specifically, the Cabinet Office Emergency Planning College has synthesized their experience coming from training a wide range of senior decision makers and their leading expertise in emergency planning and crisis management. Additionally a number of general constraints have emerged during the phase of user requirement analysis:

- *Support cooperative decision making*: it has been immediately clear how training gold commanders implies teaching them to take joint decisions after cooperation;
- *Training personalization*: it has been underscored the role of personalized teaching even though within a group decision making context;
- *Mixed-initiative interaction*: it turned out as particularly important the need of a tool that empowers the trainer with further abilities rather than a pure video-game like type of immersive experience, hence the need to create a mixed-initiative environment with the trainer fully integrated in the "lesson loop".

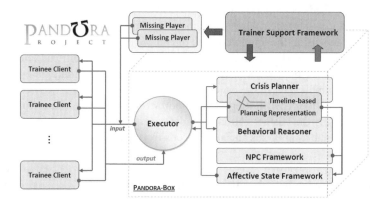

Fig. 1. *The* PANDORA-BOX *general architecture*

Figure 1 shows the software architecture of the PANDORA-BOX whose complete version has been officially demonstrated in several public events during 2011. The system is composed of three separate environments: (a) a Trainer Support Framework allows the trainer to keep run-time control of the training session and dynamically adjust the stimuli based on his/her experience; (b) distributed Trainee Clients can access the PANDORA-BOX and receive both collective and individual stimuli during a lesson; (c) a PANDORA kernel which is the main engine that generates the "lesson plan", animates it according to temporal information and continuously adjusts it to keep pace with both the evolution/evaluation of the specific group of people under training and their individual performance.

Specifically, a group of trainees, representative of different agencies involved in the resolution of a crisis (e.g., Civil Protection, Fire Rescue, Police, Transports and so on) have access to the training system. If some of the representative authorities are not present they are simulated by the PANDORA system through Non Player Character (NPC) in which case, features and decisions are synthesized either by the trainer through the system or through simplified pre-canned behaviors. As for the PANDORA-BOX kernel, its basic functioning is grounded on three of the blocks in Figure 1: a *Plan Representation* which is based on timelines as described in this paper, a *Planner* that synthesize a plan as a set of stimuli for the trainees, an *Executor* that executes a plan by dispatching subparts of it to the rendering environment. The *rendering* is not made explicit in the figure apart the two modules, called NPC Framework (developed by BFC) and the Affective State Framework (implemented by XLAB), specifically accessible by the planner to select and use realistic effects for plan segments rendering. These modules adds an additional level of "realism" to the stimuli, by customizing the appropriate presentation mode (e.g., introducing noise on a phone call report) in order to achieve a high level of realism, stress and pressure. A further aspect, relevant in PANDORA, is *personalization*. The various participants in the training are characterized according to different features, both in relation to the components closely linked to their role and responsibility, and for the particular "affective states" they may exhibit during the training experience in response to the presented stimuli. Therefore, each trainee, by interacting with the system, feeds personal data to the PANDORA-BOX, which gathers this information to build a *user model*. Also the trainees models are represented by means of time qualified facts using the timeline-based plan representation and reasoned upon by the Behavioral Reasoner (Figure 1). Such reasoner synthesizes additional temporal information used by the planner as heuristic bias to personalize training paths for each trainee (see an example later in the paper).

Overall the PANDORA-BOX supports the loop *trainer → training environment → trainee*, encouraging the customization and adaptation based on the users feedback as well as to keep in the loops goals and suggestions from the trainer.

3   Using Planning for Organizing Lesson Content

Basic goal for our learning environment is to create and dynamically adapt a four hours lesson. The pursued idea is to represent lesson's content as a *plan* composed of different "messages" to be sent to trainees which have temporal features and causal relations among them. In PANDORA a lesson master plan is first synthesized starting from an abstract specification given by the Trainer, then it is animated, expanded and updated during its execution, in relations to new information gathered form trainees and their decisions. Specifically, the lesson master plan contains time-tagged activities that trigger multimedia events presented on the Trainee Clients. A key point is represented by the reaction of trainees to lesson stimuli (e.g., the answer to a request to produce a joint decision on a specific critical point). "User reactions" are internally represented in the plan and trigger different evolutions of the current plan thus supporting dynamic adaptation.

The use of AI planning is quite natural for creating such a master plan. Previous work exists on the use of constraint reasoning for synthesizing multi-media presentations (e.g., [2]), the use of planning in story-telling (e.g., [3]), etc. The main "technological idea" we have pursued in PANDORA is to use timeline-based technology to represent and manage heterogeneous information. In particular two aspects offered an interesting challenge for timeline based technology: (a) the idea of doing planning, executing, replanning in a continuous cycle; (b) the possibility of modeling a completely different type of information with respect to the "usual" space domains in which timeline-based planning has been used (e.g., [4]).

3.1   A Timeline-Based Problem Representation

Figure 2 exemplifies the basic modeling features and introduces some terminology for the PANDORA domain modeling. The main data structure is the timeline which, in generic terms, is a function of time over a finite domain. Possible values for a timeline are generically called "events". Events are represented as a predicate holding over a time interval and characterized by start and end time[2].

---

[2] Events here are equivalent to "tokens" in other timeline-based approaches – e.g., [4]. The reason for re-naming them is for focusing attention on the PANDORA main task, namely the generation of timeline segments suitable to be "rendered" as specific multi-media *events*.

Events can be linked each other through *relations* in order to reduce allowed values for their constituting parameters and thus decreasing allowed system behaviors. In general, relations can represent any logical combination of linear constraints among event parameters. According to the number of involved events, relations can be divided into unary, binary, and *n*-ary. For example, unary relations are used in the PANDORA-BOX to

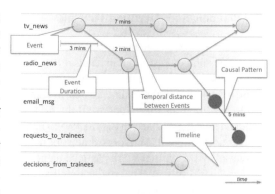

Fig. 2. The timeline-based plan data structure

fix initial scenario events' parameters placing them in time. Given an event e, an example of unary relation can be $start\text{-}at\,(e, 15, 20)$ forcing starting time of event e to be constrained inside simulation time interval $[15, 20]$. Given two events e_0 and e_1, an example of binary relation can be $after\,(e_0, e_1, 100, 120)$, forcing starting time of event e_1 constrained to be minimum 100 and maximum 120 time units after ending time of event e_0.

An "Event Network" is an hyper-graph having events as nodes and relations as hyper-edges. Through the concept of Event Network, the whole timeline-based planning procedure can be reduced to the process of reaching a target Event Network, that meets desired *goal* conditions, starting from an initial, already consistent, Event Network. In our case, goal conditions are characterized by high level scenario events representing the abstract blueprint for the master plan while the initial Event Network is, trivially, an empty Event Network. In the example of Figure 2 we see an Event Network distributed over 5 timelines (three representing different media for giving "active" information in a situation (tv_news, $radio\_news$, $email\_msg$), two more special purpose to ask input to trainees and for gathering such input ($request\_to\_trainees$, $decisions\_from\_trainees$).

A further basic ingredient in timeline modeling are the so-called "Causal Patterns" (see an example in Figure 2). They are the way to express planning domain/causal rules in the current internal representation. Any given Event Network should be consistent with respect to the set of such specified causal patterns[3]. Patterns are defined through a logic implication $reference \rightarrow requirement$ where $reference$ is the event value that demands pattern application while $requirement$ is the "consequence" of the presence of the $reference$ value in the Event Network. Making use of a recursive definition, a requirement can be a target event value, representing a new value on the same or on another timeline, a relation among reference value and target values, a conjunction of requirements or a disjunction of requirements. Being relations, in the most general case, linear constraints, causal patterns allow great expressiveness to represent quite complex behaviors.

---

[3] Causal Patters are defined with a domain description language, similar to *compatibilities* [4], that allows to specify a pattern of mixed time/causal value relations involving events.

3.2   Planning and Executing Timelines

Given these modeling ingredients, a planning domain is generically defined by a set of timelines and a set of domain Causal Patterns. From a domain representation, given a set of goals a planner generates an event network to be executed. As also shown in Figure 1 such executable plan is stored in a specific data structure, the Timeline-based Plan Representation, that supports all the back-end operations.

Indeed the planning activity in the PANDORA-BOX is initially triggered by the Trainer who acting through its support framework defines and loads a specific "Scenario", an abstract plan sketch that works as a sequence of "lesson goals" for the ground planner. A scenario fills in a particular timeline that generates sub-goaling by interacting with the set of domain causal patterns. The scenario has the double role of enabling the Trainer to reason on a high level of abstraction thus avoiding the details of the planning technology and to continuously influence a blueprint for the event network that actually implements a lesson at ground level. Furthermore the Trainer has available commands for introducing single steps in a scenario and triggering dynamic plan adaptation. It is worth highlighting how the overall system is intended as a means to empower the trainer with a more effective means to train people. Indeed the suggested crisis stimuli as well as the behavioral analysis is offered to the trainer who can influence at any moment the training session in perfect line with a *mixed-initiative* style.

The Planner works on the ground timeline representation to create the training storyboards, e.g., the set of connected "events" that are communicated to the trainees (e.g., a video news from the crisis setting, a phone call or e-mail from a field manager, and a set of temporal distances among events). Once the planner has achieved a fix-point given the abstract scenario goals from the Trainer and the domain Causal Patterns, the responsibility is left to the Executor that step-by-step executes the plan. According to their progressive start times, plan events are sent to the Rendering Environment. Some of the events are requests for trainees to make some decisions (see Figure 2), the result of which are fed back to the timeline representation as further open information for plan adaptation. In fact the Planner is able to reacts to trainees' strategic decisions, triggering consequent events to continue the training session.

A further path on the dynamic adaptation of the lesson plan is given by the personalization for each trainee which is fostered by the block named Behavioral Reasoner. Through this module psycho-physiological trainee features are modeled and updated during training and internally represented as timelines. Specifically a set of "relevant user features" has been selected among those that influence human behavior under crisis and used to build a trainee model. The timeline-based approach also supports the dynamic update of the user model. Based on this model, the Behavioral Reasoner, synthesizes specific result timelines that are used as goals by the general planner, thus introducing a continuous loop used to tailor the intensity of stimuli to individual trainees.

Planning a lesson. Starting from goal events and from the set of domain causal patterns, the planning process generates a target Event Network that is consistent with given goals, ordering events in time through scheduling features and producing proper event consequences. In general, consequences are added to current Event Network producing new goals that require pattern application in order for them to be causally justified.

Additionally, as described in the previous section, new goals can be added during crisis simulation to represent (a) decisions taken by trainees, (b) inferences made by the behavioral reasoner, (c) new scenario steps added by the Trainer. The PANDORA planner is therefore able to replan in order to make its current Event Network consistent with respect to the new dynamic input and with its consequences, namely, changing the current course of the simulated crisis. It is worth noting that disjunctions of requirements produce branches on the search tree guaranteeing varieties of presented scenarios. In particular, it may happen that some rule cannot be applied since it imposes too strict constraints resulting in an inconsistent Event Network. In such cases, a back-jumping procedure allows to go back to the highest safe decision level. The planning process is summarized in Algorithm 1, it goes on until no pattern need to be applied while providing assurance of a consistent Event Network. If the planning process succeeds, value $\top$ is returned and an Event Network, consistent with given goals, replaces current planner state.

Algorithm 1. General solving procedure

> **while** $\exists e \in events : e$ has no applied pattern **do**
>> **if** $\neg\, (applyPattern\,(e) \wedge schedule\,(e))$ **then**
>>> **if** $\neg backJump\,()$ **then**
>>>> **return** $\perp$
>>> **end if**
>> **end if**
> **end while**
> **return** $\top$

Because not all courses of actions in a crisis can be predicted at scenario design time we have endowed the Trainer with a service that allows to incrementally modify the ongoing scenario in order to adapt the simulation to unpredicted trainees' decisions. Alternatively, the trainer can manipulate ongoing crisis to bring back execution to a desired behavior having already predicted courses. This kind of scenario modifications are stored in a knowledge base providing capacity to expand and evolve the system making PANDORA "future proof". In general, the flexibility of the timeline-based underlying engine supports us in incrementally extending the services toward PANDORA-BOX users. This turned out to be particularly interesting in an application domain relatively new for timeline technology.

Trainees modeling and personalization. The trainee's profile are built by considering relevant variables known to have an influence in decision making under stress [5]. An initial assessment is made through standardized psychological tests and physiological measurements made off-line (through pre-created questionnaires), immediately before the training session begins and updated during the training session (through specific questions during the lesson). Similarly to the storyboard, we choose to model trainees variables through timelines in order to maintain a homogeneous representation of information. Therefore we will call *event* each value on any timeline.

We clarify this with a simplified example. Two trainee features that are relevant during training are:

- the *background experience*, the crisis leader' past experience in managing crisis situation. A very short questionnaire will assess leaders socio-demographic information, their previous experiences with leading public health and safety crises and their level of success in doing it. We represent *background experience* through predicates of the form $background\text{-}experience\,(x)$ where x is an integer assuming values 0 for low experience, 1 for medium experience and 2 for high experience;
- the *self efficacy* defined by [6] as the people belief in their capabilities to perform a certain task successfully. It has been shown that this variable has influence of different aspects like managing stressful situation, increase the performance as well as receive benefits from training programs. We represent the *self efficacy* through predicates of the form $self\text{-}efficacy\,(x)$ with x being an integer ranging from 0 to 10.

Let us suppose that a trainee x answers to a self-efficacy question during a training session and that, consequently, an event representing its updated self-efficacy is added to his (or her) self-efficacy timeline. The causal patterns that is applied by solving procedure will have a structure similar to the following:

$$x.self\text{-}efficacy \rightarrow \begin{cases} cat : x.category \\ during\,(this, cat, [0, +\infty]\,, [0, +\infty]) \end{cases}$$

This patterns assures that every time we have a self-efficacy update, an event, named cat locally to the rule, is added to $category$ timeline of trainee x, new self-efficacy value must appear "during" cat (triggering event's starting point is constrained to be $[0, +\infty]$ before cat's starting point while cat's ending point is constrained to be $[0, +\infty]$ before triggering event's ending point). Event cat is added to current Event Network once solving procedure is called and requires itself a pattern application.

Let's assume now that the following requirements, representing trainee categorizations, are defined inside the Behavioral Reasoner:

$$r_0 : (se.value = 0 \wedge be.value = 0 \wedge is.value = 0)$$

$$r_1 : (se.value = 1 \wedge be.value = 0 \wedge is.value = 1)$$

$$\ldots$$

These requirements basically state that, if self-efficacy value is equal to 0 and background-experience is equal to 0 than induced-stress' $value$ parameter must be equal to 0, if self-efficacy value is equal to 1 and background-experience is equal to 0 than induced-stress' $value$ parameter must be equal to 1. Exploiting given requirements, category's pattern "simply checks" values on other x's modeled timelines just imposing their values taking advantage of and/or evaluation and, then, generates consequences for Crisis Planner in terms, for example, of induced-stress for trainee x. Resulting patterns is:

$$x.category \rightarrow \begin{cases} se : (?) \, x.self\text{-}efficacy \\ be : (?) \, x.backgroung\text{-}experience \\ is : x.induced\text{-}stress \\ contains \, (this, se, [0, +\infty], [0, +\infty]) \\ contains \, (this, be, [0, +\infty], [0, +\infty]) \\ equals \, (this, is) \\ r_0 \vee r_1 \vee \ldots \end{cases}$$

where the $(?)$ symbol forces target values se and be to "unify" with an already solved event in order to close the loop and interrupt the pattern application process for the event. It is worth noticing that, with the given pattern, if unification is not possible or a self-efficacy value of 2 is given by trainee and no requirement is defined for it, there is no way of applying it thus generating an unsolvable Event Network.

Finally, induced stress pattern selects proper events from the domain knowledge and propose them to the trainee in order to generate an adequate stress level with the aim of maximizing the learning process.

Executing the lesson plan. Last crucial aspect of the PANDORA system is represented by the crisis scenario execution. Training process requires ease of temporal navigation through storyboard allowing execution speed adjustments as well as rewinding features. Simulation time t is maintained by execution module and increased of execution speed dt at each execution step. Each timeline transition that appears inside interval $[t, t + dt]$ is then dispatched to PANDORA rendering modules for creating the best effect for the target trainees.

When going back in time, two different behaviors are provided by PANDORA-BOX:

– default roll-back, intended for debriefing purposes, that simply updates current simulation time t to desired target value keeping untouched actions taken by trainees;
– heavy roll-back, intended to revert to a crucial decision point at time t, removing each event representing trainees' choices at time $t' > t$, along with their consequences, in order to allow a different simulation course.

All the information – stimuli sent to the class, trainees psycho-physiological state, their taken actions – is represented in the final Event Network. Hence, it is available as supporting material for the trainer during the debriefing session.

While default roll-back does not involve planning at all, heavy roll-back requires events removal, events generated as their consequences as well as involved relations. We need to restart from an empty Event Network and gradually add required events and relations. Although simple, this approach is computationally heavy and highlights current difficulties in removing elements from an Event Network.

Finally, maintaining information about current simulation time, the executor module is responsible for placing in time events that represent trainees' actions, adding proper relations, thus fostering replanning features in order to integrate actions' consequences inside the current Event Network.

Fig. 3. Screenshots of the current Trainer and Trainee Interfaces

4   The PANDORA-BOX Interactive Features

We close this overview of the PANDORA-BOX functionalities by describing the interfaces to real users. Figure 3 depicts some of the interaction features in the current demonstrator. As direct consequence of the choices in the architecture, the system distinguishes between two types of interactions:

- *trainer-system interaction*, indicated as *Trainer View*, which is related to the functionalities available to the trainer to create a training session, monitor, edit it and interact dynamically with the class;
- *trainee-system interaction*, indicated as *Trainee View*, implemented by XLAB, which is the interface through which trainees can connect to the PANDORA-BOX, receive stimuli and make decisions.

Additionally we have a further view, called *Expert View*, which is an inspection capability over the timeline environment and its execution functionalities.

Trainer View. This service allows to compose a *class* completing it with "missing players" to have a coverage of institutional roles in crisis strategic decision making. Created a class the trainer can load a *Scenario*, and see it in tabular form with a series of important information such as the execution time of each goal event and who is the main recipient of information. It is worth highlighting how this representation is close to the current way of working of the trainers and has been instrumental in establishing a dialogue with them, before proposing any kind of completely new solutions. Along with the scenario, the interface also contains information about available resources to resolve the crisis and the consequences of trainees' decisions, both represented through resource timelines and dynamically updated during the training. The trainer is the one to have the basic commands from executing the plan, stopping execution, resuming it and rewinding. Furthermore, a specific

requirement from user centered design has been a set of plan annotation functionalities plus a series of additional commands which allows the trainer to dynamically
add new stimuli, in perfect line with the mixed initiative interaction style.

Trainee View. The Trainee interface contains three main blocks, plus a number of features related to communication of each trainee with the rest of the class and the
trainer. The main building blocks are: (1) *Background Documents*, which represents
a set of information delivered off-line to the class in the form of maps, documents,
reports, in order to create awareness about the upcoming exercise; (2) *Dynamic
information* that represents the information dynamically scheduled and sent to the
trainee in the form of videos, maps, decision points etc.; (3) *Main Communication
Window*, which is devoted to display stimuli (possibly customized) to individual
trainees or to the class.

Expert View. In parallel with the traditional tabular view, the trainer can inspect the
more advanced view of the PANDORA module, that is the internal representation for
both the Crisis Planner and the Behavioral Reasoner. As already said, all type of
information within PANDORA is represented as a timeline and continually updated
(see different colors for timelines related to the crisis and the user model in the Expert
View). At this point, through the Execute button, the trainer can start the session.

5   Conclusions

This paper introduces features of the current demonstrator of the PANDORA project.
Main goal of the paper is to give the reader a comprehensive idea of the use of planning
technology in the PANDORA-BOX. We have seen how the representation with timelines
is the core component of the crisis simulation, and that a continuous loop of planning,
execution, plan adaptation is created to support personalized training with the Trainer
in the loop.

Acknowledgments. The PANDORA project is supported by EU FP7 joint call ICT/
Security (GA.225387) and is monitored by REA (Research Executive Agency). Authors
are indebted to all the project partners for the stimulating work environment.

References

1. Wilkins, D., Smith, S.F., Kramer, L., Lee, T., Rauenbusch, T.: Airlift Mission Monitoring and
   Dynamic Rescheduling. Eng. Appl. of AI 21(2), 141–155 (2008)
2. Jourdan, M., Layaida, N., Roisin, C.: A Survey on Authoring Techniques for Temporal Scenarios of Multimedia Documents. In: Furht, B. (ed.) Handbook of Internet and Multimedia Systems
   and Applications - Part 1: Tools and Standards, pp. 469–490. CRC Press, Boca Raton (1998)
3. Young, R.M.: Notes on the Use of Plan Structures in the Creation of Interactive Plot. In: Working
   Notes of the AAAI Fall Symposium on "Narrative Intelligence", Cape Cod, MA (1999)
4. Muscettola, N.: HSTS: Integrating Planning and Scheduling. In: Zweben, M., Fox, M.S. (eds.)
   Intelligent Scheduling. Morgan Kaufmann, San Francisco (1994)
5. Cortellessa, G., D'Amico, R., Pagani, M., Tiberio, L., De Benedictis, R., Bernardi, G., Cesta,
   A.: Modeling Users of Crisis Training Environments by Integrating Psychological and Phys-
   iological Data. In: Mehrotra, K.G., Mohan, C.K., Oh, J.C., Varshney, P.K., Ali, M. (eds.)
   IEA/AIE 2011, Part II. LNCS (LNAI), vol. 6704, pp. 79–88. Springer, Heidelberg (2011)
6. Bandura, A.: Social Foundations of Thought and Actions. A Social Cognitive Theory. Prentice
   Hall, Englewood Cliffs (1986)

Clustering and Classification Techniques for Blind Predictions of Reservoir Facies

Denis Ferraretti[1], Evelina Lamma[1], Giacomo Gamberoni[2], and Michele Febo[1]

[1] ENDIF-Dipartimento di Ingegneria, Università di Ferrara, Ferrara, Italy
{denis.ferraretti,evelina.lamma}@unife.it,
michele.febo@student.unife.it
[2] intelliWARE snc, Ferrara, Italy
giacomo@i-ware.it

Abstract. The integration of different data in reservoir understanding and characterization is of prime importance in petroleum geology. The large amount of data for each well and the presence of new unknown wells to be analyzed make this task complex and time consuming. Therefore it is important to develop reliable prediction methods in order to help the geologist reducing the subjectivity and time used in data interpretation. In this paper, we propose a novel prediction method based on the integration of unsupervised and supervised learning techniques. This method uses an unsupervised learning algorithm to evaluate in an objective and fast way a large dataset made of subsurface data from different wells in the same field. Then it uses a supervised learning algorithm to predict and propagate the characterization over new unknown wells. Finally predictions are evaluated using homogeneity indexes with a sort of reference classification, created by an unsupervised algorithm.

Keywords: Knowledge discovery and data mining, clustering, classification, application, petroleum geology.

1 Introduction

Data interpretation and understanding can be complex tasks especially when there are large amounts of data to consider and when they are of different type.

In petroleum geology, it is important task to use all the available borehole data to completely characterize both the reservoir potentials and performance of reservoirs. The large amount of data for each well and the presence of different wells also make this task complex, especially if the subjectivity of the data interpretation has to be reduced. The development of reliable interpretation methods is of prime importance regarding the understanding of the reservoirs. Additionally, data integration is a crucial step to create useful description models and to reduce the amount of time needed for each study. Those models can eventually be used to predict the characterization of unknown wells from the same field. The manual interpretation of a well can take up to one month.

R. Pirrone and F. Sorbello (Eds.): AI*IA 2011, LNAI 6934, pp. 348–359, 2011.

Artificial intelligence, data mining techniques and statistics methods are widely used in reservoir modeling. The prediction of sedimentary facies[1] from wireline logs can be performed by means of a Markov chain [1], and the approach can be improved by integrating different sources [2] from the same well. These techniques help the geologist in facies analysis [3], [4] and, when combining this with neural networks, lead to the development of new interpretative methods for reservoir characterization [5]. However, reservoir characterization is improved when information from different wells in the same field/area is taken into consideration, giving reliable support to further analysis of unknown wells in the same field.

In this paper we develop an interpretation method which uses clustering to objectively and quickly evaluate a large dataset made of subsurface data from different wells in the same field. Then it uses decision trees or regression methods to predict and propagate the characterization of new unknown wells. In particular we propose a range of techniques, i.e., clustering and learning classification algorithms, in order to, first, cluster several regions of different wells into similar groups, by applying hierarchical clustering; choose the set of most significant clusters (this is done by the experts in the domain) and finally feed a machine learning algorithm in order to learn a classifier to be applied to new instances and wells.

The paper is organized as follows: a general background on borehole logging and the available dataset are outlined in Section 2. The overall interpretation method is introduced in Section 3. A detailed explanation of experimental results over different prediction algorithms is given in Section 4. Finally, section 5 concludes the paper.

2 Background

Well logging, also known as borehole logging, is the practice of making a detailed record (a well log) of the different characteristics of the geologic formations penetrated by a borehole. There are many types of logging tools, therefore there are many type of logs: resistivity, porosity, acoustic logs and image logs or FMI[2] logs. In our work we use all the previous log types properly integrated in a large dataset. While electric logs are provided as a table with numeric values along the well depth, image logs are digital images that represent resistivity measurements of the rock formation taken from the wellbore surface. FMI log interpretation is a very complex task, due to the large number of variables and to the vast amount of data to be analyzed.

In order to integrate data from different sources, we need to convert image log observation and interpretation into numeric data. This task is done by using results from our previous work based on I^{2}AM [4].

---

[1] A body of sedimentary rock distinguished from others by its lithology, geometry, sedimentary structures, proximity to other types of sedimentary rock, and fossil content.

[2] FMI (Fullbore Formation MicroImager) is the name of the tool used to acquire image logs based on resistivity measures within the borehole.

Our interpretation method has been developed and tested using electric and image logs from 5 different wells located in the same area. Available wells and number of instances: *well1* (1799), *well2* (1023), *well3* (1214), *well4* (1041) and *well5* (953). After the testing phase, we decided to use only five of the available logs. Selected attributes are: number of sinusoid in the analysis window (SIN), spectral gamma ray (SGR), bulk density (RHOB), delta-T compressional (DTCO), neutron porosity (PHI). We also have three additional attributes: the depth of the measurement (DEPTH), the geological unit[3] (UNIT) and the name of the well (WELL-NAME). Each dataset has a sampling resolution of 10 inches.

3   Methodology

The new interpretation method we developed helps the geoscientists in their analysis, by creating a model that describe wells structures and applying it to a new unknown well.

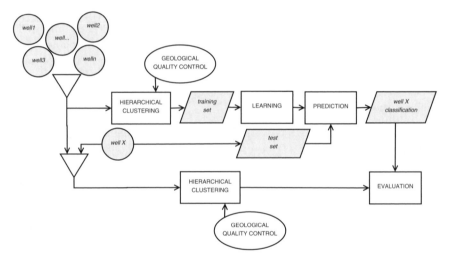

Fig. 1. Prediction of a unknown well. In the evaluation phase we use the whole dataset.

In our previous work [6] we predict a facies distribution using a classificator trained on the dataset created by merging the datasets from all the wells, including the well to be predicted. This is far from the real use: the geologist could start the analysis with some wells and then a new unknown well, from the same area, is added. In this case it is important to be able to reuse the previous validated work. Our approach involves two phases (see Figure 1 and Figure 2): first, hierarchical clustering is applied to a set of co-located wells to obtain a reliable human validated data structure that represents the facies distribution.

---

[3] A body of rock or ice that has a distinct origin and consists of dominant, unifying features that can be easily recognized and mapped.

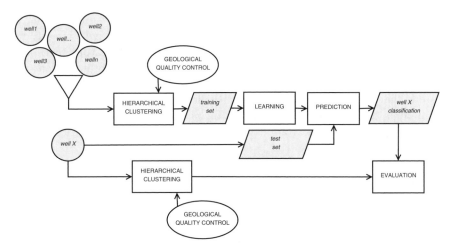

Fig. 2. Prediction of an unknown well. In the evaluation phase we use only the dataset of the unknown well

Then starting from identified clusters, a supervised learning algorithm is used to learn a model which can be applied to new wells, in order to predict the distribution of facies.

We first create a large dataset that includes data from different wells in the same area, this is the input of a clustering task. Clustering is the classification of data into different classes/clusters so that the data in each class ideally share some common characteristics. The identified clusters take into account all the characteristics and rock properties of all the known wells in the studied area.

The second phase involves the prediction of facies distribution over a new unknown well in the same area. This task is achieved learning the model of each cluster from the previous description by applying supervised learning algorithms. This technique allows the propagation of classes in new wells. Following these two phases we obtain a semi-automatic interpretation method for well logs. It is important to note that differently from [6], the dataset of the well to be predicted is not combined in the clustering with all the other datasets. This means that the test set does not contain any class or cluster information for each item. It is a sort of "blind prediction" and it involves new way of evaluation.

Data integration and clustering. The initial classification was obtained for the *starting dataset*. This dataset was built by adding all the data from 4 wells into a single table, keeping only one well as test set. The clustering process was conducted using the approach described in [4] and [7]. We applied a hierarchical agglomerative approach [8] with the following settings: Z normalization[4], Manhattan distance[5] and maximum linking.

---

[4] A linear normalization of each variable that brings mean to 0 and variance to 1.

[5] The distance algorithm used in the clustering process can handle missing data. If some attributes are missing for certain examples, the distance is computed only with the remaining ones.

Hierarchical agglomerative clustering builds the hierarchy starting from the individual elements considered as single clusters, and progressively merges clusters according to a chosen similarity measure defined in the feature space [8].

The dataset was finally cut by the geologist, in order to obtain a clear classification of rock types. In all our experiments this step generated 8 different clusters, that has been used as classes in the following step. The same clustering technique is then applied in the evaluation phase, in order to create a reliable reference classification and compare it with the predicted classification.

Supervised learning and prediction. In order to find the most reliable interpretation method and the best prediction algorithm, we tested several techniques based on different learning approaches. We use J4.8 [9], Random Forests [10], PART [11] and Rotation Forest [12] as decision tree induction and classification rule generation algorithms. For regression we use ClassificationViaRegression [13] and Logistic [14].

Evaluation of the approach. The classifier's evaluation is often based on prediction accuracy: the percentage of correct prediction divided by the total number of predictions. There are at least three techniques which are used to compute a classifier's accuracy: the one-third/two-third technique, the leave-one-out and the cross-validation [15]. In our particular case we can't apply directly none of the previous techniques because, to calculate the prediction accuracy, we miss the "correct" class information for each item. In order to evaluate the prediction algorithms we must set a reference classification of the unknown well. This will be used as an ideal result to compare performances of different prediction algorithms. We propose two different types of evaluation based on different datasets: the first technique (see Figure 1) uses a new dataset obtained by merging the *starting dataset* with the dataset of the unknown well, the second technique uses only the dataset of the unknown well (see Figure 2).

The geologist creates the cluster partition by cutting the tree at different distances. It is important to cut the tree for the same number of clusters and to use the same criteria used in the initial clustering (i.e. color mosaic observations or clustering metrics). In this way we obtain a clustering solution that will be used as reference classification comparable with the one created in the prediction[6]. First we use a visual comparison between predicted classes and reference classification. This can be done using a software that shows classes sequence with different colors along the well (see Figure 3 and Figure 4). In these results it is easy to observe classes changes and trends. Moreover with the reference classification we still can't directly calculate the accuracy of the prediction algorithm because the new clusters do not necessarily match with classes of the predicted classification. We need a measure of how two different classifications are homogeneous and consistent, regardless of the name of the classes. We use *entropy* and *purity*.

---

[6] For clarity we will refer to reference classification as *clusters*, and the predicted classification as *classes* in the rest of the paper.

Entropy, or *information entropy* [9,16], aims to highlight matching between the predicted classification and the chosen reference classification. We can define the information entropy of a single class as the uncertainty relative to the cluster attribute for its examples. Entropy for the $i - th$ class can be computed using the following equation:

$$H_i = - \sum_{j=1}^{n_c} \frac{n_{ij}}{n_i} \log \frac{n_{ij}}{n_i}$$

where n_c is the number of clusters, n_i the number of examples of the $i - th$ class, and n_{ij} the number of examples of the $j - th$ cluster in the $i - th$ class. A low entropy value reveals the "homogeneity" of a class, with respect to the cluster attribute. A class containing instances from only one cluster, will score an information entropy equal to 0. We can evaluate the entropy of the whole predicted classification by computing the weighted mean of the entropy of each class. The number of instances belonging to the class is used as weight. This equation can be written as:

$$H = \frac{1}{N} \sum_{i=1}^{n_C} n_i H_i$$

where N is the number of instances in the whole dataset, n_C the number of classes and n_i the number of examples into the $i - th$ class. A low overall entropy value represents a good matching between the predicted classification and the reference classification.

Purity [17] is a simple and transparent quality measure of classification solution. To compute *purity*, each classes is assigned to the cluster which is most frequent in the class, and then the accuracy of this assignment is measured by counting the number of correctly assigned items and dividing by N. Formally *purity* for $i - th$ class is:

$$P_i = \frac{1}{n_i} max(n_{ij})$$

The overall *purity* of the predicted solution could be expressed as a weighted sum of individual classes purities:

$$P = \sum_{i=1}^{n_c} \frac{n_i}{N} P_i$$

In general, the bigger the value of *purity* the better the solution.

4   Experimental Results

In our tests the supervised learning algorithm uses as training set the dataset created by merging 4 of the 5 wells datasets, then predicts classes in a test well, excluded from the same large dataset. It is important to note that in [6] the knowledge about the characteristics of the well that will be predicted is combined with all other wells and used in the learning phase, conversely this approach does not use any prior knowledge about the test well.

All experiments are conducted using the hierarchical clustering software developed in our previous work [4]. For the supervised learning phase we use WEKA [18], the open source data mining software written in Java. In all our experiments we use as test wells *well2* and *well4*. When the test set is *well2* the training set is made up of 5007 instances (*well1, well3, well4, well5*) and when the test is *well4* the training set counts 4989 items (*well1, well2, well3, well5*). We extend the datasets by adding two attributes: normalized depth (NORM-DEPTH) and UNIT. UNIT is the numerical ID of the geological unit and NORM-DEPTH is the depth linear normalization: its value is 0 at the top and 1 at the bottom of the analyzed section. These values are the same for all the wells although, due to the different geological description, the real depths are different. Using NORM-DEPTH, instead of DEPTH, in conjunction with UNIT in the prediction algorithm, it is possible to better consider different rock type. For every dataset we use 6 attributes: SIN, SGR, RHOB, DTCO, PHI, NORM-DEPTH, and UNIT. The geologist identified 8 different clusters, recorded as CLUSTER-NAME attribute in the dataset.

Visual comparison. Figure 3 and Figure 4 show a visual comparison between predicted classes and reference classification. The first two columns are the reference classifications: made by using the whole dataset and made by using only the test well. Due to the evaluation method, in this comparison the differences between reference and predicted color classes do not matter. More important are the changes in classes sequence. It is difficult to evaluate the algorithms in *well4* (Figure 4) because it presents rapid classes changes along the well, but in both wells UNIT IV.3 is clearly detected by predicted classification. Also the transition between UNIT IV.2 and UNIT III is correctly identified by all the algorithms. An important consideration made by the geologist is that, due to the number of classes it is very difficult to evaluate and choose the best algorithm, but looking at the reference classification, it seems that the first column (clustering made by using the whole dataset) is more readable than the second. It presents fewer details and it is less complex.

***Entropy* and *purity* comparison.** Table 1 show results of *entropy* and *purity*. In order to better understand results, making further tests, we choose to predict some interesting sections of *well2* and *well4*: UNIT IV.2 (inf+sup) and UNIT IV.3. To locate them see Figure 3 and Figure 4. For each section we create the training set extrapolating the same geological unit from all the wells. Using both the evaluation techniques, the whole dataset evaluation (see Figure 1) and the test dataset evaluation (see Figure 2), we predict and calculate *entropy* and *purity* of each well and section. Looking at these results, Logistic shows better performance than other algorithms in most cases. Logistic results for *well2 - UNIT IV.2* are not very good, but in fact this section is not very meaningful because it is short and very homogeneous. This result confirm, as expected, that regression methods are suitable for prediction of continuous numeric values.

Table 1. Result of *entropy* and *purity* for chosen wells and sections

	whole dataset eval.		test dataset eval.	
	entropy	*purity*	*entropy*	*purity*
well2				
ClassificationViaRegression	0.902	0.652	0.865	0.633
J4.8	0.946	0.625	0.948	0.603
Logistic	0.873	0.646	**0.778**	**0.668**
PART	0.944	0.628	0.943	0.608
Random Forests	0.905	0.636	0.873	0.622
Rotation Forest	**0.854**	**0.665**	0.853	0.635
well2 - UNIT IV.2				
ClassificationViaRegression	0.199	0.963	0.262	0.938
J4.8	**0.132**	**0.975**	0.195	0.951
Logistic	0.199	0.963	0.262	0.938
PART	**0.132**	**0.975**	0.195	0.951
Random Forests	0.149	0.963	**0.181**	**0.963**
Rotation Forest	0.199	0.963	0.262	0.938
well2 - UNIT IV.3				
ClassificationViaRegression	0.817	0.663	0.774	0.694
J4.8	0.869	0.641	0.842	0.665
Logistic	**0.760**	**0.679**	**0.677**	**0.719**
PART	0.889	0.647	0.859	0.679
Random Forests	0.854	0.647	0.806	0.680
Rotation Forest	0.837	0.663	0.811	0.680
well4				
ClassificationViaRegression	0.718	0.741	0.755	0.689
J4.8	0.745	0.735	0.775	0.681
Logistic	0.703	0.751	**0.737**	0.697
PART	0.694	**0.774**	0.0.742	**0.705**
Random Forests	0.728	0.748	0.779	0.689
Rotation Forest	**0.683**	0.769	0.761	0.696
well4 - UNIT IV.2				
ClassificationViaRegression	0.743	0.720	**0.702**	**0.732**
J4.8	0.799	0.701	0.805	0.720
Logistic	0.674	0.732	0.739	0.720
PART	0.643	**0.768**	0.759	0.720
Random Forests	0.690	0.750	0.742	0.701
Rotation Forest	0.640	0.739	0.743	0.726
well4 - UNIT IV.3				
ClassificationViaRegression	**0.902**	**0.671**	0.509	**0.787**
J4.8	1.004	0.606	0.559	0.740
Logistic	0.908	0.628	0.612	0.697
PART	0.998	0.599	0.564	0.711
Random Forests	0.965	0.625	0.550	0.733
Rotation Forest	0.904	0.657	0.516	0.765

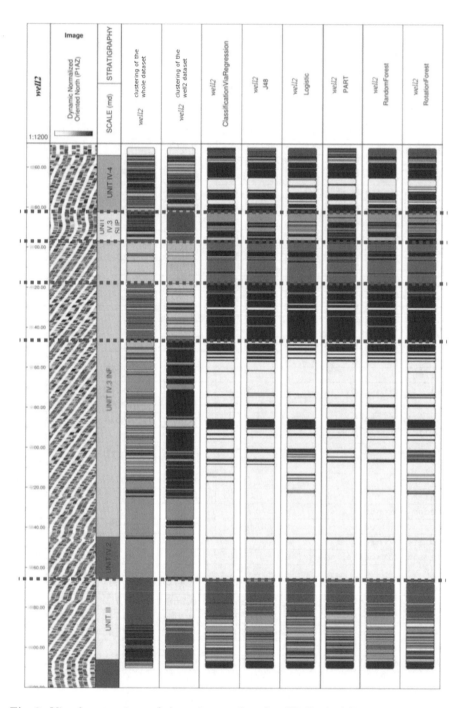

Fig. 3. Visual comparison of clustering results of *well2*. Dashed lines represent main changes in cluster distribution correctly detected by prediction algorithms.

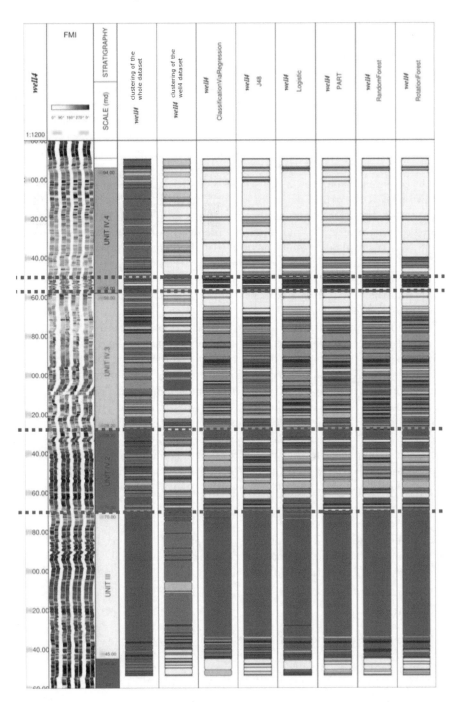

Fig. 4. Visual comparison of clustering results of *well4*. Dashed lines represent main changes in cluster distribution correctly detected by prediction algorithms.

5   Conclusions

A novel interpretation method for a large dataset made of subsurface data has been proposed and tested in a well log interpretation task. It consisted of merging the dataset from different wells in the same area, identifying facies distribution with clustering (human interpretation), learning the clustering solution in a description model and then predicting results for a new unknown well from the same area. Each well dataset was built by integrating different data: electrical logs and image logs. Image logs were automatically processed in order to obtain a numeric description of the interested features.

A large dataset made by 2 wells from a hydrocarbon reservoir was created, then we used it as a training set of decision trees or regression techniques. We finally tested the learned model by predicting the facies distribution over two unknown wells.In order to find a reliable prediction algorithm we tested several techniques. We evaluate results using a visual comparison and computing *entropy* and *purity* over a reference classification. This classification is generated using two different dataset: the starting dataset merged with the unknown well dataset and the only test well dataset. `Logistic` was a good compromise for the prediction of tested wells.

The main advantages of this approach are the extraction of realistic information about the rock properties and identify facies that can help in the reservoir characterization; the avoidance of interpretation subjectivity; and the reduction of the interpretation time by largely automating the log interpretation.The time reduction given by our approach has a great impact in costs of reservoir analysis and interpretation. The experimental results show that the approach is viable for reservoir facies prediction in real industrial contexts where it is important to reuse information about wells already analysed. This approach represents an interesting AI application, that integrates different supervised and unsupervised techniques. Furthermore, the application includes a technical quality control step, provided by human interaction, in order to extract and evaluate prediction models.

Acknowledgements. This work has been partially supported by CCIAA di Ferrara, under the project "Image Processing and Artificial Vision for Image Classifications in Industrial Applications".

References

1. Bolviken, E., Storvik, G., Nilsen, D.E., Siring, E., Van Der Wel, D.: Automated prediction of sedimentary facies from wireline logs. Geological Society, London, Special Publications 65, 123–139 (1992)
2. Basu, T., Dennis, R., Al-Khobar, B.D., Al Awadi, W., Isby, S.J., Vervest, E., Mukherjee, R.: Automated Facies Estimation from Integration of Core, Petrophysical Logs, and Borehole Images. In: AAPG Annual Meeting (2002)
3. Shin-Ju, Y., Rabiller, P.: A new tool for electro-facies analysis: multi-resolution graph-based clustering. In: SPWLA 41st Annual Logging Symposium (2000)

4. Ferraretti, D., Gamberoni, G., Lamma, E., Di Cuia, R., Turolla, C.: An AI Tool for the Petroleum Industry Based on Image Analysis and Hierarchical Clustering. In: Corchado, E., Yin, H. (eds.) IDEAL 2009. LNCS, vol. 5788, pp. 276–283. Springer, Heidelberg (2009)
5. Knecht, L., Mathis, B., Leduc, J., Vandenabeele, T., Di Cuia, R.: Electrofacies and permeability modeling in carbonate reservoirs using image texture analysis and clustering tools. In: SPWLA 44th Annual Logging Symposium, vol. 45(1), pp. 27–37 (2004)
6. Ferraretti, D., Lamma, E., Gamberoni, G., Febo, M., Di Cuia, R.: Integrating clustering and classification techniques: a case study for reservoir facies prediction. Accepted to IS@ISMIS2011: Industrial Session - Emerging Intelligent Technologies in the Industry. IS@ISMIS2011 will be held on June 2011, in Warsaw, Poland (2011)
7. Ferraretti, D., Gamberoni, G., Lamma, E.: Automatic Cluster Selection Using Index Driven Search Strategy. In: Serra, R., Cucchiara, R. (eds.) AI*IA 2009. LNCS, vol. 5883, pp. 172–181. Springer, Heidelberg (2009)
8. Theodoridis, S., Koutroumbas, K.: Pattern Recognition, 3rd edn. Academic Press, London (2006)
9. Quinlan, J.R.: Induction on Decision Trees. Machine Learning 1, 81–106 (1986)
10. Breiman, L.: Random Forests. Mach. Learn. 45(1), 5–32 (2001)
11. Frank, E., Witten, I.H.: Generating Accurate Rule Sets Without Global Optimization. In: Proceedings of the Fifteenth International Conference on Machine Learning, pp. 144–151 (1998)
12. Rodriguez, J.J., Kuncheva, L.I., Alonso, C.J.: Rotation Forest: A New Classifier Ensemble Method. IEEE Trans. Pattern Anal. Mach. Intell. 28(10), 1619–1630 (2006)
13. Frank, E., Wang, Y., Inglis, S., Holmes, G., Witten, I.H.: Using Model Trees for Classification. Machine Learning 32(1), 63–76 (1998)
14. Le Cessie, S., Van Houwelingen, J.C.: Ridge Estimators in Logistic Regression. Applied Statistics 41(1) (1992)
15. Kotsiantis, S.B.: Supervised Machine Learning: A Review of Classification Techniques. Informatica 31, 249–268 (2007)
16. Shannon, C.E.: A Mathematical Theory of Communication. CSLI Publications, Stanford (1948)
17. Zhao, Y., Karypis, G.: Criterion functions for document clustering: experiments and analysis, Technical Report. Department of Computer Science, University of Minnesota (2001)
18. Hall, M., Frank, E., Holmes, G., Pfahringer, B., Reutemann, P., Witten, I.H.: The WEKA Data Mining Software: An Update. SIGKDD Explorations 11(1) (2009)

Multi-sensor Fusion through Adaptive Bayesian Networks

Alessandra De Paola, Salvatore Gaglio, Giuseppe Lo Re, and Marco Ortolani

University of Palermo,
Viale delle Scienze, ed 6. – 90128 Palermo, Italy
{depaola,gaglio,lore,ortolani}@unipa.it

Abstract. Common sensory devices for measuring environmental data
are typically heterogeneous, and present strict energy constraints; moreover, they are likely affected by noise, and their behavior may vary across
time. Bayesian Networks constitute a suitable tool for pre-processing
such data before performing more refined artificial reasoning; the approach proposed here aims at obtaining the best trade-off between performance and cost, by adapting the operating mode of the underlying
sensory devices. Moreover, self-configuration of the nodes providing the
evidence to the Bayesian network is carried out by means of an on-line
multi-objective optimization.

Keywords: Ambient Intelligence, Bayesian Networks, Multi-objective
optimization.

1 Motivations and Related Work

Artificial reasoning in many real world scenarios relies on measurements collected from diverse sensory sources; commonly available devices are typically
affected by noise, and characterized by heterogeneity as regards their energy
requirements; moreover their behavior may vary across time.

One of the application scenarios of artificial intelligence where multi-sensor
data fusion is particularly relevant is Ambient Intelligence (AmI). The AmI
paradigm relies on the capability of sensing the environment, through the deployment of a pervasive and ubiquitous sensory infrastructure, surrounding the
user, for monitoring relevant ambient features. Among these, a high attention is
devoted to context information, such as the users' presence in monitored areas
or current users' activities [9,6,2].

In our work, we present a sample scenario of an AmI system devoted to detect users' presence through a wide set of simple and low-cost devices, possibly
affected by a non negligible degree of uncertainty, as well as devices capable of
measuring environmental features only partially related to the human presence,
and finally a limited set of more precise, though more expensive sensors. In particular, we suppose that the sensory infrastructure is embodied into a Wireless
Sensor Network (WSN) [1], whose nodes, pervasively deployed in the environment, are capable of on-board computing functionalities and are characterized
by limited, non-renewable, energy resources.

R. Pirrone and F. Sorbello (Eds.): AI*IA 2011, LNAI 6934, pp. 360–371, 2011.

In order to estimate the environmental features of interest, while keeping the sensor nodes operating costs low, we propose a system that fully exploits the intrinsic statistical dependencies in the available sensory readings and copes with their inherent uncertainty by performing a multi-sensor data fusion.

Few works in literature propose a real multi-sensor data fusion framework for Ambient Intelligence. Remarkable exceptions are works presented in [5] and [7]. The authors of [5] propose a multi-sensor fusion system for integrating heterogeneous sensory information in order to perform user activity monitoring. The authors present a comparison between two probabilistic approaches (Hidden Markov Models, and Conditional Random Fields), and point out the effectiveness of a probabilistic system for activity detection in terms of dealing with uncertainty. The authors of [7] present an activity recognition approach reinforced by information about users' location. The proposed framework uses a variety of multimodal and unobtrusive wireless sensors integrated into everyday objects; this sensory infrastructure provides data to an enhanced Bayesian Network fusion engine able to select the most informative features.

Unlike other works reported in literature, the work presented here focuses on the dynamic management of the devices providing information to the inference system, thus allowing to deal with such conflicting goals as energy saving and accuracy of the outcome. In particular, the proposed system comprises two levels of reasoning; at the low level a Bayesian network for reasoning on the relevant environmental feature (such as users' presence), merges the available sensory data, while the upper level performs a meta-reasoning on system performance and cost. This meta-level is able to trade the reliability of the Bayesian network outcome for the relative cost in terms of consumed energy, in order to steer a decision about which sensory devices are to be activated or de-activated.

The remainder of the paper is organized as follows. Section 2 presents the general architecture of the proposed system, while Section 3 details its self-configuration capability. The self-configuration process is illustrated through a running example in Section 4 and finally Section 5 reports our conclusions.

2 The Proposed System

One of the requirements characterizing AmI is the availability of a pervasive sensory infrastructure characterized by a low cost and general as much as possible. For this reason, often, reasonings about context are not be performed via specialized sensors, so that the sensed signals will only be partially correlated to the features of interest.

In order to correctly infer the presence of users from the available sensory information a Bayesian inference system for multi-sensor data fusion has been developed. Probabilistic reasoning accounts for the partial correlation between sensory signals and states, and allows to cope with noisy data. The possibility of integrating data coming from multiple sensors exploits the redundancy of such devices deployed throughout the environment.

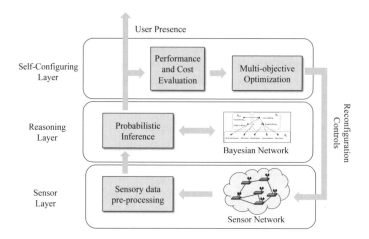

Fig. 1. Block diagram for the presence estimate system

On top of the Bayesian network, a meta-level for self-configuration is implemented, as shown in Figure 1. Such higher level component reasons about potential trade-offs between the confidence degree of the Bayesian network and the cost for using the sensory infrastructure. A plan will be produced stating which sensory devices are to be activated or de-activated.

2.1   The Sensory Infrastructure

The proposed system was developed by taking an existing AmI software architecture as a reference; the original AmI architecture has been implemented at our department and is described in [3].

The sensor network used for our AmI architecture is composed by several heterogeneous devices capable of capturing different physical phenomena. In particular, here we consider five kinds of sensory technologies: WSN, RFId readers, door sensors, sensors on actuators and software sensors.

WSN are composed by small computing devices equipped with off-the-shelf sensors for measuring ambient quantities and with wireless transceivers that enable data exchange among nearby nodes [1]. Sensor nodes in our AmI architecture have been deployed in various rooms close to "sensitive" indoor areas: by the door, by the window, and by the user's desk.

When considering the specific task of user presence detection, we will consider in particular the sound sensor, which is able to detect the amount of noise level in its proximity, thus providing some rough indication about the level of room occupancy.

Other nodes carry specific sensors, such as RFId readers, in order to perform basic access control. In the considered scenario, RFId tags are embedded into ID badges for the department personnel, while RFId readers are installed close to the main entrance and to each office door; readings from each tag are collected via the relative nodes, and forwarded by the WSN to the AmI system, that will process them and will reason about the presence of users in the different areas of the department.

Besides being equipped with a RFId reader, each entrance to the building will also be coupled to a sensor recording its status (i.e. if it is open, closed or locked), which may be used for monitoring access to the different areas, as well as for extracting information about the presence of people.

The interaction of users with the actuators may also be captured via ad-hoc "monitors"; for instance, if the user manually triggers any of the provided actuators (e.g. the air conditioning, the motorized electric curtains, or the lighting systems) via the remote controls or traditional switches, specialized sensors capture the relative IR or electric signals. Detecting some kind of interaction provides a reliable indication about the presence of at least one person in the monitored area.

Finally, a "software sensor" is installed on the users' personal computers to keep track of user login and logout, and to monitor their activity on the terminal. As long as users are actively using their terminal, such sensors will set the value for the probability of the user to be present to its maximum. On the other hand, if no activity is detected, the presence of users close to their workstations may still be inferred via a simple face recognition application, which may be triggered to refine the probability value of user presence based on a degree of confidence in the identification process.

2.2   Environmental Modeling through a Bayesian Network

A system aimed at inferring information about a specific environmental feature based on data coming from multiple sensors may be easily implemented through a rule-based approach, in cases where the sensory information is not affected by noise and uncertainty. Otherwise, the reasoning system needs to take uncertainty into account, as is the case with user's presence detection based on the sensory data mentioned earlier. In such cases, Bayesian Network theory [8] may be an optimal choice for inferring knowledge through a probabilistic process, since it provides an effective way to deal with the unpredictable ambiguities arising from the use of multiple sensors [7].

Classical Bayesian networks, however, may only provide a static model for the environment, which would not be suitable for the proposed scenario; we therefore chose dynamic Bayesian networks or, more specifically, Markov chains to implement our model, which thus allow for probabilistic reasoning on dynamic scenarios, where the estimate of the current system state depends not only on the instantaneous observations, but also on past states.

Figure 2(a) shows the Bayesian network used to infer probabilistic knowledge on a given state feature based on a set of input sensory data. Each state feature affects a set of sensory readings (we indicate each evidence node with E^i), that can be considered the perceivable manifestation of that state. The connection between the current state and its sensory manifestation is given by the probabilistic sensor model $P(E_t^i|X_t)$. Moreover the current state depends on past state according to a state transition probability $P(X_t|X_{t-1})$.

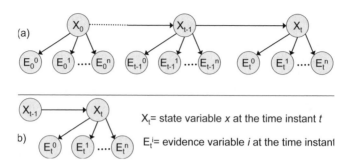

Fig. 2. Structure of a Markov chain for inferring a given state feature starting from a set of sensory data

The belief about the specific value of a state variable is the conditional probability with respect to the whole set of observations from the initial time to the current time:

$$Bel(x_t) = P(x_t|e_1^1, e_1^2, \ldots, e_1^n, \ldots e_t^1, e_t^2, \ldots, e_t^n) = \tag{1}$$
$$= P(x_t|\mathbf{E}_1, \mathbf{E}_2, \ldots, \mathbf{E}_t) = P(x_t|\mathbf{E}_{1:t}) \ .$$

Due to the simplifications introduced by the Markov assumption, and to the conditional independence among sensory measurements, once the state induced by the Bayesian network is known, the belief about the current state can be inductively defined as follows:

$$Bel(x_t) = \eta \prod_{e_t^i} P(e_t^i|x_t) \cdot \sum_{x_{t-1}} P(x_t|x_{t-1})Bel(x_{t-1}) \ . \tag{2}$$

Thanks to these simplifications, at each time step it is sufficient to consider a reduced set of variables, as shown in Figure 2(b), which results in a reduced overall computation effort.

3   Implementing Self-configuration

In some circumstances it can be useful to modify the status of the sensory infrastructure because of particular environmental conditions, or in accordance with the current system performances. In other words, it is sometimes desirable to act on the sensory devices by turning them on or off, thus modifying the flow of information feeding the probabilistic reasoner.

The first scenario occurs when the cost of some of the sensory acquisition devices increases; such variation may be caused by the enforcement of specific energy saving policies aiming to increase the WSN lifetime. In this case, each sensor may be tagged with a cost in terms of required energy, inversely proportional to its charge level, and it is preferable for the system to do without these sensory inputs, provided that the performance quality is not affected, or at least that the degradation is deemed acceptable. As a consequence, it should be possible to modify the sensory infrastructure status by allowing such devices to go into a "low consumption" mode, for instance by suspending the sensory data gathering; normal functioning would be restored only for those devices whose contribution is crucial for the probabilistic inference engine.

The second scenario occurs when the Bayesian network receives its input from a greatly reduced set of sensory devices, so that the system is forced to assign a high degree of uncertainty to the environmental states. In such situation, the adoption of an additional sensory device, although very expensive, may contribute to decease the uncertainty degree; therefore, a set of previously de-activated sensory devices may be switched back to normal functioning, so as to provide additional information for the inference process.

Following these considerations, and in order to make the system as self-sufficient as possible, we allow it to autonomously opt to modify the status of the sensory infrastructure, by suspending or restoring the data flow from some evidence nodes, in order to get the best trade-off between energy cost and precision in reasoning.

3.1   Indices for Self-configuration

An extended model of Bayesian network was adopted, where each node is tagged with additional information. In particular, evidence nodes are tagged with two additional pieces of information: cost, and operation mode.

The operation mode gives an indication about the state of the sensory devices associated to the evidence node, i.e. activated or de-activated; for the evidence node E^i, the operation mode at time instant t is indicated as op_t^i.

The cost associated to the evidence nodes is not set within the probabilistic inference system itself, rather it is set by an energy management subsystem based on the current state of the sensor nodes. Based on the costs of the evidence nodes, it is possible to compute the overall cost for a state variable as the sum of the costs of its connected evidence nodes. Such value represents the total cost necessary to infer the distribution probability for the state variable, depending on the currently used sensory information.

Furthermore, assuming that the function for computing the cost of evidence node E_t^i is indicated by $f_{cost}(E_t^i)$, then the following holds:

$$cost(E_t^i) = \begin{cases} f_{cost}(E_t^i) & \text{if } op_t^i \text{ is } on \text{ ,} \\ 0 & \text{otherwise ;} \end{cases}$$

$$cost(X_t) = \sum_{E_t^i \in \mathbf{E_t}} cost(E_t^i) \text{ .} \tag{3}$$

In order to assess the performance of the current configuration of the Bayesian network it is useful to extract information about the precision of the probabilistic reasoning, besides the overall energy cost. An uncertainty index will be used to measure the intrinsic uncertainty of the a posteriori inferred belief:

$$uncertainty(X_t) \overset{def}{=} -\sum_{x_t} Bel(x_t) \log Bel(x_t) \,. \tag{4}$$

Even though the definition of the uncertainty index is formally similar to that of the entropy for variable X, the latter is a function of the *a priori* probability distribution, whereas the uncertainty index is a function of the *a posteriori* probability distribution.

Indices $cost(X_t)$ and $uncertainty(X_t)$ give indications about the cost and effectiveness of the probabilistic reasoning about a state variable X_t.

3.2   Adapting the Sensor Network Configuration

In our model, different configurations of the sensory infrastructure may be completely expressed by indicating the operating mode of each evidence node. The configuration of the entire sensor network may thus be expressed by a vector: $s_t = [op_t^0, \ldots, op_t^i, \ldots, op_t^n]$.

By monitoring the indices for cost and uncertainty for each state variable, according to Equations 3 and 4, the system checks whether the uncertainty index is close to its maximum allowed value and whether the cost index rises unexpectedly. If one of these two events occurs, a modification of the sensory infrastructure is triggered.

In order to avoid oscillations in the configuration of the sensor network and to ensure gradual modifications of its structure, each inference step only enables atomic actions, i.e. actions operating on one evidence node at a time.

In order to select the action to be performed, a multi-objective selection system is devised, based on a Pareto-dominance criterion; the aim is to obtain the best trade-off between cost minimization and uncertainty minimization. As is evident, the two goals are conflicting and minimizing with respect to costs only would lead to deactivating all the sensory devices, whereas minimizing with respect to uncertainty only would lead to activating all of them.

The optimal action is selected with respect to the cost and the hypothetical uncertainty that the system would have obtained for each alternative configuration of sensory devices, considering the actual sensor readings. For a specific configuration s we indicate these indices as $C(s)$ and $U(s)$, respectively; they will be computed by disregarding the evidence nodes corresponding to deactivated sensory devices in Equations 3 and 2.

When dealing with the conflicting goals of minimizing $U(s)$ and $C(s)$, the traditional approach of merging them into a single objective function presents several limitations, mainly because it would require an accurate knowledge of the different objective functions, either in terms of relative priority or relevance. On the contrary, we chose to keep two independent objective functions and to manage them through a multi-objective algorithm.

We will say that a configuration s_i *Pareto-dominates* another configuration s_j if:

$$C(s_i) \leq C(s_j) \wedge U(s_i) \leq U(s_j). \tag{5}$$

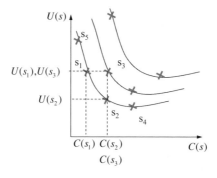

Fig. 3. Graphical example of the Pareto-dominance analysis

A configuration s^* is *Pareto-optimal* if no other solution has better values for each objective function, that is if the following holds:

$$\begin{cases} C(s^*) \leq C(s_i) \\ \quad\text{and} \qquad \forall i = 1 \ldots n \ . \\ U(s^*) \leq U(s_i) \end{cases} \tag{6}$$

Figure 3 represents an example of the Pareto-dominance analysis: configurations s_1 and s_2 belong to the same non-dominated front because $C(s_1) \leq C(s_2)$ and $U(s_2) \leq U(s_1)$, while both configurations s_1 and s_2 dominate configuration s_3; the set of optimal configurations is $\{s_1, s_2, s_4, s_5\}$.

The Pareto-dominance analysis is performed through the fast non-dominated sorting procedure proposed in [4], whose complexity is $O(mN^2)$, where m is the number of objective functions ($m = 2$ in our case) and N is the number of evidence nodes.

Within the optimal front, the configuration improving the index related to alarm triggering is chosen; namely if $cost(X_t)$ triggered the alarm, the configuration improving index $C(s)$ is selected, whereas if $U(X_t)$ triggered the alarm, the configuration improving index $U(s)$ is selected.

4   Running Example

This section describes a running example illustrating how our self-configuring Bayesian network operates. We will consider a network with one state variable X, with 2 possible different values, and three evidence nodes E^1, E^2, E^3, taking 3, 2 and 2 values respectively; the network is defined through the conditional probabilities tables reported in Tables 1 and 2.

Let us assume that the considered state variable is associated to the presence of the user in her office room and that evidence variables are associated to the software sensor, to the door-status sensor and to the set of sound sensors (see Figure 4).

Let us consider an initial a configuration where all three sensors are active; furthermore let $P(X_0) =< 0.9, 0.1 >$ represent the probability distribution for

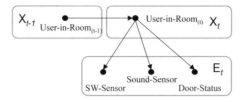

Fig. 4. Bayesian Network for the running example

the state variable at time $t = 0$ (i.e. $P(X_0 = 0) = 0.9$ and $P(X_0 = 1) = 0.1$). For the sake of the example we will assume that the cost associated to evidence variables E^1 and E^2 will not vary over time and will be equal to $cost(E^1) = cost(E^2) = 1$, whereas the cost associated to evidence variable E^3 will linearly increase over time with unitary coefficient and initial value $cost(E_1^3) = 1$. Finally, the threshold for the uncertainty index will be set to 0.9.

If the sensory readings at time $t = 1$ are

$$[E_1^1, E_1^2, E_1^3] = [1, 0, 1] ,$$

the corresponding belief for the state variable, computed according to Equation 2, will be

$$Bel(X_1) =< 0.906, 0.094 > ,$$

meaning that with high probability the user is not present in the monitored area. According to Equation 4, this belief distribution results in an uncertainty equal to $U(X_1) = 0.451$, and the cost index will be $cost(X_1) = 3$. Since the uncertainty index falls below the threshold, the sensor configuration will not be varied for the next step.

We now suppose that at the next time instant sensors produce the readings:

$$[E_2^1, E_2^2, E_2^3] = [2, 1, 1] ,$$

possibly associated to the user entering the monitored area; the belief on the state variable will be:

$$Bel(X_2) =< 0.057, 0.943 > .$$

The new sensory readings cause a dramatic change in the belief about user's presence. The uncertainty index will now amount to $U(X_2) = 0.315$, well below

Table 1. CPT for state transition: $P(X_t|X_{t-1})$.

		X_t	
		0	1
X_{t-1}	0	0.8	0.2
	1	0.2	0.8

Table 2. CPTs for sensor models: $P(E_t^1|X_t)$, $P(E_t^2|X_t)$ and $P(E_t^3|X_t)$

		E_t^1			E_t^2		E_t^3	
		0	1	2	0	1	0	1
X_t	0	0.5	0.3	0.2	0.9	0.1	0.6	0.4
	1	0.1	0.1	0.8	0.4	0.6	0.2	0.8

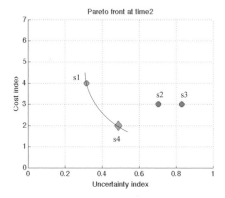

s	U(s)	C(s)
s1=[1,1,1]	0.315	4
s2=[0,1,1]	0.705	3
s3=[1,0,1]	0.830	3
s4=[1,1,0]	0.488	2

Fig. 5. Pareto-dominance analysis of the various configurations of the sensory infrastructure at time $t = 2$. The table on the right shows the indices $U(s)$ and $C(s)$ for the current configuration at time $t = 2$, and for the alternative configurations.

the threshold; however, the cost of sensor E^3 has grown up to $cost(E_2^3) = 2$, which results in a corresponding increase in the cost of the probabilistic inference $(cost(X_2) = 4)$ not ascribable to additional sensor activations. This condition triggers the re-configuration process.

During re-configuration, the hypothetical uncertainty is checked against the costs of the different sensory conditions obtained by toggling the state of one of the sensors with respect to the current configuration. In our case, the three possible configurations are (E^1, E^2), (E^1, E^3), and (E^2, E^3), corresponding to states $[1, 1, 0]$, $[1, 0, 1]$, and $[0, 1, 1]$ respectively. Indices of uncertainty and cost are computed as explained in Section 3.2, producing the values shown in the table reported in Figure 5.

The Pareto dominance analysis of the different solutions is shown in Figure 5, and allows to identify configurations $s1$, and $s4$ as belonging to the optimal front; the former corresponds to the current configuration, whereas the latter is obtained by de-activating sensors related to evidence node E^3. The solution improving the index that caused the alarm will be chosen within those in the optimal front; in this case, this will correspond to configuration $s4$, which allows to reduce the cost of the BN.

Step $t = 3$ will start with a new configuration where the considered evidence node are E^1 and E^2. Assuming that current sensory readings are

$$[E_3^1, E_3^2] = [2, 1] \ ,$$

the belief for the state variable will be

$$Bel(X_3) = < 0.013, 0.987 > \ ,$$

meaning that the new sensory readings reinforce the belief about the presence of the user. The uncertainty index is equal to $U(X_3) = 0.097$, still below the relative threshold, and the cost of the sensor relative to E^3 increases up to $cost(E_3^3 = 3)$,

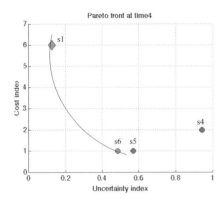

s	U(s)	C(s)
s4=[1,1,0]	0.944	2
s5=[0,1,0]	0.571	1
s6=[1,0,0]	0.484	1
s1=[1,1,1]	0.125	6

Fig. 6. Pareto-dominance analysis of the various configurations of the sensory infrastructure at time $t = 4$. The table on the right shows the indices $U(s)$ and $C(s)$ for the current configuration at time $t = 4$, and for the alternative configurations.

which however does not affect the cost of the probabilistic inference ($cost(X_3) = 2$). Since all indices fall below the relative thresholds, the configuration will not vary for the next step.

At time $t = 4$, we assume that the sensory readings are

$$[E_4^1, E_4^2] = [1, 0] \; ;$$

by looking at the CPTs, it is clear that those readings are not the ones with highest probability with respect to the user's presence. This discrepancy produces a change in the belief, which is however coupled with a high uncertainty:

$$Bel(X_4) = < 0.639, 0.361 > \; ,$$

and $U(X_4) = 0.944$. Since the uncertainty index falls over the threshold, the self-configuration process is triggered again.

The current state is $s = [1, 1, 0]$, so the three possible alternative configurations are $[0, 1, 0]$, $[1, 0, 0]$, and $[1, 1, 1]$, corresponding to activating evidence nodes (E^2), (E^1), or (E^1, E^2, E^3) respectively. Now $cost(E_4^3) = 4$, and the indices of uncertainty and cost are show in the table reported in Figure 6.

A Pareto-dominance analysis of the possible solutions identifies configurations $s1$ and $s6$ as belonging to the optimal front, as shown in Figure 6. Within such front, the solution improving the index that triggered an alarm is chosen, namely, in this case, $s1$ which allows to decrease the uncertainty.

In other words, the sensor associated to evidence node E^3 is re-activated, regardless of the high energy cost, since this is the way for the system to gather the additional information necessary to lower its uncertainty. Such costly re-activations occur when other currently activated sensors provide information not matching with the current belief, due to excessive noise or to an actual variation for the state. In both cases, it is convenient to re-activate a costly sensor just for the time necessary to decrease the system uncertainty about the state of the external world, and then deactivate it again.

5 Conclusions

This paper proposed a Bayesian networks model which includes a meta-level allowing for dynamic reconfiguration of the sensory infrastructure providing the evidence for the probabilistic reasoning. The system has been instantiated on an Ambient Intelligence scenario for the extraction of contextual information from heterogeneous sensory data. The added meta-level accounts both for the accuracy of the outcome of the system, and for the cost of using the sensory infrastructure. The provided realistic example showed that the proposed approach is promising in overcoming the difficulties arising from the inherently imprecision of sensory measurements, allowing to obtain a sufficiently precise outcome, while also minimizing the costs in terms of energy consumption.

Finally, we plan to extend the test set with a real-world scenario, in order to evaluate the scalability with respect to the number and the heterogeneity of data sources, and the sensitiveness to the variability of energy consumption functions, as well as to compare the system performances with other meta-management strategies.

References

1. Akyildiz, I., Su, W., Sankarasubramaniam, Y., Cayirci, E.: A survey on sensor networks. IEEE Communication Magazine 40(8), 102–114 (2002)
2. Bernardin, K., Ekenel, H., Stiefelhagen, R.: Multimodal identity tracking in a smart room. Personal and Ubiquitous Computing 13(1), 25–31 (2009)
3. De Paola, A., Gaglio, S., Lo Re, G., Ortolani, M.: Sensor9k: A testbed for designing and experimenting with WSN-based ambient intelligence applications. In: Pervasive and Mobile Computing. Elsevier, Amsterdam (2011)
4. Deb, K., Agrawal, S., Pratap, A., Meyarivan, T.: A fast elitist non-dominated sorting genetic algorithm for multi-objective optimization: Nsga-ii. In: Parallel Problem Solving from Nature PPSN VI, pp. 849–858. Springer, Heidelberg (2000)
5. Kasteren, T.L., Englebienne, G., Kröse, B.J.: An activity monitoring system for elderly care using generative and discriminative models. Personal and Ubiquitous Computing 14(6), 489–498 (2010)
6. Li, N., Yan, B., Chen, G., Govindaswamy, P., Wang, J.: Design and implementation of a sensor-based wireless camera system for continuous monitoring in assistive environments. Personal and Ubiquitous Computing 14(6), 499–510 (2010)
7. Lu, C., Fu, L., Meng, H., Yu, W., Lee, J., Ha, Y., Jang, M., Sohn, J., Kwon, Y., Ahn, H., et al.: Robust Location-Aware Activity Recognition Using Wireless Sensor Network in an Attentive Home. IEEE Transaction on Automation Science and Engineering 6(4), 598–609 (2009)
8. Pearl, J.: Probabilistic reasoning in intelligent systems: networks of plausible inference. Morgan Kaufmann, San Francisco (1988)
9. Pirttikangas, S., Tobe, Y., Thepvilojanapong, N.: Smart environments for occupancy sensing and services. In: Handbook of Ambient Intelligence and Smart Environments, pp. 1223–1250 (2009)

Tackling the DREAM Challenge for Gene Regulatory Networks Reverse Engineering

Alessia Visconti[1,3], Roberto Esposito[1,3], and Francesca Cordero[1,2,3]

[1] Department of Computer Science, University of Torino
[2] Department of Clinical and Biological Sciences, University of Torino
[3] Interdepartmental Centre for Molecular Systems Biology, University of Torino
{visconti,esposito,fcordero}@di.unito.it

Abstract. The construction and the understanding of Gene Regulatory Networks (GRNs) are among the hardest tasks faced by systems biology. The inference of a GRN from gene expression data (the GRN *reverse engineering*), is a challenging task that requires the exploitation of diverse mathematical and computational techniques. The DREAM conference proposes several challenges about the inference of biological networks and/or the prediction of how they are influenced by perturbations.

This paper describes a method for GRN reverse engineering that the authors submitted to the 2010 DREAM challenge. The methodology is based on a combination of well known statistical methods into a Naive Bayes classifier. Despite its simplicity the approach fared fairly well when compared to other proposals on real networks.

1   Introduction

A Gene Regulatory Network (GRN) is a complex system whose construction and understanding are major tasks in computational biology. A GRN is a model that aims at capturing the interactions among genes and transcription factors. GRNs describe the causal consequences of the regulatory processes at transcriptional level.

The inference of a GRN from gene expression data (the GRN *reverse engineering*), is a challenging task that requires the exploitation of different mathematical and computational techniques, such as methods from statistics, machine learning, data mining, graph theory, ontology analysis and more.

To construct a GRN, it is necessary to infer interactions among the genes that compose a biological system. To extract such interactions, a widespread approach consists in analysing the data from DNA microarray experiments [1]. A microarray experiment quantifies the expression levels of a large number of genes within different experimental conditions (i.e., the gene expression profiles). Each condition may refer to a different cell state (e.g. tumoral or healthy), or an environmental stress (e.g. lack of oxygen) or a specific time point.

In the last decade, a large number of methods arose for the reverse engineering of GRNs (see [2] and [3] for a review). The main techniques are based on *linear models* [4], *graphical models* [5], and *genes profiles similarity* [6].

R. Pirrone and F. Sorbello (Eds.): AI*IA 2011, LNAI 6934, pp. 372–382, 2011.

Linear models are largely employed due to their simplicity and to their low computational cost. One common drawback of this approach is that it may require too many independent measurements.

Probabilistic graphical models (e.g., Bayesian Networks [7]) explicitly state the independencies in a multivariate joint probability distribution using a graph based formalism. They provide a flexible and well-formalized language that allows one to introduce prior biological knowledge. Unfortunately, limitations in the Bayesian formalism (e.g., no-loop requirement) may hinder the correct modeling of real world dependencies, while other formalisms require one to resort to approximate techniques in order to overcome hard computational problems [8].

Approaches based on genes profiles similarity exploit measures such as the Mutual Information, the Euclidean distance, or the Pearson Correlation Coefficient, to assess the edges likelihood. In practice, edges are drawn if the assessed similarity is higher than a given threshold. These models sport low computational costs at the expense of a drastic simplification of the model.

From an applicative point of view, a complete transcriptional model may lead to important applications in areas as genetic engineering [9], drug design and medicine [10]. A GRN may help in studying hypotheses about the gene regulation functioning, e.g. by computational simulations, or it can be used to answer high level biological questions, e.g. *"how do system perturbations influence regulatory networks?"*. In summary, an accurate method for the reverse engineering of GRNs would help the development of better experimental hypotheses, thus avoiding wasting time and resources and, at the same time, it would help to improve our understanding of cellular mechanisms [11].

The annual *Dialogue for Reverse Engineering Assessments and Methods* (DREAM) conference [12] is dedicated to these topics. The DREAM conference proposes several challenges about the inference of biological networks and/or the prediction of how they are influenced by perturbations. Each method submitted to a challenge is evaluated using rigorous metrics and compared with the others. The DREAM challenges allow a fair assessment of the strengths and weaknesses of the proposed methods, and an objective judgment about the reliability of submitted models.

Four different challenges have been proposed in the last iteration of the conference. The challenge number four (the *network inference challenge*) deals with the inference of genome-scale GRNs over several organisms. The data provided to participants is derived from real and synthetic gene expression experiments. In order to allow a fair comparison of the methods, each dataset is made available in an anonymized format: genes and experimental settings are provided with meaningless identifiers so that the only meaningful information is the value of the measured gene expression levels.

Participants are provided with a list of putative Transcriptional Factors (TFs), a metadata file describing experimental conditions, and a file with experimental measurements of gene expression levels. They are requested to submit a list of edges linking TFs to their targets (either genes or TFs) sorted by a plausibility measure.

This paper is the first comprehensive description of a method for GRN reverse engineering that the authors submitted to the challenge 4 of DREAM 2010. The methodology is based on a combination of well known statistical methods into a Naive Bayes classifier. Despite its simplicity the approach fared fairly well when compared with other proposals.

2   Datasets Description

The DREAM network inference challenge consists in four reverse engineering tasks. Among the four provided datasets, one derives from *in silico* (i.e., synthetic) experiments, the others correspond to the following three organisms: *Staphylococcus aureus*, *Escherichia coli*, and *Saccharomyces Cerevisiae*. Each dataset is composed by three files containing: a list of putative TFs, the gene expression data, and some meta information.

The TFs list is of particular importance since, by problem definition, edges are required to start from a TF. The gene expression data file contains a matrix where each row corresponds to a gene and each column corresponds to an experiment. Cells contain gene expression levels. It should be noted that this is not a raw matrix of the type output by DNA microarray scanners. Instead, DREAM organizers provide values already processed to factor out specific biases due to the experimental settings. Finally, the meta information file describes the settings of each experiment.

Four broad classes of experimental settings are used in the experiments:

wild-type is specified when the values concern a cell in an unperturbed state;

deletion/overexpression is specified when one (or several) genes are switched off (deletion) or duplicated (overexpression). The meta-data also specifies the target gene(s) of the deletion/overexpression experiment.

perturbation is specified in case of multifactorial alterations. In contrast to the targeted genetic perturbations described above, a multifactorial perturbation may affect the expression levels of an unknown number of genes at the same time. For instance, a perturbation experiment may study the effects of a physical or chemical stress applied to the cells (e.g., heat shock or a change of the growth medium). In this case, the perturbation type is provided (in an anonymized form) along with its intensity.

time series is specified when the expression levels are measured at successive time points. In case of time series, more than one column of the data matrix describe the same experiment (one column for each time point). The meta-data associated with each column also reports the corresponding time point.

It is also possible that some experiments are made using a combination of different experimental settings. For instance, a perturbation could be done in conjunction of a gene deletion, or observed at different time points.

As a last remark, let us mention that in performing biological experiments it is customary to replicate the experiments several times. In this way, statistics could be evaluated using more robust figures. Information about experiments replicates is provided in the meta-data file.

3   Methods

The process of constructing a GRN is basically the process of discovering how genes influence each others. In this work, we make use of a well established approach that is based on the identification of *Differentially Expressed Genes* (DEGs). A DEG is a gene whose expression profile varies in a statistically significant way in response to a given stimuli. DEGs can be identified by comparing the expression levels of genes in wild-type (*control*) and perturbed (*treatment*) samples.

In order to interpret correctly the data, the statistical analysis tools for DEGs identification need to take into consideration the original *experimental design*, i.e., the original settings that were devised to perform the biological experiments. We processed the data using four different techniques:

Pearson correlation coefficient (PCC): a standard correlation measure from statistics. It is used to get a measure of the linkedness of all possible pairs of genes.

Limma: a linear model for the assessment of differential expressions [13]. This is one of the state-of-the-art algorithms for analyzing experiments when the number of samples is limited. The central idea is to fit a linear model to the expression data for each gene. Empirical Bayes and other shrinkage methods are used to exploit information across genes making the analysis more stable.

maSigPro: a state-of-the-art method for the analysis of experiments that include time series [14]. maSigPro follows a two steps regression strategy to find genes with significant temporal expression changes.

z-**score:** the standard *z* statistic. We used it in experiments when other tools could not be applied.

We used the R implementation of Limma and maSigPro [15]. The packages require, beside the actual data, a matrix (the *design matrix*) describing some of the features of the experimental design. Even though, ideally, this matrix should be set up following the original experimental design, in our context this information was not available and we chose the experimental design on the basis of the type of the experimental setting. For what concerns the deletion/overexpression and perturbations experimental settings, the design matrix specifies which experiments are to be considered as treated and which experiments are to be considered their wild-type counterpart. For time-series experiments, in addition to the information about wild-type experiments, the design matrix needs to specify where to find and how to deal with the time points. In our case, we specified that each time point needs to be compared with its immediate predecessor (rather than, e.g., comparing all time points against the first one).

The results of the tools analysis are used to build a set of hypothetical edges. The plausibility of the hypothetical edges will be then assessed by means of a Naive Bayes classifier that makes use of a set of statistics associated with each edge. The statistics are built on the basis of the tools results and of the experimental settings as follows:

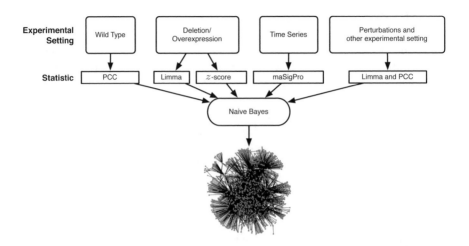

Fig. 1. System architecture

- PCCs, calculated on the basis of wild-type experiments, are associated to each possible pair of TF/gene;
- for each deletion/overexpression experiment, p-values (when smaller than 0.05) resulting from limma are associated to each deleted/overexpressed TF and all DEGs identified;
- when limma could not be used for a particular deletion/overexpression experiment, the z statistic p-value is used instead;
- for each experiment involving time-series, p-values (when smaller than 0.05) resulting from maSigPro are associated with each pair of TF/gene that can be found in the clusters returned by the tool.
- in all other cases, PCCs are calculated on the expression profiles of pairs of TF/gene that limma identified as DEGs.

The proposed method architecture containing a summary of which tools are used for each experimental setting is shown in Figure 1.

The combination of the statistics is evaluated using a Naive Bayes approach. Our goal is to compute the probability that an edge belongs to the network given the experimental evidence. Let us consider a sampling experiment where possible edges are drawn at random and denote with X a stochastic variable that assume value 1 when the drawn edge belongs to the network and 0 otherwise. Let us also define $Y_1 \ldots Y_m$ to be the values of statistics assessing the given edge. We would like to compute the probability $P(X = 1|Y_1 \ldots Y_m)$. By Bayes's theorem, this quantity could be written as:

$$P(X = 1|Y_1 \ldots Y_m) = \frac{P(Y_1 \ldots Y_m|X = 1)P(X = 1)}{\sum_{x \in \{0,1\}} P(Y_1 \ldots Y_m|X = x)P(X = x)}$$

$$= \frac{\prod_{i=1 \ldots m} P(Y_i|X = 1)P(X = 1)}{\sum_{x \in \{0,1\}} \prod_{i=1 \ldots m} P(Y_i|X = x)P(X = x)} \tag{1}$$

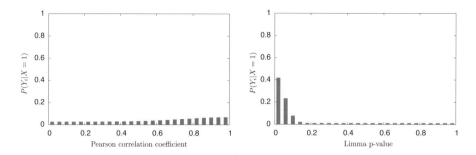

Fig. 2. $P(Y_i|X = 1)$ distribution for PCC and Limma p-value statistics

where equality holds by assuming Y_i and Y_j to be independent given X for each i, j ($i \neq j$). To compute the sought probabilities, we need to specify the distributions of X and of each Y_i given X. Since the challenge does not offer any information about the networks we are dealing with, we are not in the position of estimating X distribution from the data. Then, we exploit a widely accepted assumption: that biological networks have a scale-free topology [16]. In a scale-free network, the probability that a randomly chosen node has exactly k edges is $P(k) = k^{-\gamma}$, with $\gamma \in (2\ldots3)$. Here, as suggested in [17], we set $\gamma = 3$. It follows that the number of edges in a scale-free network with N genes can be computed as:

$$e(N) = \sum_{k=1}^{N} N \times P(k) \times k = N \sum_{k=1}^{N} \frac{1}{k^2}.$$

By approximating the quantity $\sum_{k=1}^{N} \frac{1}{k^2}$ with $\frac{\pi^2}{6}$ (its limit for $N \to \infty$), in a network of size N, the probability that "picking at random two genes they are connected" is $e(N)/N^2 = \frac{\pi^2}{6N}$. Then we set $P(X = 1) = \frac{\pi^2}{6N}$ and $P(X = 0) = 1 - \frac{\pi^2}{6N}$.

To compute formula (1) we still have the problem of setting the distributions of the Y_i given X. Since no data is provided to estimate them, we resort to assigning them on the basis of the confidence we have in the tools that generated the Y_i values. Statistics we were less confident about (PCC and z-score) were associated with uniform-like distributions, other ones (limma and maSigPro) with distributions more peaked to one of the extremes. Figure 2 shows the distributions for $P(Y_{PCC}|X = 1)$ and $P(Y_{limma}|X = 1)$.

4   Results

The format required for the DREAM 2010 network inference challenge submission is a list of (at most) $100,000$ edges sorted by edges plausibility. We generated our results by enumerating all possible pairings of TFs and genes and using formula (1) to assess the edges plausibility. The sorted edges list was then truncated to size $100,000$.

Submitted results have been compared, by the DREAM organizers, with a list of putative edges (the *gold standard*). Of the four networks under investigation: network 1 corresponds to a synthetic dataset for which a gold standard is known; network 2 corresponds to the *Staphylococcus aureus*, a bacterium for which a gold standard is not available (consequently, result sets targeting this network has not been evaluated); networks 3 and 4 correspond to well known organisms – *Escherichia coli*, and *Saccharomyces cerevisiae*– whose gold standards were built according to the literature[1].

The *Area Under the Receiver Operating Characteristic* curve (AUROC) and the *Area Under Precision/Recall* curve (AUPR) evaluation measures have been used to compare the submitted networks. The two curves are built by varying the size of edges list and measuring how the re-dimensioned lists performed. The AUROC is calculated using the ratio between the *true positive (*TP*) rate* and the *false positive (*FP*) rate* as a performance measure. The AUPR is calculated using the ratio between *precision* (i.e., $\frac{TP}{TP+FP}$) and *recall* (i.e., $\frac{TP}{TP+FN}$, with FN denoting the number of *false negative* edges).

The methods overall score is evaluated as follows. A p-value is evaluated for each AUROC and AUPR scores (additional details can be found in [18,19]). Then, an AUROC p-value (p_{AUROC}) and an AUPR p-value (p_{AUPR}) are obtained by taking the geometric mean of the scores obtained on each network. Finally, the method overall score is calculated as $-\frac{1}{2}\log_{10}(p_{AUROC} \times p_{AUPR})$.

Table 1 reports the method rankings as published by the DREAM organizers. Columns report the team id number, the overall score for the proposed method, p_{AUROC} and p_{AUPR} statistics, and the p-values for the AUPR and AUROC measure evaluated on each network. Results for the method described in this paper are typeset in a bold faced font. Our methodology obtains middle-ranking performances. Upon closer inspection however, it is apparent that our method does not perform as well as others in reconstructing network 1. Table 2 shows the resulting ranking along with the re-evaluation of the measures on networks 3 and 4 only. We note that in this new evaluation, the method ranks in the top positions. This observation is particularly interesting since network 1 is the only one based on synthetic experiments. Why this is the case it is hard to fathom.

A possible explanation is that an over-adaption of other approaches to synthetic datasets may have hindered their performances. Were this hypothesis be true, it would unveil a twofold issue. On one side it would imply that synthetic networks are still significantly different from real networks. On the other side it would imply that many current approaches are being too aggressively optimized using synthetic networks.

Another possible explanation is that the simplicity of the Naive Bayes approach pays off on real networks, where the noise level is higher and the noise model is understood to a lesser extent.

---

[1] Network 3 and 4 gold standards are based on the best information available to date. As a consequence, they may contain incorrect or missing edges.

Table 1. Results for the DREAM Network Inference Challenge. Columns 6-11 report p-values for the AUPR and AUROC measures on networks 1, 3, and 4. Columns 4 and 5 report their average scores p_{AUPR} and p_{AUROC}. Column 3 reports the overall methods score. All values correspond to those published by the DREAM organizers.

	Team	Overall	p_{AUPR}	p_{AUROC}	$p_{AUPR}1$	$p_{AUPR}3$	$p_{AUPR}4$	$p_{AUROC}1$	$p_{AUROC}3$	$p_{AUROC}4$
1	415	40.279	1.97E-41	1.83E-39	1.60E-104	5.15E-20	1.58E-01	3.06E-106	5.00E-11	1.06E-02
2	543	34.023	8.09E-30	1.37E-37	8.17E-053	1.07E-39	2.15E-02	1.34E-056	1.64E-53	1.77E-03
3	776	31.099	2.33E-41	6.78E-20	7.20E-118	4.34E-07	2.53E-01	3.48E-059	3.42E-03	2.71E-02
4	862	28.747	4.89E-44	6.40E-12	8.58E-135	9.99E-01	1.00E+00	3.82E-039	1.00E+00	1.00E+00
5	548	22.711	6.24E-36	4.24E-08	8.41E-084	4.76E-12	1.03E-16	4.09E-017	1.17E-09	2.76E-01
6	870	21.398	6.89E-32	9.06E-09	2.83E-091	1.73E-08	6.25E-01	2.46E-024	4.98E-06	1.09E-01
7	868	20.694	9.23E-31	2.64E-09	1.42E-092	5.11E-08	1.75E-01	6.97E-022	1.86E-04	4.19E-04
8	842	16.686	3.45E-14	6.81E-18	5.98E-035	4.47E-10	9.10E-01	5.13E-057	7.21E-01	8.56E-01
9	861	13.397	6.04E-13	1.03E-13	1.35E-039	2.18E-02	1.55E-01	6.95E-033	3.26E-03	4.04E-05
10	395	10.586	3.10E-10	4.80E-10	4.20E-008	1.53E-13	5.25E-12	3.42E-002	5.56E-05	4.78E-27
11	799	8.887	1.19E-03	5.00E-14	1.58E-008	3.85E-02	9.78E-01	8.18E-045	9.94E-01	9.83E-01
12	875	8.800	9.64E-03	4.13E-13	3.08E-010	3.64E-03	1.00E+00	1.50E-041	9.44E-01	1.00E+00
13	823	8.126	4.29E-05	4.16E-10	3.63E-013	3.53E-05	9.89E-01	1.59E-032	9.60E-01	9.07E-01
14	48	6.420	1.52E-08	4.55E-04	3.80E-007	3.11E-15	2.39E-04	4.63E-002	7.08E-09	3.24E-05
15	**702**	**6.332**	**2.17E-08**	**2.12E-04**	**7.98E-001**	**9.76E-06**	**1.25E-20**	**8.82E-001**	**4.32E-13**	**2.74E-01**
16	772	6.026	8.39E-08	1.34E-03	1.42E-011	2.26E-16	5.24E-01	8.78E-001	2.43E-09	1.93E-01
17	864	4.973	8.85E-09	1.00E-00	1.65E-030	8.77E-01	1.00E+00	9.99E-001	1.00E+00	1.00E+00
18	705	2.484	1.36E-02	6.84E-02	1.00E+000	7.17E-01	5.58E-07	2.34E-005	8.97E-04	1.49E-01
19	504	2.266	3.40E-04	1.00E-00	3.39E-014	7.53E-01	1.00E+00	9.99E-001	1.00E+00	1.00E+00
20	281	1.897	1.16E-00	5.33E-03	1.00E+000	9.99E-01	6.32E-01	9.99E-001	8.54E-12	7.74E-01
21	829	1.475	2.36E-02	3.79E-00	1.00E+000	1.36E-07	5.63E-01	9.99E-001	4.98E-01	3.69E-02
22	802	0.997	1.00E-00	9.89E-01	1.00E+000	9.99E-01	9.99E-01	1.12E-006	9.98E-01	9.28E-01
23	756	0.331	1.49E-00	3.08E-00	1.00E+000	9.99E-01	3.04E-01	8.21E-001	1.00E+00	4.17E-02
24	736	0.257	1.10E-00	2.97E-00	1.00E+000	9.99E-01	7.58E-01	5.69E-001	1.00E+00	6.71E-02
25	854	0.256	1.10E-00	2.96E-00	1.00E+000	9.99E-01	7.58E-01	5.71E-001	1.00E+00	6.71E-02
26	638	0.144	1.78E-00	1.09E-00	1.79E-001	9.99E-01	9.93E-01	9.52E-001	1.00E+00	8.08E-01
27	784	0.035	1.01E-00	1.17E-00	1.00E+000	9.99E-01	9.83E-01	9.99E-001	1.00E+00	6.26E-01
28	787	0.000	1.00E-00	1.00E-00	1.00E+000	9.99E-01	1.00E+00	9.99E-001	1.00E+00	1.00E+00
29	821	0.000	1.00E-00	1.00E-00	1.00E+000	9.99E-01	1.00E+00	9.99E-001	1.00E+00	1.00E+00

Table 2. Results for the DREAM Network Inference Challenge. Columns 6-9 report p-values for the AUPR and AUROC measures on networks 3 and 4. Columns 4 and 5 report their average scores p_{AUPR} and p_{AUROC}. Column 3 reports the overall methods score.

	Team	Overall	p_{AUPR}	p_{AUROC}	p_{AUPR3}	p_{AUPR4}	p_{AUROC3}	p_{AUROC4}
1	543	24.0435	4.80E-21	1.70E-28	1.07E-39	2.15E-02	1.64E-53	1.77E-03
2	395	13.6677	8.96E-13	5.15E-16	1.53E-13	5.25E-12	5.56E-05	4.78E-27
3	**702**	**9.4601**	**3.49E-13**	**3.44E-07**	**9.76E-06**	**1.25E-20**	**4.32E-13**	**2.74E-01**
4	548	9.2009	2.21E-14	1.79E-05	4.76E-12	1.03E-16	1.17E-09	2.76E-01
5	415	8.0907	9.03E-11	7.30E-07	5.15E-20	1.58E-01	5.00E-11	1.06E-02
6	48	7.6919	8.62E-10	4.79E-07	3.11E-15	2.39E-04	7.08E-09	3.24E-05
7	772	6.3140	1.09E-08	2.16E-05	2.26E-16	5.24E-01	2.43E-09	1.93E-01
8	870	3.5576	1.04E-04	7.38E-04	1.73E-08	6.25E-01	4.98E-06	1.09E-01
9	281	2.8448	7.95E-01	2.57E-06	9.99E-01	6.32E-01	8.54E-12	7.74E-01
10	868	2.7896	9.45E-03	2.79E-04	5.11E-04	1.75E-01	1.86E-04	4.19E-04
11	776	2.7481	3.31E-04	9.63E-03	4.34E-07	2.53E-01	3.42E-03	2.71E-02
12	705	2.5684	6.32E-04	1.15E-02	7.17E-01	5.58E-07	8.97E-04	1.49E-01
13	842	2.4002	2.02E-05	7.85E-01	4.47E-10	9.10E-01	7.21E-01	8.56E-01
14	861	2.3381	5.81E-02	3.63E-04	2.18E-02	1.55E-01	3.26E-03	4.04E-05
15	829	2.2131	2.77E-04	1.36E-01	1.36E-07	5.63E-01	4.98E-01	3.69E-02
16	823	1.1292	5.91E-03	9.33E-01	3.53E-05	9.89E-01	9.60E-01	9.07E-01
17	875	0.6161	6.03E-02	9.71E-01	3.64E-03	1.00E+00	9.44E-01	1.00E+00
18	756	0.4746	5.51E-01	2.04E-01	9.99E-01	3.04E-01	1.00E+00	4.17E-02
19	799	0.3587	1.94E-01	9.88E-01	3.85E-02	9.78E-01	9.94E-01	9.83E-01
20	736	0.3236	8.70E-01	2.59E-01	9.99E-01	7.58E-01	1.00E+00	6.71E-02
21	854	0.3236	8.70E-01	2.59E-01	9.99E-01	7.58E-01	1.00E+00	6.71E-02
22	784	0.0528	9.91E-01	7.91E-01	9.99E-01	9.83E-01	1.00E+00	6.26E-01
23	504	0.0309	8.68E-01	1.00E+00	7.53E-01	1.00E+00	1.00E+00	1.00E+00
24	638	0.0240	9.96E-01	8.99E-01	9.99E-01	9.93E-01	1.00E+00	8.08E-01
25	864	0.0143	9.36E-01	1.00E+00	8.77E-01	1.00E+00	1.00E+00	1.00E+00
26	802	0.0085	9.99E-01	9.63E-01	9.99E-01	9.99E-01	9.98E-01	9.28E-01
27	862	0.0002	9.99E-01	1.00E+00	9.99E-01	1.00E+00	1.00E+00	1.00E+00
28	787	0.0002	9.99E-01	1.00E+00	9.99E-01	1.00E+00	1.00E+00	1.00E+00
29	821	0.0002	9.99E-01	1.00E+00	9.99E-01	1.00E+00	1.00E+00	1.00E+00

5   Conclusion

In this paper we presented a Naive Bayes based approach to GRN reverse engineering: one of the toughest problems in systems biology. Needless to say, the presented method is one of the simplest approaches that can be developed for a problem of this complexity. Indeed, simplicity has been one of the goals we strived to attain in its design. This choice has been also motivated by the fact that past DREAM conferences emphasized that simpler methods could perform as well as others. Also, when real networks are to be analyzed, data scarcity and its quality demand for classifiers built using a small number of well understood parameters. Our method showed average performances in the 2010 DREAM network inference challenge.

An interesting facet of our methodology is that it performed remarkably better in the case of real networks than with synthetic ones: it is among the top performers (it ranks third) when the synthetic dataset is hold out from the evaluation (it ranks 15th otherwise). A number of interesting questions could be raised by this observation: what is in synthetic datasets that set them apart from natural ones? Should one strive to optimize new algorithms more aggressively on natural datasets? Could the culprit be found in the quality of real data, so that most of these methods will perform much better when this quality increases? We believe that the answers to these questions may be important to better understand current tools and to develop new ones.

References

1. Cho, K., Choo, S., Jung, S., Kim, J., Choi, H., Kim, J.: Reverse engineering of gene regulatory networks. Systems Biology, IET 1(3), 149–163 (2007)
2. He, F., Balling, R., Zeng, A.: Reverse engineering and verification of gene networks: Principles, assumptions, and limitations of present methods and future perspectives. Journal of Biotechnology 144(3), 190–203 (2009)
3. Hecker, M., Lamberk, S., Toepfer, S., van Someren, E., Guthke, R.: Gene regulatory network inference: Data integration in dynamic models - A review. BioSystems 96, 86–103 (2009)
4. Bansal, M., Gatta, G.D., di Bernardo, D.: Inference of gene regulatory networks and compound mode of action from time course gene expression profiles. Bioinformatics 22(7), 815–822 (2006)
5. Koller, D., Friedman, N.: Probabilistic graphical models: principles and techniques. MIT Press, Cambridge (2009)
6. Margolin, A.A., Nemenman, I., Basso, K., Wiggins, C., Stolovitzky, G., Favera, R.D., Califano, A.: Aracne: An algorithm for the reconstruction of gene regulatory networks in a mammalian cellular context. BMC Bioinformatics 7(S-1) (2006)
7. Friedman, N., Linial, M., Nachman, I., Pe'er, D.: Using bayesian networks to analyze expression data. Journal of Computational Biology 7(3-4), 601–620 (2000)
8. Murray, I., Ghahramani, Z.: Bayesian learning in undirected graphical models: Approximate mcmc algorithms. In: UAI, pp. 392–399 (2004)
9. Cohen, B., Mitra, R., Hughes, J., Church, G.: A computational analysis of whole-genome expression data reveals chromosomal domains of gene expression. Nature Genetics 26(2), 183–186 (2000)

10. Cutler, D., Zwick, M., Carrasquillo, M., Yohn, C., Tobin, K., Kashuk, C., Mathews, D., Shah, N., Eichler, E., Warrington, J., et al.: High-throughput variation detection and genotyping using microarrays. Genome Research 11(11), 1913 (2001)

11. Segal, E., Friedman, N., Kaminski, N., Regev, A., Koller, D.: From signatures to models: understanding cancer using microarrays. Nature Genetics 37, S38–S45 (2005)

12. DREAM: Dialogue for Reverse Engineering Assessments and Methods (2010), http://wiki.c2b2.columbia.edu/dream/index.php/The_DREAM_Project

13. Smyth, G.: limma: Linear models for microarray data. In: Gail, M., Krickeberg, K., Samet, J., Tsiatis, A., Wong, W., Gentleman, R., Carey, V.J., Huber, W., Irizarry, R.A., Dudoit, S. (eds.) Bioinformatics and Computational Biology solutions using R and Bioconductor. Statistics for Biology and Health, pp. 397–420. Springer, Heidelberg (2005)

14. Conesa, A., Nueda, M.J., Ferrer, A., Talón, M.: maSigPro: a method to identify significantly differential expression profiles in time-course microarray experiments. Bioinformatics 22(9), 1096–1102 (2006)

15. R Development core team: R: A language and environment for statistical computing. R foundation for statistical computing, Vienna, Austria (2011) ISBN 3-900051-07-0

16. Albert, R.: Scale-free networks in cell biology. Journal of Cell Science 118, 4947–4957 (2005)

17. Barabàsi, A.L., Albert, R.: Emergence of scaling in random networks. Science 286, 509–512 (1999)

18. Stolovitzky, G., Prill, R.J., Califano, A.: Lessons from the dream2 challenges. Annals of the New York Academy of Sciences 1158, 159–195 (2010)

19. Prill, R.J., Marbach, D., Saez-Rodriguez, J., Sorger, P.K., Alexopoulos, L.G., Xue, X., Clarke, N.D., Altan-Bonnet, G., Stolovitzky, G.: Towards a rigorous assessment of systems biology models: the DREAM3 challenges. PLoS One 5, e9202 (2010)

Full Extraction of Landmarks in Propositional Planning Tasks*

Eliseo Marzal, Laura Sebastia, and Eva Onaindia

Universitat Politècnica de València (Spain)
{emarzal,lstarin,onaindia}@dsic.upv.es

Abstract. One can find in the literature several approaches aimed at finding landmarks and orderings, each claiming to obtain an improvement over the first original approach. In this paper we propose a complementary view to landmarks exploitation that combines the advantages of each approach and come up with a novel technique that outperforms each individual method.

Keywords: Planning, landmarks, STRIPS.

1 Introduction

In the first work on landmarks [4,1], authors proposed an algorithm to extract landmarks and their orderings from the relaxed planning graph (RPG) of a planning task. We will call this approach LM in the following. The work presented in [3] to calculate *disjunctive landmarks*, called DL in the following, is a different and more general approximation than the one used in LM. The work in [5] adopts the LM and DL approximations to extract landmarks and orderings and introduces an additional extraction of landmarks derived from Domain Transition Graphs (DTG). On the other hand, the technique developed in [7], uses a planning graph (PG) rather than a RPG to extract landmarks. The most recent work on landmarks extraction [2] generalizes this latter method to obtain conjunctive landmarks and landmarks beyond the delete relaxation. However, authors claim this method is computationally costly and the use of conjunctive landmarks during planning does not pay off.

In this paper, we present an approach for full exploitation of landmarks in STRIPS settings by individually analyzing the existing techniques in order to determine when and how to introduce each of them in our approach.

2 Existing Approaches to the Extraction of Landmarks and Orderings

In this section, we study the existing approaches to the extraction of landmarks. First, we need some notations. A STRIPS **planning task** $\mathcal{T} = (\mathcal{A}, \mathcal{I}, \mathcal{G})$ is a

---

* This work has been partially funded by the Spanish government MICINN TIN2008-06701-C03-01, Consolider-Ingenio 2010 CSD2007-00022 and Valencian Government Project Prometeo 2008/051.

R. Pirrone and F. Sorbello (Eds.): AI*IA 2011, LNAI 6934, pp. 383–388, 2011.

triple where $\mathcal{A}$ is a set of actions, and $\mathcal{I}$ (initial state) and $\mathcal{G}$ (goals) are sets of literals. A **relaxed planning task** $\mathcal{T_R} = (\mathcal{A_R}, \mathcal{I}, \mathcal{G})$ is a triple where $\mathcal{A_R}$ is a set of actions obtained by ignoring the delete lists of all actions. In [4], **landmarks** are defined as literals that must be true at some point during the execution of any solution plan. Given that it is PSPACE-complete to decide whether a literal is a landmark or not, [1] defines a sufficient condition (denoted as *Proposition 1*) for a literal being a landmark, which states that if the RPG levels off before reaching the goals, then the planning problem is unsolvable.

The LM Approach. This work ([1], [4]) proposes a two-step process. The **first step** consists in computing a set of landmarks candidates, LC, initially set to $\mathcal{G}$. For each $l \in LC$, the literals in the intersection of the preconditions of the actions that *first achieve l* are added in turn to the set LC of landmark candidates, and this process is repeated for each literal $l \in LC$. The set of *first achievers* of a literal $l \in RPGL_i$ is defined as $fa(l) = \{a \in RPGA_{i-1} : l \in add(a)\}$. Moreover, a **disjunctive landmark** is defined as a set of literals so that at least one of the literals in the set must be true at some point on every solution plan. After the extraction of landmarks, the **second step** evaluates the literals in LC to discard those that are not provably landmarks by following *Proposition 1* (verification).

The DL Approach. In DL [3], every landmark is treated as a disjunctive set, which can contain a single landmark, or can be a set of literals representing a *resource abstracted* landmark. The process for computing landmarks is essentially the same as in the LM approach, but DL adopts a different definition of the *first achievers* of a literal l: $fa(l) = \{a \in \mathcal{A} : pre(a) \subseteq RPGL_f(\neg l) \wedge l \in add(a)\}$, where $RPGL_f(\neg l)$ is the set of literals nodes that appear when the RPG (by excluding all actions that add l) levels off. In DL, **all** the actions reachable from $\mathcal{I}$ are taken into account for computing the first achievers of a literal l, unlike LM, which only considers those actions reachable from $\mathcal{I}$ until l first appears in the RPG. This ensures the extracted landmarks in DL are valid without the need for further verification because all ways of achieving a literal are considered.

The DTG Approach. The method introduced in [5] (DTG) adapts the LM and DL approaches to the SAS^+ formalism. It uses the same algorithm than LM, but opts for the more general approach to admit disjunctive landmarks of DL. Moreover, DTG incorporates an additional step that extracts more landmarks from the information encoded in domain transition graphs (DTG). Thus, while LM only makes use of causal dependencies to extract landmarks, the DTG approach also explores dependencies among the variables values.

The Propagation Approach. The approach presented in [7] has important differences with respect to the previous approaches. First, the extraction of landmarks is performed on the planning graph rather than on the RPG, which eliminates the need of a landmark verification. Second, it uses a forward chaining process to propagate the information from the initial state to the level where the planning graph reaches a fixed point. Third, it extracts proposition and action landmarks. This algorithm is somewhat more expensive than a single landmark verification propagation but it is substantially less expensive than a large number of landmark verifications.

Ordering relations. For using landmarks during search, it is also important to partially order them so as to obtain a skeleton of the solution plan. The work in [1] also provides the following definitions for different types of ordering relations among landmarks, which have also been adopted by the other approaches: (1) there is a **dependency ordering** between l and l', written $l \leq_d l'$, if all solution plans add l before adding l'; (2) there is a **greedy necessary ordering** between l and l', written $l \leq_{gn} l'$, if l is added before l' only in those action sequences where l' is *achieved for the first time.*

A greedy necessary ordering between l and l' ($l \leq_{gn} l'$) is added when a literal l is considered as a landmark candidate for achieving another landmark l'. In the case of **dependency orderings**, which are only used by the DL and DTG approaches, the process is as follows: given two landmarks l and l', if l is not included in those literals *possibly before* l', then l cannot be achieved without achieving l' first, and hence the ordering $l' \leq_d l$ is added [5]. The calculation of orders between landmarks is more complex in the Propagation approach.

Comparison of the LM, DL, DTG and Propagation approaches We present a study over 12 planning domains from the International Planning Competitions[1] that analyzes the number of single and disjunctive landmarks and ordering relations obtained with each strategy. As Table 1[2] shows, no single landmark can be extracted for driverlog and zenotravel domains using either method. This is because there are different resources available to achieve the same task in these domains. In the domains logistics and freecell, Propagation and DTG, respectively, outperform the other approaches, whereas in the rest of domains the LM approach always obtains at least the same number of landmarks, and this number significantly increases in the domains depots, pipesworld and elevator.

From the observed results, we claim that LM is the preferable approach to obtain a larger number of landmarks. Thus, LM affords selecting a subset of landmarks candidates and subsequently applying a verification over all literals that are likely to be landmarks. In contrast, DL misses some single landmarks because it only selects the literals that are provably landmarks by considering all possible ways of achieving a particular literal. For this reason, it obtains a significative larger number of disjunctive landmarks. On the other hand, some candidate landmarks that LM discards in the verification are likely to be disjunctive landmarks in DL. Surprisingly, DTG obtains fewer disjunctive landmarks than DL despite of using the same extraction method. Once again, the reason

---

[1] These domains can be found at http://ipc.icaps-conference.org.

[2] We found difficulties in solving problems of the OpenStacks with the DTG method. The number of orderings found by DTG are not shown because the values returned by LAMA are not comparable with our implementations of LM and DL.

Table 1. Comparison among existing approaches. Each column shows the sum of the number of landmarks and orderings computed for each problem of each domain. We have used our own implementation of the LM, DL and Propagation approaches, and we have executed the LAMA planner [6] (based on the DTG approach). The ordering relations columns show the *greedy necessary orders/dependency orders*.

Domain	Single landmarks				Disjunctive landmarks			Ordering relations	
	LM	DL	DTG	Prop	LM	DL	DTG	LM	DL
Gripper	20	20	20	-	**480**	460	460	**2320/0**	940/0
Logistics	536	536	630	**665**	88	**311**	81	1331/0	**1606/426**
Blocks	523	523	523	523	0	0	0	2131/0	**1956/563**
Depots	**341**	224	283	140	277	**365**	157	**2010/0**	1376/368
Driverlog	0	0	0	-	146	**329**	120	540/0	**539/88**
ZenoTravel	2	2	1	-	123	**283**	80	367/0	**455/119**
Rovers	**76**	75	75	**76**	191	**277**	159	**998/0**	917/0
Freecell	864	596	**1032**	510	**65**	0	0	**1650/0**	676/0
PipesWorld	**149**	123	118	-	336	**362**	31	**1367/0**	835/469
Storage	**27**	25	25	-	112	**275**	123	504/0	**565/2219**
OpenStacks	1024	1024	-	-	20	**40**	-	11884/0	**11884/85**
Elevator	**264**	123	217	-	315	**749**	242	**1473/0**	1384/14

behind is that some of the disjunctive landmarks computed by DL might turn into single landmarks when using the information of the DTGs.

Regarding the Propagation approach, we only show the single landmarks for a subset of domains because of the high cost to build the planning graph. As authors claim in [7], the number of extracted landmarks is greater than with the other approaches, but this approach was able to solve only 14 problems (out of 20) in the depots domain and 12 (out of 20) in the freecell domain. Despite the overall number of landmarks is smaller, it is worth noting that, some of the landmarks found by the propagation method in these domains (8 and 113, respectively) have not been encountered by LM.

Finally, as can be observed, there is not a clear *outperformer* in terms of orderings. In general, the approach that extracts more landmarks is also the approach that obtains more orderings. In any case, it is important to remark that LM almost always obtains more greedy necessary orderings, which are the most useful orders to construct the skeleton of the solution plan.

3   The FULL Approach

The FULL approach relies on the sequential application of the LM technique, a propagation of literals and, finally, the application of DL for eliciting the set of disjunctive literals. From this initial configuration, we set up two different versions of the FULL technique, FULL-v1 and FULL-v2.

FULL-v1 simply consists in linking together the LM and DL techniques. Although LM already extracts a set of disjunctive landmarks, the method used by

Table 2. Comparison of best existing with our approaches. Columns labeled "% wrt T" indicate the percentage of landmarks found by the best existing method and FULL-v2 w.r.t. the total number of landmarks that can be found in a problem (which is calculated by applying *Proposition 1* to all the literals in a given problem).

Domain	Single landmarks					Disjunctive landmarks		
	Best existing	% wrt T	F-v1	F-v2	% wrt T	Best existing	F-v1	F-v2
Gripper	20	100%	20	20	100%	**480**	460	460
Logistics	665 (Prop)	100%	536	665	100%	311	311	311
Blocks	523	91.9%	523	523	91.9%	0	0	0
Depots	341 (LM)	72.7%	341	**353**	75.2%	365	365	365
Driverlog	0	100%	0	0	100%	329	329	329
ZenoTravel	2	100%	2	2	100%	283	283	283
Rovers	76 (LM)	85.3%	76	76	85.3%	277	277	277
Freecell	1032 (DTG)	87.9%	864	**1143**	97.3%	**65**	0	0
PipesWorld	149 (LM)	80.1%	149	149	80.1%	362	411	**411**
Storage	27 (LM)	17.7%	27	**29**	19%	275	275	275
OpenStacks	1024	98%	1024	**1044**	100%	40	40	**60**
Elevator	264 (LM)	94.6%	264	**279**	100%	749	842	**849**

DL to compute the first achievers is more complete and robust. In addition, this latter step allows us to infer new landmarks. Table 2 compares the top performer for each domain and FULL-v1. Obviously, FULL-v1 obtains the same results in all the domains where LM was the best existing method. In those domains where DTG outperformed LM, FULL-v1 was also outperformed by DTG.

FULL-v2 applies a propagation of literals similar to [7] but, in our case, the propagation is performed on a RPG rather than on a planning graph, which is a less costly process. However, this forces us to apply a landmark verification on the extracted landmarks. In summary, FULL-v2 consists of the following steps:

1. Extraction of a first set of landmark candidates through the first step of LM. The greedy-necessary orderings are also computed.
2. Addition of more landmark candidates through the propagation of literals on a RPG.
3. Verification of all the landmark candidates.
4. Computation of disjunctive landmarks through DL taking as input all the verified landmarks.
5. Computation of dependency orders: $l \leq_d l'$ (1) if l' does not belong to the literals *possibly before* l; (2) if l' is labeled with l in step 2.

FULL-v2 extracts more landmarks than the top performer for each domain as shown in Table 2. This increase in the number of landmarks is given, partly, by the forward propagation of literals along the RPG, which allows us to compute a set of literals *complementary* to those obtained by the LM approach. The extra landmarks found in the propagation of literals makes the step 4 of FULL-v2 also find more single and disjunctive landmarks. Therefore, the raise in the number

of landmarks of FULL-v2 is due to the propagation of literals but also to the
new landmarks that DL is able to find when it receives as input the landmarks
from steps 1 and 2. This can be noticed in domains OpenStacks and Elevator
where, despite using the same method to compute disjunctive landmarks, FULL-
v2 extracts a larger number of disjunctive landmarks than FULL-v1. We conclude
that FULL-v2 is able to find all the landmarks in six domains and it finds more
than 75% in (almost) the rest of domains. Regarding the **ordering relations**
(not shown in table 2), in general, both FULL-v1 and FULL-v2 elicit a larger
number of orderings compared to LM and DL as they also extract more landmarks
than these two methods. As for **performance**, we found that the computational
cost of DTG and LM are similar, whereas it is one order of magnitude slower for
DL. On the other hand, the propagation of literals in FULL-v2 is very little
time-consuming, adding only a few milliseconds over the cost of DL.

4   Final Conclusions

This paper presents a thorough analysis of all the existing methods for landmarks extraction. We studied the strengths and weaknesses of each method and
we presented an extensive experimental setup that compares the number of single
and disjunctive landmarks and orderings obtained with the existing techniques.
From the observed results, we come up with a novel landmark method as the
combination of the techniques that appear to report the best results in terms
of the extracted elements. This new method returns the best figures in landmarks extraction so far. Our next immediate work is to use these results within
a heuristic search planner and check out the benefits it brings in number of
solved planning tasks and solution quality, although as Karpas and Domshlak
say in their paper "Cost-Optimal Planning with Landmarks" *more landmarks
can never hurt.*

References

1. Hoffmann, J., Porteous, J., Sebastia, L.: Ordered landmarks in planning. Journal of
   Artificial Intelligence Research 22, 215–287 (2004)
2. Keyder, E., Richter, S., Helmert, M.: Sound and complete landmarks for and/or
   graphs. In: Proceedings of the 19th ECAI, pp. 335–340 (2010)
3. Porteous, J., Cresswell, S.: Extending landmarks analysis to reason about resources
   and repetition. In: Proc. 21st Workshop of the UK PLANSIG, pp. 45–54 (2002)
4. Porteous, J., Sebastia, L., Hoffmann, J.: On the extraction, ordering, and usage
   of landmarks in planning. In: Pre-proceedings of the 6th European Conference on
   Planning (ECP 2001), pp. 37–48. Springer, Heidelberg (2001)
5. Richter, S., Helmert, M., Westphal, M.: Landmarks revisited. In: Proc. 23rd AAAI
   Conference on Artificial Intelligence (AAAI 2008), pp. 945–982. AAAI Press, Menlo
   Park (2008)
6. Richter, S., Westphal, M.: The lama planner: Guiding cost-based anytime planning
   with landmarks. JAIR 39, 127–177 (2010)
7. Zhu, L., Givan, R.: Landmark Extraction via Planning Graph Propagation. In:
   Printed Notes of ICAPS 2003 Doctoral Consortium, Trento, Italy (2003)

Rule-Based Creation of TimeML Documents from Dependency Trees

Livio Robaldo[1], Tommaso Caselli[2], and Matteo Grella[3]

[1] Department of Computer Science, University of Turin
[2] Istituto di Linguistica Computazionale, CNR, Pisa
[3] Parsit s.r.l.
http://www.parsit.it/
robaldo@di.unito.it, tommaso.caselli@ilc.cnr.it, matteo.grella@parsit.it

1 Introduction

The access to information through content has become the new frontier in NLP. Innovative annotation schemes such as TimeML [4] have push forward this aspect by creating benchmark corpora. In TimeML, an event is defined as something that holds true, obtains/happens, or occurs and annotated with a dedicated tag, namely <EVENT>. With respect to previous annotation schemes ([2], [6], [1], among others) for event annotation, some of the most interesting and innovative aspects of TimeML concern: i.) the parts-of-speech which may be annotated and ii.) the text span of the <EVENT> tag. With respect to the first point, TimeML allows the annotation of all linguistic realization of events such as verbs (example (1.a)), complex VPs (such as light verb constructions or idioms), nouns (example (1.b)), prepositional phrases (1.c), or adjectival phrases (1.d).

(1)   a. I pompieri <EVENT>isolarono</EVENT> la sala.
         (the fireworkers <EVENT>isolated</EVENT> the room.)

      b. Al Sayed è il nuovo <EVENT>presidente</EVENT>.
         (Al Sayed is the new <EVENT>president</EVENT>).

      c. Un turista <EVENT>in</EVENT> vacanza è morto.
         (A tourist <EVENT>on</EVENT> vacation died.)

      d. Si ritiene <EVENT>furbo</EVENT>.
         (He thinks he is <EVENT>clever</EVENT>.)

As for the text span of the <EVENT> tag, TimeML implements the notion of minimal chunk, i.e. only the head of the constituent(s) which realize an event must be annotated and not the entire phrase(s). This distinction is of utmost importance, since phrases can include more than one event instance.

In the present paper, we present a new Event Detector that exploits the TULE parser. However, it differs from [5] in that all the procedural knowledge has been moved into a separated configuration file, which can be manually filled and tuned, thus facilitating the update of the rules.

R. Pirrone and F. Sorbello (Eds.): AI*IA 2011, LNAI 6934, pp. 389–394, 2011.
© Springer-Verlag Berlin Heidelberg 2011

2   The Event Detector

The Event Detector in [5] is a Java™ module that explores TULE dependency trees and, for each node that could denote an event (verbs, nouns, adjectives, and prepositions), it runs a set of ad-hoc if-then rules in order to decide if it must be annotated as <EVENT>. These rules inspect the nearest nodes of the one under examination, and check if they belong to pre-built lists of words and collocations, and if they satisfy some simple constraints.

For instance, one of the collocations in the list is (*ottenere, risultato*) [(*obtain, result*)]. This means that whenever the noun *risultato* occurs as the direct object of the verb *ottenere*, it must be annotated as event.

The TULE format allows to check such constraints even when the verb *ottenere* is a modifier of the noun, because a trace referring to the latter is inserted. For instance, the phrase in (2) correspond to the tree in fig.1.

(2)   I **risultati** ottenuti nel primo quadrimestre ...
      (*The results obtained in the first quarter ... *)

We analysed the Java™ code of the prototype in [5] and tried to improve it.

In order to keep track of the impact of all conditions and constraints, we decided to move all the procedural knowledge into a separated configuration file, which can be manually filled and tuned, thus facilitating the updating of the rules. We devised an XML format describing patterns on the dependency trees, i.e. configurations of nodes with possible other nodes in their anchestors or dependents. The configuration file executes the pattern matching with respect to the TULE trees and marks up the documents with the <EVENT> tag. A further innovation of our approach is the definition of procedures to check the *evolution* of the rules' maintainance. In other words, whenever we evaluate the system during the training phase, the TimeML documents produced are compared both with the gold standard ones in the training set, *and* to the ones that were previously generated by the system. This double-check is possible because everytime we generate the TimeML documents, they are saved and kept in a separate folder. During the following executions, they are compared with the new ones

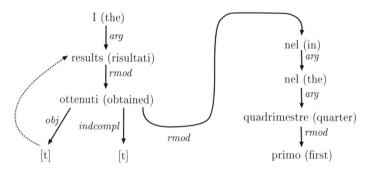

Fig. 1. Syntactic dependency tree associated with the phrase "I risultati ottenuti nel primo quadrimestre" (*The results obtained in the first quarter*)

in order to detect all effects made by the modified XML configuration file. Suppose, for instance, that a certain event is not recognized. Then, we modify the rules to get it. However, it could be the case that the new modification leads to a failure in the annotation of a previously recognized event. By means of this double-check procedure, we can make further modifications to the rules in order to identify both events. In this way, the Event Detector converges to the 100% of recall and precision, minus the percentages of gaps due to parsing errors and to cases where more complex disambiguation rules, possibly involving further techniques based on semantic and pragmatic information, are needed. The results shown in section 5 provide evidence to these hypotheses and claims.

In the next subsection, we describe the XML format of the rules, and how it is interpreted to harvest the events occurring in the text.

2.1   The XML Configuration File

The XML format used in the configuration file is rather simple. Each "keyword", i.e. each word that *could* be annotated as event or whose anchestors/dependents could be annotated as such, is associated with one or more **constraint** tags. The tag specifies which words, and under what conditions, must be annotated as <EVENT>. The conditions are specified in terms of other subtags:

- **headAlternatives**: it specifies the morphological features that must be satisfied by a word. The word must be annotated, provided that all other constraints are satisfied, just in case **headAlternatives** includes the attribute **tagIt="yes"**. **headAlternatives** includes one or more **head** tags whose subtags correspond to the TULE morphological features. **headAlternatives** is satisfied iff at least one of its **head**s matches the input word.

- **governorAlternatives**: it specifies the possible governors of the word, in terms of one or more **governor** tags. Each **governor** can specify the attribute **maxDistance** that indicates what is the maximal number of steps that could be done in order to find a word matching with the **governor**. The tag includes a tag **headAlternatives**, and possibly a **labelAlternatives** tag, specifying the possible labels linking the two words, and another (recursive) tag **governorAlternatives**.

- **dependentsAlternatives**: it specifies the possible dependents of the word, in exactly the same way **governorAlternatives** does, except that, while the governor is unique, we can specify tuples of dependents that must be found under the word. Thus, **dependentsAlternatives** includes one or more tag **dependents**, and the latter one or more tag **dependent**. The tag is satisfied iff at least one tuple of **dependents** matches the dependents of the word.

For instance, with respect to the two collocations *(risultato, ottenere)* [*(result, obtain)*], and *(risultato, conseguire)* [*(result, achieve)*], the script produces the constraint in fig.2. The maximal distance of the governor has been set to 3 intervening items. This allows to account for all the variants "*ottenere un risultato*', "*ottenere dei risultati*", etc. via a single rule.

```
<keyword value="risultato">
  <constraint>
    <headAlternatives tagIt = "yes">
      <head>
        <Lemma>risultato</Lemma>
        <POS>noun</POS>
        <CatType>common</CatType>
      </head>
    </headAlternatives>
    <governorAlternatives>
      <governor maxDistance="3">
        <labelAlternatives>
          <label>*OBJ*</label>
        </labelAlternatives>
        <headAlternatives tagIt = "yes">
          <head>
            <Lemma>ottenere</Lemma>
            <POS>verb</POS>
          </head>
          <head>
            <Lemma>conseguire</Lemma>
            <POS>verb</POS>
          </head>
        </headAlternatives>
      </governor>
    </governorAlternatives>
  </constraint>
</keyword>
```

Fig. 2. Example of constraints included in the configuration file

In order to account for exceptions to the rules, we added two attributes to the tag `constraint`, namely `not` and `priority`. If `not` has value 'yes' (default value is 'no'), the words identified by the rule must *not* be annotated as <EVENT>. `priority` marks the "importance" of the rule (default value is 0). Each word is annotated depending on the rule with highest priority that identifies it.

For instance, the verb *riguardare*, i.e. *to concern*, denotes an <EVENT> in most cases, except when it occurs in the syntactic construction *"per quanto riguarda"* [as far as it concerns], that is a paraphrase of the preposition *"about"* and so it must not be tagged as <EVENT>. Nevertheless, a further exception to this latter applies: the verb denotes an event if it occurs in this syntactic construction and if either a personal pronoun or a proper name occurs among its dependents. Two examples are *"per quanto mi riguarda"* and *"per quanto riguarda Luigi"*, i.e. *"as far as I/Luigi am/is concerned"*. In order to account for these cases, the configuration file includes the `constraints` reported in fig.3.

```
<keyword value="riguardare">
  <constraint>
    <headAlternatives tagIt = "yes">
      <head><Lemma>riguardare</Lemma>
      <POS>verb</POS></head>
    </headAlternatives>
  </constraint>
  <constraint priority="1" not="yes">
    <headAlternatives tagIt = "yes">
      <head><Lemma>riguardare</Lemma>
      <POS>verb</POS></head>
    </headAlternatives>
    <governorAlternatives>
      <governor maxDistance="1">
        <headAlternatives>
          <head><Form>quanto</Form></head>
        </headAlternatives>
        <governorAlternatives>
          <governor maxDistance="1">
            <headAlternatives>
              <head><Lemma>per</Lemma>
              <POS>Preposition</POS></head>
            </headAlternatives></governor>
        </governorAlternatives></governor>
    </governorAlternatives>
  </constraint>
  <constraint priority="2">
    <headAlternatives tagIt = "yes">
      ... the same as the second constraint ...
    </governorAlternatives>
    <dependentAlternatives>
      <dependents>
        <dependent maxDistance="1">
          <headAlternatives>
            <head><POS>Pronoun</POS>
            <CatType>Pers</CatType></head>
            <head><POS>Noun</POS>
            <CatType>Proper</CatType></head>
          </headAlternatives>
        </dependent>
      </dependents>
    </dependentAlternatives>
  </constraint>
</keyword>
```

Fig. 3. Example of constraints included in the configuration file

3   Evaluation

Table 1 shows the results of our system on TempEval-2 corpus for Italian.

The performance are rather satisfactory. Notice that the precision remains rather high both in the training and in the test set. The reason is that the XML configuration file mainly behaves as a "filter", as it is designed to specify which words must be annotated as <EVENT>. These values can be improved by enhancing TULE's accuracy and with a better integration with lexical resources for word-sense disambigation such as MultiWordNet [3].

Table 1. Results of the Event Detector on TempEval-2 corpus for Italian.

Data set	Precision	Recall	F-measure
TempEval-2 TRAINING	0.9438	0.8773	0,9092
TempEval-2 TEST	0.9131	0.7827	0,8428

References

1. Filatova, E., Hovy, E.: Assigning time-stamps to event-clauses. In: Proceedings of the Workshop on Temporal and Spatial Information Processing, TASIP 2001, vol. 13, pp. 13:1–13:8. Association for Computational Linguistics, Stroudsburg (2001)
2. Katz, G., Arosio, F.: The annotation of temporal information in natural language sentences. In: Proceedings of the Workshop on Temporal and Spatial Information Processing, TASIP 2001, vol. 13, pp. 15:1–15:8. Association for Computational Linguistics, Stroudsburg (2001)
3. Pianta, E., Bentivogli, L., Girardi, C.: Multiwordnet: Developing and aligned multilingual database. In: Proceedings of the First International Conference on Global WordNet, Mysore, India, pp. 293–302 (January 2002)
4. Pustejovsky, J., Castao, J., Saurì, R., Ingria, R., Gaizauskas, R., Setzer, A., Katz, G.: TimeML: Robust specification of event and temporal expressions in text. In: Fifth International Workshop on Computational Semantics (IWCS-5)
5. Robaldo, L., Caselli, T., Russo, I., Grella, M.: From italian text to timeml document via dependency parsing. In: Proc. of the 12th International Conference on Intelligent Text Processing and Computational Linguistics (2011)
6. Setzer, A., Gaizauskas, R.: A pilot study on annotating and extracting temporal information. In: Proceedings of the Workshop on Temporal and Spatial Information Processing, TASIP 2001, vol. 13, pp. 11:1–11:8. Association for Computational Linguistics, Stroudsburg (2001)

Handling Partial Preferences in the Belief AHP Method: Application to Life Cycle Assessment

Amel Ennaceur[1], Zied Elouedi[1], and Eric Lefevre[2]

[1] University of Tunis, Institut Supérieur de Gestion de Tunis, 41 Avenue de la
liberté, cité Bouchoucha, 2000 Le Bardo, Tunis, Tunisia
amel_naceur@yahoo.fr, zied.elouedi@gmx.fr
[2] Univ. Lille Nord of France, UArtois EA 3926 LGI2A, France
eric.lefevre@univ-artois.fr

Abstract. This paper proposes a novel multi-criteria decision making
method under uncertainty that combines the Analytic Hierarchy Process
(AHP) with the belief function theory. Our method, named belief AHP,
allows the expert to express incomplete and imprecise information about
groups of alternatives instead of single ones. On the other hand and in
order to judge the importance of criteria, he can also present his opinions
on groups of criteria. Then, the uncertainty will be taken into account in
the final decision. Finally, another purpose of this paper is also to solve a
real application problem which deals with the PVC life cycle assessment.

1 Introduction

Within the framework of Multi-Criteria Decision Making (MCDM) problems, a
decision maker often needs to make judgments on decision alternatives that are
evaluated on the basis of its preferences (criteria) [11]. Amongst the most well
known methods is the Analytic Hierarchy Process (AHP) [5] [6]. In fact, the
strength of this method is that it is easier to understand and it can effectively
handle both qualitative and quantitative data. In spite of its popularity, this
method is often criticized [3] because, in real-life decision making situation, the
decision maker may encounter several difficulties when building the pair-wise
comparison. These difficulties arise due to different situations: the lack of data for
making decisions, the inability to compare separate alternatives and/or criteria
between each other, etc. As a result, several extensions of AHP method were
proposed such as the Fuzzy AHP [4], the probabilistic AHP [1]. In particular in
the belief function framework, the DS/AHP method [2] was proposed.

The objective of this paper is to develop what we call a belief AHP, a MCDM
method adapted to imprecise and incomplete preferences, where the uncertainty
is represented by the belief function theory. Our aim through this work is to
allow the decision maker to give subjective judgments in two levels: the criterion
and alternative levels. On the one hand, our method offers a formalism allowing
the expert to express his ranking even over subgroups of alternatives. On the
other hand, to judge the importance of criteria, the belief AHP method will
be able to compare on groups of criteria instead of single criterion. Finally, to

R. Pirrone and F. Sorbello (Eds.): AI*IA 2011, LNAI 6934, pp. 395–400, 2011.
© Springer-Verlag Berlin Heidelberg 2011

illustrate the feasibility of our approach, we have applied our proposed method on a real application problem.

This paper is organized as follows: we start by introducing the AHP method, then we give an overview of the basic concepts of the belief function theory. In the main body of the paper, we present our new approach: the belief AHP which is based on the belief function theory. Finally, our method will be illustrated on a real application problem in order to understand its real unfolding.

2   Analytic Hierarchy Process

The AHP approach is a decision-making technique developed by Saaty [5] [6] to solve complex problems of choice and prioritization. The basic idea of the approach is to convert subjective assessments of relative importance to a set of overall scores or weights. Its first step is to set up a hierarchy consisting of the final goal of the problem or the decision to be made, a number of criteria, and a number of alternatives to select. Once the hierarchy is built, the decision maker starts the prioritization procedure. Elements of a problem on each level are paired (with respect to their upper level decision elements) and then compared using a nine-point scale [5] [6]. This semantic scale is used to translate the preferences of a decision maker into crisp numbers. After filling the pair-wise comparison matrices, the relative importance (priority) of the elements on each level of the hierarchy are determined by using the eigenvalue method. Finally, AHP aggregates all local priorities from the decision table by a simple weighted sum. The global priorities thus obtained are used for final ranking of the alternatives and selection of the best one.

3   Belief Function Theory

In this section, we briefly review the main concepts underlying the belief function theory as interpreted by the TBM. Details can be found in [8], [10].

Let Θ be the frame of discernment representing a finite set of elementary hypotheses related to a problem domain. We denote by 2^{Θ} the set of all the subsets of Θ.

The impact of a piece of evidence on the different subsets of the frame of discernment Θ is represented by the so-called basic belief assignment (bba) (denoted by m). It quantifies the impact of a piece of evidence on the different subsets of the frame of discernment [10].

The belief function theory offers many interesting tools. To combine beliefs induced by distinct pieces of evidence, we can use the conjunctive rule of combination [8]. Also, the discounting technique allows to take in consideration the reliability of the information source that generates the bba m [9].

It is necessary when making a decision, to select the most likely hypothesis. One of the most used solutions within the belief function theory is the pignistic probability [7].

4   Belief AHP Approach

Belief AHP aims at performing a similar purpose as AHP. In fact, its main purpose is to find the preferences' rankings of the decision alternatives in an uncertain environment. Within this context, a first work has been tackled by Beynon et al. [2], they developed a method, called DS/AHP. Despite all the advantages of this method, which allows different comparisons to be made for certain group of alternatives, they do not take into account the uncertainty in the criterion level. Thus, we propose a more general method for solving complex problems under the condition that it tolerates imprecision and uncertainty when the expert expresses his preferences between criteria and also alternatives. In other words, our approach will be able to compare groups of criteria and also groups of alternatives. Hence, the computational procedure of our proposed approach is summarized in the following steps.

1. **Identification of the candidate criteria:** By nature, the importance of criteria is relative to each other. Therefore, a decision maker may encounter some difficulties to compare separate ones. In our work, a new method for judging the importance of these criteria is proposed. In fact, we suggest to extend the AHP method to an imprecise representation rather than forcing the decision maker to provide precise representations of imprecise perceptions. We suppose that there is a set of criteria $\Omega = \{c_1, ..., c_m\}$ consisting of m elements. Denote the set of all subsets of C by 2^Ω, and let C_k be the short notation of a subset of C, i.e., $C_k \subseteq C$ and $C_k \in 2^\Omega$. An expert chooses a subset $C_k \subseteq C$ of criteria from the set C and compares this subset with another subset $C_j \subseteq C$. Thus, criteria that belong to the same group have the same degree of preferences. Since we are not performing pair-wise comparisons of criterion but relating groups of criteria, these sets of criteria should not consider a criterion in common, because if one criterion is included in two groups, then each group will give a different level of favorability. By generalization, the subsets of criteria can be defined as:

$$C_k \succ C_j, \forall \ k, j | C_k, C_j \in 2^\Omega, C_k \cap C_j = \emptyset \ . \tag{1}$$

2. **Identification of the candidate alternatives:** As mentioned, under this approach we suggest to compare groups of alternatives instead of single one. The decision maker has to identify the subsets of favorable alternatives from all the set of the possible ones. One of the possible solutions of this task is to use the DS/AHP method [2]. Similarly to the criterion level, we assume that there is a set of alternatives $\Theta = \{a_1, ..., a_n\}$ consisting of n elements. Denote the set of all subsets of A by 2^Θ, and let A_k be the short notation of a subset of A, i.e., $A_k \subseteq A$ and $A_k \in 2^\Theta$. The main aim behind this method was explained in [2].

3. **Computing the weight of considered criteria and the alternative priorities:** After constructing the hierarchical structure of the problem, what is left is setting priorities of the subsets of alternatives and criteria. At this point, standard pair-wise comparison procedure is made to obtain these priorities.

4. **Updating the alternatives priorities:** Once the priorities of decision alternatives and criteria are computed, we have to define a rule for combining
   them. The problem here is that we have priorities concerning criteria and
   groups of criteria instead of single ones, whereas the sets of decision alternatives are generally compared pair-wise with respect to a specific single
   criterion. In order to overcome this difficulty, we choose to apply the belief
   function theory because it provides a convenient framework for dealing with
   individual elements of the hypothesis set as well as their subsets.

   At the decision alternative level, we propose to follow the main idea of
   the DS/AHP method. In fact, we have the priority vector corresponding to
   each comparison matrix sums to one. So, we can assume that $m(A_k) = w_k$,
   where w_k is the eigen value of the kth sets of alternatives.

   The next step is to update the obtained bba with the importance of their
   respective criteria. In this context, our approach proposes to regard each
   priority value of a specific set of criteria as a measure of reliability. In fact,
   the idea is to measure most heavily the bba evaluated according to the most
   importance criteria and conversely for the less important ones. If we have
   C_k a subset of criteria, then we get β_k its corresponding measure of reliability, given by the division of the importance of criteria by the maximum of
   priorities. As a result, two cases will be presented: First, if the reliability factor represents a single criterion, then the corresponding bba will be directly
   discounted, and we get:

   $$m_{C_k}^{\alpha_k}(A_j) = \beta_k.m_{C_k}(A_j), \ \forall A_j \subset \Theta \ . \tag{2}$$

   $$m_{C_k}^{\alpha_k}(\Theta) = (1 - \beta_k) + \beta_k.m_{C_k}(\Theta) \ . \tag{3}$$

   where $m_{C_k}(A_j)$ the relative bba for the subset A_j, and we denote $\alpha_k = 1 - \beta_k$.
   Second, if this factor represents a group of criteria, their corresponding bba's
   must be combined. Based on the belief function framework, our proposed
   approach assumes that each pair-wise comparison matrix is considered as
   a distinct source of evidence, which provides information on opinions towards the preferences of particular decision alternatives. Then, we apply the
   conjunctive rule of combination and we get:

   $$m_{C_k} = \bigcirc m_{c_i}, \ \ i = \{1, ..., h\} \ . \tag{4}$$

   where h is the number of element of a specific group of criteria C_k and
   $c_i \in C_k$ (c_i a singleton criterion). Finally, these obtained bba's (m_{C_k}) will
   be discounted by their corresponding measure of reliability (the same idea
   used in Equation 2 and 3).

5. **Synthetic utility and decision making:** After updating the alternatives
   priorities', we must compute the overall bba. An intuitive definition of the
   strategy to calculate these bba's will be the conjunctive rule of combination
   $(m_{final} = \bigcirc m_{C_k}^{\alpha_k})$.

   To this end, the final step is to choose the best alternative. In this context,
   we choose to use the pignistic transformation to help the expert to make his
   final choice.

5   Application

The problem in this application is not to use or not the Polyvinyl chloride (PVC) in general, but to know in which country the environmental impact is less important for the destruction of a kilogram of PVC?

1. **Identification of the candidate criteria and alternatives:** In this application problem, the environmental criteria are playing the role of multiple criteria, and it was decided to restrict them to four areas: $\Omega = \{$abiotic depletion (C1), eutrophication (C2), toxicity infinite (C3), Fresh water aquatic ecotoxicity infinite (C4)$\}$. Apart from the four criteria, the initial interview also identified three selected countries on the set of alternatives: $\Theta = \{\text{France}(FR), \text{USA}(US), \text{England}(EN)\}$.
2. **Computing the weights of considered criteria:** Now the expert is asked to express the intensity of the preference for one criterion versus another. By using the eigenvector method, the final priorities values can be obtained as shown in Table 1, and a normalized vector is given.

Table 1. The weights assigned to the criteria according to the expert's opinion

Criteria	$\{C1\}$	$\{C4\}$	$\{C2,C3\}$	Priority	Nomalized vector
$\{C1\}$	1	2	6	0.58	1
$\{C4\}$	$\frac{1}{2}$	1	4	0.32	0.55
$\{C2,C3\}$	$\frac{1}{6}$	$\frac{1}{4}$	1	0.1	0.17

3. **Computing the alternatives priorities:** Similarly to the standard AHP, comparison matrices are constructed, and we suppose that each priority vector is considered as a bba (see Table 2).

Table 2. Priorities values

C1	m_{C1}	C2	m_{C2}	C3	m_{C3}	C4	m_{C4}
$\{EN,US\}$	0.896	$\{EN\}$	0.526	$\{EN\}$	0.595	$\{US\}$	0.833
$\{EN,US,FR\}$	0.104	$\{US,FR\}$	0.404	$\{FR\}$	0.277	$\{EN,US,FR\}$	0.167
		$\{EN,US,FR\}$	0.07	$\{EN,US,FR\}$	0.128		

4. **Updating the alternatives priorities:** Firstly, this step concerns the groups of criteria $\{C2,C3\}$. Our aim is to combine the bba relative to the criteria C2 and C3. Then, the obtained bba's is discounted by their measure of reliability $\beta_{C2,C3} = 0.17$. After that, this step concerns the single criterion $\{C1\}$ and $\{C4\}$. The relative bba's are directly discounted by their reliability measure $\beta_{C1} = 1$ and $\beta_{C4} = 0.55$ and we get the following Table 3.
5. **Synthetic utility and decision making:** After updating the sets of priority value, the conjunctive rule of combination can be applied, this leads us to get a single bba denoted by $m_{final} = m_{C1}^{\alpha_{C1}} \otimes m_{C2,C3}^{\alpha_{C2,C3}} \otimes m_{C4}^{\alpha_{C4}}$.

Table 3. The adjusted Priority values

	$\emptyset$	$\{EN\}$	$\{FR\}$	$\{US, FR\}$	Θ		$\{EN, US\}$	Θ		$\{US\}$	Θ
$m_{C2,C3}^{\alpha C2,C3}$	0.066	0.0718	0.0224	0.0088	0.831	$m_{C1}^{\alpha C1}$	0.896	0.104	$m_{C4}^{\alpha C4}$	0.458	0.542

Then, these obtained bba is transformed into pignistic probabilities, we get $BetP_{final}(EN) = 0.2911$, $BetP_{final}(US) = 0.6894$ and $BetP_{final}(FR) = 0.0195$. Consequently, USA is the recommended country since it has the highest values.

6   Conclusion

Our objective through this work is to develop a new MCDM method providing a formal way to handle uncertainty in AHP method within the belief function framework. Moreover, our approach has reduced the number of comparisons because instead of using single elements, we have used subsets. At the end, we have shown the flexibility and feasibility of our proposed approach by applying it on a real application problem related to "the end of life phase" of PVC product.

This work calls for several perspectives. One of them consists in comparing our proposed approach with other MCDM methods. In addition, the proposed method will be more flexible, if it will be able to handle uncertainty in the Saaty's scale.

References

1. Basak, I.: Probabilistic judgments specified partially in the Analytic Hierarchy Process. European Journal of Operational Research 108, 153–164 (1998)
2. Beynon, M., Curry, B., Morgan, P.: The Dempster-Shafer theory of evidence: An alternative approach to multicriteria decision modelling. OMEGA 28(1), 37–50 (2000)
3. Joaquin, P.: Some comments on the analytic hierarchy process. The Journal of the Operational Research Society 41(6), 1073–1076 (1990)
4. Laarhoven, P.V., Pedrycz, W.: A fuzzy extension of Saaty's priority theory. Fuzzy Sets and Systems 11, 199–227 (1983)
5. Saaty, T.: A scaling method for priorities in hierarchical structures. Journal of Mathematical Psychology 15, 234–281 (1977)
6. Saaty, T.: The Analytic Hierarchy Process. McGraw-Hill, New York (1980)
7. Smets, P.: The application of the Transferable Belief Model to diagnostic problems. International Journal of Intelligent Systems 13, 127–158 (1998)
8. Smets, P.: The combination of evidence in the Transferable Belief Model. IEEE Pattern Analysis and Machine Intelligence 12, 447–458 (1990)
9. Smets, P.: Transferable Belief Model for expert judgments and reliability problems. Reliability Engineering and System Safety 38, 59–66 (1992)
10. Smets, P., Kennes, R.: The Transferable Belief Model. Artificial Intelligence 66, 191–234 (1994)
11. Zeleny, M.: Multiple Criteria Decision Making. McGraw-Hill Book Company, New York (1982)

Combining Description Logics and Typicality Effects in Formal Ontologies

Marcello Frixione and Antonio Lieto

University of Salerno,
Via Ponte Don Melillo, 84084, Fisciano, Italy
{mfrixione,alieto}@unisa.it

Abstract. In recent years, the problem of concept representation received great attention within knowledge engineering because of its relevance for ontology-based technologies. However, the notion of concept itself turns out to be highly disputed and problematic. In our opinion, one of the causes of this state of affairs is that the notion of concept is in some sense heterogeneous, and encompasses different cognitive phenomena. This results in a strain between conflicting requirements, such as, for example, compositionality on the one side and the need of representing prototypical information on the other. AI research in some way shows traces of this situation. In this paper we propose an analysis of this state of affairs and sketch some proposal for concept representation in formal ontologies, which takes into account suggestions coming from psychological research.

Keywords: Ontologies, knowledge representation, reasoning, linked data.

1 Introduction

Computational representation of concepts is a central problem for the development of ontologies and for knowledge engineering. However, the notion of concept itself results to be highly disputed and problematic. One of the causes of this state of affairs is that the notion itself of concept is in some sense heterogeneous, and encompasses different cognitive phenomena. This has several consequences for the practice of knowledge engineering and for the technology of formal ontologies. In this paper we propose an analysis of this situation. In section 2, we point out some differences between the way concepts are conceived in philosophy and in psychology. In section 3, we argue that AI research in some way shows traces of the contradictions individuated in sect. 2. In particular, the requirement of compositional, logical style semantics conflicts with the need of representing concepts in the terms of typical traits that allow for exceptions. In section 4 we individuate some possible suggestions coming from different aspects of cognitive research: namely, the proposal to keep prototypical effects separate from compositional representation of concepts, and the possibility to develop hybrid, prototype and exemplar-based representations of concepts. In section 5 we give some tentative suggestion to implement our proposals within the context

R. Pirrone and F. Sorbello (Eds.): AI*IA 2011, LNAI 6934, pp. 401–406, 2011.

of semantic web languages, in the framework of the linked data perspective. Finally, in section 6, we discuss some expected results.

2   Compositionality vs. Prototypes

Within the field of cognitive science, conceptual representation seems to be constrained by conflicting requirements, such as, for example, compositionality on the one side and the need of representing prototypical information on the other. A first problem (or, better, a first symptom that some problem exists) consists in the fact that the use of the term concept in the philosophical tradition is not homogeneous with the use of the same term in empirical psychology [1]. Briefly, we could say that in cognitive psychology a concept is essentially intended as the mental representations of a category, and the emphasis is on such processes as categorisation, induction and learning. According to philosophers, concepts are above all the components of thoughts. This fact brought a great emphasis on *compositionality*, and on related features, such as productivity and systematicity, that are often ignored by psychological treatments of concepts. On the other hand, it is well known that compositionality is at odds with *prototypicality effects*, which are crucial in most psychological characterisations of concepts. Prototypical effects are a well established empirical phenomenon. However, the characterisation of concepts in prototypical terms is difficult to reconcile with the requirement of compositionality. According to a well known argument by Jerry Fodor [2], prototypes are not compositional. In synthesis, Fodor's argument runs as follows: consider a concept like PET FISH. It results from the composition of the concept PET and of the concept FISH. But the prototype of PET FISH cannot result from the composition of the prototypes of PET and of FISH. For example, a typical PET is furry and warm, a typical FISH is greyish, but a typical PET FISH is not furry and warm neither greyish.

3   Concept Representation in Artificial Intelligence

The situation sketched in the section above is in some sense reflected by the state of the art in Artificial Intelligence (AI) and, more in general, in the field of computational modelling of cognition. This research area seems often to hesitate between different (and hardly compatible) points of view. In AI the representation of concepts is faced mainly within the field of knowledge representation (KR). Symbolic KR systems (KRs) are formalisms whose structure is, in a wide sense, language-like. This usually involves that KRs are assumed to be compositional. In a first phase of their development (historically corresponding to the end of the 60s and to the 70s) many KRs oriented to conceptual representations tried to keep into account suggestions coming from psychological research, and allowed the representation of concepts in prototypical terms. Examples are early semantic networks [3] and frame systems [4]. However, such early KRs where usually characterised in a rather rough and imprecise way. They lacked a clear formal definition, and the study of their meta-theoretical properties was almost impossible. When AI practitioners tried to provide a stronger formal foundation to concept oriented KRs, it turned out to be difficult to reconcile compositionality and prototypical representations. As a consequence, they often choose to sacrifice the latter. In particular, this is the solution adopted in a class of concept-oriented KRs which had (and still have) wide diffusion within AI, namely the class of formalisms that stem from the so-called structured inheritance networks and from the KL-ONE system. Such systems were subsequently called terminological logics, and today are usually known as description logics (DLs) [5]. A standard inference mechanism for this kind of networks is inheritance. Representation of prototypical information in semantic networks usually takes the form of allowing exceptions to inheritance. Networks in this tradition do not admit exceptions to inheritance, and therefore do not allow the representation of prototypical information. Indeed, representations of exceptions can be hardly accommodated with other types of inference defined on these formalisms, concept classification in the first place [6]. In more recent years, representation systems in this tradition have been directly formulated as logical formalisms (the above mentioned description logics [5]), in which Tarskian, compositional semantics is straightly associated to the syntax of the language. Logical formalisms are paradigmatic examples of compositional representation systems. As a consequence, this kind of systems fully satisfy the requirement of compositionality. This has been achieved at the cost of not allowing exceptions to inheritance. By doing this we gave up the possibility of representing concepts in prototypical terms. From this point of view, such formalisms can be seen as a revival of the classical theory of concepts (according to which concepts can be defined in terms of necessary and sufficient conditions [7]), in spite of its empirical inadequacy in dealing with most common-sense concepts. Nowadays, DLs are widely adopted within many application fields, in particular within the field of the representation of ontologies. For example, the OWL (Web Ontology Language) system[1] is a formalism in this tradition that has been endorsed by the World Wide Web Consortium for the development of the semantic web.

4   Some Suggestions from Cognitive Science

The empirical results from cognitive psychology show that most common-sense concepts cannot be characterised in terms of necessary/sufficient conditions [7]. Classical, monotonic DLs seem to capture the compositional aspects of conceptual knowledge, but are inadequate to represent prototypical knowledge. But a non classical alternative, a general DL able to represent concepts in prototypical terms, does not still emerge. As a possible way out, we sketch a tentative proposal that is based on some suggestions coming from cognitive science. In particular, we individuate some hints that, in our opinion, could be useful for the development of artificial representation systems, namely: (i) the proposal to keep prototypical effects separate from compositional representation of concepts

---

[1] http://www.w3.org/TR/owl-features/

(sect. 4.1); and (ii) the possibility to develop hybrid, prototype and exemplar-based representations of concepts (sect. 4.2).

4.1 A "Pseudo-Fodorian" Proposal

As seen before (section 2), according to Fodor, concepts cannot be prototypical representations, since concepts must be compositional, and prototypes do not compose. On the other hand, in virtue of the criticisms to classical theory, concepts cannot be definitions. Therefore, Fodor argues that (most) concepts are atoms, i.e., are symbols with no internal structure. Their content is determined by their relation to the world, and not by their internal structure and/or by their relations with other concepts [8]. Of course, Fodor acknowledges the existence of prototypical effects. However, he claims that prototypical representations are not part of concepts. Prototypical representations allow to individuate the reference of concepts, but they must not be identified with concepts. Consider for example the concept DOG. Of course, in our minds there is some prototypical representation associated to DOG (e.g., that dogs usually have fur, that they typically bark, and so on). But this representation does not coincide with the concept DOG: DOG is an atomic, unstructured symbol. We borrow from Fodor the hypothesis that compositional representations and prototypical effects are demanded to different components of the representational architecture. We assume that there is a compositional component of representations, which admits no exceptions and exhibits no prototypical effects, and which can be represented, for example, in the terms of some classical DL knowledge base. In addition, a prototypical representation of categories is responsible for such processes as categorisation, but it does not affect the inferential behaviour of the compositional component.

4.2 Prototypes and Exemplars

Within the field of psychology, different positions and theories on the nature of concepts are available. Usually, they are grouped in three main classes, namely prototype views, exemplar views and theory-theories [7]. All of them are assumed to account for (some aspects of) prototypical effects in conceptualisation. Theory-theory approach is more vaguely defined if compared to the other two points of view. As a consequence, in the following we focus on prototype and exemplar views. According to the prototype view, knowledge about categories is stored in terms of prototypes, i.e. in terms of some representation of the best instances of the category. For example, the concept CAT should coincide with a representation of a prototypical cat. In the simpler versions of this approach, prototypes are represented as (possibly weighted) lists of features. According to the exemplar view, a given category is mentally represented as set of specific exemplars explicitly stored within memory: the mental representation of the concept CAT is the set of the representations of (some of) the cats we encountered during our lifetime. These approaches turned out to be not mutually exclusive. Rather, they seem to succeed in explaining different classes of cognitive phenomena, and many researchers hold that all of them are needed to explain psychological data. In this perspective, we propose to integrate some of them in computational representations of concepts. More precisely, we try to combine a prototypical and an exemplar based representation in order to account for category representation and prototypical effects (for a similar proposal, see [9]).

5 Some Suggestion for the Implementation

In the field of web ontology languages, the developments sketched above appear nowadays, technologically possible. Within the Semantic Web research community, in fact, the Linked Data perspective is assuming a prominent position [10]. Consider for example the opposition between exemplar and prototype theories (see sect. 4.2 above). Both theories can be implemented in a re-presentation system using the Linked Data perspective. Let us consider first the case of prototype theory. A dual representation of concepts and reasoning mechanisms appears to be possible trough the following approach: a concept is represented both in a formal ontology (based on a classical, compositional DL system), and in terms of a prototypical representation, implemented using the Open Knowledge-Base Connectivity (OKBC) protocol[2]. The knowledge model of the OKBC protocol is supported and implemented in the so called Frame Ontologies. Since it is possible to export (without losing the prototypical information) the Frame Ontologies in classical Ontology Web Language, the connection between these two types of representation can be done using the standard formalisms provided by the Semantic Web community within the Linked Data perspective (e.g. using the owl:sameAs construct). In a similar way, an exemplar based representation of a given concept can be expressed in a Linked Data format, and connected to a DL ontological representation. In this way, according to our hypothesis, different types of reasoning processes (e.g., classification and categorization) can follow different paths. For example, classification could involve only the DL ontology, while the non monotonic categorization process could involve exemplars and prototypical information. A possible solution to perform non monotonic categorization of instances could be based on the PEL-C algorithhm, Prototype-Exemplar Learning Classifier [9]. The PEL-C is a hybrid machine learning algorithm able to account for typicality in the categorization process, using both prototype and exemplar based representations. The application of this algorithm requires the choice of a metric of semantic similarity between concepts within the prototype and exemplar based component of the architecture.

6 Expected Results

It is our intention to evaluate our proposal by comparing its performance with that of a traditional ontology representing the same domain. The evaluation tasks could consist mainly in two types of controls: property checking and instance

---

[2] http://www.ai.sri.com/ okbc/

checking running SPARQL query on the different knowledge bases. Instance checking aims to answer at such questions as is a particular instance member of a given concept?. In this case we expect that, the processes running on prototypical and exemplar based representations, could provide a different answer if compared to that of a traditional DL ontology. This result does not cause inconsistencies or create any problem because of the separation of representation and reasoning process. The second evaluation task is based on property checking. It consists answering such questions as does the class A have the property b?. We expect that the query-answering mechanism should take advantage from the integration of different types of information provided for the same concept. For example: let us suppose that an user runs an informational query on a dual knowledge base representing information concerning fruit in order to know which kind of citrus is yellow (that is an indirect formula to ask: does any citrus have the property of being yellow?). The expected answer that fits the informational needs of the user is lemon. However, does not exist in the knowledge base any kind of citrus that has the property of being yellow as a defining condition because being yellow is not a necessary condition for being a lemon. However the property to be yellow is relevant from a cognitive point of view to characterize the concept lemon, and can be represented into the prototypical component of the class lemon. In this way is possible to retrieve the desired information from the prototypical and/or exemplar part of the representation. So, given a SPARQL query such as:

SELECT? citrus
WHERE ?citrus :has colour : YELLOW

the result returned from the DL representation should be null, while the correct answer (correct with respect to the intention of the user) will be generated from the prototypical component of the representation.

References

1. Frixione, M., Lieto, A.: The computational representation of concepts in formal ontologies: Some general considerations. In: Proc. KEOD, Valencia (2010)
2. Fodor, J.: The present status of the innateness controversy. In: Fodor, J. (ed.) Representations. MIT Press, Cambridge (1981)
3. Brachman, R., Levesque, H. (eds.): Readings in Knowledge Representation. Morgan Kaufmann, Los Altos (1985)
4. Minsky, M.: A framework for representing knowledge (1975); Now in [3]
5. Baader, F., Calvanese, D., McGuinness, D., Nardi, D., Patel-Schneider, P.: The Description Logic Handbook. Cambridge University Press, Cambridge (2003)
6. Brachman, R.: I lied about the trees. The AI Magazine 3(6), 80–95 (1985)
7. Murphy, G.L.: The Big Book of Concepts. The MIT Press, Cambridge (2002)
8. Fodor, J.: Psychosemantics. MIT Press, Cambridge (1987)
9. Gagliardi, F.: Cognitive Models of Typicality in Categorization with Instance-Based M.L., in Practices of Cognition, pp. 115–130. Univ. of Trento Press (2010)
10. Bizer, C., Heath, T., Berners-Lee, T.: Linked Data - The Story So Far. Int. J. on Semantic Web and Inf. Syst. 5(3), 1–22 (2009)

Italian Anaphoric Annotation with the Phrase Detectives Game-with-a-Purpose

Livio Robaldo[1], Massimo Poesio[2,4], Luca Ducceschi[3],
Jon Chamberlain[4], and Udo Kruschwitz[4]

[1] University of Turin
robaldo@di.unito.it
[2] University of Trento
poesio@essex.ac.uk
[3] University of Utrech/Verona
lucaducceschi@gmail.com
[4] University of Essex
{jchamb,udo}@essex.ac.uk

1 Introduction

Recently, web collaboration (also known as crowd sourcing) has started to emerge as a viable alternative for building the large resources that are needed to build and evaluate NLP systems.

In this spirit, the Anawiki project (http://anawiki.essex.ac.uk/) [8] aimed at experimenting with Web collaboration and human computation as a solution to the problem of creating large-scale linguistically annotated corpora. So far, the main initiative of the project has been Phrase Detectives (PD) [2], a game designed to collect judgments about anaphoric annotations. To our knowledge, Phrase Detectives was the first attempt to exploit the effort of Web volunteers to annotate corpora (subsequent efforts include [1] and [5]).

In this paper, we discuss how we developed a version of the game to annotate anaphoric relations in Italian, focusing in particular on the methods we developed to annotate anaphoric phenomena that are present in Italian but not English, like clitics and zeros, and on the development of a pipeline to convert Italian texts into the XML format used in the game. This convertion is simpler than that for English, and produces better results.

2 Phrase Detectives

Phrase Detectives offers a simple graphical user interface for non-expert users to learn how to annotate text and how to make annotation decisions [3]. The goal of the game is to identify relationships between words and phrases in a short text. An example of a task would be to highlight an anaphor-antecedent relation between the markables (sections of text) 'This parrot' and 'He' in 'This parrot is no more! He has ceased to be!'.

There are two ways to annotate within the game: by selecting a markable that corefers to another one (**Annotation Mode**); or by validating a decision

R. Pirrone and F. Sorbello (Eds.): AI*IA 2011, LNAI 6934, pp. 407–412, 2011.

Fig. 1. Screenshots of the game: Annotation Mode (left) and Validation Mode (right)

previously submitted by another player (`Validation Mode`). `Annotation Mode` (Figure 1, on the left) is the simplest way of collecting judgments. The player has to locate the closest antecedent markable of an anaphor markable, i.e. an earlier mention of the object. By moving the cursor over the text, markables are revealed in a bordered box. To select it the player clicks on the bordered box and the markable becomes highlighted. They can repeat this process if there is more than one antecedent markable (e.g. for plural anaphors such as *'they'*). They submit the annotation by clicking the `Done!` button. The player can also indicate that the highlighted markable has not been mentioned before (i.e. it is not anaphoric), that it is non-referring (for example, *'it'* in *'Yeah, well it's not easy to pad these Python files out to 150 lines, you know.'*) or that it is the property of another markable (for example, *'a lumberjack'* being a property of *'I'* in *'I wanted to be a lumberjack!'*. In `Validation Mode` (Figure 1, on the right) the player is presented with an annotation from a previous player. The anaphor markable is shown with the antecedent markable(s) that the previous player chose. The player has to decide if he agrees with this annotation. If not he is shown the Annotation Mode to enter a new annotation.

In the game, groups of players work on the same task over a period of time as this is likely to lead to a collectively intelligent decision [9]. An initial group of players are asked to annotate a markable. If all the players agree with each other then the markable is considered complete.

So far, 4428 players registered to the game, and 1,5 million of annotations and validations have been collected. 324 documents have been completed, for a total of 108k words and 60,000 markables[1]. As shown in [4], the overall agreement is superior to 94%, although it refers to a smaller set of markables.

---

[1] To make a comparison, OntoNotes 3.0 (`http://www.ldc.upenn.edu`) includes 45,000 markables.

3   Porting Phrase Detectives to Italian

In order to use Phrase Detectives to annotate Italian data, the following pipeline was created using the TULE dependency parser [7]:

- The input is analysed by the Tule Parser [7], which is a Dependency parser.
- A Java module identifies markables on the basis of the dependency links among words. The Java module produces the MAS-XML format corresponding to the input text.
- MAS-XML is converted to SGF via Saxon.

Phrase Detectives stores its data in SGF format, which is obtained via Saxon starting from a MAS-XML file. The latter is a xml-based format encoding a very basic form of constituency structure: sentences (tag <s>), noun expressions (tag <ne>), nominal modifiers (tag <mod>), nominal heads (<head>) and verbal expressions (tag <ve>). In Figure 2, we show the MAS-XML representation of sentences[2] (1.a-b), which are taken from a Wikipedia entry loaded in the game. In the box on the left we show all the tags that should be annotated in the format, while in the box on the right we show the noun expressions only, i.e. the only tags that are considered in Phrase Detectives. During each round, the game highlights a noun expression, and asks the player to select (at least one) previous noun expression that co-refers it.

(1) a. La battaglia di Austerlitz fu l'ultima battaglia che si attuò durante le guerre napoleoniche della terza coalizione e che portò alla vittoria l'esercito napoleonico.
     b. Avvenne il 2 dicembre 1805 nei pressi della cittadina di Austerlitz tra l'armata francese composta da circa 65.000 uomini comandati da Napoleone ed una armata congiunta, formata da russi e austriaci ...

It is quite easy to harvest the noun expressions from the dependency trees. We check if certain "key" words (nouns, articles, etc.) could be the governors of a noun expression via some simple test, then we take (and tag) the noun expression as the union of all contiguous words occurring in the subtrees of these governors. In the next subsection, we discuss the process in some more detail.

3.1   Main Syntactic Phenomena

Nouns and Traces

As pointed out above, noun expressions are primarily denoted by nouns and (non-relative) pronouns. For instance, the phrases *La battaglia di Austerlitz* and *Avvenne il 2 dicembre 1805* are represented as in fig.3. In the figure, the part-of-speech of each word is indicated next to it.

---

[2] English translation of (1.a): *The battle of Austerlitz was the last battle that took place during the napoleonic wars of the third coalition and that led to the victory the napoleonic army.* English translation of (1.b): *(It) took place the 2nd of December 1805 around the village of Austerlitz between the French army composed by 65.000 men commanded by Napoleon and a joint army, composed by Russians and Austrians.*

```
<ne>La battaglia                    <ne>(esso/essa)</ne>
  <mod>di                           Avvenne
    <ne>Austerlitz</ne>             <ne>il 2 dicembre 1805</ne>
  </mod>                            nei pressi
</ne>                               <ne>della cittadina di
<ve>fu</ve>                           <ne>Austerlitz</ne>
<ne>l'ultima battaglia             </ne>
  <mod>                             tra
    che si <ve>attuò</ve>           <ne>l'armata francese composta da
    <mod>durante                      <ne>circa 65.000 uomini
      <ne>le guerre napoleoniche         comandati da
        <mod>                             <ne>Napoleone</ne>
          <ne>della terza              </ne>
                coalizione</ne>     </ne>
        </mod>                      ed
      </ne>                         <ne>una armata congiunta</ne>
    </mod>                          , formata da
    e che <ve>portò</ve>           <ne>
    <ne>alla vittoria</ne>           <ne>russi</ne>
    <ne>l'esercito napoleonico</ne>  e
  </mod>                             <ne>austriaci</ne>
</ne> .                            <ne> ...
```

Fig. 2. MAS-XML representation of sentences (1.a) and (1.b)

The rule to identify such noun expressions is rather simple: we scan the nodes in the trees and, for each determiner, we create a noun expression including it, its argument (DET-ARG) and every other modifier or apposition in its subtree. Then, we recursively apply the rule to identify nested noun expressions.

Finally, since the TULE parser introduces traces in the trees to mark pro-dropped arguments, it is possible to create noun expressions referring to the corresponding zero-pronouns. In the final MAS-XML they appear among round brackets, and they are morphologically realized via pronouns having the same gender and number. For instance, the subject of *Avvenne* in fig.3 is pro-dropped. The tree includes a trace indicating a neutral and singular missing argument. In the MAS-XML, we output a noun expression whose content is *(esso/essa)*, where *esso* and *essa* are, respectively, the male and the female Italian translations of the English pronoun *it*.

Modifiers

In TUT dependency trees, modifiers (aka adjuncts) are optional dependents appearing as subtrees of the head they modify. Therefore, whether an head denotes a noun expression, all words occurring in these subtrees should be included therein. For instance, in fig.3, the prepositional modifier *di Austerlitz* has been included in the noun expression headed by the determiner *La*, while in fig.4, the adverb *circa* is included in the noun expression headed by the determiner *65.000* as well as the adjective *congiunta* is part of the noun expression headed by *una*.

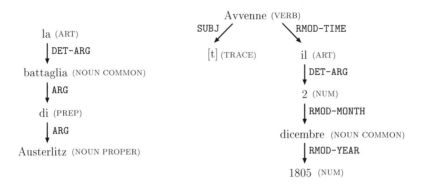

Fig. 3. TUT sub-trees for: *La battaglia di Austerlitz* and *Avvenne il 2 dicembre 1805*

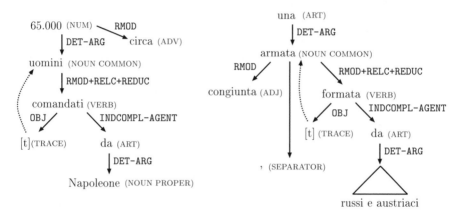

Fig. 4. TUT sub-trees for: *Circa 65.000 uomini comandati da Napoleone* and *Una armata congiunta, formata da russi e austriaci*

However, there is an exception to this general rule for relative clauses. In fact, many Italian sentences contain several embeddings of relative clauses, that would lead to rather long noun expressions. As a result, an annotator could have more doubts and difficulties while selecting the previous co-referent(s). When relative clauses tend to be very long or to have several levels of nesting, they are usually separated from the main sentence via punctuation. Therefore, in order to enhance readability, we include the relative clauses in the noun expression associated with the main sentence only in absence of punctuation. For instance, with respect to the trees in fig.4, we include the relative clause *comandati da Napoleone* in the noun expression headed by *65.000*, while we do not include the clause *formata da russi e austriaci...* in the sentence headed by *una armata congiunta*.

4   Evaluation and Discussion

The pipeline required to process Italian texts has two relevant features, namely:

- a smaller number of independent tools is needed (only the Tule parser and Saxon);
- the syntactic analysis is obtained via a dependency parser;

Both these features contribute to the quality of the processed text that is marked up by the players.

With respect to that, PD allows expert players to correct the markables resulting after the pipeline processing. This feature enables us to observe and record the performances of the whole process: the English pipeline of PD has a score of 4.56 errors per text [6], while the Italian version has only 0.67.

The pipeline used for Italian texts has been designed with the aim of minimizing errors in the markables and, consequently, further review process. The dependency parser used for Italian has the advantage of arranging the words in a structure rather close to the recursive embedding of the markables. In case of error, it is rather easy and quick to directly correct the dependency tree, as in most cases it is sufficient to change the pointer of one of its nodes. In addition to that the TULE parser can easily detect pro-dropped arguments and insert traces to mark them.

References

1. Attardi, G., the Galoap Team: Phratris. Demo presented at INSEMTIVES 2010 (2010)
2. Chamberlain, J., Kruschwitz, U., Poesio, M.: Constructing an anaphorically annotated corpus with non-experts: Assessing the quality of collaborative annotations. In: Proc. of the 2009 Workshop on The People's Web Meets NLP: Collaboratively Constructed Semantic Resources. Collocated at ACL 2009, Singapore (2009)
3. Chamberlain, J., Poesio, M., Kruschwitz, U.: Phrase detectives - a web-based collaborative annotation game. In: Proc. of I-Semantics (2008)
4. Chamberlain, J., Poesio, M., Kruschwitz, U.: A new life for a dead parrot: Incentive structures in the phrase detectives game, Madrid (2009)
5. Hladká, B., Mírovský, J., Schlesinger, P.: Play the language: play coreference. In: Proceedings of the ACL-IJCNLP 2009 Conference Short Papers, pp. 209–212. Association for Computational Linguistics (2009)
6. Kruschwitz, U., Chamberlain, J., Poesio, M.: (linguistic) science through web collaboration in the anawiki project, Athens (2009)
7. Lesmo, L., Lombardo, V.: Transformed subcategorization frames in chunk parsing. In: In Proc. of the 3rd Int. Conf. on Language Resources and Evaluation (LREC 2002), Las Palmas, pp. 512–519 (2002)
8. Poesio, M., Kruschwitz, U., Chamberlain, J.: Anawiki: Creating anaphorically annotated resources through web cooperation. In: Proc. of the Sixth International Language Resources and Evaluation, Marrakech, Morocco (2008)
9. Surowiecki, J.: The Wisdom of Crowds. Anchor, New York (2005)

A New Possibilistic Clustering Method: The Possibilistic K-Modes

Asma Ammar and Zied Elouedi

LARODEC, Institut Supérieur de Gestion de Tunis,
Université de Tunis,
41 Avenue de la Liberté, 2000 Le Bardo, Tunisie
asma.ammar@voila.fr, zied.elouedi@gmx.fr

Abstract. This paper investigates the problem of clustering data pervaded by uncertainty. Dealing with uncertainty, in particular, using clustering methods can be of great interest since it helps to make a better decision. In this paper, we combine the k-modes method within the possibility theory in order to obtain a new clustering approach for uncertain categorical data; more precisely we develop the so-called possibilistic k-modes method (PKM) allowing to deal with uncertain attribute values of objects where uncertainty is presented through possibility distributions. Experimental results show good performance on well-known benchmarks.

Keywords: Clustering, possibility theory, uncertainty, categorical data, k-modes method.

1   Introduction

Clustering aims to group together data that are similar into subsets called clusters. The k-modes method [6], [7] is among the most successful clustering methods for dealing with categorical data on which we focus in this paper.

When we model knowledge, uncertainty related to attribute values of objects can occur. Dealing with this imperfection can improve the results. Thus, handling uncertain data using uncertainty theories has attracted a lot of attention; we can mention methods combining clustering approaches with possibility theory [4], [11], [14]. As possibility theory has shown its effectiveness in these works. In this paper, we propose to apply possibility theory to the k-modes method to obtain the possibilistic k-modes method (PKM) dealing with another kind of uncertainty related to attribute values of objects.

2   The K-Modes Method

The k-modes method [6], [7], denoted in the following by SKM, has been developed to deal with large categorical data sets. It uses the simple matching method, the modes and a frequency-based function. The simple matching measure is computed by d(X, Y)=$\sum_{j=1}^{m} \delta(x_j, y_j)$ with X and Y two objects with

R. Pirrone and F. Sorbello (Eds.): AI*IA 2011, LNAI 6934, pp. 413–419, 2011.
© Springer-Verlag Berlin Heidelberg 2011

m categorical attributes, $0 \leq d \leq m$ and $\delta(x_j, y_j)$ is equal to 0 if $x_j = y_j$ and is equal to 1 if $x_j \neq y_j$. The optimization problem minimizing the clustering cost function is illustrated by Minimizing D(W, Q)=$\sum_{j=1}^{k} \sum_{i=1}^{n} w_{i,j} d(X_i, Q_j)$ with W an n×k partition matrix, Q the set of k modes and $w_{i,j} \in \{0, 1\}$ is the membership degree of the object X_i in the cluster C_j.

3   Possibility Theory: An Overview

The possibility theory has been introduced by Zadeh in [16] and developed by several researchers mainly Dubois and Prade [1].

Let $\Omega = \{\omega_1, \omega_2, ..., \omega_n\}$ be the universe of discourse [15]. The *possibilistic scale* L can be defined by $[0, 1]$ in the quantitative setting and by $(L, \prec)$ in the qualitative setting. *The possibility distribution function π is defined as the mapping* from the set Ω to L. π is *normalized* when $max_i \{\pi(\omega_i)\} = 1$. Moreover, based on π the complete knowledge $(\exists \omega_0, \pi(\omega_0) = 1$ and $\pi(\omega) = 0$ otherwise) and the total ignorance $(\forall \omega \in \Omega, \pi(\omega) = 1)$ are defined. In addition to that, the information fusion consists in the conjunctive fusion, applied on reliable and in agreement sources [2], and the disjunctive fusion, defined where sources are conflicting [2]. They combine possibility distributions of the same object. In the possibility theory, possibilistic similarity/dissimilarity measures aim to compute the similarity degree between two normalized possibility distributions. They are characterized by several properties [9], [10] (symmetry, upper and lower bounds, etc).

The Information Closeness [5]	Sanguesa et al. Distance [13]
$G(\pi_1, \pi_2) = g(\pi_1, \pi_1 \ Disj \ \pi_2) + g(\pi_2, \pi_1 \ Disj \ \pi_2)^1$	$Distance(\pi_1, \pi_2) = U(|\pi_1 - \pi_2|)$
The Information Affinity [8]	
$InfoAff(\pi_1, \pi_2) = 1 - 0.5[D(\pi_1, \pi_2) + Inc(\pi_1, \pi_2)]^2$	

4   The Possibilistic K-Modes Method

4.1   The Possibilistic K-Modes Parameters

Notations. We define the following notations that will be used in this paper:

- Ts is the training set and S=$(X_1, X_2, ..., X_n)$ is the set of n objects.
- A_j=$(a_{j1}, a_{j2}, ..., a_{jt})$ is the set of t values and $1 \leq j \leq m$ (m attributes).
- π_{ij} is the possibility distribution defined for A_j related to the object X_i.

---

[1] $g(\pi_i, \pi_j) = |U(\pi_j) - U(\pi_i)|$, U is the U-uncertainty measure and the disjunctive Disj is the maximum operator.

[2] $D(\pi_1, \pi_2) = \frac{1}{n} \sum_{i=1}^{n} |\pi_1(\omega_i) - \pi_2(\omega_i)|$ is the Manhattan distance, Inc is the inconsistency measure with $Inc(\pi_1, \pi_2) = 1 - max(\pi_1(\omega) \ Conj \ \pi_2(\omega))$ and the conjunctive Conj is the minimum operator.

Parameters. The PKM differs from the SKM by these parameters:

1. *The structure of the training set.* we define a new training set for the PKM where we assume that training objects may contain both certain and uncertain values of attributes. A possibility distribution will be defined for each attribute relative to each object expressing the possibility degree that the actual attribute value of the object is true.

2. *The updating of modes.* The PKM handles possibility distributions defined for the different attributes related to the objects. So, in order to update the mode, we need an adequate method able to fuse different possibility distributions of attributes related to several objects. From methods that seem to be adequate to our need, we can mention the conjunctive fusion, the disjunctive fusion and the mean operator. However, both conjunctive and disjunctive fusion operators cannot be applied because they are only used to combine possibility distributions of the same object. Nevertheless, the mean operator can be easily applied, using this formula, by computing the average of the possibility degrees of objects belonging to the same cluster (C).

$$\forall \omega \in A_j, \pi_{jC}(\omega) = \frac{\sum_{i=1}^{p} \pi_{ij}(\omega)}{p}, \ p = |C|. \tag{1}$$

3. *The similarity measure.* We need to adapt a different possibility similarity measure suitable to the PKM in order to handle uncertain attributes' values. Based on the satisfaction of the possibilistic similarity measure properties [9] that can influence the clustering result, we select the adequate similarity measure to the PKM. Note that, the information closeness violates the property of lower bound [10], whereas Sanguesa et al. distance does not satisfy the upper bound [10]. However, the InfoAff satisfies all properties in [9]. So, it is expected that it gives the most accurate results. The InfoAff between modes and objects is computed as follows:

$$InfoAff_{PKM}(X_1, X_2) = \frac{\sum_{j=1}^{m} InfoAff(\pi_{1j}, \pi_{2j})}{m}. \tag{2}$$

4.2   The Possibilistic K-Modes Algorithm

1. *Select randomly the k initial modes, one mode for each cluster.*
2. *Allocate objects to the cluster whose mode is the most similar to them after computing the distance measure (see Equation (2))*
3. *Update the cluster mode using the mean operator (see Equation (1)).*
4. *Retest the similarity between objects and modes. Reallocate an object to the cluster whose mode is more similar to it then update the modes.*
5. *Repeat (4) until all objects are stable i.e. they do not change clusters.*

5   Experiments

5.1   The Framework

We have used nine data sets from the UCI Repository [12] to evaluate the PKM results namely Shuttle Landing Control (SL.DS), Balloons (B.DS),

Post-Operative Patient (PO.DS), Congressional Voting Records (C.DS), Balance Scale (BS.DS), Tic-Tac-Toe Endgame (T.DS), Solar-flare (S.DS), Car Evaluation (CE.DS) and Nursery data sets (N.DS).

5.2   Artificial Creation of Uncertain Data Sets

In order to have attribute values labeled by possibility degrees in the Ts of the PKM, we have modified UCI databases.

1. *In the case of certain attributes' values.* First, the true attribute value $r_{a_{jl}}$ takes the possibility degree 1. Then, we assign a possibility degree 0 to the remaining attributes' values ($\neq r_{a_{jl}}$).
2. *In the case of uncertain attributes' values.* We assign to the true attribute value $r_{a_{jl}}$ the possibility degree 1. Then, we assign random possibility degrees from $[0, 1]$ to the remaining attributes' values ($\neq r_{a_{jl}}$).

5.3   Evaluation Criteria

To evaluate the PKM, we focus on three primary criteria of evaluation namely the accuracy, the iteration numbers and the time of execution.

We use the accuracy [6] $AC = \frac{\sum_{l=1}^{k} a_l}{n}$ in the PKM to determine the rate of objects correctly classified. n and a_l are respectively the total number of objects in the Ts and the number of objects correctly classified in the cluster l.

5.4   Experimental Results

We have run our PKM algorithm several times then we have calculated the mean of all obtained results in order to avoid biased results.

Results of Certain Databases. As follows, the SKM and the PKM (based respectively on the information affinity ($PKM_{InfoAff}$), the information closeness (PKM_G) and Sanguesa et al. distance (PKM_D)) results are compared and analyzed using the AC, the iteration numbers and the execution time criteria.

Looking at Table 1, the $PKM_{InfoAff}$ has the highest AC and it is the fastest. Through the $PKM_{InfoAff}$, we have obtained better quality of clustering results.

Results of Uncertain Databases. Let "A" be the percentage of attributes with uncertain values and "d" the possibility degree of the different values except $r_{a_{jl}}$. Table 2 presents a comparison between the three versions of the PKM. Through the results provided in this table, we can remark that the AC of the $PKM_{InfoAff}$ is higher than the AC of both the PKM_G and PKM_D. Table 3 details the execution time and the iteration numbers criteria and they prove that the $PKM_{InfoAff}$ converges after few iterations.

As expected and based on these criteria (the AC, the execution time and the iteration numbers), the InfoAff has proved a high performance in both certain and uncertain framework. As a result, the infoAff is the similarity measure the most adequate to our PKM.

Table 1. The SKM vs the PKM based on the evaluation criteria

Data sets	SL.DS	B.DS	PO.DS	C.DS	BS.DS	T.DS	S.DS	CE.DS	N.DS
SKM									
AC	0.61	0.52	0.684	0.825	0.785	0.513	0.87	0.795	0.896
Iteration numbers	8	9	11	12	13	12	14	11	12
Elapsed time/s	12.431	14.551	17.238	29.662	37.819	128.989	2661.634	3248.613	8379.771
$PKM_{InfoAff}$									
AC	0.69	0.694	0.72	0.896	0.789	0.564	0.876	0.84	0.91
Iteration numbers	2	2	6	3	2	5	6	3	4
Elapsed time/s	0.017	0.41	0.72	3.51	8.82	37.118	51.527	75.32	5473.98
PKM_G									
AC	0.633	0.601	0.682	0.804	0.701	0.532	0.811	0.80	0.86
Iteration numbers	4	2	10	4	6	9	8	7	8
Elapsed time/s	0.989	0.567	1.331	5.566	10.854	40.238	58.665	81.982	6410.331
PKM_D									
AC	0.654	0.675	0.715	0.894	0.716	0.539	0.87	0.837	0.901
Iteration numbers	3	2	6	4	5	8	10	4	7
Elapsed time/s	0.51	0.46	1.75	4.42	9.398	38.995	52.183	78.01	5829.094

Table 2. The AC of $PKM_{InfoAff}$ vs PKM_G and PKM_D

Similarity measures	SL.DS	B.DS	PO.DS	C.DS	BS.DS	T.DS	S.DS	CE.DS	N.DS
A < 50% and 0<d<0.5									
$PKM_{InfoAff}$	0.652	0.632	0.713	0.851	0.783	0.566	0.864	0.821	0.895
PKM_G	0.614	0.601	0.665	0.809	0.605	0.503	0.796	0.779	0.784
PKM_D	0.634	0.632	0.696	0.851	0.78	0.564	0.861	0.812	0.861
A < 50% and 0.5≤d≤ 1									
$PKM_{InfoAff}$	0.638	0.634	0.69	0.843	0.785	0.569	0.868	0.791	0.899
PKM_G	0.593	0.604	0.613	0.803	0.714	0.512	0.839	0.68	0.812
PKM_D	0.629	0.63	0.682	0.864	0.783	0.512	0.867	0.751	0.895
A ≥ 50% and 0<d<0.5									
$PKM_{InfoAff}$	0.713	0.774	0.758	0.896	0.795	0.572	0.877	0.875	0.908
PKM_G	0.681	0.719	0.692	0.825	0.70	0.51	0.872	0.779	0.788
PKM_D	0.697	0.75	0.712	0.857	0.76	0.554	0.87	0.806	0.882
A ≥ 50% and 0.5≤d≤ 1									
$PKM_{InfoAff}$	0.698	0.693	0.737	0.843	0.78	0.558	0.864	0.859	0.897
PKM_G	0.663	0.723	0.639	0.801	0.694	0.504	0.813	0.715	0.756
PKM_D	0.679	0.731	0.682	0.827	0.759	0.537	0.849	0.764	0.816

Table 3. The iteration number and the execution time/second of the PKM

Databases	SL.DS	B.DS	PO.DS	C.DS	BS.DS	T.DS	S.DS	CE.DS	N.DS
$PKM_{InfoAff}$									
Iteration numbers	3	2	8	4	2	8	6	5	4
Elapsed time	0.02	0.437	0.794	3.881	8.914	36.219	51.766	76.221	5671.406
PKM_G									
Iteration numbers	6	4	8	8	6	12	10	8	8
Elapsed time	1.01	0.532	1.739	5.937	11.623	39.661	59.772	80.611	7912.663
PKM_D									
Iteration numbers	4	4	8	6	4	10	10	8	6
Elapsed time	0.27	0.522	1.629	5.061	8.321	37.351	52.712	77.768	5989.276

6   Conclusion

In this paper, we have proposed the possibilistic k-modes (PKM) based on the standard k-modes method (SKM) within the possibility theory having the objective to cluster objects characterized by either certain or uncertain attribute values. PKM was tested on several datasets soiled by uncertainty, and the induced results have shown its efficiency. Based on the accuracy, the iteration numbers and the execution time criteria, PKM based on information affinity as a similarity measure has given better results than using the other measures namely the information closeness and Sanguesa et al. distance.

References

1. Dubois, D., Prade, H.: Possibility theory: An approach to computerized processing of uncertainty. Plenium Press, New York (1988)
2. Dubois, D., Prade, H.: Possibility theory and data fusion in poorly informed environments. Control Engineering Practice 25, 811–823 (1994)
3. Dubois, D., Prade, H.: Possibility theory: Qualitative and quantitative aspects. In: Gabbay, D.M., Smets, P. (eds.) Handbook of Defeasible Reasoning and Uncertainty Management Systems, vol. I, pp. 169–226. Kluwer Academic Publishers, Netherlands (1998)
4. Haghighi, M.S., Yazdi, H.S., Vahedian, A.: A hierarchical possibilistic clustering. International Journal of Computer Theory and Engineering 1, 465–472 (2009)
5. Higashi, M., Klir, G.J.: On the notion of distance representing information closeness: Possibility and probability distributions. International Journal of General Systems 9, 103–115 (1983)
6. Huang, Z.: Extensions to the k-means algorithm for clustering large data sets with categorical values. Data Mining Knowl. Discov. 2(2), 283–304 (1998)
7. Huang, Z., Ng, M.K.: A note on k-modes clustering. Journal of Classification 20(2), 257–261 (2003)
8. Jenhani, I., Ben Amor, N., Elouedi, Z., Benferhat, S., Mellouli, K.: Information Affinity: a new similarity measure for possibilistic uncertain information. In: Mellouli, K. (ed.) ECSQARU 2007. LNCS (LNAI), vol. 4724, pp. 840–852. Springer, Heidelberg (2007)

9. Jenhani, I., Benferhat, S., Elouedi, Z.: Properties Analysis of Inconsistency-based Possibilistic Similarity Measures. In: Proceedings of IPMU 2008, pp. 173–180 (2008)
10. Jenhani, I., Benferhat, S., Elouedi, Z.: Possibilistic similarity measures. STUD-FUZZ 249, 99–123 (2010)
11. Krishnapuram, R., Keller, J.M.: A possibilistic approach to clustering. IEEE Trans. Fuzzy Syst. 1, 98–110 (1993)
12. Murphy, M.P., Aha, D.W.: Uci repository databases (1996), http://www.ics.uci.edu/mlearn
13. Sanguesa, R., Cabos, J., Cortes, U.: Possibilistic conditional independence: A similarity based measure and its application to causal network learning. International Journal of Approximate Reasoning, 145–167 (1997)
14. Yang, M.S., Wu, K.L.: Unsupervised possibilistic clustering. Pattern Recognition 39, 5–21 (2006)
15. Zadeh, L.A.: Fuzzy sets. Inform. And Control 8, 338–353 (1965)
16. Zadeh, L.A.: Fuzzy sets as a basis for a theory of possibility. Fuzzy Sets and Systems 1, 3–28 (1978)

Over-Subscription Planning with Boolean Optimization: An Assessment of State-of-the-Art Solutions

Marco Maratea[1] and Luca Pulina[2]

[1] DIST, University of Genova, Viale F. Causa 15, Genova, Italy
marco@dist.unige.it
[2] DEIS, University of Sassari, Piazza Università 11, Sassari, Italy
lpulina@uniss.it

Abstract. In this work we present an assessment of state-of-the-art Boolean optimization solvers from different AI communities on over-subscription planning problems. The goal of the empirical analysis here presented is to assess the current respective performance of a wide variety of Boolean optimization solvers for solving such problems.

1 Introduction

Over-Subscription Planning Problems (OSPPs) [1,2] are planning problems containing quantitative preferences expressed on goals in case not all the goals can be satisfied. In particular, a cost is associated to the violation of goals, and the aim is to find a plan whose metric maximizes the rewards of satisfied goals. OSPPs are thus suitable to model a wide set of practical applications – see, e.g., [2]. The increasing interest of the AI planning community on OSPPs is also witnessed by recent editions of the International Planning Competition (IPC), where all the "optimization" tracks of IPC'08 consider these problems. Furthermore, in IPC'08 also quantitative preferences expressed on action's preconditions are taken into account for plan metrics.

Considering a fixed plan horizon, i.e. a *makespan*, a recently adopted approach used to deal with such problems is to reduce them to Boolean propositional problems with linear optimization functions [3,4] – e.g., Max-SAT and Pseudo-Boolean (PB) problems. Both Max-SAT and PB are extensions of the well-known propositional satisfiability (SAT) problem. These formalisms allow the user to naturally reason with integers, which is one of the main limitation of SAT, instead of relying on complicated and/or space consuming encodings – see, e.g., [5].

In this paper we present an assessment of state-of-the-art systems coming from different scientific AI communities. We show the result of an experimental analysis involving all the best performing Max-SAT and PB solvers and other systems that, even if designed to mainly deal with other formalisms, can solve Boolean optimization problems. In particular, we also consider Answer Set Programming (ASP), Integer Programming (IP), Constraint Integer Programming (CIP), and Interval Constraints Propagation (ICP) systems. All the solvers are tested on domains comprised in both IPC'06 and IPC'08. Our results reveal that the ASP solver CLASP and the PB solver MINISAT+ are currently the overall best systems on these instances.

R. Pirrone and F. Sorbello (Eds.): AI*IA 2011, LNAI 6934, pp. 420–425, 2011.

2   Instances and Solvers

For our analysis, we considered the domains OPENSTACKS, OPENSTACKS-IPC08, PATH-WAYS, PEGSOL, STORAGE, TPP, and TRUCKS. They were used in past IPCs, in par-ticualar in the "SimplePreferences" track of IPC'06 and in the "netben-opt" track of IPC'08. In these domains, the plan metrics, in terms of quantitative preferences, are ex-pressed on goals and/or on actions preconditions. In the following, we provide a short overview about the modeling of the problems of interest – details can be found in [3]. Considering a fixed makespan, benchmarks are reduced to Boolean optimization prob-lems with different formalisms. This is done by using a modified version of the SAT-PLAN planner [6] on the STRIPS problems formulation at fixed makespan. Given that SATPLAN can only handle STRIPS domains, while IPC'06 domains are non-STRIPS, and some ADL [7] constructs are used, we have first adapted the non-STRIPS problems in the following way: (i) Preferences[1] on actions preconditions are expressed with two actions that do not contain preferences. For both actions, the related preference formula is treated as hard, further negated in the second. The second action also achieves a new dummy literal; and (ii) the goal preferences are imposed as preconditions of dummy ac-tions, which achieve new dummy literals defining the new problem goals. The treatment of actions preferences is inspired by the ones used in [8]. The metrics of the planning problems are expressed with (linear) optimization function.

Concerning the instances generation, we have modified SATPLAN at each makespan of the SATPLAN's approach, until the optimal. Thus, our compilation allows to find plans with optimal metrics at fixed makespan. Further, note that while literals related to goal preferences can be implicitly considered to hold only "at the end" modality [9] – i.e., at the final makespan – this is not the case for the ones related to preconditions that can, in general, hold at any makespan, unless we know that, instead, STRIPS actions can be only executed once (e.g., this is the case for well known real-world planning domain like blocks-world and logistics). The changes in SATPLAN were mainly related to the creation of formulas in the formats accepted by the various solvers employed instead of the DIMACS format for SAT formulas in CNF. To mention the more widely used input formats, our approach generates Weighted Partial Max-SAT problem – a further extension of Max-SAT. In particular, in the Weighted Partial Max-SAT problem, a positive integer weight is associated to each soft clause, and the goal is to satisfy all hard clauses and maximize the sum of weights associated to satisfied soft clauses.

In our experimental evaluation, we focus on the results obtained by the various solvers on the first satisfiable formula following the SATPLAN approach, augmented with optimization issues defined by the metric of the problem. Further, we consider the case where actions can be executed at most once. Globally, we will focus on the 71 in-stances that we could compile with the ADL2STRIPS tool, and then can be solved by at least one of the considered solvers. In particular, some PATHWAYS, TPP from #11 to #16, TRUCKS from #3 to #7, and OPENSTACKS #1 (as numbered in IPC'06) instances could

---

[1] We consider that at most one preference formula in expressed on the preconditions of an action: This is the case for all domains we consider in this paper. If this would not be the case, we should consider their power set.

Table 1. Solvers involved in the evaluation. The table is structured as follows. Column "Solver" reports the name of the solver, while column "Version" indicates the version used in the experiments. For AKMAXSAT indicates the version submitted to the 2010 Max-SAT Competition. Finally, column "Formalism" reports the input formalisms accepted by the solver.

Solver	Version	Formalism	Solver	Version	Formalism
AKMAXSAT	2010	Max-SAT	MINIMAXSAT	1.0	Max-SAT, PB
BSOLO	3.0.17	PB	MINISAT+	1.14	PB
CLASP	1.3.6	ASP	MSUNCORE	1.2	Max-SAT
CPLEX	12.0	IP	SAT4J	2.1.0	Max-SAT, PB
GLPPB	0.2	PB	SCIP	1.2.0	CIP
HYSAT	0.8.6	ICP	WBO	1.4	Max-SAT, PB
INCWMAXSATZ	1.2	Max-SAT	WMAXSATZ	2.5	Max-SAT

be compiled but not solved by any system. They thus provide challenging benchmarks for state-of-the-art solvers.

Table 1 summarizes the solvers involved in the evaluation. Looking at the table, we can see that the selected systems come from different scientific AI communities, namely ASP, CIP, ICP, IP, Max-SAT, and PB. Moreover, we can see that, in some cases, a solver is able to deal with problems expressed with different formalisms, e.g. MINIMAXSAT. Concerning PB and Max-SAT solvers, we selected the best solvers that have participated to Max-SAT and PB evaluations along the years [10,11], with emphasis on the "Weighted Partial" and "OPT-SMALL-INT" categories, the last being part of PB evaluations, and where (i) there is no constraint with a sum of coefficients greater than 2^{20} (20 bits), and (ii) the objective function is linear. CLASP [12] is the overall winner of the 2009 ASP Competition [13], CPLEX is a well-known linear arithmetic solver that can solve IP problems, while HYSAT [14] is the best solver based on ICP.

3   Experimental Analysis

The experiments reported in this section ran on 10 identical PCs equipped with a processor Intel Pentium IV running at 3.2GHz with 1GB of RAM, and running GNU Linux Debian 2.4.27-2. For each run, the CPU time limit was set to 900 CPU seconds and, in order to prevent memory swapping, we set a memory limit at 900MB.

Table 2 shows the results of our analysis for each single domain. Notice that we drop the domain OPENSTACKS, for which we report that no solver solved the single instance contained. Looking at Table 2 (top-left), domain OPENSTACKS-IPC08, we can see that only CLASP was able to deal with all domain instances, while MINIMAXSAT, MINISAT+, and WBO top to 1. The bad performance related to the remaining 10 solvers is mainly to ascribe to the instances size. Considering now domain PATHWAYS (Table 2, top-right), we report that no solver was able to solve all instances. The best solver is MINISAT+ that tops to 15 out of 20 instances, followed by CLASP. Also BSOLO and SAT4J solved the same amount of instances, but they spent one order of magnitude more of CPU time. Overall, 8 solvers were able to solve at least 50% of the instances, and we also report that all solvers solved at least 30% of the domain dataset. Looking yet at Table 2 (middle-left), domain PEGSOL, the reported results highlight how these

Table 2. Evaluation results by domain. We report the number of instances solved within the time limit ("#") and the total CPU time ("Time") spent on solved instances. Solvers are sorted according to the number of solved instances, and, in case of a tie, according to CPU time. A dash means that a solver did not solve any instance in the related domain.

Domain	Solver	Solved #	Solved Time	Domain	Solver	Solved #	Solved Time
OPENSTACKS-IPC08 (2)	CLASP	2	623.81	PATHWAYS (20)	MINISAT+	15	21.00
	MINIMAXSAT	1	54.20		CLASP	15	24.79
	MINISAT+	1	62.58		BSOLO	15	127.24
	WBO	1	645.73		SAT4J	15	721.99
	AKMAXSAT	–	–		MINIMAXSAT	14	69.20
	BSOLO	–	–		CPLEX	14	943.86
	GLPPB	–	–		INCWMAXSATZ	13	499.80
	HYSAT	–	–		WMAXSATZ	12	1343.86
	INCWMAXSATZ	–	–		MSUNCORE	8	14.54
	MSUNCORE	–	–		SCIP	8	138.48
	SAT4J	–	–		AKMAXSAT	8	387.17
	SCIP	–	–		HYSAT	8	1537.65
	WMAXSATZ	–	–		WBO	7	1.59
	CPLEX	–	–		GLPPB	6	389.46
PEGSOL (28)	INCWMAXSATZ	28	0.35	STORAGE (7)	CPLEX	7	20.18
	MINISAT+	28	1.19		WMAXSATZ	7	36.62
	WMAXSATZ	28	1.23		MINISAT+	7	43.86
	MINIMAXSAT	28	1.35		INCWMAXSATZ	7	58.18
	BSOLO	28	1.53		CLASP	7	91.96
	CLASP	28	1.55		SCIP	7	197.90
	CPLEX	28	3.4		SAT4J	7	457.71
	AKMAXSAT	28	15.04		MINIMAXSAT	6	10.84
	SAT4J	28	68.64		AKMAXSAT	5	5.39
	SCIP	28	70.40		BSOLO	5	64.54
	HYSAT	28	207.70		WBO	5	101.66
	WBO	24	92.43		MSUNCORE	4	0.32
	MSUNCORE	21	694.43		HYSAT	4	56.86
	GLPPB	15	1817.76		GLPPB	2	1.17
TPP (10)	CLASP	10	106.2	TRUCKS (3)	MSUNCORE	2	18.67
	MINISAT+	10	124.97		WBO	2	23.04
	BSOLO	10	422.92		CLASP	2	48.46
	MINIMAXSAT	8	48.95		MINIMAXSAT	2	57.95
	WBO	8	168.15		MINISAT+	2	390.36
	SAT4J	8	424.54		BSOLO	1	119.69
	CPLEX	8	2042.1		SAT4J	1	359.17
	WMAXSATZ	7	687.19		AKMAXSAT	–	–
	MSUNCORE	4	0.34		CPLEX	–	–
	INCWMAXSATZ	4	1.32		GLPPB	–	–
	SCIP	4	15.21		HYSAT	–	–
	AKMAXSAT	4	51.56		INCWMAXSATZ	–	–
	HYSAT	4	63.45		SCIP	–	–
	GLPPB	–	–		WMAXSATZ	–	–

instances seem to be easier for the solvers. All considered systems but 3 were able to solve all instances, and we can conjecture that these results are mainly due to the fact that such instances are composed of a small number of variables and constraints. Considering now domain STORAGE (Table 2, middle-right), we can see that 50% of solvers are able to solve all instances in the domain. Looking at the results, CPLEX is the solver having best performance, and it is faster than both MINISAT+ and CLASP, by a 2x and a 4x factor, respectively. Looking now at the results related to the domains at the bottom of Table 2, we can see that, concerning TPP domain (bottom-left), CLASP and

Table 3. Classification of instances by domain. For each domain we report the name (column "Domain"), the total amount of instances in the domain, and the number of solved instances (group "Overall", columns "N" and "#", respectively). It follows column "Time", which report the CPU time taken to solve the instances. Finally, group "Hardness" reports the total amount of easy, medium, and medium-hard instances (columns "EA", "ME", and "MH", respectively).

Domain	Overall		Time	Hardness			Domain	Overall		Time	Hardness		
	N	#		EA	ME	MH		N	#		EA	ME	MH
OPENSTACKS-IPC08	2	2	623.81	–	1	1	PATHWAYS	20	15	17.77	6	9	–
PEGSOL	28	28	0.32	11	17	–	STORAGE	7	7	14.25	2	5	–
TPP	10	10	105.68	–	10	–	TRUCKS	3	2	18.67	–	2	–

MINISAT+ confirm their good performance. We also notice the performance of BSOLO in this domain, that is the only other solver that solves all comprised instances. Six solvers are not able to solve 50% of the total amount of instances. Concluding the analysis of Table 2, about TRUCKS domain (bottom-right) we can first report that no solver solved all instances. We also can see that the best two solvers, MSUNCORE and WBO did not perform very well in the other domains. If we consider statistics related to the 2 solved instances, we report that their structure in terms of relationship between variables, soft, and hard constraints is quite different from instances in the other domains. We can conjecture that heuristics in MSUNCORE and WBO are effective in such cases.

Globally, we report that CLASP is the best solver, able to solve more than 90% of the dataset, and that MINISAT+ performance are very close. On the other hand, one solver only – namely GLPPB– was not able to solve 50% of the whole dataset, and 7 solvers were able to deal with at least 75% (53 instances) of the whole dataset. Notice that these 7 solvers were designed to solve 4 (out of 6 taken into account) different problem formalisms, namely ASP (CLASP), PB (MINISAT+ and BSOLO), Max-SAT (MINIMAXSAT, SAT4J, and WMAXSATZ), and IP (CPLEX). We also notice that SCIP, a CIP solver, tops to about 66% of the dataset, while the ICP solver (HYSAT) tops to 62% of the dataset.

In Table 3 we report a "domain-centered" classification. In the table, the number of instances solved and the cumulative time taken for each domain is computed considering the "State Of The Art" (SOTA) solver, i.e., the ideal solver that always fares the best time among all the solvers. Thus, an instance is solved if at least one of the solvers solves it, and the considered time is the best among the times of the solvers that solved the instances. The instances are also classified according to their empirical hardness as follows: an instance is called "easy" if it has been solved by all the considered solvers; "medium" are those non-easy that can be solved by at least two solvers; "medium-hard" are those solved by only one solver; "hard" are the ones remained unsolved. Looking at Table 3, we can see that the SOTA solver is able to solve 64 instances, resulting in 19 easy, 44 medium, 1 medium-hard, and 7 hard instances. Considering OPENSTACKS-IPC08, performances of SOTA solver are to ascribe to CLASP, that also solves uniquely 1 instance. Looking now at PATHWAYS, we notice that 5 (out of 20) instances are very challenging for the whole pool of solvers. We also notice that the SOTA solver major contributors are MINIMAXSAT and INCWMAXSATZ, with 6 and 4 instances, respectively. Finally, considering the whole dataset, we report that the major contributors to the SOTA solver are INCWMAXSATZ and MINIMAXSAT (33%), followed by MIN-ISAT+ and CLASP (11%). We also report that the remaining 7 solvers (out of 14) do not contribute to the SOTA solver.

References

1. van den Briel, M., Nigenda, R.S., Do, M.B., Kambhampati, S.: Effective approaches for partial satisfaction (over-subscription) planning. In: Proc. of AAAI 2004, pp. 562–569 (2004)
2. Smith, D.E.: Choosing objectives in over-subscription planning. In: Proc. of ICAPS 2004, pp. 393–401. AAAI, Menlo Park (2004)
3. Maratea, M.: Planning as satisfiability with IPC simple preferences and action costs. Technical report (2011),
   http://www.star.dist.unige.it/~marco/Data/TR-planning.pdf
4. Robinson, N., Gretton, C., Nghia Pham, D., Sattar, A.: Partial weighted maxsat for optimal planning. In: Zhang, B.-T., Orgun, M.A. (eds.) PRICAI 2010. LNCS, vol. 6230, pp. 231–243. Springer, Heidelberg (2010)
5. Bailleux, O., Boufkhad, Y.: Efficient CNF encoding of boolean cardinality constraints. In: Rossi, F. (ed.) CP 2003. LNCS, vol. 2833, pp. 108–122. Springer, Heidelberg (2003)
6. Kautz, H., Selman, B.: Unifying SAT-based and graph-based planning. In: Dean, T. (ed.) Proc. of IJCAI 1999, pp. 318–325. Morgan Kaufmann, San Francisco (1999)
7. Pednault, E.: ADL and the state-transition model of action. Journal of Logic and Computation 4, 467–512 (1994)
8. Gazen, B.C., Knoblock, C.A.: Combining the expressivity of UCPOP with the efficiency of Graphplan. In: Steel, S. (ed.) ECP 1997. LNCS, vol. 1348, pp. 221–233. Springer, Heidelberg (1997)
9. Gerevini, A., Haslum, P., Long, D., Saetti, A., Dimopoulos, Y.: Deterministic planning in the 5th IPC: PDDL3 and experimental evaluation of the planners. Artificial Intelligence 173(5-6), 619–668 (2009)
10. Argelich, J., Li, C.M., Manyà, F., Planes, J.: The first and second Max-SAT evaluations. Journal od Satisfiability, Boolean Modeling and Computation 4(2-4), 251–278 (2008)
11. Manquinho, V.M., Roussel, O.: The first evaluation of pseudo-Boolean solvers (PB'05). Journal on Satisfiability, Boolean Modeling and Computation 2, 103–143 (2006)
12. Gebser, M., Kaufmann, B., Neumann, A., Schaub, T.: Conflict-driven answer set solving. In: Proc. of IJCAI 2007, pp. 386–392 (2007)
13. Denecker, M., Vennekens, J., Bond, S., Gebser, M., Truszczynski, M.: The second Answer Set Programming Competition. In: Erdem, E., Lin, F., Schaub, T. (eds.) LPNMR 2009. LNCS, vol. 5753, pp. 637–654. Springer, Heidelberg (2009)
14. Franzle, M., Herde, C., Teige, T., Ratschan, S., Schubert, T.: Efficient solving of large non-linear arithmetic constraint systems with complex boolean structure. Journal of Satisfiability, Boolean Modeling and Computation 1, 209–236 (2007)

Multi-platform Agent Systems with Dynamic Reputation Policy Management

Vincenzo Conti[1], Marco Sardisco[1], Salvatore Vitabile[2], and Filippo Sorbello[3]

[1] Facoltà di Ingegneria e Architettura,
Università degli Studi di Enna Kore,
Cittadella Universitaria, 94100 Enna, Italy
[2] Dipartimento di Biopatologia e Biotecnologie Mediche e Forensi,
Università degli Studi di Palermo,
Via del Vespro, 90127 Palermo, Italy
[3] Dipartimento di Ingegneria Chimica, Gestionale, Informatica e Meccanica,
Università degli Studi di Palermo,
Viale delle Scienze, 90128 Palermo, Italy

Abstract. Open, distributed multi-platform agent systems require new management approaches for resources and data secure access. In this paper a Jade-S based multi-platform agent system implementing dynamic reputation policy management is proposed. The implemented extension deals with biometrics, X-Security, DES cryptography and agent reputation. With more details, the proposed reputation management system helps to assess the agent's behavior and reliability, in order to select trusted agents. This is made possible by the knowledge that agents are able to acquire, over time, and that allows them to choose the best solution using own intelligence in total autonomy.

Keywords: Multi-agents system, security levels, dynamic reputation management.

1 Introduction

The increasing use of distributed systems highlights the need for a new approach allowing the management and use of all resources and information requested by users in an efficient and safe way [7] [8]. In this context, mobile agent systems have a number of characteristics that make them particularly suitable for the realization of distributed applications, such as, robustness, reliability and computational efficiency.

The use of such systems raises a set of new problems related to agents data protection and of the system itself that must be adequately addressed [2] [4] [9].

In some contexts, such as, systems for electronic commerce or financial transactions, each user should be held accountable for actions that the processes or agents, related to it, play within a system or a multi-agent platform. The first step requires that each user or agent is uniquely identified and monitored [1] [10]. Examples of actions for which to define responsibilities include: accesses to particular resources such as local files, changes to platform security policies, and the responsibility of communicative acts between agents.

R. Pirrone and F. Sorbello (Eds.): AI*IA 2011, LNAI 6934, pp. 426–431, 2011.

The agents should be able to provide their identity both to the platform and to other agents, while the platforms should provide their identity to own agents and to other platforms. So that responsibilities can be assigned in precise and unambiguous way, it need to keep track of significant events from the point of view of safety and to list the people or the processes responsible for that event [4] [9] [11] [12]. Several approaches and systems, particularly in recent years, have adapted the multi-agent platforms to the solution of the problems exposed above. Most of this works have identified the user identifying process as a starting point for the solution of these problems. A useful tool for the individual identification has always been the fingerprint. Unlike the username and password, a fingerprint cannot be copied, reproduced, and counterfeit [13]. Fingerprints, in fact, are unique in each individual, there aren't individuals with the same fingerprint, and thus represent a possible means of individual recognition without error possibility.

With regard to user identification systems exists JADE-S that is one of the most popular platform in multi-agent systems field. JADE-S is born from the fusion of JADE and a security plug-in that provides for the user authentication, licensing and secure communication between agents within the same platform and between different platforms [3].

In this paper a framework of an open, distributed and multi-platform agent system has been implemented. For its implementation, the platform JADE of TILab, JADE-S add-on, to which was added a block "Fingerprint Verification System" which represents the biometric authentication system, described in [6], and X-Security plug-in were used. Besides, the system use the DES, a symmetric key encryption algorithm, to protect communications between community agents. Besides, the main characteristic of the proposed system is related to the dynamic management of reputation and policy of user/agent of the system.

The paper has been structured as follows. Section 2 gives a description of the system. In the Section 3 the dynamic reputation management is described. Finally, Section 4 reports the conclusion of the proposed work.

2   System Description

The implemented system is an open, distributed and multi-platform system because for its operation requires the use of two or more platforms JADE resident on different computers and/or networks. With more details, many platforms were used to realize the proposed system. The system was created with devices and countermeasures able to guarantee the communications and transactions security between agents of the different platforms.

The system, in a possible scenario, is composed of two parts (Fig. 1): the first, called *Server*, represents the credit institution that manages and makes available to investors the securities of the stock market, the second, called *Client*, represents, instead, investors that require to the credit institution securities on which to invest. Therefore, on many platforms the Client systems, composed by agents that allow to manage the interaction with the user, have been realized.

These systems request securities and communications with agents of the Server platform. On another platform the system that represents the credit institution has been realized with the agents that enable the securities management and communication with investors to respond to their requests.

In the next subsections, the implemented system and the proposed agents community will be described, starting from a description of the platforms and the explanation of the security system characterized by different levels.

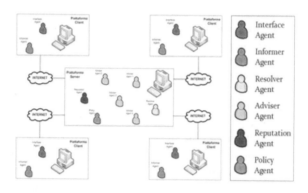

Fig. 1. Graphic representation of the proposed open distributed system. Agent's colors identify different services.

2.1   The Implemented Security Levels

The proposed system aims to prevent that a multi-agent platform may be started by unauthorized users: unauthorized users can not launch agents within it. The security level already present in JADE-S provides that the authorization is identified by username and password that are required to the start of the platform. Every agent can be started only by users that successfully completed the authentication process. The JADE platform provides a security level based on the user recognition by entering username and password. This security level ensures that users that have not these items cannot access to the platform. To increase this authentication phase, each individual could be identified by its biometric characteristics that uniquely distinguish him from other individuals, for example one of these features could be the fingerprint [5] [12]. For this reason, a block called "Fingerprint Verification System", described in [6], has been added to the existing security system to perform the verification step and to assign the ownership to user/agent. Therefore, in the system is possible to identify two platform protection levels: a first security level that provides the insert of username and password, the second consists in verifying the user fingerprint requesting access to the platform. Within the system other two security levels were realized: the first to protect the information flow exchanged between agents within the same platform and between different platforms, the second to ensure access to services, offered by agents registered on the platform, only to agents that possess a certificate. To protect the information exchanged between agents have been used the cryptography. All communications were encrypted through the use of DES, a symmetric key encryption algorithm. Finally, to realize the second security level mentioned above the X-Security has been used, a security plug-in for JADE platform that implements a public key infrastructure based on X.509v3 certificates [4]. The X-Security heart is constituted by an agent, called Security Certification Authority (SCA), dedicated to the issue of identity certificates to other agents [6]. In the proposed system all agents acting on the platform have a certificate issued by the SCA.

2.2   Agents community description

To prove/describe the system goodness and validity, the picture of a very common scenario, for example banks investments, could be used. This possible scenario introduces the following tasks:

- a demand for investment will be performed;
- all banks of the system will be checked and all investments with a relative budget, will be proposed;
- user will perform a choice on the basis of proposed investments.

The agents community used to resolve this transaction type is composed of the following actors.

The *Interface agent*: it receives the requests made by users through a GUI (the proposed investments list by the credit institution) and forwarding those requests to the Resolver agent. The *Informer agent* has the responsibility to communicate to the Resolver, and then to the credit institution, the choice made by the user on the basis of proposed investments by the Adviser agents. The *Resolver agent*, upon receipt message with the user request received from the Interface agent, performs a search in the DF to obtain useful information about the service requested and about the Advisers that can help it. The Resolver agent forwards them the request including, in addition, a parameter that represents the minimum reputation level required to perform the research. Performed this operation, the Resolver waits for research results that as soon as will be communicated by the Advisers, will be displayed in a GUI. *The Adviser agent* is the agent responsible for managing securities and represents the single bank of the system to which it belongs. This agent, first to replay to Resolver agent, check own reputation level, which is communicated to it by the Policy agent, and based on parameters that are transmitted it by the Resolver. *The Reputation agent* has the task of monitoring developments on the stock market to assess, for each month, changes of securities on which users have invested. This agent reads the file that contains a list of all made investments and verifies if the interest rates of investments made by users are increased, decreased or remained stable compared to those dating back to date on which such investments are been carried. Successively, the Reputation agent assigns a penalty or a reward point to each proposed investment by the single Adviser through a dedicated mathematical function. *The Policy agent* evaluates the Advisers reliability through the reputation level updated by the Reputation agent. Besides, this agent defines the credibility and confidence level that the entire credit institution builds itself against the investors through the work of their Advisers. The Policy agent managed dynamically the Adviser agents reputation through the use of a particular mathematical function. The resulting value will be written in the policy file and communicated to each Adviser agent.

3 Dynamic Reputation Management

The Adviser agents are one of the key points of the proposed open system, because they represent the banks to which they belong and because they personally respond about the investments that offer to the users. The correct functioning of these agents, however, is governed by their reputation level. Reputation means credibility, confidence that the credit institution builds itself against the investors through the work of their Advisers. A mechanism to manage the agents reputation, in a common context as bank transactions, will be now described.

The Resolver agent, first, keeps memory of all investments made by users during the time writing them on a special file. At the end of each month the Reputation agent reads the file containing the list of all made investments and verifies if the interest investments rates, made by users, are increased, decreased or remained stable. The Reputation agent compares the updated interest rate of each investment with the original interest rate, in particular, if the interest rate is increased compared to the original, the Reputation assigns a reward point, otherwise, it assigns a penalty point to the Adviser agent. These scores are then written to separate files, for each specific Adviser. Successively, the Policy Agent reads the files (wrote, previously, by Reputation agent) containing the scores of the users investments, and calculates their sum to obtain a value that will be used to dynamically manage the Adviser agents reputation. The Advisers reputation update is performed through the following function:

$$f = (\alpha * reputation + (1 - \alpha) * availability) + k$$

where α is a real value in [0,1], k is a binary value equal to 1 for all agents whose meets the user constraints otherwise its value is 0. *Reputation* is a value in [0,100] and it is calculated by Reputation agent in according to EigenTrust algorithm [14].

Finally, *availability*, value in [0,100], represents the user past experience obtained a particular service in the multi-platform system. Completed the updating reputation process of all Adviser agents, the resulting value will be written on the policy file of the specific Adviser. Finally. When the Adviser agent reaches a reputation level lower than a threshold prescribed by the Resolver, will no longer be contacted to search new investments. The credit institution will decide whether to delete this Adviser, or to report, after a certain time period, the reputation of this Adviser to a level that would allow it to be recalled by the Resolver to respond to the users requests.

The proposed reputation management system helps to assess the Advisers behavior and reliability, in order to distinguish trusted than unreliable Advisers.

This is made possible by the knowledge that agents are able to acquire, over time, and that allows them to choose the best solution using own intelligence in total autonomy.

4   Conclusions

In this paper a Jade-S based multi-platform agent system implementing dynamic reputation policy management has been proposed. To test the goodness of the proposed approach an enhanced JADE-S based framework has been developed, adding biometrics, X-security plug-in and DES encryption algorithm to protect platform and agent's data. Experimental trials have confirmed that the proposed reputation management system assesses trusted agent's behavior and reliability.

References

1. Jansen, W.A.: Countermeasures for mobile agent security, Computer Communications. Special Issue on Advanced Security Techniques for Network Protection, Elsevier Science (2000)
2. Poggi, A., Rimassa, G., Tomaiuolo, M.: Multi-user and security support for multi-agent systems. In: Proc. of WOA 2001 Workshop, Modena, Italy (September 2001)
3. Vitaglione, G.: Jade Tutorial Security Administrator Guide
4. Novak, P., Rollo, M., Hodik, J., Vlcek, T., Pechoucek, M.: X-Security architecture in agentcities, http://agents.felk.cvut.cz/security/main/index.php
5. Jain, A., Hong, L., Bolle, R.: On-Line Fingerprint Verification. IEEE Transactions on Pattern Analysis and Machine Intelligence 19(4) (April 1997)
6. Vitabile, S., Conti, V., Militello, C., Sorbello, F.: An Extended JADE-S Based Framework for Developing Secure Multi-Agent Systems. Computer Standard and Interfaces Journal 31(5), 913–930 (2009) ISSN: 0920-5489, doi:10.1016/j.csi.2008.03.017
7. Pilato, G., Vitabile, S., Conti, V., Vassallo, G., Sorbello, F.: A Concurrent Neural Classifier for HTML Documents Retrieval. In: Apolloni, B., Marinaro, M., Tagliaferri, R. (eds.) WIRN 2003. LNCS, vol. 2859, pp. 210–217. Springer, Heidelberg (2003)
8. Farmer, W.M., Guttman, J.D., Swarup, V.: Security for Mobile Agents: Authentication and State Appraisal. In: Martella, G., Kurth, H., Montolivo, E., Hwang, J. (eds.) ESORICS 1996. LNCS, vol. 1146, pp. 118–130. Springer, Heidelberg (1996)
9. Chi Wong, H., Sycara, K.: Adding Security and Trust to Multi-Agent System. In: Proc. of Autonomous Agents (1999)
10. Jansen, W.A., Karygiannis, T.: NIST Special Publication 800-19 - Mobile Agent Security. National Institute of Standards and Technology (2000)
11. Corradi, A., Montanari, R., Stefanelli, C.: Security Issues in Mobile Agent Technology, Distributed Computing Systems, 1999. In: Proc. of 7th IEEE Workshop on Future Trends of Distributed Computing Systems, pp. 3–8 (1999)
12. Vitabile, S., Conti, V., Pilato, G., Sorbello, F.: A Fingerprint Based Authentication System for the JADE-S Platform. In: Agentcities ID3, Barcelona, Spain, February 6-8 (2003)
13. JADE home page, http://jade.tilab.com
14. Kamvar, S., Schlosser, M., Garcia-Molina, H.: The eigentrust algorithm for reputation management in p2p networks. In: Proc. of 12th International WWW Conference, pp. 640–651 (May 2003)

Integrating Built-in Sensors of an Android with Sensors Embedded in the Environment for Studying a More Natural Human-Robot Interaction

Giuseppe Balistreri[1,2], Shuichi Nishio[2], Rosario Sorbello[1],
and Hiroshi Ishiguro[2,3]

[1] DICGIM Università degli Studi di Palermo,
Viale delle Scienze, 90128 Palermo, Italy
balistreri@dinfo.unipa.it
[2] ATR Intelligent Robotics and Communication Laboratory,
2-2-2 Hikaridai Seikacho Sourakugun Kyoto, Japan
[3] Graduate School of Engineering Science, Osaka University,
1-3 Machikaneyama Toyonaka Osaka, Japan

Abstract. Several studies supported that there is a strict and complex relationship between outer appearance and the behavior showed by the robot and that a human-like appearance is not enough for give a positive impression. The robot should behave closely to humans, and should have a sense of perception that enables it to communicate with humans. Our past experience with the android "Geminoid HI-1" demonstrated that the sensors equipping the robot are not enough to perform a human-like communication, mainly because of a limited sensing range. To overcome this problem, we endowed the environment around the robot with perceptive capabilities by embedding sensors such as cameras into it. This paper reports a preliminary study about an improvement of the controlling system by integrating cameras in the surrounding environment, so that a human-like perception can be provided to the android. The integration of the development of androids and the investigations of human behaviors constitute a new research area fusing engineering and cognitive sciences.

Keywords: Android, gaze, sensor network.

1 Introduction

The ultimate goal of robot development from the human-robot interaction perspective is to build a robot that exhibits comprehensible behaviours and that supports a rich and multimodal interaction [1], and the robot appearance must be sufficiently anthropomorphic to elicit natural reactions from people interacting with it [2]. A robot which realizes a very human-like appearance is called an "android". However, the android system of prof. Hiroshi Ishiguro [3] is not yet sufficient to realize a natural human-robot interaction, because the android's

R. Pirrone and F. Sorbello (Eds.): AI*IA 2011, LNAI 6934, pp. 432–437, 2011.

perceptional functions are not implemented or are substituted by a human controller. Some perception functions can be provided to the robot by embedding sensors in the environment [4] [5] [6]. The resulting architecture is different from the human perception system, but the modalities are not important for a natural communication. A robot can obtain the position of someone else by using sensors embedded in its surrounding environment, but it can use the obtained information as if it comes from a complex vision system like the human one.

Supporting the hypothesis that face-to-face is the best model of interaction, we want to leverage people's ability to recognize the subtleties in the eye focusing as a feedback, making the conversation with the robot richer and more effective. To establish an effective eye focusing, the robot needs to know the exact position of the head of the conversational partner, as people are known to be highly sensitive in distinguish gaze directions and identifying eye contacts.

This paper reports the development of a communication system which integrates sensors embedded in the environment with an android.

2   Geminoid System

This section briefly describes our robot system used for eye-contact in a sensor network study.

Geminoid HI-1 was developed to closely resemble the outer appearance of its creator Prof. Hiroshi Ishiguro [7] to pursue research in the field of "Android Science" [8], in which these special robots are seen as "a key testing ground for social, cognitive, and neuroscientific theories" [9]. Figure 1 on the left side shows the Geminoid HI-1 and its real counterpart.

The robot is equipped with teleoperation functionality for avoid the current limitations in AI technologies (Figure 1 on the right side). A webcam points the face of the operator and drives Geminoid's head orientation using the "FaceAPI" software from Seeing Machines, and microphones and a headphone are used to capture and transmit utterances. The robot has been designed to work also with tactile and floor sensors but for this study only cameras have been used.

Fig. 1. Left side: Geminoid HI-1 with its creator Hiroshi Ishiguro. **Right side**: control room for the Geminoid HI-1

3   Related Work

Laser range finder (LRF) and RFID have been used for accurate people tracking, but they cannot obtain positions of heads as people heights are not same.
Methods for tracking people have been proposed using multi-camera. Suzuki [10]
and Matsumoto [11] have realized a people head tracking using hypotheses in a
3D space and likelihood estimation of projected hypotheses on each image captured by cameras. These methods are less robust than the system using LRFs.
Recently, several robots have been proposed that utilize gaze for meta-communi-
cation. ROBITA turns to the specific person speaking at the moment in a group
conversation. With Robovie [12] has been investigated the relationship between
a robot's head orientation and its gaze in a Robot Mediated Round Table (RM-
RT) experimental setup. A similar task has been addressed with Cog [13], but to
our knowledge, none seems to address the problem using an android robot. We
believe that use an android robot can give us a more detailed knowledge about
the "uncanny valley" hypothesis.

4   System

This section describes the idea behind the human head detection and tracking
using a network of cameras embedded in the environment surrounding the android. This is basically a slightly modified implementation of the work of [14],
using only cameras for the tracking system, and taking advantage of the regular
light conditions of the room where the robot is placed.

We employ the PF [14] for tracking the human head. The algorithm estimates
a posterior probability with random sampling and its evaluation from likelihood.

In the configuration tested we prepared four cameras placed in the four corners
of the Geminoid's room and a total of three TCP/IP networked pc's: one "master

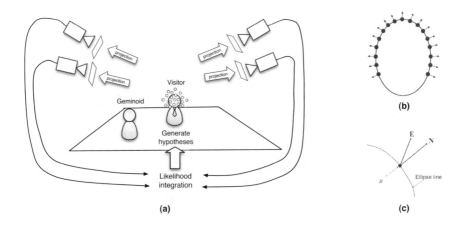

Fig. 2. (a) The distributed PF system; (b) the normals on the ellipse that approximates
the human head; (c) the parts of the normal

pc" and two "slave pc", which verify the hypotheses generated from the "master pc". Each "slave pc" runs two instances of a developed client, and each client is connected to one of the cameras placed in the room. The software is written in C++, the machines were equipped with an Intel I7 processor and 8Gb of RAM. The algorithm uses 100 particles that are initially placed in the entrance of the door so that a new comer into the room can be easily detected. The system was able to run in real time.

5   Expected Results and Improvements of the Control System

The solution previously shown adds to the human-like appearance of the android an efficient perceptual functions useful for communicate through the typical human communication channels. Using the cameras placed in the four corners of the room is possible to cover the area in front of the robot detecting when someone is very near to the robot (short range interaction) or when someone is moving in the room (long range interaction). In figure 3 are shown the map of the room and pictures of the setup.

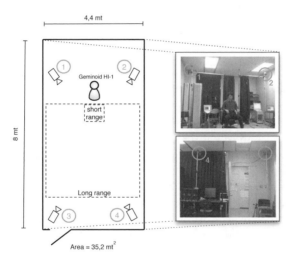

Fig. 3. On the **left side**: map of the Geminoid room from the top showing the cameras and approximately the two ranges of interaction, the short one and the long one. On the **right side**: pictures of the setup.

A possible way to evaluate the human-robot interaction in a short range would be to couple non-conscious human responses together with a complementary source of information such as a questionnaire, so that the gaze behavior in the human-robot interaction can be compared to the one in human-human interaction. A typical behavior is to establish an eye contact looking with the right eye to the interlocutor's right eye. This contact is often lost, especially when one of the conversational partners is thinking. This behavior has been explained by three main theories: arousal reduction theory [15]; the different cortical activation hypothesis [16] and the social signal theory [17].

Our hypothesis is that if we allow the Geminoid to produce the same kinds of eye movements following the same social rules of humans, subjects will consider the robotic interlocutor as if it were a person, or at least a social agent. We believe that eye movements act as signals about whether the subject is, for example, thinking or listening, and an android must use this kind of non-verbal communication in order to increase its human-likeness. To test the validity of this speculation, is it needed to consider the duration and the timing together with the focus of gaze. The same kind of conclusions have been reached in [18].

In [19] has been observed that in a long range interaction scenario consisting of a quietly sitting person (subject B) and another person free to act in a 2.0 m x 2.0 m area (subject A), typical behaviors of B were: look toward the direction that A looks, look at A a few times, look down to the ground, or keep looking at subject A (following him). Results from experiments like the one in [19] suggest that reactions for a human-like presence are efficient for achieving the subjective human-like nature in an android communication. The system described in this paper will allow us to tackle issues focusing on appearance and perception, although such studies have not started yet. The described system can be a test bed for cognitive science, and some research approaches in cognitive science used robots for experiments.

6   Conclusion and Future Works

This paper proposed a hypothesis for the improvement of the human likeness of an android by establishing, breaking and recovering the eye contact with a human interlocutor. Those social signals can be used in both face-to-face interactions and long range interactions, and they can reinforce our expectation of androids as a responsive agent. However, this study is only preliminary and a more comprehensive one is required to contribute to the study of android science and human nature. Our next work will include further investigation on the effect of the gaze for an android in order to confirm the psychological effects.

References

1. Kanda, T., Hirano, T., Eaton, D., Ishiguro, H.: Interactive Robots as Social Partners and Peer Tutors for Children: A Field Trial. Human Computer Interaction (Special Issues on Human-Robot Interaction) 19, 61–84 (2004)
2. Shimada, M., Minato, T., Itakura, S., Ishiguro, H.: Evaluation of android using unconscious recognition. In: Proceedings of the IEEE-RAS International Conference on Humanoid Robots, pp. 157–162 (2006)
3. Nishio, S., Ishiguro, H., Hagita, N.: Geminoid: Teleoperated Android of an Existing Person. In: Humanoid Robots, New Developments, I-Tech, pp. 343–352 (2007)

4. Morishita, H., Watanabe, K., Kuroiwa, T., Mori, T., Sato, T.: Development of robotic kitchen counter: A kitchen counter equipped with sensors and actuator for action-adapted and personally assistance. In: Proceedings of the 2003 IEEE/RSJ International Conference on Intelligent Robots and Systems, pp. 1839–1844 (2003)
5. Mori, T., Noguchi, H., Sato, T.: Daily life experience reservoir and epitomization with sensing room. In: Proceedings of Workshop on Network Robot Systems: Toward Intelligent Robotic Systems Integrated with Environments (2005)
6. Ishiguro, H.: Distributed vision system: A perceptual information infrastructure for robot navigation. In: International Joint Conference on Artificial Intelligence (IJCAI 1997), pp. 36–41 (1997)
7. Becker-Asano, C., Ogawa, K., Nishio, S., Ishiguro, H.: Exploring the uncanny valley with Geminoid HI-1 in a real-world application. In: IADIS Intl. Conf. Interfaces and Human Computer Interaction, pp. 121–128 (2010)
8. Ishiguro, H.: Android Science: Toward a new cross-interdisciplinary framework. In: Proc. of the CogSci 2005 Workshop Toward Social Mechanisms of Android Science, Stresa, Italy, pp. 1–6 (2005)
9. MacDorman, K.F., Ishiguro, H.: The uncanny advantage of using androids in cognitive and social science research. Interaction Studies, 297–337 (2006)
10. Suzuki, T., Iwasaki, S., Kobayashi, Y., Sato, K., Sugimoto, A.: Incorporating environmental models for improving vision-based tracking of people. Syst. Comput. Jpn. 38(2), 1592–1600 (2007)
11. Matsumoto, Y., Kato, T., Wada, T.: An occlusion robust likelihood integration method for multi-camera people head tracking. In: Proceedings of International Conference on Networked Sensing Systems (INSS), pp. 235–242 (2007)
12. Imai, M., Kanda, T., Ono, T., Ishiguro, H., Mase, K.: Robot mediated round table: Analysis of the effect of robot's gaze. In: Proceedings of 11th IEEE International Workshop on Robot and Human Interactive Communication, pp. 411–416 (2002), doi:10.1109/ROMAN.2002.1045657
13. Brooks, R.A., Breazeal, C., Marjanović, M., Scassellati, B., Williamson, M.M.: The cog project: building a humanoid robot. In: Nehaniv, C.L. (ed.) CMAA 1998. LNCS (LNAI), vol. 1562, pp. 52–87. Springer, Heidelberg (1999)
14. Matsumoto, Y., Wada, T., Nishio, S., Miyashita, T., Hagita, N.: Scalable and Robust Multi-people Head Tracking by Combining Distributed Multiple Sensors. Intelligent Service Robotics 3(1) (2010)
15. Argyle, M., Cook, M.: Gaze and Mutual Gaze. Cambridge University Press, Cambridge (1976)
16. Previc, F.H., Murphy, S.: Vertical eye movements during mental tasks: A reexamination and hypothesis. Percept. Motor Skills 84, 835–847 (1997)
17. McCarthy, A., Muir, D.: Eye movements as social signals during thinking: Age differences. In: Biennial Meeting of the Society for Research in Child Development (2003)
18. MacDorman, K.F., Minato, T., Shimada, M., Itakura, S., Cowley, S.J., Ishiguro, H.: Assessing human likeness by eye contact in an android testbed. In: Proceedings of the XXVII Annual Meeting of the Cognitive Science Society, Stresa, Italy, July 21-23 (2005)
19. Chikaraishi, T., Minato, T., Ishiguro, H.: Development of an android system integrated with sensor networks. In: IEEE/RSJ International Conference on Intelligent Robots and Systems, IROS 2008, September 22-26, pp. 326–333 (2008)

Probabilistic Anomaly Detection for Wireless Sensor Networks

Alfonso Farruggia, Giuseppe Lo Re, and Marco Ortolani

DICGIM, University of Palermo,
Viale delle Scienze, ed.6 - Palermo

Abstract. Wireless Sensor Networks (WSN) are increasingly gaining popularity as a tool for environmental monitoring, however ensuring the reliability of their operation is not trivial, and faulty sensors are not uncommon; moreover, the deployment environment may influence the correct functioning of a sensor node, which might thus be mistakenly classified as damaged. In this paper we propose a probabilistic algorithm to detect a faulty node considering its sensed data, and the surrounding environmental conditions. The algorithm was tested with a real dataset acquired in a work environment, characterized by the presence of actuators that also affect the actual trend of the monitored physical quantities.

Keywords: Autonomic Computing, Probabilistic Reasoning, Wireless Sensor Networks.

1 Introduction

Wireless Sensor Networks (WSNs) are composed of a set of interconnected devices equipped with sensors for measuring various physical quantities [1] and as such, they may be used in AI systems for acquiring knowledge about an application domain; clearly, it is important that they are not affected by faults. The present work describes a probabilistic approach for the detection of such anomalies in WSN by exploiting statistical information extracted from data gathered by the nodes themselves. The present work extends our previous work [2], and it aims at modeling the overall behavior of a sensor node, as well as the external factors potentially affecting its operations by a Bayesian network, so that belief propagation may be used to infer the overall health status of the node.

2 Related Work

The topic of anomaly detection for WSN has already been addressed in current literature, but many works fail to consider peculiar characteristics that may lead to a wrong anomaly detection. In [3] an approach for identifying regions of faulty sensor nodes is presented, with good performance for large faulty sets, but which focuses only on hardware faults. Our approach is not sensitive to the nature of the faults, and the probability of a correct diagnosis is independent of the amount

R. Pirrone and F. Sorbello (Eds.): AI*IA 2011, LNAI 6934, pp. 438–444, 2011.

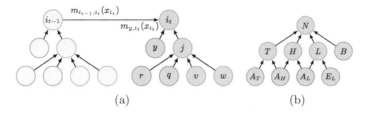

Fig. 1. (a) A Bayesian network, highlighting message passing between two showing hidden and observed nodes. (b) The proposed Bayesian network.

of nodes. In [4] a method for detecting faulty sensor nodes is presented. The method uses Principal Component Analysis (PCA), and wavelet decomposition for analyzing historical data for small-scale WSNs. The faulty sensor nodes can be detected by extracting high-frequency coefficients of wavelet decomposition, but the presence of actuators is not considered, which may alter the outcome of the proposed algorithm. In [5], the authors address potential errors in sensor measurements due to faults, and develop a distributed Bayesian algorithm for detecting and correcting such faults. They present a sample scenario where sensors detect concentrations of some chemical agent may exceed some pre-defined threshold, and propose a Bayesian approach based on the assumption that measurement errors due to faulty equipment are likely to be uncorrelated, whereas environmental measurements are spatially correlated. A similar assumption is used in our work, although our application domain is different, and our approach is applicable to generic off-the-shelf sensors, and takes the specific operational context into account.

3   Detecting Anomalies in Wireless Sensor Nodes

In order to assess the operational good standing of a sensor node we represent its behavior through a Bayesian network (BN) able to model the influence of external factors. The target application domain is an indoor environment; the WSN nodes are supposed to be powered by non renewable energy sources and the user is allowed to influence the environment by operating actuators.

3.1   Modeling Sensor Nodes Behavior

We are interested in modeling the behavior of each sensor node by way of a BN capturing the influence of the surrounding environmental conditions over the sensors on board of the node using the belief propagation (BP) [6] to infer on graphical models expressed by the BN. We refer to the rightmost BN in Figure 1(a) to introduce the theoretical grounds; let S indicate the set of nodes of the BN, with $|S| = s$. Each variable can assume a discrete number of states, and we will indicate one of the different states of node i as x_i. To compute the message between hidden nodes j and i, let H_j and B_j indicate the sets of hidden and observed variables connected to node j; in the depicted case $H_j = \{v, w\}$, and $B_j = \{q, r\}$, with $H_j, B_j \subset S$, at the end the messages are of the form:

$$m_{ji}(x_i) \leftarrow \sum_{b \in B_j} \phi_j(x_j, x_b) \psi_{ji}(x_j, x_i) \cdot \prod_{k \in Ngh(j) \setminus i} m_{kj}(x_j) \qquad (1)$$

where $\phi_j(x_j, x_b)$ and $\psi_{ji}(x_j, x_i)$ represent the *potential functions* between pairs of variables of the graphical model. The former controls the relationship between observed and hidden variables, whereas the latter controls the relationship among hidden variables of the graphical model. $Ngh(j)$ represents the set of neighbors of node j. The "belief" x_i assumed by node i is expressed as follows:

$$b_i(x_i) = \frac{1}{z_i} \phi_i(x_i, y_i) \cdot \prod_{k \in Ngh(j)} m_{ki}(x_i) \qquad (2)$$

where z_i is a normalization factor. Finally, we consider that the values of the variables in the model may change over time, so the beliefs are actually re-computed for each instant. Figure 1(a) shows the relationship between two instances of the model at consecutive time instants; a message will also be exchanged between two consecutive instances of node i, and represents an estimate of the state node i will assume at time t, computed at time $t - 1$.

3.2   Specializing the Model for Indoor Environmental Monitoring

Figure 1(b) shows the structure of the BN we used in our context. Each sensor node is assumed to be equipped with three sensors for measuring light exposure, temperature and relative humidity, respectively, and its operating status is modeled as a binary stochastic variable. Variable N represents the health status of a sensor nodes, and it has to be ultimately inferred; in our model it is influenced by variables L, H, and T which represent the estimators of the operating status of the three on-board sensors. Each of them models the status of the corresponding sensor also taking into account the operating context, which in our case is represented by the surrounding environmental conditions, as well as the potential influence of actuators over the readings of each sensor. Variables E_L, E_H, and E_T represent the raw estimators of the health status of the three sensor with respect to their surrounding environment, and they are computed via the technique described in our previous work [2], that labels a healthy sensor as GOOD, and a faulty sensor as DAMAGED. Variables A_L, A_H, and A_T model the influence of the actuators for light, humidity and temperature respectively. The probabilities associated with such variables are computed with respect to the acquired readings; if the actuator is turned on, the probabilities are computed on the fly by applying Gaussian regression that allows to estimate $p(x_N = \text{GOOD} \,|\, x_{A_T} = \text{ON})$. Whenever the actuator is turned off, we assume a uniform distribution for the corresponding variable. Finally, the operating status of a sensor node is also influenced by the charge level of its battery; our model captures it through variable B using an approach based in the correlation computed between a node having low power and one with sufficient power.

3.3   Inferring the Health Status of an Environmental Sensor Node

As previously mentioned, the overall health status of a sensor node is inferred by computing the belief $b_{N(t)}(x_N)$ of the corresponding node N in the BN; in our scenario x_N is a 2-dimensional vector containing the probabilities associated to the two labels, GOOD and DAMAGED. As the model evolves over time, it takes on a configuration depending on the acquired measurements as well as on external perturbing factors. Eventually, the belief about x_N at time t will indicate which of the two possible states is the correct inference for the operating status of the sensor node. In our model, variables $E_L, E_H, E_T, A_L, A_H, A_T, B$ are the observed variable, whereas variables N, L, H, T are hidden. The marginal probability of the hidden variable N_t, in particular is estimated via BP, by applying Equation 2, which in the specific case becomes:

$$b_{N(t)}(x_N) = \frac{1}{z_N}\phi_t(x_N, x_B) \cdot \prod_{j \in \{H,L,T,N_{t-1}\}} m_{jN}(x_N) \tag{3}$$

where $\phi_t(x_N, x_B)$ is the probability $P(N|B)$, whereas messages are computed by Equation 1. As shown in Equation 3, the state of node N at time t requires a message from node N_{t-1}; we assume that such message is null for N_0.

4   Experimental Results

In order to assess the performance of our method we set up a typical work environment, and we deployed 5 sensor nodes. The setting also included an actuator that influences temperature and humidity, and an artificial lighting. For our experiments, we collected a dataset acquired during the period ranging from March, 24th to April, 14th 2011. The sampling period of each node is 3 minutes; each of the following test scenarios considered an overall time span of 24 hours. Due to its central location, node 5 has been specifically considered as representative for the evaluation of the performance of the proposed algorithm; as will be shown, the influence of all kinds of actuators is more noticeable as compared to the remaining nodes. Three sample scenarios are presented follow to better explain the performance of our approach. The performance of the proposed approach was quantified by computing two metrics: the *accuracy*, measuring the reliability of the classifier with respect to the detection of GOOD and DAMAGED nodes, and the *precision*, which specifically considers the detection of faulty node; they are computed as follows:

$$Ac = \frac{Tn + Tp}{Tp + Fn + Fp + Tn} \quad (4) \qquad Pr = \frac{Tp}{Tp + Fp} \quad (5)$$

where Tp measures the amount of nodes whose health status is GOOD and are actually detected as such (i.e. true positives), Fp measures the amount of nodes whose health status is GOOD, but are erroneously detected as DAMAGED (i.e. false positives), and analogously for the two remaining parameters.

4.1 Scenario 1: Dataset influenced by actuators

In this scenario, the BN processes data influenced solely by the action of the actuators, it identifies data where such influence is relevant, and succeeds in classifying the relative sensors as healthy, even when the underlying MRF-based classifier would trigger an alarm.Our BN-based classifier provides better performance thanks to the additional information extracted from the environmental context, like that the actuators are turned-on. The outcome of the proposed algorithm is shown in the topmost plot of Figure 2(a); the three other plots in the same Figure show the status of the individual sensors for humidity, temperature, and light as computed by the MRF-based algorithm. The reported plots specifically consider node 5. Figure 2(a) highlights that the proposed algorithm outperforms the basic MRF-based classifier. The performance of this scenario is reported in the first row of Table 1.

4.2 Scenario 2: Dataset influenced by a simulated fault

In this scenario, the dataset corrupted by an artificial error only is processed. Figure 3 shows the original dataset; the dotted rectangle highlights the presence of errors. In this case the accuracy value for the proposed algorithm is lower than the classifier based on the MRF, due to a transition phase necessary for the algorithm to converge on the exact state. In the first plot of Figure 2(b), the evolution of the belief about state GOOD for node 5 is shown. In the others plots, the dotted rectangle surrounds the interval containing the errors for the sensors, which are thus regarded as DAMAGED. The transition is due to the fact that the network is time dependent, so that the previous state of a node influences the estimation of next value (through message passing). Just for this scenario the performance is shown in the second row of the Table 1.

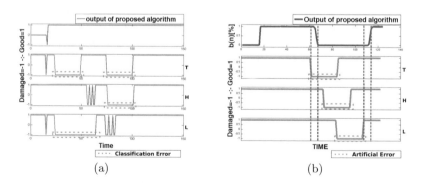

Fig. 2. (a) Environmental information accounted in the Bayesian classifier, the errors of classification are committed by the classifiers MRF based; (b) Progress of belief of the node 5 during the errors occurred in its sensors, the last three charts indicate the period which the error occur respectively on the sensor of temperature, humidity, and light

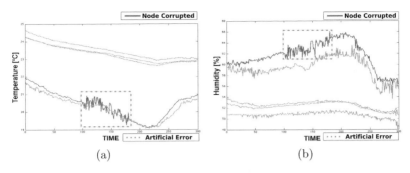

Fig. 3. (a) Real dataset perturbed by a Gaussian error: (a): temperature; (b): humidity

4.3   Scenario 3: Dataset Influenced by Actuators and a Simulated Error

In this scenario, the dataset used is influenced by the action of the actuators, and by an artificial error. As in the first scenario, classifier accounts for the environmental information in its reasoning, and correctly identifies the action of the actuators, but similarly to the second scenario, it singles out the artificial error. The first plot in Figure 4(a) shows the evolution of the belief when the artificial errors occurred on the sensors, and that the belief of the node decreases only in the proximity of errors, so that the status of the node switching toward DAMAGED value as shown in the first plot of Figure 4(b). The other plots of Figure 4(b) show that the MRF-based classifier approximately identifies the faulty sensor, signaling the error for a longer time than the Bayesian classifier, which detects the error upon its occurrence. On the third row of Table 1, the performance of both kind of classifiers are presented for this scenario.

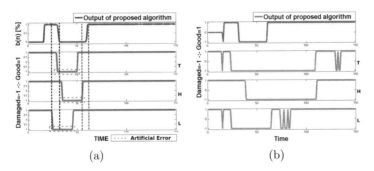

Fig. 4. (a) Real dataset of humidity perturbed by the air conditioner and by a fault. (b) Dynamics of the estimate of the status for the classifiers in scenario 3

Table 1. Performance summary of the experimental scenarios

	BN Classifier		MRF-based Classifiers					
			T		H		L	
	Ac[%]	Pr[%]	Ac[%]	Pr[%]	Ac[%]	Pr[%]	Ac[%]	Pr[%]
Scenario 1	89	90	77	78	63	64	63	63
Scenario 2	70	93	88	99	78	87	–	–
Scenario 3	78	78	52	51	50	42	80	77

5   Conclusion

In this paper we proposed a Bayesian classifier for the health status of sensor nodes for environmental monitoring considering the external factors that surrounding a node, like actuators. A possible future use of the our work might be in a wider and more complex architecture, such as an autonomic system.

References

[1] Yick, J., Mukherjee, B., Ghosal, D.: Wireless sensor network survey. Computer Networks 52, 2292–2330 (2008)

[2] Farruggia, A., Re, G.L., Ortolani, M.: Detecting faulty wireless sensor nodes through stochastic classification. In: 2011 IEEE International Conference on Pervasive Computing and Communications Workshops, pp. 148–153 (2011)

[3] Chen, J., Kher, S., Somani, A.: Distributed fault detection of wireless sensor networks. In: Proceedings of the 2006 Workshop on Dependability Issues in Wireless Ad Hoc Networks and Sensor Networks, pp. 65–72. ACM, New York (2006)

[4] Zhang, X.-L., Zhang, F., Yuan, J., Weng, J.-l., Zhang, W.-h.: Sensor fault diagnosis and location for small and medium-scale wireless sensor networks. In: 2010 Sixth International Conference on Natural Computation, pp. 3628–3632 (2010)

[5] Krishnamachari, B., Iyengar, S.: Distributed bayesian algorithms for fault-tolerant event region detection in wireless sensor networks. IEEE Transactions on Computers 53, 241–250 (2004)

[6] Yedidia, J., Freeman, W., Weiss, Y.: Understanding belief propagation and its generalizations. Exploring Artificial Intelligence in the New Millennium 8, 236–239 (2003)

From Natural Language Definitions to Knowledge Bases Axioms

Matteo Casu and Armando Tacchella

Università di Genova
{matteo.casu,armando.tacchella}@unige.it

Abstract. In this paper we propose to assist ontology development in the case of simple dictionary-like definitions discussing a system which translates natural language definitions into logical formulas and providing a partial translation from first-order formulas into Description Logics formulas. We compared our approach with other existing state-of-the-art tools using different sets of dictionary-like sentences, obtaining some preliminary but encouraging results.

1 Introduction

Being a W3C recommendation, the ontology language *par excellence* is OWL [1] – currently on its second revision – whose fragments, called "profiles", are based on specific Description Logics (DLs) [2]. Since most DLs can be embedded in First-Order-Logic (FOL) (see, e.g., [3] for a survey) FOL is sufficient to express ontological knowledge expressible in OWL. Therefore, designing knowledge bases in OWL can be considered as a – perhaps more intuitive – way of formalizing logical theories that could have been equivalently expressed in FOL. Often, general-purpose reference ontologies undergo several revisions that improve their quality, in terms of correctness, completeness and usefulness. On the other hand, special-purpose ontologies used in small-scale applications do not undergo any form of review by ontology experts. Since it is unlikely that domain experts drafting such ontologies have prior knowledge of logic, the chance of getting low quality theories is relatively high [4] and costly in terms of human effort.

In this paper, we contribute to the problem of developing domain-specific ontologies by translating natural language sentences provided by a domain expert into first-order logic and description logics. We implemented our approach in a prototypical system platform based on off-the-shelf tools which uses Montague semantics [5] and that can handle dictionary-like definitions – text that can be found in legal or industrial specifications and product taxonomies – which tend to be short, explicit and with a consistent usage of terms. Extracting T-box and A-Box axioms is performed by extracting DL sentences from a FOL-based representation. Since the embedding of DLs into FOL cannot be "reversed", i.e., there are FOL sentences for which an equivalent DL sentence cannot be written, we obtain DL axioms by applying heuristic simplifications and partial translation techniques.

2 System

Architecture As parser, we used a chart parser from Nltk that handles feature-based context-free grammars (FCFG) in which the semantic feature of a word is given as

R. Pirrone and F. Sorbello (Eds.): AI*IA 2011, LNAI 6934, pp. 445–450, 2011.

a λ-term. The grammar we defined implements a pure, extensional (i.e. non modal), Montague grammar for English, and can be easily extended with Discourse Representation Theory (DRT, [6]) to handle anaphoras and to parse several sentences in a unique batch. The grammar covers some natural language phenomena, such as relative clauses, cross-categorial connectives, and genitive construction. We assume a one-to-one correspondence between syntactic structures and semantic interpretation, and we do not deal with the lexical meaning of basic expressions. However, we are interested in ambiguity deriving from multiple parse trees of a sentence. Among these trees, some will produce the same semantic reading, some others will not. In the applications we target, having *genuine* different readings in output is a good feature, because they represent different understandings of the same definition. We are interested in pruning logically equivalent readings. Summing up, we have readings that define (i) the very same (syntactically equal) formula; (ii) different but logically equivalent formulas; and (iii) different and non-equivalent formulas. For this reason, a theorem prover (Prover9 [7]) can be called in order to return only one formula for each class of logically equivalent formulas.

A translation FOL $\rightarrow$ DL. The translation proceeds top-down in the formula tree, without propagating failures bottom-up. This approach permits to have some partial but possibly useful translations. The DL currently covered by the translation is $\mathcal{ALCIO}$ plus negation of roles. We report the syntax of the target language (A is an atomic concept, C, D are concepts, S is an atomic role, R is a role):

$$C, D ::= A \mid \top \mid \bot \mid \neg C \mid C \sqcap D \mid C \sqcup D \mid \forall R.C \mid \exists R.C$$
$$R ::= S \mid \neg S$$

The FOL fragment we are interested in is FOL with equality and without functional symbols, except for constants. We assume the set of *terms* and the set $\mathrm{Wff}_{\mathrm{FOL}}$ of well-formed-formulas to be defined in the standard way. We will use the following conventions in using metavariables: $x, y, z, \ldots$ will denote individual variables, $a, b, c, \ldots$ individual constants, P, Q unary relational symbols, R, S binary relational symbols. The notion of closed formula (or statement), and of bound, free, fresh, mute variable will be assumed. The translation is presented as two functions, Ts and Tc.

Ts Translates FOL statements into DL axioms. Ax_{DL} is the set of axioms for a certain DL (for now intentionally left undeclared). We assume the functions:

- *fresh_skolem()* – returns a fresh constant
- *fresh_var()* – returns a fresh individual variable
- *instantiate_exists(ϕ)* – takes a formula ϕ of form $\exists x.\psi$ and returns $\psi[x :=$ $fresh\_skolem()]$
- *instantiate_equality(X)* – makes sure that formulas of the form $\exists x.(lawyer(x)\wedge$ $x = John)$ are simplified in $lawyer(John)$
- *indiv_const_leaves(ψ)* – returns the list of individual constants in ψ
- *length(list)* – returns thelength of the list $list$

Ts is presented in algorithm 1.

Tc Translates FOL formulas in one free variable in DL concepts. We assume functions:

- deal_with_nominal(ψ) – takes a predicate $R(x, a)$ and returns a concept $\exists R.\{a\}$
- choose_concept(X,Y) – takes two formulas one of which is a concept, and choose which one is the concept
- choose_role(X,Y) – takes two formulas one of which is a role, and choose which one is the role
- extract_predicate(ψ) – takes a predicate $P(x_1, \ldots, x_n)$ and returns the relation symbol P
- is_concept(X) – returns True if X is a 1-ary predicate
- is_role(X) – returns True if X is a 1-ary predicate
- on a role $P(x_1, x_2)$ the functions $P(x_1, x_2).first$ – returns x_1 – and $P(x_1, x_2).second$ – returns x_2 can be called
- free(X) returns the free variable of the concept X

Tc is presented in algorithm 2.

Algorithm 1. $Ts : \text{Wff}_{\text{FOL}} \rightarrow \mathbb{P}(Ax_{DL})$

Input: A FOL sentence ϕ.

```
 1: if φ = P(a₁, ... aₙ) then
 2:      if n = 1 then
 3:           return {a₁ : P}          // concept assertion
 4:      else if n = 2 then
 5:           return {(a₁, a₂) : P}    // role assertion
 6:      else if φ = (a = b) then
 7:           return {(a = b)}         // equality assertion
 8:      else
 9:           return ∅                 // catch-all
10: else if φ = ¬ψ and (ψ = P(a) or ψ = P(a₁, a₂) or ψ = (a₁ = a₂)) then
11:      return {¬Tc(ψ)}               // negated assertion
12: else if φ = ∀x.ψ and ψ = X → Y then
13:      return {Tc(X, x) ⊑ Tc(Y, x)}  // concept inclusion
14: else if φ = ∃x.ψ then
15:      return Ts(instantiate_exists(φ))  // anonimous individual
16: else if φ = X ∧ Y then
17:      return {Ts(X)} ∪ {Ts(Y)}
18: else if φ = X ∨ Y then
19:      if length(indiv_const_leaves(X)) = 1 and indiv_const_leaves(X) = indiv_const_leaves(Y) then
20:           c := indiv_constants_leaves(X)[0]
21:           concept := instantiate_equality(X) ∨ instantiate_equality(Y)
22:           return {c : Tc(concept[c := fresh_var()], fresh_var())}
23: else
24:      return ∅
```

2.1   APE, C&C and SystemOne

The tool coming closer to ours is APE, the parser of the AttempTo project [8]. APE uses grammatical methods for natural language understanding and provides output in OWL, but it is restricted to deal with a subset of English known as ACE (AttempTo Controlled English). Moreover, ACE's grammar is unambigous. This can be a feature in some context and a weakness in others (e.g. in rendering alternative interpretations of the sentences provided by the user).

Algorithm 2. $Tc(\alpha, free)$

Input: An FOL formula α with one free variable, and the free variable.

```
 1: if α = ¬ψ then
 2:     return ¬Tc(ψ, free)
 3: else if α = X ∧ Y then
 4:     return Tc(X, free) ⊓ Tc(Y, free)
 5: else if α = X ∨ Y then
 6:     return Tc(X, free) ⊔ Tc(Y, free)
 7: else if α = P(a₁, . . . , aₙ) and n = 2 and any([a₁, a₂]) is constant or function variable then
 8:     return deal_with_nominal(α)
 9: else if α = ∃x.ψ and ψ = X ∧ Y then
10:     role=choose_role(X,Y)
11:     concept=choose_concept(X,Y)
12:     if is_binary(role) and role.first=free and role.second=x and x=concept.free then
13:         return ∃extract_predicate(role).Tc(concept, free)
14:     else if is_binary(role) and role.second=free and role.first=x and x=concept.free then
15:         return ∃extract_predicate(role⁻¹).Tc(concept, free)
16:     else
17:         return Fail
18: else if α = ∀x.ψ and ψ = X → Y then
19:     if is_role(X) and X = P(a₁, a₂) and is_concept(Y) and a₁ = free and a₂ = x and Y.free=x then
20:         return ∀extract_predicate(X).Tc(Y, free)
21:     else if X = P(a₁, a₂) and is_concept(Y) and a₁ = x and a₂ = free and free(Y) = x then
22:         return ∀P⁻¹.Tc(Y, free)
23:     else if Y = P(a₁, a₂) and is_concept(X) and a₁ = free and a₂ = x and free(X) = x then
24:         return ∀¬P.Tc(X, free)
25:     else if Y = P(a₁, a₂) and is_concept(X) and a₁ = x and a₂ = free and free(X)=x then
26:         return ∀¬P⁻¹.Tc(X, free)
27: else if α = P(a) then
28:     return P
29: else
30:     return Fail
```

C&C is a suite of tools that performs syntactic analysis using Combinatory Categorial Grammars[9]. It comes with Boxer, a tool that takes a CCG tree and returns a DRT structure (or, equivalently, a FOL formula). Using our current setting, competing with C&C in coverage is provably impossible, for it is based on a class of grammars (CCG) that recognize mildly-context-sensitive languages (a superset of the context-free languages). Despite this, we claim that our simple framework is more easily extensible, and is more fit to the class of sentences we are interested in. A simple example is given by the following C&C/Boxer translation:

"Every car is owned by someone." is translated in (here in FOL syntax):

$$\forall x.(car(x) \rightarrow \exists y.\exists z.(person(y) \wedge event(z) \wedge own(z) \wedge patient(z, x) \wedge agent(z, y)))$$

The translation is correct, but it is in some sense too much for the task: Boxer implements a neo-Davidsonian treatment of events, which can be a burden on simple sentences like the one cited.

3 Experiments and Results

We have considered three sets of sentences in our experiments. The first one is taken from AttempTo website, and is comprised of sentences respecting ACE constraints. The second set of sentences comes from WikipediaMiner [10], a web tool that, given

Table 1. Our system vs. APE on ACE sentences (left), and our system on WikiMiner and Mimosa sentences. The time is in seconds (s) or minutes (m), and it refers to the whole set of sentences considered

	APE	Our system
coverage:	36/36	31/36
to-DL:	22	18
time:	$7s$	$25s$
readings:	1	1.9
different:	1	1.2

	WikiMiner	Mimosa
coverage:	20/25	21/28
to-DL:	19	14
time:	$2m\ 28s$	$5m\ 3s$
readings:	6.5	6.2
different:	5.25	5.4

a search entry, returns a short descriptive sentence extracted from Wikipedia. These sentences are a good example of what we would like to address, i.e., generally descriptive and "realistic" sentences, not confined in general in a controlled fragment of English. The third set of sentences is extracted from the Terminology Dictionary of the Mimosa OSA-EAI specification (Open System Architecture for Enterprise Application Integration, [11]). This terminology dictionary constitutes another example of industrial specification text that we would like to conceptualize in an automated way.

Overall, we consider 36 sentences from the AttempTo website (ACE sentences); 25 sentences from WikipediaMiner; 28 sentences from the Mimosa Terminology Dictionary. Due to space constraints we do not show detailed output for all the sentences.[1] The results of our tests are reported in Table 1. In the Table we consider:

- The coverage (in number of sentences translated - **coverage**);
- The number of sentences translated to DL (**to-DL**);
- The total time used to perform the translation (**time**) on the set of sentences;
- The average number of readings per sentence (**readings**);
- the average number of different readings per sentence (**different**);

We do not take into account the time needed to prune equivalent formulas. As reported in Table 1 (left), the performances of our system on ACE sentences are comparable to those of APE: 86% of APE coverage can be obtained by our system with a $\times 3.5$ increase in overall CPU time. The APE parser can translate the ACE sentences but not the other test sets, therefore in Table 1 (right) we report only our system's performance. As we can see, considering that these sentences are not controlled, the results obtained are encouraging: 80% coverage is obtained on WikiMiner sentences – with most sentences translated into DL as well – and 75% coverage is obtained on Mimosa sentences – which turn out to be more difficult to translate to DL.

4   Conclusions and Future Work

Summing up, in this paper we presented a translator from natural language definitions to FOL and DL sentences which is meant to assist special-purpose ontology development and some related experimental results. Possible extensions to our system include

---

[1] All the tests described in the following were performed on a MacBook running MacOSX 10.6.5, CPU 2.13 GHz, RAM 2 GB, APE version 6.6.

reducing number of readings, named-entity recognition, POS tagging, and adding background and lexical information. The above points, plus the usage of discourse representation theory to handle anaphoras and the investigation of different grammar formalisms, will be the subject of our future work.

References

1. Mcguinness, D.L., van Harmelen, F.: OWL web ontology language overview. W3C recommendation, W3C (2004)
2. Baader, F., Calvanese, D., McGuinness, D., Nardi, D., Patel-Schneider, P.F. (eds.): The Description Logic Handbook: Theory, Implementation, and Applications. Cambridge University Press, Cambridge (2003)
3. Hustadt, U., Schmidt, R.A., Georgieva, L.: A Survey of Decidable First-Order Fragments and Description Logics. JoRMiCS 1, 251–276 (2004)
4. Rector, A., Drummond, N., Horridge, M., Rogers, J., Knublauch, H., Stevens, R., Wang, H., Wroe, C.: OWL pizzas: Practical experience of teaching OWL-DL: Common errors & common patterns. Engineering Knowledge in the Age of the SemanticWeb, 63–81 (2004)
5. Montague, R.: The proper treatment of quantification in ordinary English. In: Hintikka, K.J.J., Moravcsic, J., Suppes, P. (eds.) Approaches to Natural Language, pp. 221–242. Reidel, Dordrecht (1973)
6. Kamp, H., Reyle, U.: From Discourse to Logic: Introduction to Model-theoretic Semantics of Natural Language. In: Formal Logic and Discourse Representation Theory. Studies in Linguistics and Philosophy, vol. 42. Kluwer, Dordrecht (1993)
7. McCune, W.: Prover9 and mace4 (2005-2010),
   http://www.cs.unm.edu/~mccune/prover9/
8. Fuchs, N., Höfler, S., Kaljurand, K., Rinaldi, F., Schneider, G.: Attempto Controlled English: A Knowledge Representation Language Readable by Humans and Machines. In: Eisinger, N., Małuszyński, J. (eds.) Reasoning Web. LNCS, vol. 3564, pp. 213–250. Springer, Heidelberg (2005)
9. Steedman, M.: The syntactic process. MIT Press, Cambridge (2000)
10. Milne, D., Witten, I.: An open-source toolkit for mining wikipedia (2009)
11. Open System Architecture for Enterprise Application Integration,
    http://www.mimosa.org/?q=resources/specs/osa-eai-v321

A Multimodal People Recognition System for an Intelligent Environment

Salvatore M. Anzalone[1], Emanuele Menegatti[1], Enrico Pagello[1],
Rosario Sorbello[2], Yuichiro Yoshikawa[3], and Hiroshi Ishiguro[3]

[1] Intelligent Autonomous Systems Laboratory, Dep. of Information Engineering,
Faculty of Engineering, Padua University, Padua, Italy
{anzalone,emg,epv}@dei.unipd.it
[2] Robotic Laboratory, Dep. of Computer Science,
Faculty of Engineering, University of Palermo, Palermo, Italy
sorbello_rosario@unipa.it
[3] Intelligent Robotics Laboratory, Dep. of Systems Innovation,
Graduate School of Engineering Science, Osaka University, Osaka, Japan
{yoshikawa,ishiguro}@sys.es.osaka-u.ac.jp

Abstract. In this paper, a multimodal system for recognizing people
in intelligent environments is presented. Users are identified and tracked
by detecting and recognizing voices and faces through cameras and microphones spread around the environment. This multimodal approach
has been chosen to develop a flexible and cheap though reliable system,
implemented through consumer electronics. Voice features are extracted
through a short time spectrum analysis, while face features are extracted
using the eigenfaces technique. The recognition task is achieved through
the use of some Support Vector Machines, one per modality, that learn
and classify the features of each person, while bindings between modalities are also learnt through a cross-anchoring learning rule based on
the mutual exclusivity selection principle. The system has been developed using NMM, a middleware software capable of splitting the sensors
processing in several software nodes, making the system scalable in the
number of cameras and microphones.

1 Introduction

Distributing intelligence among agents and objects of the environment is the
main concept of Ambient Intelligence. These kinds of systems are based on a
network of sensors and actuators distributed around the environment, that try
to reveal important information and to act consequently in an intelligent way[6].
Typical applications involve the real-time logging of the information coming
from several sensors to accomplish various kind of missions, such as recognizing
allowed people, signalling warning state or hazardous situations, thus acting
accordingly[1]. The extraction of relevant information from the raw data streams
collected by the sensors is essential: usually they are full of unimportant data
and important information is "hidden" in it.

R. Pirrone and F. Sorbello (Eds.): AI*IA 2011, LNAI 6934, pp. 451–456, 2011.

The system presented focuses on the recognition of human users inside an indoor, closed environment, through a multimodal approach. It can be used in several domains in which an user based customization of the behaviours is needed. Humans are tracked using several consumer microphones and cameras, spread around an intelligent environment: identification occurs through their voices and faces. This multimodal approach permits the use of small and cheap cameras and microphones, allowing the implementation of a great variety of coverage topologies and solving situations in which a single source of information fails. Cameras can track and recognize people inside the environment while microphones improve the robustness of the whole system by recognizing people hidden to all the cameras field of view.

The approach used assesses identities trough a characterization based on voices and faces. However, in long-term daily-life applications, sensors add noise and people features are not static, altering the effectiveness of the recognition. In order to take in account these changes, the system is trained to estimate a model of the voices and of the faces for each person, together with their correct bindings. Binding in a correct way the information coming from different modalities to the respective real objects, the problem of "anchoring" as defined in [3] is one of the main problems in multimodal systems.

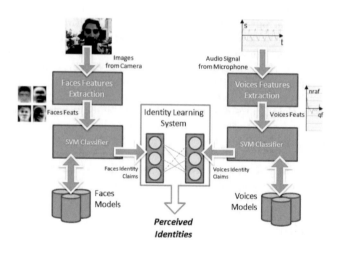

Fig. 1. A sketch of the system

2   System Overview

The system recognizes people using both the features from their voices and faces (fig. 1). Voice identification is based on a noise resistant short time spectrum analysis (NRAF)[7], while facial recognition is based on Eigenfaces technique[9]. Features are then classified using some Support Vector Machine[8], one for each channel. Voice clusters and face clusters are then "anchored" to their high level symbols representing the identifiers of each person that the system recognizes[[11].

This supervised learning approach needs a training setup phase in which voices and faces models, jointly with their bindings with the high level ids are learnt. The system has been developed using a distributed framework called Network-Integrated Multimedia Middleware (NMM)[5]. According to the distributed design of this framework, the system is able to perceive input streams coming from various sensors on different sides, rooms or areas of the intelligent environment, allowing the scalability of the system to different environments and domains.

3   Voices Detection and Recognition

The classification of people's voices is one of the main targets of the system. While distinguishing sound from silence has been easily achieved by threshold-ing the average value of the perceived sound, recognizing the speaker's identity is not a trivial problem that can be achieved through different ways[2]. Focusing on noisy daily-life intelligent environment, real-time user recognition should be per-formed by the system regardless of what users are saying, so passwords cannot be used: the system needs to recognize voices in a text independent way. Accord-ingly, speaker recognition system is based on a short time spectrum analysis. The voice recognition system is performed in several steps (fig. 2): firstly, voice activ-ity is detected; then NRAF, Noise Robust Audio Features, are extracted; finally, the speaker identity claim is calculated, using the SVM classifier. In particular, NRAF are inspired to the human auditory system[7]: features of the current perceived sound are extracted according to a short time analisys, by applying to it a succession of filters. Then, in a training stage, collected NRAF are used to build an SVM model capable of describing the voice of each human user. During the running phase, the trained SVM is used to find the best identity claim. As last stage of refinement, to obtain more stable results, the mode of the last 10 voices identity claims is calculated and given as final voice claim.

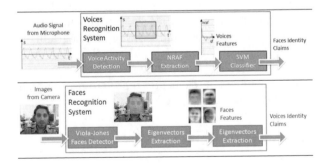

Fig. 2. The recognition systems for voices and faces

4   Faces Detection and Recognition

Image frames coming from the camera are processed to detect human faces and, eventually, recognized by associating them to a face claim identifier. There are three stages on face recognition (fig. 2): in the first step, the system tries to find faces in the camera frames, using a Viola-Jones classifier[10] previously trained to recognize frontal faces. Then, faces found are processed to calculate a vector of features that best describe them. PCA theory offers concepts to achieve this, through the use of the technique of the Eigenfaces. A small set of sample pictures is used to generate a space that best encodes the variations between faces, called Eigenspace. This eigenspace has been built using different generic faces coming from the database "YaleFaces"[4], obtaining a generalized features space for frontal faces. Facial features are calculated by projecting on this space the cropped image of each face. Vectors of face features are then classified by a SVM, properly trained to classify frontal faces of the human users present in the environment, giving for them a face claim identifier. To obtain stable results, the mode of last 3 claims is performed to obtain the current face identity claim.

5   Identity Anchoring

Claims of voices and faces coming from the SVM classifiers are referred to clusters that better encode the similarities between features. However, they do not inform in an explicit way about the persons identities. During a long-term use of the system, the daily changes of human features, the changing of the environment condition or other unexpected reason can bring the system to model several different clusters belonging to the same identity. An anchoring system capable to link in a correct way the features from each modality to their high level people identity symbols is needed. An Hebbian network has been used to represent the connection between the clusters from each modality[11]. In an Hebbian network if two connected neurons are repeatedly actived at the same time, they will strengthen their connection. According to this, during the learning stage, each user will speak in front of the camera, while the system will learn and improve its bindings between the perceived voice and face, forming a graph of links between the two modalities. It is possible to explore this graph to find groups of strongly connected nodes: each of these groups will be associated to an identifier that will be unique for all its membership elements, representing the high level symbol related to the identity of the human user.

6   Experimental Results

The system has been tested at the IAS-Lab of Padova University in a room equipped as an intelligent environment (fig. 3), with several computers, microphones, an Omnidome(R) system, an Ulisse(R) PTZ camera, and two Panasonic cameras. In particular, the Omnidome system is composed by an omnidirectional camera and a pan-tilt camera besides it.

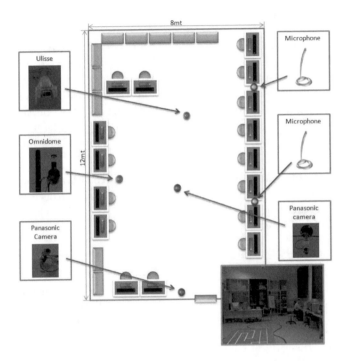

Fig. 3. The test room of the IASLab equipped as an intelligent environment

A training stage is performed before users enter the intelligent environment by asking them to read in front of an equipped computer 60 seconds of text. Once a person comes inside the intelligent environment, cameras will track him using a temporary identifier "not yet identified" until one of them will detect and recognize his face. If the cameras fail this recognition because occlusion or other unpredictable reasons, the system will try to identify him using microphones, by his voice. The first sensor able to recognize the person will give him his id. If the tracking system loses a person, again he will be tracked as "not yet identified" until a sensor is able to classify again, starting a new tracking with the correct user identifier. The system has been tested 20 times using a population of 3 people chosen randomly. The faces identification system based on eigenfaces was able to recognize people in the 82% of cases, while the voices recognition system based on NRAF features was able to identify speakers in the 87% of cases. However, thanks to the multimodal approach, the system has been able to outperform the performances of the single modalities, giving a correct evaluation of 91%: when the system is not capable of recognizing people through one modality, a second one will recover, improving the total performances of the whole system. Causes of error come from the nature of signals perceived. Emotions, unconsciously changing of voice tonality, variation on lighting conditions can introduce important differences compared to the training set, degrading the performances of the system.

7   Conclusions and Future Works

In this paper a multimodal people recognition system has been showed. The system has been used as a part of an intelligent environment to recognize and track people, using both faces and voices bound together. Results presented show that the use of multiple modalities on recognition outperforms the results of a single modality, and encourage us to pursuit researches based on this approach, to simultaneous learn the clustering and the mapping of different perceptions: an automatic, unsupervised, multimodal learning system of human identities.

Acknowledgments. Authors would like to thank Prof. D. Anderson and Ph.D. S. Ravindran for their help with NRAF features. This project was partially founded by Regione Veneto Cod. progetto: 2105/201/5/2215/2009 approved with DGR n. 2215, 21/07/2009.

References

1. Anzalone, S.M., Menegatti, E., Pagello, E., Yoshikawa, Y., Ishiguro, H., Chella, A.: Audio-video people recognition system for an intelligent environment. In: 2011 IEEE, 4th International Conference on Human System Interactions (HSI). IEEE, Los Alamitos (2011)
2. Bimbot, F., Bonastre, J.F., Fredouille, C., Gravier, G., Magrin-Chagnolleau, I., Meignier, S., Merlin, T., Ortega-García, J., Petrovska-Delacrétaz, D., Reynolds, D.A.: A tutorial on text-independent speaker verification. EURASIP Journal on Applied Signal Processing, 430–451 (2004)
3. Coradeschi, S., Saffiotti, A.: An introduction to the anchoring problem. Robotics and Autonomous Systems 43(2-3), 85–96 (2003)
4. Hurkens, C., Van Iersel, L., Keijsper, J., Kelk, S., Stougie, L., Tromp, J., Dolech, D., Eindhoven, A.: Face Image Database (2008), publicly available for non-commercial use http://cvc.yale.edu/projects/yalefaces/yalefaces.html
5. Lohse, M., Repplinger, M., Slusallek, P.: Network-Integrated Multimedia Middleware. Services, and Applications, Department of Computer Science, Saarland University, Germany, Diss. (2005)
6. Preuveneers, D., Van den Bergh, J., Wagelaar, D., Georges, A., Rigole, P., Clerckx, T., Berbers, Y., Coninx, K., Jonckers, V., De Bosschere, K.: Towards an extensible context ontology for ambient intelligence. Ambient Intelligence, 148–159 (2004)
7. Ravindran, S., Schlemmer, K., Anderson, D.V.: A physiologically inspired method for audio classification. EURASIP Journal on Applied Signal Processing, 1374–1381 (2005)
8. Steinwart, I., Christmann, A.: Support vector machines. Springer, Heidelberg (2008)
9. Turk, M., Pentland, A.: Face recognition using eigenfaces. In: Proc. IEEE Conf. on Computer Vision and Pattern Recognition, vol. 591, pp. 586–591 (1991)
10. Viola, P., Jones, M.: Robust real-time object detection. International Journal of Computer Vision 57(2), 137–154 (2002)
11. Yoshikawa, Y., Nakano, T., Ishiguro, H., Asada, M.: Multimodal joint attention through cross facilitative learning based on μx principle. In: Proceedings of the 7th International Conference on Development and Learning, pp. 226–231 (2008)

Instance-Based Classifiers to Discover the Gradient of Typicality in Data

Francesco Gagliardi

Department of Philosophy, "*La Sapienza*" University of Rome,
Via Carlo Fea, 00161 Roma, Italy
fnc.ggl@gmail.com

Abstract. One of the aims of machine learning and data mining regards the problem of discovering useful and interesting knowledge from data. Usually instance-based (IB) classifiers are considered unsuitable for knowledge extraction tasks. Conversely in this paper we consider the families of IB classifiers based on prototype methods and on nearest-neighbours and we show that some hybrid IB classifiers can infer a mixture of representative instances, varying from abstracted prototypes to previous observed atypical exemplars, which can be used to discover the "typicality structure" of learnt categories. Experimental results show that one of the proposed hybrid classifiers "the Prototype exemplar learning classifier", detects a concise and meaningful set of representative instances varying from prototypical ones to atypical ones, which form a gradient of typicality. This kind of class representations cohere with theories developed in cognitive science about how human mind classifies.

Keywords: Instance-based Learning, Knowledge Discovery.

1 Introduction

The problem of supervised classification [3] concerns the construction of classifier systems that opportunely trained can assign to each instance or object presented in input, the proper class among a set of possible classes. A classifier system can have two mayor purposes: to predict class of new observations and to extract knowledge [1] from past experiences. In the latter case, the internal representation of classes, that a classifier infers from dataset play a central role. Instance-based (*IB*) classifier systems [3] [7] constitute a family of classifiers which main distinctive characteristic is to use the instances themselves as classes representation. Within this family we can identify two sub-families [7]: the first is based on prototype methods and the second on Nearest-Neighbours. The prototype methods build representative instances as abstracted centroids of classes or sub-classes, often using iterative clustering algorithms; conversely the nearest-neighbours methods use exemplars filtered from dataset as representative instances. The instance-based classifiers and in particular the *k-NNC* (*k-Nearest Neighbour Classifier*) achieve such performances to be used in real-life problems, but they usually are not used in the knowledge discovery process, because it is commonly assumed that *"instances do not really «describe» the patterns in data"* [3, p.79] and that the instance-based learning *"in a sense [...] violates the*

R. Pirrone and F. Sorbello (Eds.): AI*IA 2011, LNAI 6934, pp. 457–462, 2011.
© Springer-Verlag Berlin Heidelberg 2011

notion of «learning»" [3, p.79]. Some recent developments show that some hybrid *IB* classifiers, such as the *Prototype exemplar learning classifier* (*PEL-C*) [6] and the *Total recognition by adaptive classification experiments* (*T.R.A.C.E.*) [2], which generalize both prototype-based and exemplar-based classifiers, can be used to discover the "typicality structure" of learnt category, detecting how the class is decomposable in some subclasses and their typicality grade within the class. The concept descriptions obtained from these classifiers are composed of a mixture of instances varying from prototypical ones to atypical ones, which form the *"gradient of typicality"*, which has been proposed to explain the human mind representation of categories. In fact this kind of representation of classes cohere with the evidence of experimental psychology about how humans classify data [4][5][8] and that are summarized in the well-known *Prototype-theory* and the *Exemplar-theory* developed in cognitive science [8]. In this work we investigate the capability of *IB* classifiers to extract the typicality structure in data and so we apply them to the well-know problem of classification of *Iris* plant proposed by R.A. Fisher.

2 Hybrid Instance-Based Classifier Systems

The problem of supervised classification, concerns the construction of classifier systems which can assign to every presented input instance, the proper class among a set of possible classes. In *IB* systems the concept description extracted from the training set, consists of the storage of directly observed or abstracted instances belonging to the set of all possible observations. Classification is performed comparing a new instance, for which the class is unknown, with the labelled instances saved in memory, using the nearest neighbour rule (*NN rule*). We focus on the learning algorithm introduced in [2] (see Algorithm 1 in the following) because it has particular formal characteristics which are explained in detail in [2; sect. 3]. In particular it is possible to demonstrate that the representative instance set inferred by this learning algorithm can vary from that of the Nearest Prototype Classifier (*NPC*), which is completely based on prototypes, to one of the Nearest Neighbour Classifier (*NNC*), which is completely based on exemplars according to the number of learning iterations and to the particular dataset (see proof of theorem 3.2 in [2]). So the hybrid classifiers considered here are a generalization of *NPC* and *NNC* and in fact we use the term "hybrid" to refer to the type of representative instances set which can be obtained from classifiers that use this learning algorithm and hence it does not refer to a kind of classifier obtainable with a simple joining of classifiers *NPC* and *NNC*.

Concerning the learning phase of hybrid *IB* classifiers, the starting step is, as the *NPC*, the calculation of one prototype for every class, then for any new learning iteration, the misclassified instance in the previous iteration, which results the farthest from the nearest prototype of its own class (viz. the most atypical one), is added as candidate instance. This candidate instance may or may not undergo an abstraction process by a re-calculation of the prototypes positions of its class according to k-means algorithm. If the abstraction takes place, the considered instance "generates" a new prototype and consequently the old ones adapt; otherwise it is stored as an exemplar of that class in the set of representative instances.

We present in the following the learning algorithm[1]; we indicate *TS* as the training set, *RI* as the representative instances set and C_k as the items of the *k-th* class:

Algorithm 1. Learning algorithm of hybrid instance-based classifiers

```
1. Initialize RI with the barycentres of the classes Cₖ
2. WHILE NOT (Termination Condition)
%% (Find a new candidate representative instance)
    2.1 Calculate the distances between every instance of TS and every
        instance of RI
    2.2 Among the misclassified instances of TS, find the instance which
        is the farthest from the nearest instance of RI belonging to its
        own class. Call it X and assume that it belongs to the class Cₖ
    2.3 Add X to RI.
%% (Update RI)
    2.4 Consider only instances of RI and TS belonging to Cₖ Call them as
        RIₖ and TSₖ, respectively
    2.5 Update the positions of RI using the k-means clustering algorithm
        applied only to TSₖ with starting conditions RIₖ:
        2.5.1 Apply the NN-rule to the items of TSₖ respect to the RIₖ
        2.5.2 Iteratively re-calculate the locations of instances of RIₖ
              by updating the barycentres calculated respect to the
              subclasses determined with the NN-rule.
3. END
```

The behaviour of these classifier systems can vary from the one of the *NPC*, completely based on prototypes, to the one of *NNC*, completely based on exemplars, in an adaptive way and according to the chosen termination condition and to the particular classification problem. In intermediate cases the number and the kind of the representative instances is dynamically determined as a combination of prototypes, exemplars and instances of an intermediate abstraction level.

In general, we can think about different possible termination conditions of this learning algorithm, such as the following:

- *Training accuracy.* The accuracy percentage in classification of the training set is fixed and it can be equal to or less than *100%*. In the case it is set to *100%*, the system is forced to classify correctly all the training set, and the obtained classifier is the one proposed by Nieddu and Patrizi [2] and it is known as *T.R.A.C.E.* (acronym of Total Recognition by Adaptive Classification Experiments). This classifier is the first one proposed in literature which basically uses the iterative learning algorithm showed above.

- *Predictive accuracy.* The system can estimate its own performance on new instances by using a technique of cross validation [3, p.149] as varying the number of iterations. Therefore, the system is able to find the minimum number of iterations to obtain the maximum capability of generalizing (predictive accuracy on new instances), even if it worsens the training accuracy. In fact, beyond a certain value of the iteration number, the accuracy on the training set increases while the accuracy on new instances used for validation does not: this is the so called over-fitting phase. This termination condition is the one we have chosen for the *Prototype exemplar learning classifier (PEL-C)*.

---

[1] A similar version of this algorithm which is more formal but with an assigned termination condition can be found in [2, p.481].

3 Experimental Results

We have considered the problem of discriminating among three types of the Iris plant, based on the dataset introduced by R.A. Fisher. The aim of performing these tests is twice: to compare the classification performances of *IB* classifiers as black-boxes and to analyze the kind of internal representations of classes. We carry out different test suites for every classifier system: *NPC, NNC, k-NNC, T.R.A.C.E.* and *PEL-C*. Each test suite is prepared by using as cross-validation technique the *leave-one-out* procedure. In addition, for the *k-NNC* and *PEL-C* we execute runs, constituted of more test suites as varying *k* or the stop condition, respectively. The *k-NNC* is applied to *Iris* dataset by carrying out a run composed of *12* test suites as *k* varies between *1* and *12*. The best predictive accuracy on test set is obtained for *k=7*. In order to analyze the behaviour of the *PEL-C* and the *T.R.A.C.E.* a run composed of more test suites is carried out by varying the stop criterion of their iterative learning algorithm. To compare classification performance and the kind of class representations obtained by different *IB* classifiers we have considered, as varying the used classifier, the following indexes: the accuracy on test set, the cardinality of concepts description (the number of representative instances), the number of pure prototypes (single instances representing the entire class) and exemplars (instances representing only themselves) which are present in the representative instances set, and finally the mean percentage of representativeness, inside their own class, of representative instances. The representativeness of an instance is computed as the ratio between the number of observations assigned to a class using a given representative instance, and the total of the instances assigned to that class. In the table 1 we compare these indexes.

Table 1. Performance indexes obtained by different classifier systems on *Iris* dataset

	NPC	PEL-C	T.R.A.C.E.	NNC	k-NNC
Representation	*Prototype-based*	*Hybrid*		*Exemplar-based*	
Accuracy (%)	92.7	**96.7**	93.3	95.3	**96.7**
Representative instances	3	7	15	150	150
Pure-prototypes	3	1	1	0	0
Exemplars	0	1	3	150	150
Mean percentage of representativeness (%)	100	42.9	20.0	2.0	n/a

We observe that the *PEL-C* reaches the best classification performance equal to one of the *k-NNC*, that has the *k* optimized in order to maximize performances. Moreover, *PEL-C* obtains it with only *7* instances against the *150* of the *k-NNC*.

The *NPC* uses only *3* instances, which are all pure prototypes, but its accuracy is lower than *PEL-C* and *k-NNC*. The *k-NNC* and *NNC* use a classes representation entirely composed of exemplars, which are the *150* instances of the whole training set. The *T.R.A.C.E.* and the *PEL-C* infer from the training set a hybrid representation of classes, for example, regarding the *7* patterns used by the *PEL-C* to build the classes description: *1* is a pure prototype, *1* is a pure exemplar and the remaining *5* are representative instances with an intermediate abstraction. Knowledge discovery has been defined as *"a non-trivial process of identifying valid, novel, potentially useful and ultimately understandable patterns from collections of data"* [1]; its main aim is to reveal some new and useful information from the data. In the table 2 we show in detail the concepts description found by the *PEL-C* for the *Iris* dataset.

Table 2. The *concepts description* obtained by *PEL-C* and relative representativeness

Sepal length (mm)	Sepal width (mm)	Petal length (mm)	Petal width (mm)	Class	Cardinality	representativeness (%)	
50.06	34.28	14.62	2.46	*Iris Setosa*	50	100.0	*Pure-prototype*
62.11	30.00	45.15	14.30	*Iris Versicolour*	27	52.9	
54.89	25.16	38.00	11.47		19	37.3	*Intermediate*
62.00	24.25	47.25	14.75		5	9.8	*abstraction*
70.00	31.59	58.74	21.74	*Iris Virginica.*	27	55.1	*prototypes.*
61.59	27.68	52.05	18.59		21	42.9	
49	25	45	17		1	2.0	*Exemplar.*

We observe that the representative instances type varies from a pure prototype (the first row) to a pure exemplar (last row), while the classes descriptions vary from a class totally based on a prototype (*Iris Setosa*) to a class based on prototypes plus *1* exemplar (*Iris Virginica*). The inferred representative instances detect how the classes are decomposable in some subclasses and their typicality grade within the own class: for the linearly separable class *Iris Setosa* there is no need to define subtypes; while *Iris Versicolour* class presents three subtypes (composed respectively of the *52.9%, 37.3%,* and *9.8%* of the entire class); and the *Iris Virginica* class presents two subtypes (composed respectively of the *55.1% 42.9%* of the entire class) and one atypical exemplar. This "gradient of typicality" can be showed as in the figure 1.

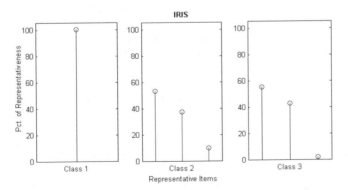

Fig. 1. The "gradient of typicality" extracted by *PEL-C* for the *Iris* dataset

Summarizing, the set of representative instances inferred by *PEL-C* with its typicality gradient, is the most interesting from a knowledge extraction viewpoint, because it produces good classification accuracy and discovers the typicality structure of learnt classes thanks to a meaningful mixture of prototypical instances, with graded abstraction, and atypical exemplars.

4 Concluding Remarks

We have compared the performances of five *IB* classifiers based on different types of internal representations which vary among prototypes-based, exemplars-based and hybrid ones, in order to analyze their behaviour regarding classifying performances and the kind of knowledge extracted. The *PEL-C* and the optimized *k-NNC* obtain the best classifying accuracy, but the first with a very lower number of representative instances. From the viewpoint of knowledge extraction, the *PEL-C*, results to be the most interesting because it obtains representations of classes very concise and composed of a meaningful mixture of representative instances (varying from prototypical ones to atypical ones) which detect how the classes are decomposable in subclasses and their typicality grade within the own class. Achieved results are a first confirmation outside the medical domain [6] of the suitableness of *PEL-C* as tool for knowledge extraction, so we plan to apply it and other *IB* classifier systems, on other classification problems to confirm further the present findings.

Acknowledgments. I wish to thank Roberto Cordeschi for the valuable discussions about matters illustrated in the present paper.

References

1. Fayyad, U., Piatetsky-Shapiro, G., Smyth, P., Uturusamy, R.: Advances in Knowledge Discovery and Data Mining. MIT Press, Cambridge (1996)
2. Nieddu, L., Patrizi, G.: Formal methods in pattern recognition: A review. European Journal of Operational Research 120, 459–495 (2000), doi:10.1016/S0377-2217(98)00368-3
3. Witten, I.H., Frank, E.: Data Mining: Practical Machine Learning Tools and Techniques with Java Implementations, 2nd edn. Morgan Kaufmann, San Francisco (2005)
4. Cordeschi, R., Frixione, M.: Rappresentare i concetti: filosofia, psicologia e modelli computazionali. Sistemi Intelligenti XXIII(1), 25–40 (2011), doi:10.1422/34610
5. Gagliardi, F.: The Necessity of Machine Learning and Epistemology in the Development of Categorization Theories: a Case Study in Prototype-Exemplar Debate. In: Serra, R., Cucchiara, R. (eds.) AI*IA 2009. LNCS, vol. 5883, pp. 182–191. Springer, Heidelberg (2009), doi:10.1007/978-3-642-10291-2_19
6. Gagliardi, F.: Instance-based classifiers applied to medical databases: diagnosis and knowledge extraction. Artificial Intelligence in Medicine 52(3), 123–139 (2011), doi:10.1016/j.artmed.2011.04.002
7. Hastie, T., Tibshirani, R., Friedman, J.: Prototype Methods and Nearest-Neighbors. In: The Elements of Statistical Learning. Data Mining, Inference, and Prediction, 2nd edn., pp. 459–484. Springer, New York (2009), doi:10.1007/b94608_13
8. Murphy, G.L.: The big book of concepts. The MIT Press, Cambridge (2002)

Author Index